Advances in Experimental Medicine and Biology

Advances in Experimental Medicine and Biology presents multidisciplinary and dynamic findings in the broad fields of experimental medicine and biology. The wide variety in topics it presents offers readers multiple perspectives on a variety of disciplines including neuroscience, microbiology, immunology, biochemistry, biomedical engineering and cancer research. Advances in Experimental Medicine and Biology has been publishing exceptional works in the field for over 30 years and is indexed in Medline, Scopus, EMBASE, BIOSIS, Biological Abstracts, CSA, Biological Sciences and Living Resources (ASFA-1), and Biological Sciences. The series also provides scientists with up to date information on emerging topics and techniques. 2015 Impact Factor: 1.953

More information about this series at http://www.springer.com/series/5584

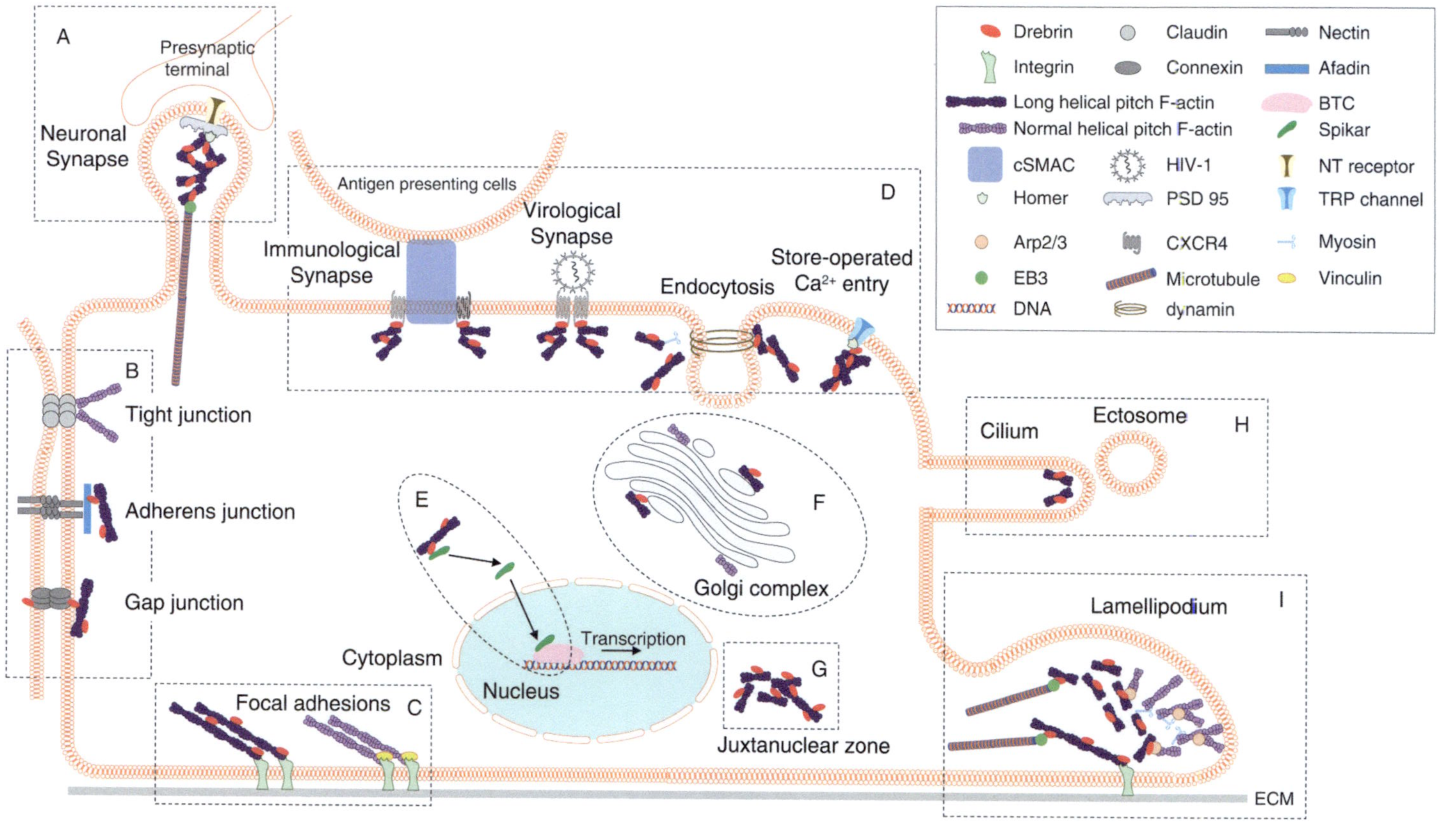

Drebrin forms unique stable F-actin with a long helical crossover and plays a pivotal role in intercellular communication at neuronal synapses, adherens and gap junctions, and immunological and virological synapses. Additionally, drebrin is involved in the cellular mechanisms of cell migration, cell process formation, cancer metastasis, and gene transcription through a transcription co-activator spikar. Drebrin is also found on the Golgi complex, at the juxtanuclear zone, and at a tip of cilia

Tomoaki Shirao • Yuko Sekino
Editors

Drebrin

From Structure and Function to Physiological and Pathological Roles

Springer

Editors
Tomoaki Shirao
Department of Neurobiology and Behavior
Gunma University Graduate School
of Medicine
Maebashi, Gunma
Japan

Yuko Sekino
Laboratory of Chemical Pharmacology
Graduate School of Pharmaceutical Sciences
The University of Tokyo, Tokyo
Japan

ISSN 0065-2598 ISSN 2214-8019 (electronic)
Advances in Experimental Medicine and Biology
ISBN 978-4-431-56548-2 ISBN 978-4-431-56550-5 (eBook)
DOI 10.1007/978-4-431-56550-5

Library of Congress Control Number: 2017951869

Printed on acid-free paper

This Springer imprint is published by Springer Nature
The registered company is Springer Japan KK
The registered company address is: Chiyoda First Bldg. East, 3-8-1 Nishi-Kanda, Chiyoda-ku, Tokyo 101-0065, Japan

This book is dedicated to our mentor Dr. Kunihiko Obata whose wise counsel has led us to the discovery and a deeper understanding of drebrin.

Preface

This book consolidates drebrin studies that have accumulated over three decades, since the first identification of drebrin by our group in 1985. Although in the 1980s we could not envision exactly how the study of drebrin would develop, the progress reflected in the chapters presented here was beyond our wildest expectations. This book begins with a general introduction of drebrin from a historical perspective, and then the chapters in the second part provide the molecular characterization of drebrin and drebrin-decorated F-actin. The third and fourth parts discuss its function in the nervous and non-nervous system, respectively.

This review will appeal to researchers who are interested in synapse formation and synaptic plasticity, as well as subcellular local morphogenesis, such as cell protrusion formation, cell migration, intercellular junction formation, and endocytosis. The book will also appeal to researchers who use drebrin as a tool, such as a marker of synaptic function or a disease marker. This book was kept as concise as possible, to be understood by readers from diverse scientific disciplines. Because of the clarity of its presentations, it can also serve as a textbook in graduate courses.

We wish to express our gratitude to the authors who so willingly contributed to this book. We would also like to thank the staff of Springer Japan, in particular Ms. Momoko Asawa and Dr. Yasutaka Okazaki.

Maebashi, Japan — Tomoaki Shirao
Tokyo, Japan — Yuko Sekino

The original version of this book was revised.
An erratum to this book can be found at DOI 10.1007/978-4-431-56550-5_24

Contents

Part I
History of Drebrin Discovery as an General Introduction

Chapter 1
General Introduction to Drebrin

Tomoaki Shirao and Yuko Sekino

Abstract Drebrin was first discovered by our group as "developmentally regulated brain protein" from the chicken optic tectum. Drebrin is an actin-binding protein, which is classified into two major isoforms produced by alternative splicing from a single *DBN1* gene. The isoform predominantly expressed in the adult brain (drebrin A) is neuron specific, containing a neuron-specific sequence (Ins2) in the middle of the molecule. Drebrin A is highly concentrated in dendritic spines, and its accumulation level is regulated by synaptic activity. In contrast, drebrin E, which lacks Ins2, is found in widespread but not ubiquitous cell types in various tissues. The isoform conversion from drebrin E to drebrin A occurs in parallel with synaptogenesis. Drebrin decorating F-actin is found at the recipient side of cell-cell communication systems, such as gap junctions, adherens junctions, immunological synapses, and neuronal synapses. In addition, it is involved in the cellular mechanisms of cell migration, cell process formation, cancer metastasis, and spermatogenesis. Lack of drebrin leads to the dysfunction of cell-cell communication, resulting in aberrant migration of metastatic cancer cells, aberrant synaptic function in dementia, and rupture of endothelial integrity. Because drebrin forms a unique F-actin with a longer helical crossover, drebrin may create an F-actin platform for molecular assembly and play a pivotal role in intercellular communication.

Keywords Alternative splicing • Cancer • Cell migration • Intercellular communication • Physical property of actin filament • Synaptogenesis • Synaptic plasticity

1.1 Introduction

Drebrin was first discovered by our group as "developmentally regulated brain protein" from the chicken optic tectum in 1985 (Shirao and Obata 1985). In the first 15 years after the discovery, no other groups except us were interested in drebrin,

T. Shirao (✉)
Department of Neurobiology and Behavior, Gunma University Graduate School of Medicine, Maebashi, Gunma 371-8511, Japan
e-mail: tshirao@gunma-u.ac.jp

Y. Sekino
Laboratory of Chemical Pharmacology, Graduate School of Pharmaceutical Sciences, The University of Tokyo, Tokyo 113-0033, Japan

T. Shirao, Y. Sekino (eds.), *Drebrin*, Advances in Experimental Medicine and Biology 1006, DOI 10.1007/978-4-431-56550-5_1

which is expressed in the nervous tissue. During these years, we purified chicken and rat drebrins, raised polyclonal and monoclonal antibodies, and cloned *DBN1* cDNAs. Consequently, we have identified major isoforms of drebrin in chicken, rodent, feline, and human expressed in the nervous tissue. We further clarified the genetic and biochemical properties of drebrin, such as actin-binding activity and phosphorylation. The expression of each isoform depends on the developmental stage. Because the isoform predominantly expressed in adult brain (drebrin A) is neuron specific, our later studies were mainly focused on drebrin A (Shirao et al. 2017).

In 1996, we found that drebrin A is highly concentrated in dendritic spines in adult rat brain, forms a complex with actin and myosin, and inhibits the actin-activated ATPase activity of myosin II (Hayashi et al. 1996). Thus, we proposed that drebrin may play a role in the structure-based plasticity of synapses through the actin-linked control of the actomyosin interaction in dendritic spines. In 1999, we successfully showed that exogenously expressed drebrin A specifically elongates dendritic spines of primary cultured neurons (Hayashi and Shirao 1999). This was the first report demonstrating that the manipulation of a single actin-binding protein in a neuron alters spine morphology. After these epoch-making findings, drebrin and the actin cytoskeleton in dendritic spines were thrown into the limelight. Since then we have shown the pivotal roles of drebrin in spine formation (Takahashi et al. 2003; Aoki et al. 2005) and synaptic plasticity (Takahashi et al. 2006; Mizui et al. 2014; Sekino et al. 2006). Nowadays, hundreds of spine-resident proteins have been found, but drebrin is still a key protein in modulating the actin cytoskeleton in dendritic spines (Sekino et al. 2007; Koganezawa et al. 2017).

Actin-binding proteins modulate the characteristics of the actin cytoskeleton and consequently regulate cell structures or produce the motile force of cells. Drebrin isoforms other than drebrin A are widely distributed in nonnervous tissues as well as the nervous tissue, not only in avian (Shirao and Obata 1986) and mammals (Shirao et al. 1994; Peitsch et al. 1999) but also in the soil amoebae (Luna et al. 1997). Furthermore, drebrin has been found at the recipient side of various intercellular communication systems, such as gap junctions, adherens junctions, immunological synapses, and neuronal synapses. This suggests the universal role of drebrin as an actin modulator.

How does drebrin change F-actin structures? Why does drebrin appear at the cell-cell communication sites? More generally, what is the physiological function of drebrin? This chapter will briefly introduce the key discoveries and proposals contributing to elucidating the above questions.

1.2 Historical Orientation

1.2.1 Background of Drebrin Study

The development of the brain is achieved by a combination of several fundamental processes, such as the proliferation and migration of neurons, the directed extension of nerve fibers, and synapse formation. Before 1960 classic morphological techniques were used for the study of brain development, because morphological structures of the brain dramatically change when each process occurs. In the 1960s and 1970s, developmental studies were accelerated by the progression of new technologies such as the autoradiography using tritiated thymidine. These new methods disclosed in detail the birth date of each neuron, the layer formation, and subsequent maturation in mammalian cerebral and cerebellar cortices and in the chicken optic tectum. However, the molecular mechanism of each process was not yet clarified.

To disclose the molecular mechanism of the brain development, the identification of the master proteins that govern each fundamental process was eagerly pursued. One approach was to select a key function in each developmental process and to look for the protein(s) that mediates that function. Adopting this approach, Edelman and his collaborators developed a specific immunological assay for molecules involved in cell adhesion (Brackenbury et al. 1977) and discovered cell adhesion molecules (CAMs) as key molecules in brain development (Hoffman et al. 1982).

Another approach was based on the conjecture that the master proteins are expressed at limited developmental stages in a restricted region of the brain. Sperry hypothesized the presence of two orthogonal gradients of molecules on retinal ganglion neurons that determine specific connections between retinal and tectal neurons (Sperry 1963), and Nirenberg's group identified an antigen that is distributed in a dorsal-ventral topographic gradient in chick embryo retina by screening a library of monoclonal antibodies in 1981 (Trisler et al. 1981).

1.2.2 Discovery of Drebrin by Proteomics

In January 1982, we started seeking for yet-to-be-discovered master proteins in the developing brain. We surveyed the changes in the proteome of the developing brain using O'Farrell's two-dimensional gel electrophoresis (2DGE) (O'Farrell 1975). The chicken optic tectum was chosen as the target region, because it is a uniform and regularly layered structure that develops correctly on a timetable, as revealed by Cowan and colleagues (LaVail and Cowan 1971a, b). After the electrophoresed gel was stained with Coomassie brilliant blue, 54 proteins were counted (Fig. 1.1). Most of them were found at the beginning (4-day embryo) and remained unchanged until adulthood. There were eight proteins that remarkably changed their staining intensities during embryonic development (Shirao and Obata 1985). These eight proteins were further classified into three groups. The first group was monotonically increasing proteins, including neurofilament proteins and drebrin A (adult-type isoform). The second group was monotonically decreasing proteins. The third group was intensely stained only at embryonic stages and was later named chicken drebrin E1 and E2 isoforms. Note that in mammals there is only one embryonic isoform named drebrin E, while chickens have two embryonic isoforms. The developmental changes in the amount of drebrins in the optic tectum are shown in Fig. 1.2. Drebrin isoforms were found with similar developmental changes in other brain regions. However, the time course of their changes varied from region to region. Even within the optic tectum, developmental changes in drebrin occur earlier in the rostral

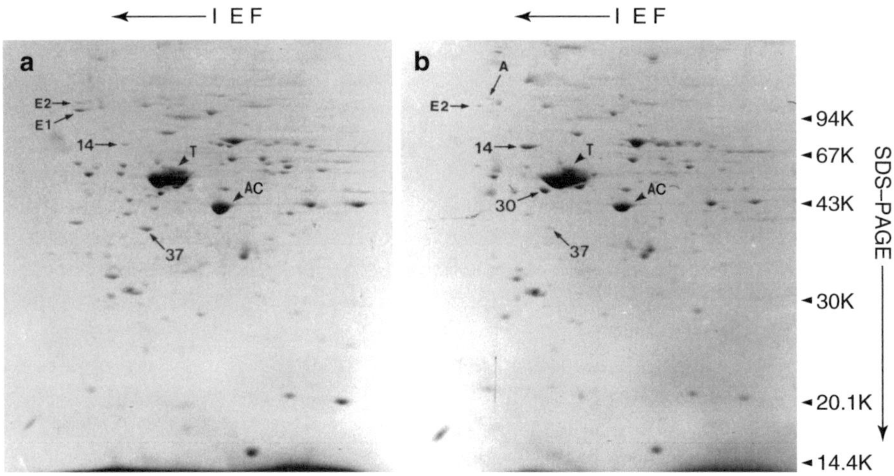

Fig. 1.1 Two-dimensional patterns of proteins of optic tecta. (**a**) Seven-day chick embryo. (**b**) Newly hatched chicken. *A* drebrin A, *E2* drebrin E2, *E1* drebrin E1, *T* tubulin, *Ac* actin. Coomassie Brilliant Blue staining

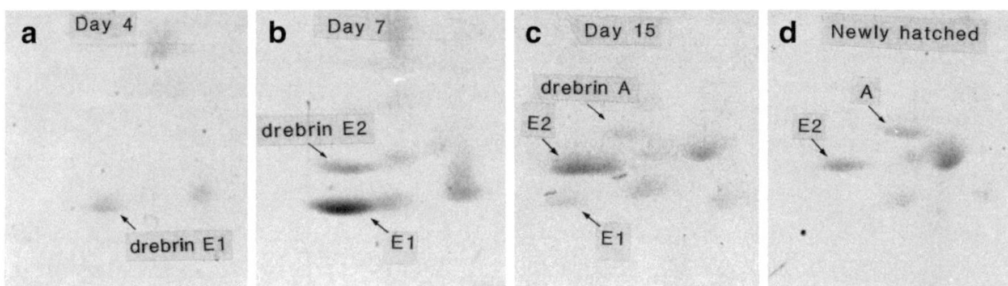

Fig. 1.2 Developmental changes of drebrin isoforms in the chick optic tectum. Panels are regions of interest in two-dimensional gel electrophoresis. (**a**) Day 4. (**b**) Day 7. (**c**) Day 15. (**d**) Newly hatched chick. Coomassie Brilliant Blue staining

portion than in the caudal portion, which corresponds to the rostro-caudal gradient of histological development (LaVail and Cowan 1971a). Together, these results suggest that the changes in drebrin expression are paralleled with brain development, which are explained in detail in Part III of this book.

1.2.3 Purification of Drebrin

In 1985, we succeeded to purify drebrin E1 and E2 from embryonic day (ED) 11 chicken brains (Shirao and Obata 1985). We used the 2DGE assay and found that all drebrins were recovered in the same fractions by various purification methods such as isoelectric precipitation, ammonium sulfate precipitation, and ion-exchange chromatography. Gel electrophoresis peptide mapping using the *Staphylococcus aureus* V8 protease demonstrated structural homology of drebrins. Using radiolabeled methionine, we confirmed that changes in the amount of drebrin in 2DGE are due to changes in drebrin synthesis, but not in drebrin degradation.

During these purification steps, contamination of actin in the drebrin fraction seemed constant at each step. In addition, all drebrin isoforms are acidic proteins (isoelectric point around 4.2), while one of the common features of actin cytoskeleton-related proteins such as calmodulin and troponin is a low isoelectric point. Furthermore, drebrins are heavily phosphorylated proteins on ED5 and ED11 in the optic tectum. Collectively, the results suggested that drebrin isoforms are closely related phosphoproteins associated with the actin cytoskeleton. It is known that the actin-binding activity is the most important function of drebrin, which is discussed below. The actin-binding activity has recently been suggested to be modified by phosphorylation (Worth et al. 2013), which plays a role in cell migration (Tanabe et al. 2014) and neuritogenesis (Geraldo et al. 2008).

1.2.4 Antibodies Against Drebrin

In 1986, we succeeded to produce polyclonal and monoclonal antibodies against electrophoretically purified drebrin E1 and E2. The resulting antibodies included monoclonal antibody (mAb) M2F6, which specifically recognizes all drebrins (E1, E2, and A) (Shirao and Obata 1986). Ever since, mAb M2F6 has been used for drebrin studies as the standard antibody, because its epitope is located in the common sequence of all drebrin isoforms and is conserved from avian to mammals.

We immunohistochemically stained frozen sections of chick embryo at various developmental stages using mAb M2F6 (Shirao and Obata 1986). Drebrin first appeared on ED2 in the myotome. In the early developmental stages of the optic tectum, drebrin is widely distributed in the neuronal somata and processes. As developmental stages proceed, drebrin highly accumulates in the tip of the growing cell processes, namely, the axonal and dendritic growth cones. On the other hand, in the adult, the drebrin immunostaining pattern is dot-like in the neuropil region, suggesting that drebrin is localized at the synapses. Then, we further analyzed the subcellular localization of drebrin using immunoelectron microscopy and found that drebrin is localized in the dendritic spine in the adult brain (Shirao et al. 1987). However, at this point, it was not yet clarified whether the developmental changes in drebrin subcellular distribution depend on each isoform or not.

1.2.5 Cloning of **DBN1** *cDNAs*

Using the antibodies, we then screened a λgt11 cDNA library from a 10-day-old chicken embryo and isolated a cDNA clone of *DBN1* (*gDcw1*) (Shirao et al. 1988). Then, using *gDcw1* we isolated the full-length cDNAs of drebrin E1, E2, and A

(Kojima et al. 1988; Kojima et al. 1993). In 1993, we further elucidated that chicken drebrin E1, E2, and A are translated from three mRNAs that are produced from a single *DBN1* gene by alternative splicing (Kojima et al. 1993). Among the drebrin isoforms, drebrin E1 (564 amino acids) is expressed in the earliest developmental stages. Drebrin E2 (607 amino acids) has an additional 43-amino-acid insertion sequence (Ins1) in the middle of drebrin E1. Drebrin A (653 amino acids) has another 46-amino-acid sequence (Ins2) inserted right in front of Ins1.

In parallel with the above study, we cloned the cDNAs of rat drebrin A (707 amino acids) (Shirao et al. 1992) and human drebrin E (649 amino acids) (Toda et al. 1993). The overall amino acid identity between chicken and rat drebrin A is 60%. In particular, the homology of the N-terminal half including the Ins2 sequence (1–364 amino acid residues in rat) and two short regions in the C-terminus (581–599 and 650–707 residues) is greater than 80% (Kojima et al. 1993). The regions are also well conserved in human. It has been shown that the rat, mouse, and human Ins2 sequence is almost identical. So far, the drebrin A isoform, which contains the Ins2 sequence, including s-drebrin A (Jin et al. 2002), has only been detected in neurons from chicken to human, and its expression depends on brain maturation. The schematic structure of drebrin E and A is shown in Fig. 1.3.

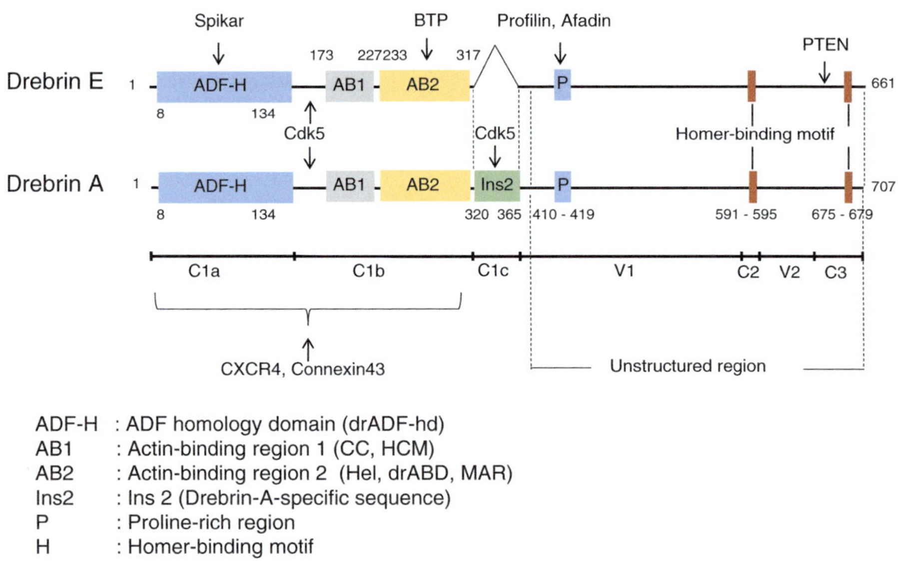

Fig. 1.3 Schematic representation of domain structures of rat drebrin. Drebrin A has "Ins2" insertion that is not present in drebrin E. Rat drebrin E corresponds to chick drebrin E2. ADF-H is described as drADF-hd in Chap. 5. Actin-binding region 1 (AB1) is comparable to coiled-coil domain (CC) and helical-charged motif (HCM) in Chaps. 4 and 5, respectively. Actin-binding region 2 is comparable to helical domain (Hel), drebrin actin-binding domain (drABD) and minimal actin-remodeling region in Chaps. 3, 4 and 21, respectively. C1, C2, and C3 are conserved regions between chicken and rodent, and V1 and V2 are variable regions. C1 is further subdivided into C1a, C1b, and C1c (see Chap. 2). Spikar binds to ADF-H. 3,5-Bistrifluoromethyl pyrazole (BTP) binds to AB2 region. Cyclin-dependent kinase 5 (CDK5) phosphorylates S142 and S342. Phosphatase and tensin homologue (PTEN) dephosphorylates S647

It is still unclear why rodents have one embryonic isoform, while chickens have two. The chicken *Dbn1* gene is over 15.4 kb, and the exons encoding the Ins1 and Ins2 sequences are separate (Kojima et al. 1993). In contrast, the mouse *Dbn1* gene contains exon 12a encoding the Ins2 sequence, which directly connects to exons12b and 12c (see details in Chap. 2). Either amino acid sequence of the chicken E2 or mouse E different from comparable drebrin A is only Ins2 (Jin et al. 2002). Therefore, mammalian drebrin E is thought to be homologous to chicken drebrin E2, but not drebrin E1.

1.2.6 Actin-Binding Activity of Drebrin

In 1994, the first solid proof for the actin-binding activity of drebrin E was reported (Ishikawa et al. 1994). We improved the purification method of drebrin E from rat brain to get enough protein for biochemical characterization. We have shown that drebrin E does not bind to actin monomers (G-actin) but binds to F-actin with a stoichiometry of 1:5 (drebrin E: actin) and an apparent dissociation constant (K_d) of 1.2×10^{-7} M.[1] Drebrin E competitively binds to F-actin with tropomyosin, fascin, and α-actinin.

On the other hand, drebrin A was not successfully purified from brain tissue for biochemical analyses, although we knew that drebrin A precipitated with F-actin. To examine whether drebrin A binds to F-actin similarly to drebrin E, we transfected various cell lines with GFP-tagged drebrin A cDNA and found that drebrin A and F-actin colocalize within a transfected cell and that tropomyosin disappears from drebrin A-bound F-actin (Shirao et al. 1994). This suggested that drebrin A also competitively binds to F-actin with tropomyosin. However, we had to wait until Ishikawa's group succeeded in purifying bacterially expressed drebrin A (Ishikawa et al. 2007) (more than 10 years) before we could elucidate the exact actin-binding property of drebrin A.

1.2.7 Drebrin Modifies the Actin Cytoskeleton in Dendritic Spines

As mentioned previously in the chicken optic tectum and cerebellum, drebrin A is mainly localized in dendritic spines (Shirao et al. 1987; Shirao and Obata 1986). In 1996, we studied the detailed localization of drebrin in the rat brain (Hayashi et al. 1996). Drebrin immunostaining is distributed in a dot-like pattern in the rat brain similarly to the chicken brain, suggesting a dendritic spine localization of drebrin.

---

[1] This K_d was quantified by densitometry of a Coomassie blue-stained SDS-PAGE gel for F-actin co-sedimentation assay. Therefore, we cannot exclude the possibility that the K_d of drebrin and F-actin is less than 10 nM (see details in Chap. 3).

However, the drebrin immunostaining intensity did not exactly correlate with the intensity of synaptophysin, a synaptic marker protein. Interestingly, immunoblot analysis showed that drebrin E was uniformly distributed at a low level throughout the brain, but drebrin A was expressed differentially, as it is abundant in the forebrain but present only at low levels in the cerebellar cortex. The high concentration of drebrin A in the forebrain suggests that the role of drebrin in the adult brain is related to learning and memory, which is characteristic of the forebrain. Electron microscopy showed that drebrin is localized at postsynaptic sites. Confocal microscopy of double labeling of drebrin and synaptophysin at the cerebral cortex showed a clear separation between synaptophysin-labeled presynaptic terminals and drebrin-labeled postsynaptic spines (Fig. 1.4a) (Hayashi et al. 1996).

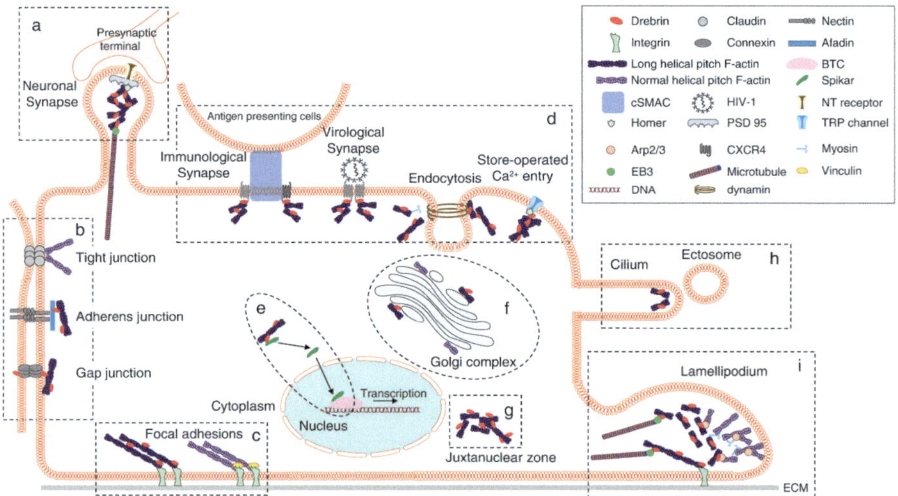

Fig. 1.4 Cellular functions of drebrin at a glance. (**a**) Drebrin forms stable F-actin pool at postsynaptic sites of excitatory synapses (Aoki et al. 2005). (**b**) Drebrin stabilizes nectin via afadin at adherens junctions (Rehm et al. 2013) and regulates gap junction via connexin (Butkevich et al. 2004). (**c**) Drebrin is not localized at vinculin-positive focal adhesions (Peitsch et al. 1999), but some focal adhesions are stabilized by drebrin (Ikeda et al. 1996). (**d**) Association of drebrin and chemokine receptor CXCR4 is enhanced by antigenic stimulation at immunological synapses (Perez-Martinez et al. 2010) and inhibits the entry of HIV-1 at virological synapses (Gordon-Alonso et al. 2013) and endocytosis (Li et al. 2017). Drebrin is also necessary for the store-operated Ca^{2+} channel function (Mercer et al. 2010) but inhibits TRP channel activity (Stiber et al. 2016). (**e**) Spikar, a transcription coactivator, enters into nucleus when it dissociates from drebrin (Yamazaki et al. 2014). (**f**) Drebrin is a specific component of small GTP-binding protein ARF-dependent actin pool on the Golgi complex (Fucini et al. 2000). (**g**) Drebrin is localized at the juxtanuclear drebrin-enriched zone, which is purportedly concerned with cell migration (Peitsch et al. 2006). (**h**) Drebrin mediates ectosome release from the tip of cilia (Nager et al. 2017). (**i**) Drebrin is localized at the transitional zone between lamellipodia and microtubules at the tip of cell processes (Geraldo et al. 2008; Mizui et al. 2009). Abbreviations: *ARF* ADP-ribosylation factor, *BTC* basal transcription complex, *cSMAC* central supramolecular activation cluster, *ECM* extracellular matrix, *HIV* human immunodeficiency virus, *NT* neurotransmitter, *TRP* transient receptor potential. Long and normal helical pitches of F-actin are 40 nm and 36 nm, respectively

Next, we examined the association of drebrin with the spine cytoskeleton. Actin, gelsolin, and myosin I, II, and V were immunoprecipitated as drebrin-containing cytoskeleton. On the other hand, microtubule-associated protein 2 (MAP2), fodrin, caldesmon, α-actinin, fascin, and tropomyosin, which are all detectable in the brain, were not detected within the immunoprecipitates. Thus, drebrin forms a distinct cytoskeletal complex with actin, gelsolin, and myosins in dendritic spines. Furthermore, we found that drebrin inhibits the actin-activated ATPase activity of myosin II (Hayashi et al. 1996). This suggests that drebrin regulates the actin-myosin interaction by the actin-linked control. It is known that caldesmon and tropomyosin, which have inhibitory effects on the actomyosin interaction similarly to drebrin, carry out the actin-linked control in smooth muscles (Ngai and Walsh 1984; Yamaguchi et al. 1984; Bugyi et al. 2010). Drebrin may take the place of caldesmon or tropomyosin in the actin-linked control of the actomyosin interaction in dendritic spines and may play a role in the structure-based plasticity of synapses.

To examine this hypothesis, we introduced excessive drebrin A into primary cultures of cortical neurons and analyzed its effect on the spine shape. In 1999, we reported that exogenous GFP-tagged drebrin A localized at the spines and induced spines to elongate (Hayashi and Shirao 1999). This was the first direct evidence that manipulation of a single actin-binding protein in a neuron specifically alters dendritic spine morphology. Although the underlining molecular mechanism was not clarified, we proposed the hypothesis that drebrin gives a unique character to the actin cytoskeleton bound to it (Shirao and Sekino 2001).

After these epoch-making findings, drebrin and the actin cytoskeleton in dendritic spines were thrown into the limelight, and in 2000 several papers focusing on drebrin localization were published by other groups (Allison et al. 2000; Conroy et al. 2000; Shoop et al. 2000) using the anti-drebrin monoclonal antibody, M2F6, developed in our laboratory. Since then, mAb M2F6 has been used in drebrin studies as the standard antibody worldwide.

1.2.8 Advances in Drebrin Studies in the Twenty-First Century

In the last 18 years, numerous drebrin functions related to various physiological phenomena, such as cell migration, synaptogenesis, and synaptic plasticity, have been elucidated extensively.

In 2006, we found the *N*-methyl-D-aspartate (NMDA) receptor activity-dependent change in drebrin localization (Sekino et al. 2006) and proposed the hypothesis that the drebrin localization change modifies the association of F-actin with other spine-resident actin-binding proteins, consequently leading to changes in spine morphology during synaptic plasticity (Sekino et al. 2006). However, it was difficult to discriminate neuronal F-actin from the glial one in a cortical mixed culture. To achieve precise studies about drebrin A and F-actin distribution in neurons, we innovated Banker's method (Kaech and Banker 2006) for our neuronal culture technique. In this method embryonic hippocampal neurons plated on a coverslip were cultured over a glial sheet in a dish. By immunocytochemically analyzing the pure neuronal culture on the coverslip, we could analyze the distribution of drebrin and F-actin without interference of the glial F-actin.

In two days in vitro (DIV) neurons, drebrin E but not A was concentrated in the growth cone, where F-actin predominantly localized in the peripheral half and microtubules localized in the central half. Drebrin E is localized only in the proximal area of the F-actin-rich region, which is the transitional zone to the central microtubule-rich region (Geraldo et al. 2008). Interestingly, the retrograde flow of drebrin-decorated F-actin was slower than that of other F-actins (Mizui et al. 2009).

In seven DIV neurons, drebrin immunostaining and F-actin were observed discontinuously at the submembranous region of the dendritic shaft and as a fibrous pattern in filopodia, except for the tip where only F-actin was observed. In 21 DIV neurons, strong drebrin immunostaining was colocalized with F-actin at the dendritic spines, but the colocalization at the dendritic shafts was decreased (Takahashi et al. 2003).

In 2008, we found that migrating neurons contain drebrin in their cell body even in the adult, which is different from the typical synaptic localization of drebrin A. Therefore, we hypothesized that migrating neurons contain drebrin E, but not drebrin A. Unfortunately, a drebrin E-specific antibody has not been successfully produced, yet, although a drebrin A-specific antibody (DAS2) was available. Therefore, it was difficult to determine whether the drebrin isoform distributed in migrating cell somata is drebrin E. Thus, we developed an image subtraction method using mAb M2F6 and DAS2 to identify the subcellular distribution of drebrin E (Song et al. 2008). Using this method, we found that drebrin E was widely distributed within the migrating neuron until drebrin A appeared. Once drebrin A is expressed in a developing neuron, drebrin E is no longer observed in the cell soma but is distributed in cell processes (Song et al. 2008).

As we mentioned previously, drebrin is also expressed in nonnervous tissues (Shirao and Obata 1986). The drebrin isoforms expressed in non-neuronal cells are only the embryonic types of drebrin, which do not have the Ins2 sequence in the molecule; hence, drebrin A, which has the Ins2 sequence, is not expressed in these cells. Many studies have been recently reported in relation to drebrin's role in various cellular functions, such as cell migration, cell process formation, intercellular communication, cancer metastasis, and spermatogenesis. These studies are discussed in more detail in specific chapters of this book, although some of them are also highlighted in the following section of this chapter (Fig. 1.4).

1.3 Physiological Roles of Drebrin

1.3.1 Morphogenetic Activity of Drebrin

One of the most striking features of drebrin function is the cell morphogenetic activity. In 1992, we demonstrated that overexpression of exogenous drebrin A in cultured fibroblasts induced many filopodia-like processes around the cell body, and some of them became highly branched long processes similar to neuronal dendrites (Fig. 1.5a) (Shirao et al. 1992). Time-lapse recording analysis of the transfected cells indicated that microvilli protruded from the cell bodies seldom retracted, but elongated and fused side-by-side, consequently forming neurite-like cell processes.

We examined the ultrastructural changes associated with process formation in drebrin A cDNA-transfected L cells by electron microscopy. Non-transfected L cells were large and epitheloid or spindle shaped with lamellipodia on the cell borders. Many short microvilli were present diffusely on cell surfaces, but not on lamellipodia. Surface and intracellular structures showed a diffused arrangement and had no special organization, except for stress fibers. On the other hand, transfected cells had round or polygonal cell bodies and neurite-like cell processes. The surfaces of

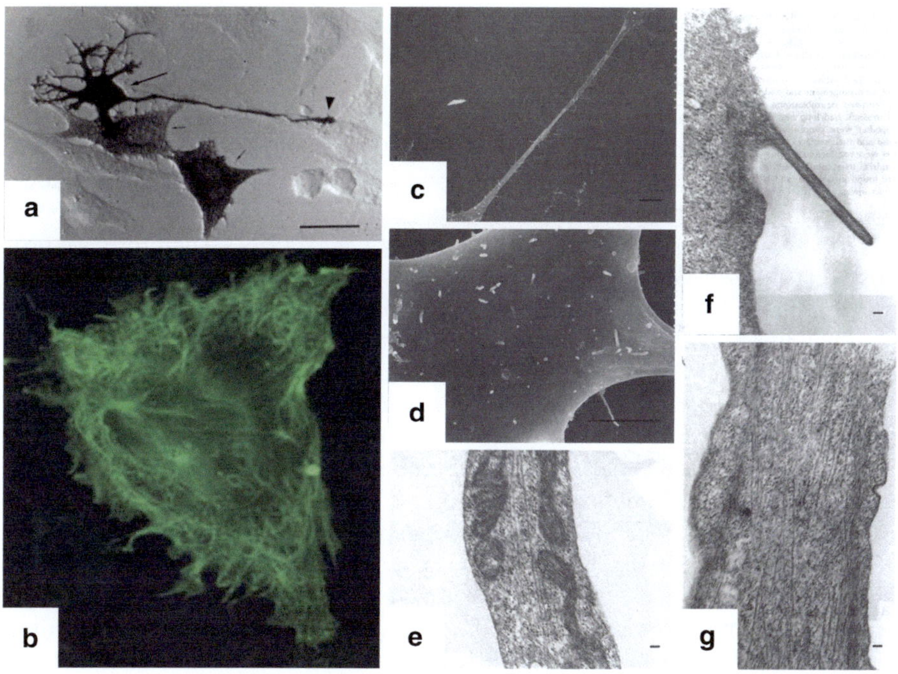

Fig. 1.5 Immunocytochemistry and immunoelectron microscopy of drebrin A transfected fibroblasts. (**a**) and (**b**) Transfected cells were subjected to immunostaining with anti-drebrin antibody (Mab M2F6). Intensely stained cell was neuron-like with round thick cell bodies and highly branched processes extending from the cell perimeter (**a**). Within a transfected cell, immunofluorescence of drebrin staining colocalized with F-actin forms thick curving bundles (**b**). C–G, Fine structures of drebrin A transfected fibroblasts. A neurite-like process of the transfectant showing microspikes (**c**). The round cell body shows occasional microvilli on its surface (**d**). Cell process containing microtubules and intermediate filaments in the central region (**e**). A microspike containing microfilaments connected to those in the cell process (**f**). Bundles of microfilaments were shown in the submembranous region of the cell process (**g**). Scale bars: (**a**), 40 μm; (**c**) and (**d**), 5 μm; (**e**), (**f**), and (**g**), 0.1 μm. (**c**) to (**g**) were reprinted from Inoue and Shirao (1997)

the cell bodies were smooth overall, although some microvilli were observed. The neurite-like cell processes had numerous microspikes on them. In the cell processes, mitochondria, microtubules, and intermediate filaments, as well as the submembranous F-actin bundles, were observed. In contrast, F-actins were only predominant structures in the microspikes (Fig. 1.5c–f) (Inoue and Shirao 1997).

We then immunocytochemically examined the distribution of drebrin A and F-actin within the cell. In the transfected cells, tropomyosin disappeared from F-actin, and drebrin A-decorated F-actin formed thick curving bundles that were different from the straight tropomyosin-binding stress fibers (Fig. 1.5b) (Shirao et al. 1994). How are thick curving F-actin bundle formed? Does drebrin A have F-actin-bundling activity?

Although drebrin E purified from brain tissue did not exhibit any actin-bundling activity (Ishikawa et al. 1994), bacterially expressed drebrin E and A did exhibit actin-bundling activity. In addition, molecular dissection of drebrin revealed two actin-binding regions (AB1 and AB2 in Fig. 1.3), and the fragment containing both regions had actin-bundling activity (Worth et al. 2013). These results suggest that drebrin has actin-bundling activity, and the bundled drebrin-decorated F-actin forms thick curving structures.

However, many cell lines express drebrin E without forming the thick curving F-actin bundles and keep straight tropomyosin-binding stress fibers. In addition, overexpression of AB2 region is enough to induce the thick actin bundles in Chinese hamster ovary (CHO) cells (Hayashi et al. 1999). Therefore, we propose that drebrin has actin-bundling activity under specific conditions. In the transfected CHO cells, overexpressed drebrin AB2 region may induce the specific conditions and allow endogenous drebrin E to bundle drebrin-decorated F-actins.

In 2009, we reported that drebrin-decorated F-actin shows slow treadmilling (Mizui et al. 2009). In 2011, Sharma et al. reported that drebrin induces remodeling of F-actin (Sharma et al. 2011). Drebrin-decorated F-actin shows unexpected unique physical and chemical characteristics, such as a longer helical crossover (40 nm), higher heat stability, and slower depolymerizing speed (Sharma et al. 2011; Mikati et al. 2013). Furthermore, the cooperative binding activity of drebrin to F-actin has been elucidated (Sharma et al. 2012). This suggests that the binding of drebrin to F-actin leads to the exclusion of other existing actin-binding proteins from F-actin and consequently the formation of F-actin fully covered by drebrin.

Thus, one possible underlining mechanism of the drebrin morphogenetic activity is that drebrin-decorated F-actin responds differently to various actin-regulating signals, resulting in the morphological change of cells. On the other hand, it is known that drebrin-decorated F-actin can link to microtubules (see below). Therefore, another possible mechanism is that submembranous drebrin-decorated F-actin bundles in the drebrin A cDNA-transfected cells enable microtubules to intrude into microvilli and consequently form cell processes.

1.3.2 Drebrin's Function in Synaptogenesis

During development, the expression of drebrin A in the brain is paralleled with synapse formation. Electron microscopy analysis of the developing rat brain has demonstrated that drebrin A locates at dendritic membranes at the initial contact sites between axons and dendrites (Aoki et al. 2005). In addition, we have shown using primary hippocampal cultured neurons that inhibition of drebrin A expression resulted in the delay of synapse formation as well as the inhibition of postsynaptic protein accumulation, such as postsynaptic density protein 95 (PSD-95), Ca^{2+}/calmodulin-dependent protein kinase II (CaMKII), spikar, and glutamate receptors (Takahashi et al. 2003; Yamazaki et al. 2014). It has been indicated that spikar (Yamazaki et al. 2014) and CaMKII (Yamazaki et al. submitted) accumulate by their direct binding activity to drebrin, while PSD-95 is likely to accumulate through the drebrin-Homer-Shank complex (Shiraishi-Yamaguchi et al. 2009; Tu et al. 1999).

How does drebrin A accumulate at postsynaptic sites? We have shown that synaptic activity governs drebrin A accumulation at postsynaptic sites (Takahashi et al. 2009). We analyzed drebrin dynamics within individual spines using the fluorescence recovery after photobleaching (FRAP) technique and found that α-amino-3-hydroxy-5-methyl-4-isoxazolepropionic acid (AMPA) receptor activity promotes drebrin A stabilization. This suggests the AMPA receptor-dependent appearance of the drebrin A-decorated F-actin complex at postsynaptic sites, which may function as a platform for molecular assembly of other postsynaptic proteins such as PSD-95, CaMKII, spikar, and glutamate receptors (Fig. 1.6).

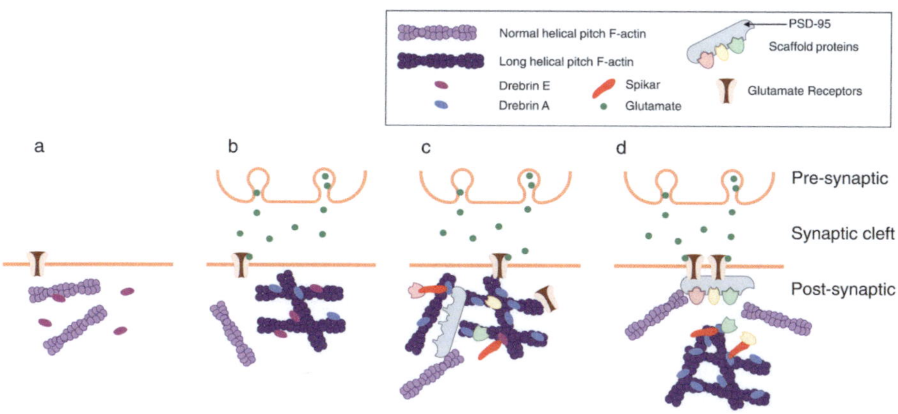

Fig. 1.6 Diagram of excitatory postsynaptic maturation. (**a**) Dynamic F-actin (normal helical pitch F-actin) is transported at cortical cytoplasm of dendritic shaft and dendritic filopodia. (**b**) Drebrin A accumulates at presumptive postsynaptic sites through AMPA receptor activity and forms drebrin A-decorated F-actin (long helical pitch F-actin). Note that drebrin E frequently binds to the long helix F-actin, but rarely to normal F-actin in vivo. (**c**) Long helical pitch F-actin builds up a platform for assembly of glutamate receptors, CaMKII, spikar, and scaffold proteins such as PSD-95. (**d**) Postsynaptic density structure is constructed in a mature spine, and drebrin A-decorated F-actin is located relatively far from the postsynaptic membrane in a fully mature spine

1.3.3 Drebrin's Function in Synaptic Plasticity

It has been shown that stable and dynamic F-actins localize at the core and periphery of dendritic spines, respectively (Honkura et al. 2008). As we mentioned, drebrin-decorated F-actin is stable with slow treadmilling and a longer helical crossover. Super resolution microscopy and electron microscopy (see details in Chap. 8) studies have shown that drebrin is located at the center of dendritic spines, indicating that stable F-actin at the core of dendritic spines consists of drebrin-decorated F-actin.

We have recently demonstrated that NMDA receptor-dependent Ca^{2+} influx activates myosin II ATPase and consequently induces the exodus of drebrin A-decorated F-actin from dendritic spines, resulting in the initiation of morphological synaptic plasticity (Mizui et al. 2014).

Taken together, we propose that synapses show morphological stability and reactivity when drebrin A-decorated F-actin is present and absent, respectively. When NMDA receptor is activated by a strong input that will induce synaptic plasticity (Hebbian plasticity), the spine gets the reactivity, because of the drebrin exodus from dendritic spines. After the spines have changed their morphology, drebrin reforms drebrin A-decorated F-actin at the core region and gives the morphological stability to the spines. This idea is consistent with our recent finding that synaptic plasticity is impaired in drebrin A-specific knockout (DAKO) mice (Kojima et al. 2016).

In addition to the aforementioned Hebbian synaptic plasticity, the brain has a homeostatic synaptic plasticity, which is a compensatory adjustment of synaptic excitability relative to the network activity. In 2001, we proposed that NMDA receptor accumulates through a specific trafficking system associated with drebrin and is anchored to F-actin via α-actinin at the postsynaptic membrane (Shirao and Sekino 2001). Moreover, 5 years later we showed that drebrin is involved in the mechanism of activity-dependent synaptic targeting of NMDA receptors using cultured neurons (Takahashi et al. 2006). Furthermore, Aoki et al. showed using DAKO mice that the lack of drebrin A results in the loss of homeostatic plasticity of NMDA receptor in vivo (Aoki et al. 2009).

In the brain of Alzheimer's disease patients, drebrin disappears from dendritic spines even though the synapses are still there (Harigaya et al. 1996). Moreover, the decrease in drebrin is observed in relation to the cognitive impairment associated with normal aging (Hatanpaa et al. 1999). The loss of drebrin A means loss of stable drebrin A-decorated F-actin, suggesting that the compatibility between the stability and the reactivity of synapses dissociates in Alzheimer's disease.

1.3.4 Drebrin's Function in Intercellular Communication

Various drebrin-associated proteins have been found thick and fast. Interestingly, drebrin and its partner proteins are often found in the structure for cell-cell communication in recipient cells as well as the postsynaptic structure of neurons. Connexin 43 binds to drebrin E and forms a supramolecular complex of drebrin-decorated F-actin at the gap junction in astrocytes (Butkevich et al. 2004) and in migrating neurons (Ambrosi et al. 2016). At adherens junctions, drebrin E binds to afadin and acts as a stabilizer of nectins by linking the nectin-afadin complex to cortical F-actin and thus preserves endothelial integrity under shear stress (Fig. 1.4b) (Rehm et al. 2013).

In $CD4^+$ T cells, the chemokine receptor, CXCR4, is associated with drebrin E (Fig. 1.4C) (Perez-Martinez et al. 2010). This association is enhanced by antigenic stimulation and is necessary for complete activation of $CD4^+$ T cells. Furthermore, drebrin negatively regulates human immunodeficiency virus (HIV)-1 infection of CD^{4+} T lymphocytes, where CXCR4 is known as an HIV-1 co-receptor (Gordon-Alonso et al. 2013). Interestingly, drebrin E is a target of the immunosuppressant 3,5-bistrifluoromethyl pyrazole (BTP), which inhibits calcium influx into cells. Drebrin is necessary for the store-operated Ca^{2+} entry (Fig. 1.4c) (Mercer et al. 2010). Particularly, in mast cells, drebrin E regulates allergic responses, namely, the FcεRI-mediated increase in intracellular Ca^{2+} and the IgE-mediated histamine release (Law et al. 2015).

In the testis, drebrin E plays a role through association with the Arp2/3 complex at the ectoplasmic specialization, conferring plasma membrane plasticity, accommodating changes in spermatid shape, promoting germ cell transport, and inducing junction restructuring (Li et al. 2011).

Taken together, drebrin E functions as a dynamic linker between membrane proteins and submembranous F-actin structures, to regulate the intercellular communication.

1.3.5 Drebrin-Decorated F-Actin Can Link to Microtubules

Dynamic interactions between F-actin and microtubules underlie various cellular processes; however, the molecular mechanism involved was unknown. Drebrin-decorated F-actins in neurons are localized in adjoining zones to the microtubule-rich region; e.g., the dendritic spines (Fig. 1.4a) are in contact with the microtubule-rich dendritic shaft, and the transitional zone of the axonal growth cone (Fig. 1.4i) is between F-actins and microtubules. This characteristic localization suggests that drebrin may link microtubules to F-actin. In 2008, it was reported that drebrin binds to the microtubule plus-end binding protein, EB3, indicating that drebrin-decorated F-actin can link to microtubules via EB3 (Geraldo et al. 2008).

1.3.6 Drebrin and Cancer

Cancer cells break the intercellular connection found in normal cells and show motility ability as well as a dramatic change in cellular morphology. Because drebrin activity is involved in these cellular processes, the role of drebrin in carcinogenesis has been actively studied. In 2005, the change in drebrin E expression in skin cancers and their precursors was reported (Peitsch et al. 2005). Of particular note is the fact that drebrin E is markedly overexpressed in metastatic cancer cells (Lin et al. 2014). Recently, it has been reported that drebrin is critical for progranulin-dependent activation of the Akt and MAPK pathways and modulates motility, invasion, and anchorage-independent growth in bladder carcinomas (Xu et al. 2015).

1.4 Perspective and Future Directions

The compatibility between the stability and the reactivity of synapses plays a pivotal role in synaptic plasticity. Dendritic filopodia seem too motile to achieve this compatibility. On the other hand, dendritic spines are suitable for keeping this compatibility. Drebrin is present both in filopodia and spines, and the drebrin isoform change from E to A occurs in parallel with the change of filopodia into spines. Furthermore, inhibition of drebrin isoform conversion forms aberrant synapses, which impairs long-term potentiation (LTP). The difference between the two isoforms is the 46-amino-acid sequence, Ins2, which is inserted in the middle of the drebrin molecule by alternative splicing. It is of great interest to elucidate the role of this sequence in drebrin function. In addition, elucidation of the alternative splicing mechanism of drebrin may make an even greater impact on developmental neurobiology, because many brain proteins change their isoform by alternative splicing simultaneously with drebrin around the synaptogenesis period.

Drebrin forms slow treadmilling F-actin with a longer helical crossover. It is now believed that the structural heterogeneity of F-actin has an important physiological role by affecting the recruitment and occupancy of actin regulators. Drebrin-decorated F-actin is often found at the recipient side of various cell-cell communication systems, and the lack of drebrin leads to the dysfunction of cell-cell communication. In the brain, drebrin is postsynaptic, and its deficiency leads to dementia. In the immune system, drebrin is in the immunological synapses, regulating viral infection and modifying allergic responses. In carcinogenesis, drebrin is at intercellular adherens junctions, and its expression level is associated with cancer metastasis. Therefore, we propose that drebrin-decorated F-actin creates a unique platform for molecular assembly necessary for intercellular communication.

References

Allison DW, Chervin AS, Gelfand VI, Craig AM (2000) Postsynaptic scaffolds of excitatory and inhibitory synapses in hippocampal neurons: maintenance of core components independent of actin filaments and microtubules. J Neurosci 20:4545–4554

Ambrosi C, Ren C, Spagnol G, Cavin G, Cone A, Grintsevich EE, Sosinsky GE, Sorgen PL (2016) Connexin43 forms Supramolecular complexes through non-overlapping binding sites for drebrin, tubulin, and ZO-1. PLoS One 11:e0157073

Aoki C, Sekino Y, Hanamura K, Fujisawa S, Mahadomrongkul V, Ren Y, Shirao T (2005) Drebrin A is a postsynaptic protein that localizes in vivo to the submembranous surface of dendritic sites forming excitatory synapses. J Comp Neurol 483:383–402

Aoki C, Kojima N, Sabaliauskas N, Shah L, Ahmed TH, Oakford J, Ahmed T, Yamazaki H, Hanamura K, Shirao T (2009) Drebrin a knockout eliminates the rapid form of homeostatic synaptic plasticity at excitatory synapses of intact adult cerebral cortex. J Comp Neurol 517:105–121

Brackenbury R, Thiery JP, Rutishauser U, Edelman GM (1977) Adhesion among neural cells of the chick embryo. I. An immunological assay for molecules involved in cell-cell binding. J Biol Chem 252:6835–6840

Bugyi B, Didry D, Carlier MF (2010) How tropomyosin regulates lamellipodial actin-based motility: a combined biochemical and reconstituted motility approach. EMBO J 29:14–26

Butkevich E, Hulsmann S, Wenzel D, Shirao T, Duden R, Majoul I (2004) Drebrin is a novel connexin-43 binding partner that links gap junctions to the submembrane cytoskeleton. Curr Biol 14:650–658

Conroy WG, Ogden LF, Berg DK (2000) Cluster formation of alpha7-containing nicotinic receptors at interneuronal interfaces in cell culture. Neuropharmacology 39:2699–2705

Fucini RV, Navarrete A, Vadakkan C, Lacomis L, Erdjument-Bromage H, Tempst P, Stamnes M (2000) Activated ADP-ribosylation factor assembles distinct pools of actin on golgi membranes. J Biol Chem 275:18824–18829

Geraldo S, Khanzada UK, Parsons M, Chilton JK, Gordon-Weeks PR (2008) Targeting of the F-actin-binding protein drebrin by the microtubule plus-tip protein EB3 is required for neuritogenesis. Nat Cell Biol 10:1181–1189

Gordon-Alonso M, Rocha-Perugini V, Alvarez S, Ursa A, Izquierdo-Useros N, Martinez-Picado J, Munoz-Fernandez MA, Sanchez-Madrid F (2013) Actin-binding protein drebrin regulates HIV-1-triggered actin polymerization and viral infection. J Biol Chem 288:28382–28397

Harigaya Y, Shoji M, Shirao T, Hirai S (1996) Disappearance of actin-binding protein, drebrin, from hippocampal synapses in Alzheimer's disease. J Neurosci Res 43:87–92

Hatanpaa K, Isaacs KR, Shirao T, Brady DR, Rapoport SI (1999) Loss of proteins regulating synaptic plasticity in normal aging of the human brain and in Alzheimer disease. J Neuropathol Exp Neurol 58:637–643

Hayashi K, Shirao T (1999) Change in the shape of dendritic spines caused by overexpression of drebrin in cultured cortical neurons. J Neurosci 19:3918–3925

Hayashi K, Ishikawa R, Ye LH, He XL, Takata K, Kohama K, Shirao T (1996) Modulatory role of drebrin on the cytoskeleton within dendritic spines in the rat cerebral cortex. J Neurosci 16:7161–7170

Hayashi K, Ishikawa R, Kawai-Hirai R, Takagi T, Taketomi A, Shirao T (1999) Domain analysis of the actin-binding and actin-remodeling activities of drebrin. Exp Cell Res 253:673–680

Hoffman S, Sorkin BC, White PC, Brackenbury R, Mailhammer R, Rutishauser U, Cunningham BA, Edelman GM (1982) Chemical characterization of a neural cell adhesion molecule purified from embryonic brain membranes. J Biol Chem 257:7720–7729

Honkura N, Matsuzaki M, Noguchi J, Ellis-Davies GC, Kasai H (2008) The subspine organization of actin fibers regulates the structure and plasticity of dendritic spines. Neuron 57:719–729

Ikeda K, Kaub PA, Asada H, Uyemura K, Toya S, Shirao T (1996) Stabilization of adhesion plaques by the expression of drebrin A in fibroblasts. Brain Res Dev Brain Res 91: 227–236

Inoue HK, Shirao T (1997) Neurite formation induced in neuroblastoma cells and genetically altered non-neuronal cells. J Electron Microsc 46:497–502

Ishikawa R, Hayashi K, Shirao T, Xue Y, Takagi T, Sasaki Y, Kohama K (1994) Drebrin, a development-associated brain protein from rat embryo, causes the dissociation of tropomyosin from actin filaments. J Biol Chem 269:29928–29933

Ishikawa R, Katoh K, Takahashi A, Xie C, Oseki K, Watanabe M, Igarashi M, Nakamura A, Kohama K (2007) Drebrin attenuates the interaction between actin and myosin-V. Biochem Biophys Res Commun 359:398–401

Jin M, Tanaka S, Sekino Y, Ren Y, Yamazaki H, Kawai-Hirai R, Kojima N, Shirao T (2002) A novel, brain-specific mouse drebrin: cDNA cloning, chromosomal mapping, genomic structure, expression, and functional characterization. Genomics 79:686–692

Kaech S, Banker G (2006) Culturing hippocampal neurons. Nat Protoc 1:2406–2415

Koganezawa N, Hanamura K, Sekino Y, Shirao T (2017) The role of drebrin in dendritic spines. Mol Cell Neurosci. doi:10.1016/j.mcn.2017.01.004

Kojima N, Kato Y, Shirao T, Obata K (1988) Nucleotide sequences of two embryonic drebrins, developmentally regulated brain proteins, and developmental change in their mRNAs. Brain Res 464:207–215

Kojima N, Shirao T, Obata K (1993) Molecular cloning of a developmentally regulated brain protein, chicken drebrin A and its expression by alternative splicing of the drebrin gene. Brain Res Mol Brain Res 19:101–114

Kojima N, Yasuda H, Hanamura K, Ishizuka Y, Sekino Y, Shirao T (2016) Drebrin A regulates hippocampal LTP and hippocampus-dependent fear learning in adult mice. Neuroscience 324:218–226

LaVail JH, Cowan WM (1971a) The development of the chick optic tectum. I. Normal morphology and cytoarchitectonic development. Brain Res 28:391–419

LaVail JH, Cowan WM (1971b) The development of the chick optic tectum. II. Autoradiographic studies. Brain Res 28:421–441

Law M, Lee Y, Morales JL, Ning G, Huang W, Pabon J, Kannan AK, Jeong AR, Wood A, Carter C et al (2015) Cutting edge: drebrin-regulated actin dynamics regulate IgE-dependent mast cell activation and allergic responses. J Immunol 195:426–430

Li B, Ding S, Feng N et al (2017a) Drebrin restricts rotavirus entry by inhibiting dynamin-mediated endocytosis. Proc Natl Acad Sci U S A 114:E3642–E3651

Li MW, Xiao X, Mruk DD, Lam YL, Lee WM, Lui WY, Bonanomi M, Silvestrini B, Cheng CY (2011) Actin-binding protein drebrin E is involved in junction dynamics during spermatogenesis. Spermatogenesis 1:123–136

Lin Q, Tan HT, Lim TK, Khoo A, Lim KH, Chung MC (2014) iTRAQ analysis of colorectal cancer cell lines suggests Drebrin (DBN1) is overexpressed during liver metastasis. Proteomics 14:1434–1443

Luna EJ, Pestonjamasp KN, Cheney RE, Strassel CP, Lu TH, Chia CP, Hitt AL, Fechheimer M, Furthmayr H, Mooseker MS (1997) Actin-binding membrane proteins identified by F-actin blot overlays. Soc Gen Physiol Ser 52:3–18

Mercer JC, Qi Q, Mottram LF, Law M, Bruce D, Iyer A, Morales JL, Yamazaki H, Shirao T, Peterson BR et al (2010) Chemico-genetic identification of drebrin as a regulator of calcium responses. Int J Biochem Cell Biol 42:337–345

Mikati MA, Grintsevich EE, Reisler E (2013) Drebrin-induced stabilization of actin filaments. J Biol Chem 288:19926–19938

Mizui T, Kojima N, Yamazaki H, Katayama M, Hanamura K, Shirao T (2009) Drebrin E is involved in the regulation of axonal growth through actin-myosin interactions. J Neurochem 109:611–622

Mizui T, Sekino Y, Yamazaki H, Ishizuka Y, Takahashi H, Kojima N, Kojima M, Shirao T (2014) Myosin II ATPase activity mediates the long-term potentiation-induced exodus of stable F-actin bound by drebrin A from dendritic spines. PLoS One 9:e85367

Nager AR, Goldstein JS, Herranz-Perez V, Portran D, Ye F, Garcia-Verdugo JM, Nachury MV (2017) An actin network dispatches ciliary GPCRs into extracellular vesicles to modulate signaling. Cell 168(252–263):e214

Ngai PK, Walsh MP (1984) Inhibition of smooth muscle actin-activated myosin Mg2+−ATPase activity by caldesmon. J Biol Chem 259:13656–13659

O'Farrell PH (1975) High resolution two-dimensional electrophoresis of proteins. J Biol Chem 250:4007–4021

Peitsch WK, Grund C, Kuhn C, Schnolzer M, Spring H, Schmelz M, Franke WW (1999) Drebrin is a widespread actin-associating protein enriched at junctional plaques, defining a specific microfilament anchorage system in polar epithelial cells. Eur J Cell Biol 78:767–778

Peitsch WK, Hofmann I, Bulkescher J, Hergt M, Spring H, Bleyl U, Goerdt S, Franke WW (2005) Drebrin, an actin-binding, cell-type characteristic protein: induction and localization in epithelial skin tumors and cultured keratinocytes. J Invest Dermatol 125:761–774

Peitsch WK, Bulkescher J, Spring H, Hofmann I, Goerdt S, Franke WW (2006) Dynamics of the actin-binding protein drebrin in motile cells and definition of a juxtanuclear drebrin-enriched zone. Exp Cell Res 312:2605–2618

Perez-Martinez M, Gordon-Alonso M, Cabrero JR, Barrero-Villar M, Rey M, Mittelbrunn M, Lamana A, Morlino G, Calabia C, Yamazaki H et al (2010) F-actin-binding protein drebrin regulates CXCR4 recruitment to the immune synapse. J Cell Sci 123:1160–1170

Rehm K, Panzer L, van Vliet V, Genot E, Linder S (2013) Drebrin preserves endothelial integrity by stabilizing nectin at adherens junctions. J Cell Sci 126:3756–3769

Sekino Y, Tanaka S, Hanamura K, Yamazaki H, Sasagawa Y, Xue Y, Hayashi K, Shirao T (2006) Activation of N-methyl-D-aspartate receptor induces a shift of drebrin distribution: disappearance from dendritic spines and appearance in dendritic shafts. Mol Cell Neurosci 31:493–504

Sekino Y, Kojima N, Shirao T (2007) Role of actin cytoskeleton in dendritic spine morphogenesis. Neurochem Int 51:92–104

Sharma S, Grintsevich EE, Phillips ML, Reisler E, Gimzewski JK (2011) Atomic force microscopy reveals drebrin induced remodeling of f-actin with subnanometer resolution. Nano Lett 11:825–827

Sharma S, Grintsevich EE, Hsueh C, Reisler E, Gimzewski JK (2012) Molecular cooperativity of drebrin1-300 binding and structural remodeling of F-actin. Biophys J 103:275–283

Shiraishi-Yamaguchi Y, Sato Y, Sakai R, Mizutani A, Knopfel T, Mori N, Mikoshiba K, Furuichi T (2009) Interaction of Cupidin/Homer2 with two actin cytoskeletal regulators, Cdc42 small GTPase and drebrin, in dendritic spines. BMC Neurosci 10:25

Shirao T, Obata K (1985) Two acidic proteins associated with brain development in chick embryo. J Neurochem 44:1210–1216

Shirao T, Obata K (1986) Immunochemical homology of 3 developmentally regulated brain proteins and their developmental change in neuronal distribution. Brain Res 394:233–244

Shirao T, Sekino Y (2001) Clustering and anchoring mechanisms of molecular constituents of postsynaptic scaffolds in dendritic spines. Neurosci Res 40:1–7

Shirao T, Inoue HK, Kano Y, Obata K (1987) Localization of a developmentally regulated neuron-specific protein S54 in dendrites as revealed by immunoelectron microscopy. Brain Res 413:374–378

Shirao T, Kojima N, Kato Y, Obata K (1988) Molecular cloning of a cDNA for the developmentally regulated brain protein, drebrin. Brain Res 464:71–74

Shirao T, Kojima N, Obata K (1992) Cloning of drebrin A and induction of neurite-like processes in drebrin-transfected cells. Neuroreport 3:109–112

Shirao T, Hayashi K, Ishikawa R, Isa K, Asada H, Ikeda K, Uyemura K (1994) Formation of thick, curving bundles of actin by drebrin A expressed in fibroblasts. Exp Cell Res 215:145–153

Shirao T, Hanamura K, Koganezawa N, Ishizuka Y, Yamazaki H, Sekino Y (2017) The role of drebrin in neurons. J Neurochem 141(6):819–834

Shoop RD, Yamada N, Berg DK (2000) Cytoskeletal links of neuronal acetylcholine receptors containing alpha 7 subunits. J Neurosci 20:4021–4029

Song M, Kojima N, Hanamura K, Sekino Y, Inoue HK, Mikuni M, Shirao T (2008) Expression of drebrin E in migrating neuroblasts in adult rat brain: coincidence between drebrin E disappearance from cell body and cessation of migration. Neuroscience 152:670–682

Sperry RW (1963) Chemoaffinity in the orderly growth of nerve fiber patterns and connections. Proc Natl Acad Sci U S A 50:703–710

Stiber JA, Wu JH, Zhang L, Nepliouev I, Zhang ZS, Bryson VG, Brian L, Bentley RC, Gordon-Weeks PR, Rosenberg PB et al (2016) The actin-binding protein drebrin inhibits neointimal hyperplasia. Arterioscler Thromb Vasc Biol 36:984–993

Takahashi H, Sekino Y, Tanaka S, Mizui T, Kishi S, Shirao T (2003) Drebrin-dependent actin clustering in dendritic filopodia governs synaptic targeting of postsynaptic density-95 and dendritic spine morphogenesis. J Neurosci 23:6586–6595

Takahashi H, Mizui T, Shirao T (2006) Down-regulation of drebrin A expression suppresses synaptic targeting of NMDA receptors in developing hippocampal neurones. J Neurochem 97(Suppl 1):110–115

Takahashi H, Yamazaki H, Hanamura K, Sekino Y, Shirao T (2009) Activity of the AMPA receptor regulates drebrin stabilization in dendritic spine morphogenesis. J Cell Sci 122: 1211–1219

Tanabe K, Yamazaki H, Inaguma Y, Asada A, Kimura T, Takahashi J, Taoka M, Ohshima T, Furuichi T, Isobe T et al (2014) Phosphorylation of drebrin by cyclin-dependent kinase 5 and its role in neuronal migration. PLoS One 9:e92291

Toda M, Shirao T, Minoshima S, Shimizu N, Toya S, Uyemura K (1993) Molecular cloning of cDNA encoding human drebrin E and chromosomal mapping of its gene. Biochem Biophys Res Commun 196:468–472

Trisler GD, Schneider MD, Nirenberg M (1981) A topographic gradient of molecules in retina can be used to identify neuron position. Proc Natl Acad Sci U S A 78:2145–2149

Tu JC, Xiao B, Naisbitt S, Yuan JP, Petralia RS, Brakeman P, Doan A, Aakalu VK, Lanahan AA, Sheng M et al (1999) Coupling of mGluR/Homer and PSD-95 complexes by the shank family of postsynaptic density proteins. Neuron 23:583–592

Worth DC, Daly CN, Geraldo S, Oozeer F, Gordon-Weeks PR (2013) Drebrin contains a cryptic F-actin-bundling activity regulated by Cdk5 phosphorylation. J Cell Biol 202:793–806

Xu SQ, Buraschi S, Morcavallo A et al (2015) A novel role for drebrin in regulating progranulin bioactivity in bladder cancer. Oncotarget 6:10825–10839

Yamaguchi M, Ver A, Carlos A, Seidel JC (1984) Modulation of the actin-activated adenosinetriphosphatase activity of myosin by tropomyosin from vascular and gizzard smooth muscles. Biochemistry 23:774–779

Yamazaki H, Kojima N, Kato K, Hirose E, Iwasaki T, Mizui T, Takahashi H, Hanamura K, Roppongi RT, Koibuchi N et al (2014) Spikar, a novel drebrin-binding protein, regulates the formation and stabilization of dendritic spines. J Neurochem 128:507–522

Yamazaki H, Sasagawa Y, Yamamoto H, Shirao T. CaMKIIβ is localized in dendritic spines as both drebrin-dependent and drebrin-independent pools (submitted)

Part II
Basic Information About Drebrin

Chapter 2
Molecular Cloning of Drebrin: Progress and Perspectives

Nobuhiko Kojima

Abstract Chicken drebrin isoforms were first identified in the optic tectum of developing brain. Although the time course of protein expression was different in each drebrin isoform, the similarity between their protein structures was suggested by biochemical analysis of purified protein. To determine their protein structures, the cloning of drebrin cDNAs was conducted. Comparison between the cDNA sequences shows that all drebrin cDNAs are identical except that the internal insertion sequences are present or absent in their sequences. Chicken drebrin are now classified into three isoforms, namely, drebrins E1, E2, and A. Genomic cloning demonstrated that the three isoforms are generated by an alternative splicing of individual exons encoding the insertion sequences from single drebrin gene. The mechanism should be precisely regulated in cell-type-specific and developmental stage-specific fashion. Drebrin protein, which is well conserved in various vertebrate species, although mammalian drebrin has only two isoforms, namely, drebrin E and drebrin A, is different from chicken drebrin that has three isoforms. Drebrin belongs to an actin-depolymerizing factor homology (ADF-H) domain protein family. Besides the ADF-H domain, drebrin has other domains, including the actin-binding domain and Homer-binding motifs. Diversity of protein isoform and multiple domains of drebrin could interact differentially with the actin cytoskeleton and other intracellular proteins and regulate diverse cellular processes.

Keywords Neural development • cDNA cloning • Alternative splicing • Homology Protein family • Actin-binding • ADF-H domain

2.1 Introduction

The organization of the nervous system is accomplished by the events which are precisely and genetically programmed during neural development. These events include neurogenesis, cell migration, axon guidance, and synapse formation. These

N. Kojima
Faculty of Life Sciences, Toyo University, Itakura-machi, Ora-gun, Gunma 374-093, Japan
e-mail: kojima033@toyo.jp

T. Shirao, Y. Sekino (eds.), *Drebrin*, Advances in Experimental Medicine and Biology 1006, DOI 10.1007/978-4-431-56550-5_2

events are likely to depend on developmental stage-dependent and cell-type-specific expression of key molecules. We demonstrated that drebrin is one of these molecules and may contribute in part to the molecular mechanisms of neural development.

Chicken drebrin isoforms were first identified as three developmentally regulated proteins (S5, S6, and S54, later renamed as E1, E2, and A) in chicken developing optic tectum by two-dimensional gel electrophoresis (Shirao and Obata 1985). Peptide mapping of partial digested proteins by a protease revealed that the digestion pattern of these proteins resembles among drebrins E1, E2, and A, suggesting that these three proteins are structurally related with each other.

Drebrin-like immunoreactive proteins exist in rat brain (Shirao et al. 1989). The proteins also showed heterogeneity, and their expression is developmentally regulated. Their independent changes in expression during neural development suggest that each play a unique role at a particular developmental stage. The distinct functions should be caused by their distinct molecular structures and developmental stage- and cell-type-specific control of expression of these proteins. To clarify their structural relationship and biological functions, the structural differences among these drebrin proteins, as well as their genomic organization, should be clarified. Cloning of drebrin cDNAs and gene revealed that the structures of these three proteins were identical except for the presence or absence of the internal sequences, and these are derived from a single gene by alternative splicing mechanism (Kojima et al. 1988, 1993). Interspecies comparison of the deduced amino acid sequences of drebrin shows a high degree of similarity, although the number of isoforms is different between chicken and mammals (Kojima et al. 1993). This suggests that functions of drebrin are important beyond species. We describe here a progress in 30 years and perspectives through the molecular cloning of drebrin.

2.2 cDNA Structure of Each Drebrin

Three isoforms of drebrin have been first identified in chicken developing optic tectum as S5, S6, and S54 of those spots are developmentally changed in their protein contents in two-dimensional gel electrophoresis (Shirao and Obata 1985). Peptide mapping using *Staphylococcus aureus* V8 protease, which selectively cleaves peptide bonds on the carboxyl side of glutamate (and in some case aspartate) residues, revealed that the digestion pattern of these proteins resembles between S5, S6, and S54, suggesting that these three proteins are structurally related with each other. Thereafter we renamed these proteins S5, S6, and S54 to drebrins E1, E2, and A, respectively.

To determine the protein structure of them, we tried to isolate the cDNA clones encoding drebrin mRNA using anti-drebrin antibody. For this purpose, we screened a lambda gt11 cDNA library prepared from chick 10-day-old embryo mRNAs. The lambda gt11 phage produces stable fusion proteins of β-galactosidase and the products of mRNAs having the same orientation and reading frame as lacZ. Thus, only

1/6 clones containing the desired sequence will express the protein. Also, incomplete cDNA clone missing important epitopes would not be recognized by the antibodies. We successfully obtained one cDNA clone (Dcw1) from a lambda gt11 chick 10-day-old embryo cDNA library by plaque hybridization with anti-drebrin antiserum (Shirao et al. 1988). The fusion protein of Dcw1 reacted with only three out of five individual monoclonal antibodies, and the length of inserted cDNA of Dcw1 was about half of that of drebrin mRNA detected by Northern blot analysis, indicating that the Dcw1 does not cover the entire cDNA for drebrin. To isolate the cDNA clones that cover the full-length drebrin sequence, Dcw1 was employed as a probe to rescreen the same cDNA library. Screening using cDNA probe instead of antibodies allows us to get the cDNA clones carrying the drebrin sequences, even though they are not inserted in the vector with the same orientation and in-flame with lacZ's reading frame. Consequently, 74 positive clones in which the cDNA inserts were strongly hybridized with the Dcw1 probe (Kojima et al. 1988). They were divided into two groups by restriction mapping. Nucleotide sequences of cDNA of Dcw6 and Dcw17, each of which was longest in each group, were separately determined (Fig. 2.1a). They were identical except that an internal 129-bp sequence, named Ins1, was present in Dcw17 and absent in Dcw6. To determine the drebrin isoform to which each cDNA clone corresponded, we raised antiserum (named anti-pep1) against the synthetic polypeptide pep1 that was the last 25 amino acid sequence of deduced Ins1. Immunoblot analysis showed that the anti-pep1 antiserum specifically reacted with drebrins E2 and A, but not drebrin E1 (Kojima et al. 1988), suggesting that drebrins E2 and A have Ins1 sequence, whereas E1 does not. Using a 80-bp sequence within Ins1 as a probe, we further isolated another group of cDNA (Dcb21) encoding the drebrin mRNA from the newly hatched chicken brain cDNA library (Kojima et al. 1993). Dcb21 had another internal 138-bp sequence, named Ins2, in the 5' direction of immediately upstream from Ins1, and the remaining part of the sequence was identical to Dcw17 (Fig. 2.1a). Calculated molecular weights of the proteins encoded in the full-length sequences of the three cDNAs Dcw6, Dcw17, and Dcb21 are 62,165, 66,553, and 71,532, respectively. Although they are about two thirds of the molecular weights estimated by sodium-dodecyl-sulfate polyacrylamide gel electrophoresis (SDS-PAGE), immunoblot analysis showed that protein generated by in vitro transcription of each full-length cDNA, Dcw6, Dcw17, and Dcb21, and translation of their mRNAs in a rabbit reticulocyte lysate co-migrates with drebrins E1, E2, and A, respectively (Kojima et al. 1993), indicating that each of the three cDNAs corresponds to drebrins E1, E2, and A (Fig. 2.1a). The discrepancy between the molecular weights of drebrins predicted from the cDNA sequences and the apparent molecular weights estimated by SDS-PAGE seems to result from the anomalous migration of drebrins on SDS polyacrylamide gel, i.e., drebrins do not bind a constant weight ratio of SDS due to the unusual features of their amino acid sequences, such as a paucity of hydrophobic regions and an abundance of negatively charged amino acid residues. We raised antisera against various synthetic peptides, which covered a part of the deduced amino acid sequence of Ins2 from rat drebrin A cDNA (Shirao et al. 1992) described below, and two of them, namely DAS1 and DAS2, specifically reacted

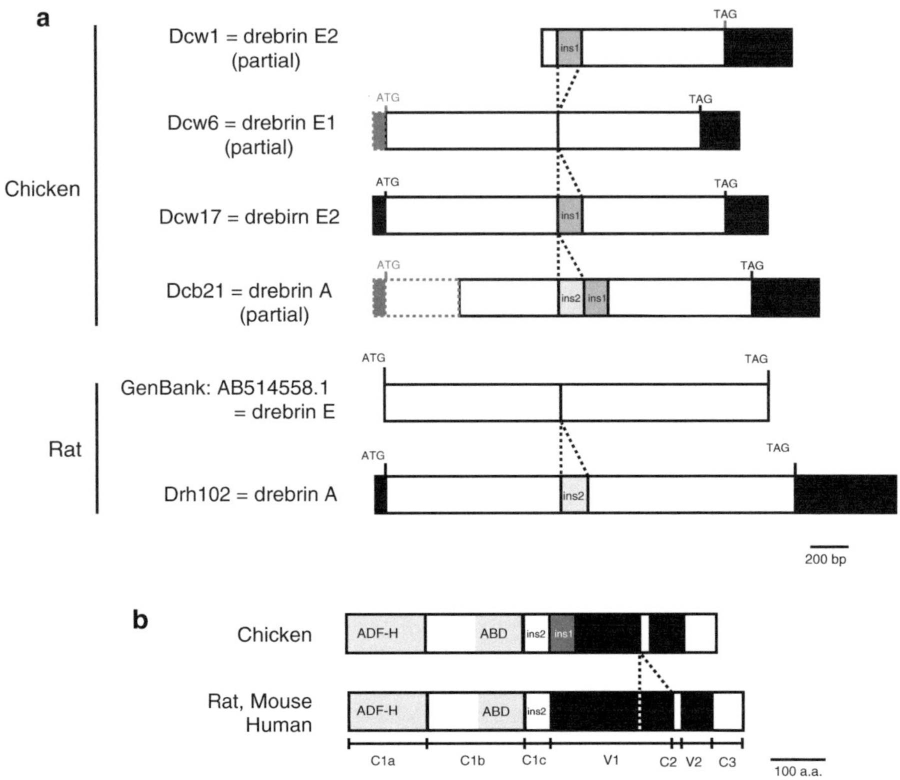

Fig. 2.1 Structure of drebrin cDNAs and domain structure of the deduced drebrin proteins. (**a**) Alignment of cDNA sequences of chicken and rat drebrin. The open rectangles and the black rectangles indicate coding regions and noncoding regions, respectively. The putative initiation codon (ATG) and the stop codon (TAG) are indicated. The shaded rectangles indicate the insertion sequences (Ins1 and Ins2). (**b**) Comparison of the protein structure of chicken drebrin A (top) and mammalian drebrin A (bottom). A diagram of the conserved (C1a, C1b, C1c, C2, and C3) and variable (V1 and V2) domains in the deduced amino acid sequence is indicated. ADF-H actin-depolymerizing factor homology. *ABD* actin-binding domain

with only drebrin A (Aoki et al. 2005). This result is consistent with the above conclusion that Dcw6, Dcw17, and Dcb21 carry the cDNA corresponding to drebrins E1, E2, and A, respectively.

Immunoblot and immunohistochemical analyses of rat tissues using an anti-drebrin antibody (mAb M2F6) raised against chicken drebrins suggest that drebrin also exists in mammalian brain (Shirao et al. 1989). Immunoblot analysis indicates that at least two distinct protein bands react to mAb M2F6 in rat brain. Their apparent molecular weights are slightly higher (approx. 130 and 140 kDa) than those of chicken drebrins (approx. 95, 100 and 110 kDa), and their isoelectric points (pI = 4.5) are similar. As in chicken the expression of each drebrin isoforms is developmentally changed in rat (Shirao et al. 1989). To obtain the rat drebrin cDNA, we constructed

```
INS2
CHICKEN ins2:  GRLHCPFIKTADSGPPSSSSSSSSPPRTPFPYITCHRTPNLSSFFP
RAT     ins2:  GRPYCPFIKASDSGPSSSSSSSSSPPRTPFPYITCHRTPNLSSSLP
MOUSE   ins2:  GRPYCPFIKASDSGPSSSSSSSSSPPRTPFPYITCHRTPNLSSSLP
HUMAN   ins2:  GRPYCPFIKASDSGPSSSSSSSSSPPRTPFPYITCHRTPNLSSSLP

INS1
CHICKEN ins1:  GSQSD-YRKVSAAG---CSPCESSPASTPLGEQRTRAPAEETPATPK
RAT   ins1-H:  GSHLDSHRRMAPTPIPTRSPSDSSTASTPITEQIERALDEVTSSQPP
MOUSE ins1-H:  GSHLDSHRRMAPTPIPTRSPSDSSTASTPIAEQIERALDEVTSSQPP
HUMAN ins1-H:  GSHLDSHRRMAPTPIPTRSPSDSSTASTPVAEQIERALDEVTSSQPP

Identical Amino Acid
Substitutable Amino Acid
```

Fig. 2.2 Comparison of amino acid sequence between species. Identical and substitutable amino acids compared among various species are marked with *yellow* and *green*, respectively. Ins1-H, Ins1-homology amino acid sequence

rat hippocampal cDNA library and screened the positive cDNA clones using chicken drebrin cDNA Dcw17 (encoding chicken drebrin E2) as a probe. We isolated the cDNA clone Drh102 encoding the full length of drebrin mRNA including the homologous sequence of chicken Ins2 (Shirao et al. 1992) and later a non-published cDNA variant (GenBank: AB514558.1) encoding another drebrin mRNA that lacks Ins2 in its sequence (Fig. 2.1a). Ins2 sequence is composed of 46 amino acids and exhibits 85% identity between chicken and rat (Fig. 2.2), suggesting that the rat cDNAs with and without the Ins2-homologous insertion sequence are corresponding to rat drebrin A and drebrin E, respectively. Unlike Ins2 sequence, there is no drebrin cDNA that lacks Ins1 sequence in its sequence. However, the alignment of the sequence just behind the Ins2 sequence shows some similarity to the chicken Ins1 sequence. Although identity is low (44%), the sequence comparison including substitutable amino acids in addition of identical amino acids shows the higher similarity between chicken and mammals (81%) (Fig. 2.2). We further tried to isolate another cDNA variant which does not have Ins1 in its sequence by reverse transcription-polymerase chain reaction (RT-PCR) of mouse brain mRNAs, but we have not isolated such mouse version of drebrin E1. Instead, besides the two isoforms drebrin E and drebrin A in rodent, we isolated the truncated form of drebrin, named by s-drebrin A, which lacks C-terminal half of drebrin A (Jin et al. 2002), as described below.

Interspecies comparison of the deduced amino acid sequences of drebrin between chicken and rodent shows a high degree of similarity (Fig. 2.1b; see also Fig. 1.3 in Chap. 1) (Kojima et al. 1993), suggesting that biological functions of drebrin are important beyond species. In particular, the homology of the N-terminal half (named domain C1) and two short regions in the C-terminal region (domains C2 and C3) is greater than 85%. These regions are also conserved in human (Toda et al. 1993). The secondary structure prediction by the method of Kyte and Doolittle (1982) suggests that the large homologous region domain C1 can be subdivided into three. The most N-terminal domain C1a is relatively hydrophobic and has potential to form repeating stretches of β-sheet that are separated by short stretches of α-helix. This region

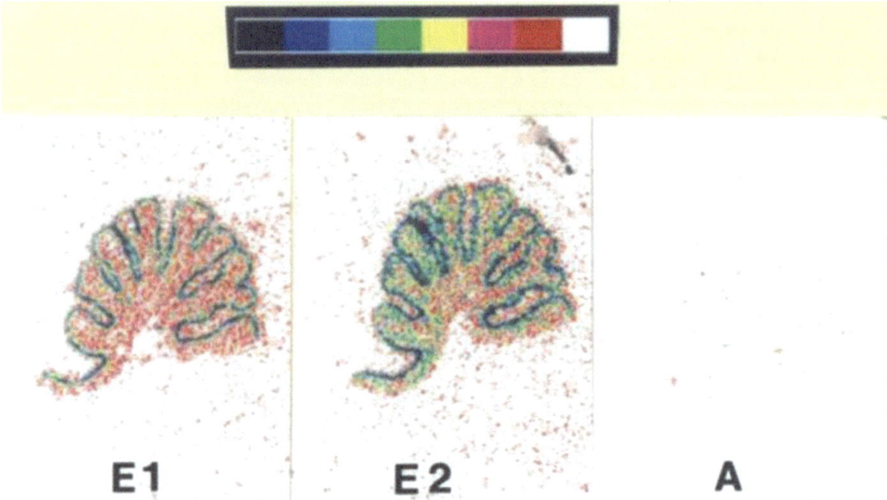

Fig. 2.3 Isoform-specific distribution of drebrin in the developing chicken cerebellum. Chicken drebrin E1 probe (GCAGCCATAGGACAGACTCCCCCAGCCCCA) hybridized strongly to the external granular layer, but not to the internal granular layer. Chicken drebrin E2 probe (GCAGCCATAGGACAGGCAGCCAGTCCGACT) hybridized to both the external and internal granular layers. There was no significant hybridization with the A probe (CGTCTGCACTGTCCTTTCATAAAGACAGCT). (Adopted from Shirao et al. 1990)

is followed by a hydrophilic, elongated α-helical region (domain C1b). Ins2 sequence is included in domain C1c, in which β-turns are concentrated. A repeat of seven to nine serine residues within a middle part of Ins2 is conserved between chicken and mammals. This highly conserved domain C1 may contribute to expression and regulation of biological function of drebrin. By analogy with other proteins, domain C1a may organize into a relatively stable β-sheet and form a globular core, named by actin-depolymerizing factor homology (ADF-H) domain (Lappalainen et al. 1998; Poukkula et al. 2011). The systematic analysis using various fragments of drebrin cDNA reveals that 85-amino acid sequence within the domain C1b (also called as coiled-coil domain) contributes to the binding to and remodeling of F-actin (Hayashi et al. 1999; Grintsevich et al. 2010). In contrast to the homologous domains, the sequence of the C-terminal half is diverged and shows low homology among species. This variable region is separated into two domains V1 and V2 by a short conserved domain C2. The domain V1 has a potential to form repeating stretches of α-helix with several β-turns. Although a repeat of ten proline residues observed in this domain of rat drebrin does not exist in that of chicken drebrin, and positioning of proline residue in domain V1 is not identical, proline-rich feature is retained among species. Proline residue generally facilitates the formation of protein secondary structure elements such as turns and destabilizes α-helix and β-sheet conformation and serves a site for protein-protein interactions. Mammoto et al. reported that a proline-rich region of this domain may contribute to the interaction with other proteins such as actin-binding protein profilin (Mammoto et al. 1998). The sequence of Ins1 in chicken is included in this domain. Although these conserved domains in the C-terminal region do not exhibit domain structure, a binding motif (PPXXF or PXXF) for Homer (Tu et al. 1998), a scaffolding protein of metabotropic glutamate receptors, exists in domain C2 of both chicken and rodents, and in domain C3 of rodents but not of chicken, suggesting that the conserved domains in the C-terminal region serve a function in interaction with intracellular proteins other than F-actin.

The structural difference between the adult and embryonic isoforms is a 46-amino acid sequence (Ins2). This sequence is absent from embryonic isoforms of drebrin (drebrin E in mammals and drebrins E1 and E2 in chicken) but present in drebrin A and s-drebrin A. As we mentioned previously, rat drebrin E is more similar to chicken drebrin E2 than drebrin E1 because drebrin E has Ins1 homologous sequence (1ns1-H in Fig. 2.2). Then does not rat need the other drebrin isoform that is similar to chicken drebrin E1?

In order to analyze the isoform-specific distribution of drebrins E1 and E2 in the chicken brain, we produced the 30-mer oligonucleotide probes corresponding to the joining region of each drebrin mRNA and successfully identified the localization of each drebrin mRNA in embryonic day 16 cerebellum by in situ hybridization (Shirao et al. 1990). Chicken drebrin E1-specific probe hybridized strongly to the external granular layer, but not to the internal granular layer. Chicken drebrin E2-specific probe hybridized to both the external and internal granular layers. There was no significant hybridization with the drebrin A-specific probe in the developing cerebellum (Fig. 2.3). Thus, the drebrin E1 is expressed in newly generated neurons but not in migrating neurons, and drebrin E2 is expressed in migrating and neurite growing neurons in addition to newly generated neurons, suggesting that drebrin E1 plays a part of drebrin E2 roles. In other words, drebrin E1 is not necessary for optic tectum development, and this might be a reason why rat does not need the other drebrin isoform similar to chicken drebrin E1 in the brain.

The drebrin isoforms exhibit differences in (1) the time course of their expression during neuronal development (Shirao et al. 1989), (2) the type of cells or tissues expressing drebrin (Luna et al. 1997; Keon et al. 2000; Peitsch et al. 1999; Shirao and Obata 1986; Shirao et al. 1987, 1994), and (3) their intracellular distribution (Shirao and Obata 1986; Shirao et al. 1987; Aoki et al. 2005; Sasaki et al. 1996; Song et al. 2008). We have, therefore, hypothesized that drebrin E and drebrin A are functionally different in the brain. To know the genomic relationship between drebrin isoforms, we next isolated and analyzed the drebrin gene.

2.3 Drebrin Isoforms Produced by Alternative Splicing

Ribonuclease protection assay is a highly sensitive technique for detecting and quantifying the specific mRNA in RNA samples. Using this assay with a radioisotope-labeled antisense-stranded cRNA probe of the drebrin fragment containing the insertion sequences, we revealed that stage- and tissue-specific expression of each drebrin isoform is regulated at the transcriptional level (Kojima et al. 1993). Drebrin E is widely expressed in the embryonic brain and other various non-neuronal tissues including the intestine, heart, and skeletal muscle. On the other hand, drebrin A is expressed in a neural tissue-specific fashion and is predominant in the postnatal brain.

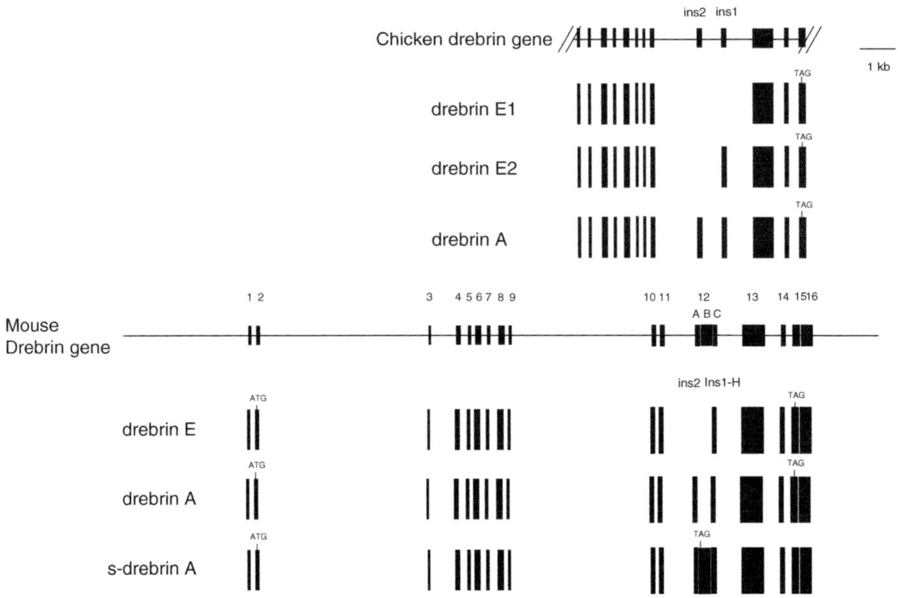

Fig. 2.4 Structural organization of drebrin gene. The genomic region is partial in chicken and covered full-length transcripts in mouse. Exons are indicated by rectangles in each genome. The selected exons in each drebrin mRNA isoform are indicated. The putative initiation codon (ATG) and the stop codon (TAG) are also indicated. See text for details

To clarify the genetic correlation of each drebrin isoform, a drebrin genomic clone was isolated from the chicken and mouse genomic library by drebrin cDNA as a probe (Kojima et al. 1993, 2010; Jin et al. 2002). The exact location of the exons was confirmed through comparison with the cDNA sequence in partial chicken drebrin gene and mouse drebrin gene covering full-length transcripts (Fig. 2.4). The exon boundaries in cDNA sequence are well conserved between chicken and mouse, that is, the cDNA sequence for domain C1a is located in exons 2 to 5; the sequence for domain C1b is in exons 6 to 11; the sequence for domain C1c (Ins2) is in exon 12A[1]; the sequences for domains V1, C2, and a half part of C3 are in exons 12C to 13; and the sequence for the remaining half part of C3 is in exons 14 to 15. Striking feature is that the sequences of Ins2 and Ins1 (Ins1-homology in mouse) are independently encoded in the individual exons of chicken genome. This fact demonstrates that the mechanism for generating drebrin isoform mRNAs from a single drebrin gene is likely to be according to a common mechanism for an alternative splicing that is the process by which the exons of primary transcripts (pre-mRNAs) from genes can be spliced in different combinations to produce structurally and functionally distinct mRNA and protein variants (Roy et al. 2013). Unidentified trans-acting factor(s) expressed in a tissue-specific and a stage-dependent manner

[1]Exon 12 was referred to as exon 11 in the original reports published before the discovery of the initial (5' terminal) exon (Jin et al. 2003; Kojima et al. 2010).

should regulate the alternative splice site selection of the drebrin gene. Similarly, in the mouse drebrin gene, Ins2 sequence is encoded in the 5'-terminal region of exon 12 (exon12A). Ins1-H sequence is encoded in the 3'-terminal region of exon 12 (exon 12C). Further, sequence analysis in mouse drebrin gene has shown that the 319-nucleotide sequence in the middle region of exon 12 (exon 12B) is excluded as an intron in drebrin E and drebrin A but included as an exon in s-drebrin A (Fig. 2.4). The fact that there is no significant sequence homology of exon 12B in the intron located between exons encoding Ins2 and Ins1 in chicken genome may suggest the difference in isoform constitution between chicken and rodent, that is, the existence of s-drebrin A and the absence of drebrin E1 in rodent. Because of the existence of a stop codon located in-frame at the nucleotide 7 in exon 12B, s-drebrin A possesses only the N-terminal half of drebrin A (including Ins2) but lacks the C-terminal half. We confirmed the actin-remodeling activity of this isoform in s-drebrin A-transfected fibroblasts (Jin et al. 2002). We, however, have not yet known the protein level and cellular distribution of s-drebrin A in the brain as a comparison with those of drebrin A.

2.4 Protein Family to Which Drebrin Belongs

As described above, drebrin protein is well conserved in chicken and mammals including human with a slight variation in isoform constitution. What is the origin of drebrin protein? The current knowledge for the ancestor protein and homologous family proteins of drebrin is overviewed below.

First reported protein which shows a significant similarity to drebrin is SH3P7 (Sparks et al. 1996). SH3P7 (also known as Abp1) is an actin-binding protein identified as a SH3 domain-containing protein (Sparks et al. 1996). The N-terminal half of SH3P7 is highly homologous to the N-terminal region of drebrin, and the SH3 domain of SH3P7 is similar to that of another actin-binding protein cortactin (Kessels et al. 2000). SH3P7 is located primarily in dendrites and co-localized with drebrin in dendritic spines (Yamazaki et al. 2001). The N-terminal region of drebrin and SH3P7 is now known as the ADF-H domain (Lappalainen et al. 1998; Poukkula et al. 2011). The ADF-H domain is an actin-binding module, which is present in members of a protein family that consists of five phylogenetically distinct classes: ADF/cofilins, twinfilins, coactosins, glia maturation factors (GMFs), and drebrin/Abp1s (Fig. 2.5) (Poukkula et al. 2011). Based on the structural homology, it has been proposed that these five classes of ADF-H domain proteins have evolved from a common ancestral protein. Although ADF-H domains between different protein classes share less than 20% sequence identity at the amino acid level, the predicted secondary structure is well conserved throughout the entire family. It is likely that all ADF-H domains have a similar three-dimensional fold composed of five internal β-strands surrounded by at least four α-helices (Poukkula et al. 2011). Such conserved three-dimensional structures of ADF-H domain proteins appear to promote cytoskeletal dynamics by interacting with actin. However, differences in

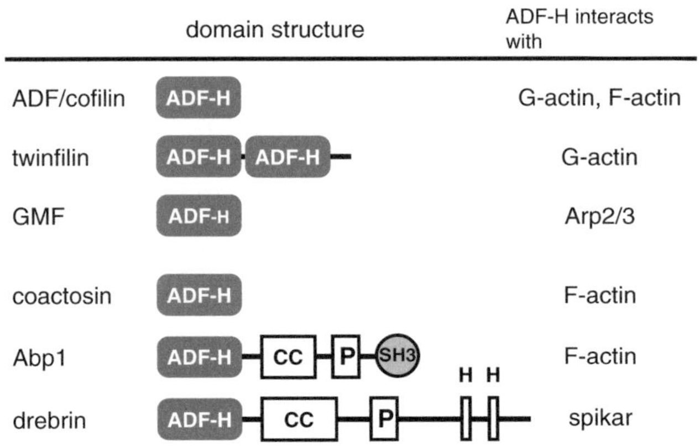

Fig. 2.5 A diagram comparing the domain structures of the ADF-H domain proteins and their ability of ADF-H domain to bind other proteins. ADF/cofilins, glia maturation factor (GMF), and coactosin consist of single ADF-H domain. Twinfilin is composed of two tandem repeats of ADF-H domain. Abp1 and drebrin are multidomain proteins that have single ADF-H domain in addition of coiled-coil domain (CC), proline-rich domain (P), SH3 domain (SH3), and Homer-binding motifs (H). (Modified from Poukkula et al. 2011)

the chemical properties of surface-exposed residues and small structural variations are responsible for the diversity in domain architectures and biochemical functions between distinct classes of ADF-H domain proteins. Of the ADF-H domain proteins, twinfilin interacts with G-actin, and ADF/cofilin binds both G-actin and F-actin, whereas drebrin/Apb1 and coactosin bind F-actin but not G-actin. GMF does not bind actin but interacts directly with Arp2/3 complex, a key regulator of F-actin nucleation. The binding of drebrin to F-actin may not be achieved via the ADF-H domain but via the actin-binding domain located downstream of ADF-H domain (Hayashi et al. 1999; Grintsevich et al. 2010). Therefore, drebrin ADF-H domain is unlikely to make a significant contribution to actin binding. We have recently discovered novel drebrin partner, named spikar, which directly interacts with drebrin ADF-H domain using yeast two-hybrid system (Yamazaki et al. 2014). Thus, ADF-H domain of drebrin is likely to have a distinct role from other members of ADF-H domain proteins.

2.5 Conclusions and Perspectives

A class of ADF-H domain protein family, drebrin/Abp1 shows higher homology with each member. After drebrin was first identified in vertebrate (Shirao and Obata 1985), Abp1 was identified as an actin-interacting protein in budding yeast (Drubin et al. 1988). Later Abp1 was also found in mammals (Sparks et al. 1996; Kessels et al. 2000; Yamazaki et al. 2001). Drebrin is predominantly expressed in the brain (Shirao and Obata 1985). Drebrin A is a neuron-specific isoform and enriched in dendritic spines, whereas drebrin E (E1 and E2 in chicken) is expressed predominantly in the embryonic brain and in a wide range of non-neuronal tissues. All the isoforms are generated from a single gene by alternative splicing. Unidentified trans-acting factor(s) expressed in a tissue-specific and a stage-dependent manner should regulate the alternative splice site selection of the drebrin gene. Although in vitro study does not show the difference in F-actin-binding activity between two isoforms (Ishikawa et al. 1994), in vivo interaction of F-actin and drebrin E might not be stronger than that of F-actin and drebrin A (Aoki et al. 2005; Kojima et al. 2010). Recently our studies using drebrin A-specific knockout mice also indicate fundamentally different biological activities between drebrin E and drebrin A (Kojima et al. 2010, 2016). It is important to note how the internal insertion sequence Ins2 influences the biological activities of drebrin. Since Ins2 has several unique features in its amino acid sequence (i.e., a long stretch of serine residue and repeats of β-turn), the insertion of Ins2 into drebrin may result in a change of the tertiary structure by phosphorylation or by binding with certain specific proteins. Also, it will be important to elucidate whether its activity is complemented by other functionally similar proteins, since mice homozygous for null mutation of drebrin does not show overt abnormalities in many aspects of brain anatomy and animal behavior.

References

Aoki C et al (2005) Drebrin A is a postsynaptic protein that localizes in vivo to the submembranous surface of dendritic sites forming excitatory synapses. J Comp Neurol 483:383–402

Drubin DG et al (1988) Yeast actin-binding proteins: evidence for a role in morphogenesis. J Cell Biol 107(Pt 2):2551–2561

Grintsevich EE et al (2010) Mapping of drebrin binding site on F-actin. J Mol Biol 398:542–554

Hayashi K et al (1999) Domain analysis of the actin-binding and actin-remodeling activities of drebrin. Exp Cell Res 253:673–680

Ishikawa R et al (1994) Drebrin, a development-associated brain protein from rat embryo, causes the dissociation of tropomyosin from actin filaments. J Biol Chem 269:29928–29933

Jin M et al (2002) A novel, brain-specific mouse drebrin: cDNA cloning, chromosomal mapping, genomic structure, expression, and functional characterization. Genomics 79:686–692

Keon BH et al (2000) Isoform specific expression of the neuronal F-actin binding protein, drebrin, in specialized cells of stomach and kidney epithelia. J Cell Sci 113(Pt 2):325–336

Kessels MM, Engqvist-Goldstein AEY, Drubin DG (2000) Association of Mouse Actin-binding Protein 1 (mAbp1/SH3P7), an Src Kinase Target, with Dynamic Regions of the Cortical Actin Cytoskeleton in Response to Rac1 Activation. Mol Biol Cell 11:393–412

Kojima N et al (1988) Nucleotide sequences of two embryonic drebrins, developmentally regulated brain proteins, and developmental change in their mRNAs. Brain Res 464:207–215

Kojima N et al (1993) Molecular cloning of a developmentally regulated brain protein, chicken drebrin A and its expression by alternative splicing of the drebrin gene. Brain Res Mol Brain Res 19:101–114

Kojima N et al (2010) Genetic disruption of the alternative splicing of drebrin gene impairs context-dependent fear learning in adulthood. Neuroscience 165:138–150

Kojima N et al (2016) Drebrin A regulates hippocampal LTP and hippocampus-dependent fear learning in adult mice. Neuroscience 324:218–226

Kyte J, Doolittle RF (1982) A simple method for displaying the hydropathic character of a protein. J Mol Biol 157:105–132
Lappalainen P et al (1998) The ADF homology (ADF-H) domain: a highly exploited actin-binding module. Mol Biol Cell 9:1951–1959
Luna EJ, Pestonjamasp KN, Cheney RE, Strassel CP, Lu TH, Chia CP, Hitt AL, Fechheimer M, Furthmayr H, Mooseker MS (1997) Actin-binding membrane proteins identified by F-actin blot overlays. Soc Gen Physiol Ser 52:3–18
Mammoto A et al (1998) Interactions of drebrin and gephyrin with profilin. Biochem Biophys Res Commun 243:86–89
Peitsch WK, Grund C, Kuhn C, Schnölzer M, Spring H, Schmelz M, Franke WW (1999) Drebrin is a widespread actin-associating protein enriched at junctional plaques, defining a specific microfilament anchorage system in polar epithelial cells. Eur J Cell Biol 78:767–778
Poukkula M et al (2011) Actin-depolymerizing factor homology domain: a conserved fold performing diverse roles in cytoskeletal dynamics. Cytoskeleton 68:471–490
Roy B et al (2013) Review: alternative Splicing (AS) of genes as an approach for generating protein complexity. Curr Genomics 14:182–194
Sasaki Y, Hayashi K, Shirao T, Ishikawa R, Kohama K (1996) Inhibition by drebrin of the actin-bundling activity of brain fascin, a protein localized in filopodia of growth cones. J Neurochem 66:980–988
Shirao T, Obata K (1985) Two acidic proteins associated with brain development in chick embryo. J Neurochem 44:1210–1216
Shirao T, Obata K (1986) Immunochemical homology of 3 developmentally regulated brain proteins and their developmental change in neuronal distribution. Brain Res 394:233–244
Shirao T, Inoue HK, Kano Y, Obata K (1987) Localization of a developmentally regulated neuron-specific protein S54 in dendrites as revealed by immunoelectron microscopy. Brain Res 413:374–378
Shirao T et al (1988) Molecular cloning of a cDNA for the developmentally regulated brain protein, drebrin. Brain Res 464:71–74
Shirao T et al (1989) Two forms of drebrins, developmentally regulated brain proteins, in rat. Proceedings of the Japan Academy 65:169–172
Shirao T et al (1990) Expression of three drebrin isoforms in the developing nervous system. Neurosci Res Suppl 13:S106–S111
Shirao T et al (1992) Cloning of drebrin A and induction of neurite-like processes in drebrin-transfected cells. Neuroreport 3:109–112
Shirao T, Hayashi K, Ishikawa R, Isa K, Asada H, Ikeda K, Uyemura K (1994) Formation of thick, curving bundles of actin by drebrin A expressed in fibroblasts. Exp Cell Res 215:145–153
Song M et al (2008) Expression of drebrin E in migrating neuroblasts in adult rat brain: coincidence between drebrin E disappearance from cell body and cessation of migration. Neuroscience 152:670–682
Sparks AB et al (1996) Cloning of ligand targets: systematic isolation of SH3 domain-containing proteins. Nat Biotechnol 14:741–744
Toda M et al (1993) Molecular cloning of cDNA encoding human drebrin E and chromosomal mapping of its gene. Biochem Biophys Res Commun 196:468–472
Tu JC et al (1998) Homer binds a novel proline-rich motif and links group 1 metabotropic glutamate receptors with IP3 receptors. Neuron 21:717–726
Yamazaki H et al (2001) Molecular cloning and dendritic localization of rat SH3P7. Eur J Neurosci 14:998–1008
Yamazaki H et al (2014) Spikar, a novel drebrin-binding protein, regulates the formation and stabilization of dendritic spines. J Neurochem 128:507–522

Chapter 3
Biochemistry of Drebrin and Its Binding to Actin Filaments

Ryoki Ishikawa

Abstract Drebrin is an actin-binding protein mainly expressed in developing neurons and dendritic spine in mature neurons. To understand the functions of drebrin in vivo, we must understand its molecular properties. In this chapter, I will focus on the purification and characterization of drebrin in vitro. Drebrin binds to F-actin with a stoichiometry of 1:5~6 with a K_d of 1~3 × 10^{-7} M and strongly inhibits the binding of other actin-binding proteins such as tropomyosin, caldesmon, fascin, α-actinin, and cofilin. It also inhibits the activities of myosin-II and myosin-V. These results are discussed in terms of the possible roles of drebrin in the stability, dynamics, and organizations of actin structures in neuronal cells.

Keywords Actin • Myosin-II • Myosin-V • Tropomyosin • Caldesmon • Fascin α-Actinin • Cofilin

3.1 Purification of Drebrin

In the early 1990s, we started to search for actin-binding proteins specifically expressed in the embryonic rat brain using the F-actin co-sedimentation method. The total cell extract of embryonic day 18 (E18) or adult rat brain was ultracentrifuged at 100,000 × g for 1 h. Many structures composed of multi-molecules such as F-actin sediment under this centrifugal force. Exogenous F-actin was added to the supernatant, and the sample was ultracentrifuged again. Proteins co-sedimented with F-actin were analyzed by SDS-PAGE and stained with Coomassie Brilliant Blue. As shown in Fig. 3.1, two bands with masses of approximately 140 and 55 kDa were obtained from E18 brain samples. Using immunoblotting, we identified the 55 kDa band as fascin, a filopodial actin-bundling protein, but we could not find a good candidate for the 140 kDa band, and so we purified the protein. We

R. Ishikawa
School of Nursing, Gunma Prefectural College of Health Science,
Maebashi, Gunma 371-0052, Japan
e-mail: ryoki@gchs.ac.jp

T. Shirao, Y. Sekino (eds.), *Drebrin*, Advances in Experimental Medicine and Biology 1006, DOI 10.1007/978-4-431-56550-5_3

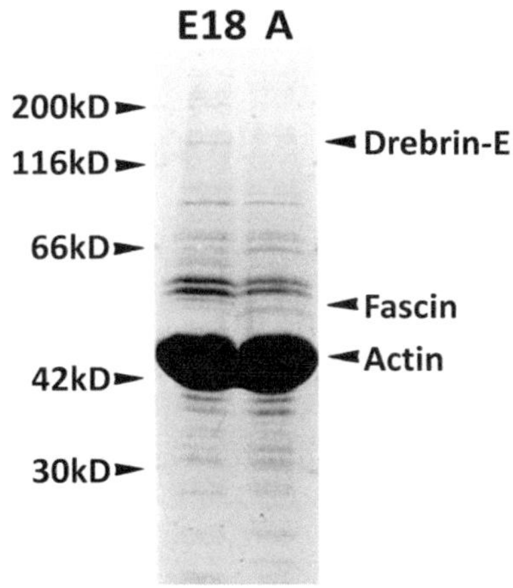

Fig. 3.1 Actin-binding fractions from rat brain. Brains were homogenized in 100 mM KCl, 1 mM $MgCl_2$, 5 mM EGTA, 5 mM DTT, and 20 mM Tris-Cl (pH 7.5) and centrifuged at 100,000 × *g* for 1 h. The supernatant was recovered, mixed with 24 μM chicken skeletal muscle F-actin for 30 min, then centrifuged again at 100,000 × *g* for 1 h. The sediment was analyzed by SDS-PAGE. E18, fraction from embryonic day 18 rat. *A* fraction from adult rat

found that this protein was heat stable: after 5 min of incubation in a boiling water bath followed by centrifugation, the protein was recovered in the supernatant. The protein was further purified by ammonium sulfate fractionation, DEAE column chromatography, and superose 6HR column chromatography. The purity of the 140 kDa band after passage through the superose 6HR column was more than 95% as judged by SDS-PAGE. Twenty grams of E18 rat brain yielded 0.05 mg of purified 140 kDa protein (Ishikawa et al. 1994). The protein was identified by partially digesting it with V8 protease in SDS-PAGE, blotting it onto a PVDF membrane, and cutting out and sequencing the resulting 30 kDa band. The partial amino acid sequence thus obtained, LSGHFENQKVMYGF, was identical to the amino acid sequence of rat drebrin (residues 54–67). Furthermore, antibody raised against chicken drebrin (M2F6; Shirao and Obata 1986) reacted with this protein. The precise position of the protein band on an SDS gel coincided with drebrin-E and not with drebrin-A. Thus, we concluded that the 140 kDa protein was drebrin-E.

We failed to purify drebrin-A from adult rat brain. As shown in Fig. 3.1, the drebrin-A band was not detected in adult rat brain extract when the SDS-PAGE gel was stained with Coomassie Brilliant Blue. Does drebrin-A lose the ability to bind to F-actin? When drebrin-A was detected by immunoblotting, centrifugation of the heat-stable fraction of adult rat brain at 140,000 × *g* resulted in the recovery of drebrin-A to the supernatant. In contrast, drebrin-A co-sedimented with F-actin under the same centrifugation conditions (Shirao et al. 1994), indicating that drebrin-A has actin binding activity following its extraction in soluble form. In the absence of exogenous F-actin, most of the drebrin-A in the sample sedimented after centrifugation at 100,000 × *g* for 1 h (Hayashi et al. 1996). It seems likely that only a small fraction of drebrin-A was extracted in the soluble form; therefore, purifying drebrin-A from brain tissue would require identifying conditions under which most drebrin-A would be extracted in the soluble form. Consequently, we decided to express drebrin in a bacterial expression system and purify it. We first tried using a normal expression system, pET19b (Novagen) as a vector and BL21(DE3) as a host, but the expression levels of both drebrin-A and drebrin E were very low, and we failed to purify either protein. Because drebrin has a domain containing ten consecutive proline residues (residues 364–373 in drebrin-E and 410–419 in drebrin-A), it is probable that tRNAs recognizing proline codons may be a limiting factor. We therefore changed the host to BL21(DE3) codon plus RP that contained an extra copy of the poly *pro-L* gene. This gene encodes the tRNA that recognizes the proline codon CCC. Using this expression system, we obtained sufficient quantities of drebrin-A in the bacterial cell extract (Ishikawa et al. 2007). Other hosts such as Rossetta 2™ are also available for expression (Sharma et al. 2011). The protein was purified using the same methodology as used to purify drebrin from brain tissue. An 8 liter bacterial culture provided 2~3 mg of purified drebrin-A.

3.2 Actin-Binding and Actin-Bundling Properties of Drebrin

We reported that drebrin-E purified from brain tissue bound to F-actin with a stoichiometry of 1:5 (drebrin-E/actin) and an apparent dissociation constant (K_d) of 1.2×10^{-7} M (Ishikawa et al. 1994). Bacterially expressed drebrin-A also bound to F-actin with a stoichiometry of 1:5~6 with a K_d of 2.7×10^{-7} M (Ishikawa et al. 2007). The bacterially expressed N-terminal half of drebrin (residues 1–300, common to drebrin-E and drebrin-A) contains the actin-binding domain and showed similar affinity for F-actin, with a K_d of 2.0×10^{-7} M, although with a different stoichiometry of 1:3 (Grinstevich et al. 2010). Therefore, it seems likely that the F-actin-binding properties of drebrin-E and drebrin-A are the same or very similar. As discussed below, however, 6 nM of drebrin strongly inhibited the sliding velocity of F-actin on myosin-II (Hayashi et al. 1996), and 10 nM of drebrin-E strongly inhibited the binding of myosin-V to F-actin (Kubota et al. 2010). Furthermore, the addition of 10 nM tetramethylrhodamine-labeled drebrin-E to 23 nM F-actin fixed on a glass surface revealed that F-actin (23 nM) strongly and uniformly binds fluorescently labeled drebrin-E (Kubota et al. 2010). All these results suggest that the K_d of drebrin from F-actin is less than 10 nM. Our F-actin co-sedimentation assay was quantified by SDS-PAGE and subsequent Coomassie Brilliant Blue staining, followed by densitometry; however, it is difficult to detect nanomolar levels of protein using this method. Therefore, the true affinity of drebrin to F-actin might be much higher than that indicated by the F-actin co-sedimentation assay.

Does drebrin have actin-bundling activity? Drebrin-E purified from brain tissue did not exhibit any actin-bundling activity, as confirmed by electron microscopy and a low-speed centrifugation assay (Ishikawa et al. 1994). In contrast, some bacterially expressed drebrin-E and drebrin-A did exhibit actin-bundling activity. SDS-PAGE indicated that the drebrin fraction collected after DEAE column chromatography was more than 95% pure. Application of this fraction to a superose 6HR column provided two major peaks which both showed as a single drebrin band on SDS-PAGE. The first peak, which was collected in the void volume, had actin-bundling activity, while the second peak, which was eluted at 24~28 min at the flow rate of 0.5 mL/min, did not have actin-bundling activity, as confirmed by a low-speed centrifugation assay (unpublished result). Because drebrin corresponding to the first peak did not sediment after centrifugation at 100,000 × *g* for 40 min, it seems likely that drebrin forms oligomers but not aggregate. We do not currently know whether oligomer formation by drebrin is an artifact of being synthesized in a bacterial expression system. Therefore, the second peak has been used as "purified drebrin" in the experiments we have published to date.

Molecular dissection of drebrin revealed two actin-binding domains, and the fragment containing both domains had actin-bundling activity as confirmed by low-speed centrifugation and electron microscopy (Worth et al. 2013). Furthermore, the exposure of F-actin to drebrin reacted with cdk5, an enzyme that phosphorylates S142 in drebrin, or the S142D mutant of drebrin, resulted in the formation of tight bundles of F-actin, indicating that drebrin has actin-bundling activity under specific conditions (Worth et al. 2013; also see Sect. 3.3 for details).

3.3 Competitive Binding of Drebrin for F-actin with Other Actin-Binding Proteins

Tropomyosin is a rod-shaped protein consisting of two α-helical polypeptides that form a coiled coil structure. Tropomyosin binds to thin filaments in skeletal muscle cells and to stress fibers in smooth muscle and non-muscle cells and may stabilize actin structures by protecting F-actin from actin-destabilizing proteins such as gelsolin (Ishikawa et al. 1989a, b) and ADF/cofilin (Bernstein and Bamburg 1982). We found that drebrin strongly inhibited tropomyosin binding to F-actin, and this was confirmed by an F-actin co-sedimentation assay (Ishikawa et al. 1994). Similar concentrations of drebrin and tropomyosin (1.1 μM drebrin-E, 1.7 μM tropomyosin, and 7.1 μM actin) resulted in a 95% reduction in the amount of tropomyosin bound to F-actin, whereas drebrin-E binding to F-actin was only weakly inhibited by tropomyosin. Even a ten times higher concentration of tropomyosin (0.67 μM drebrin-E versus 7.6 μM tropomyosin at an actin concentration of 7.1 μM) reduced the amount of drebrin-E bound to F-actin by only 70%. These results may reflect the relatively higher affinity of drebrin for F-actin compared to the affinity of tropomyosin for F-actin. We concluded that drebrin and tropomyosin competitively bind to F-actin in vitro. Furthermore, transfection of drebrin-E into CHO-K1 cells caused the dissociation of tropomyosin from microfilaments (Ishikawa et al. 1994), consistent with the results obtained in in vitro experiments.

Caldesmon is a rod-shaped, calmodulin-binding protein expressed in smooth muscle and in some non-muscle cells such as neurons. Because tropomyosin and caldesmon mutually enhance F-actin binding (Ishikawa et al. 1989a, b, 1998), we also examined whether drebrin affects the actin-binding of caldesmon in the presence or absence of tropomyosin. As shown in Fig. 3.2a, the amount of caldesmon bound to F-actin gradually decreased as the concentration of drebrin-A increased (open circles). This inhibition was partially negated in the presence of tropomyosin

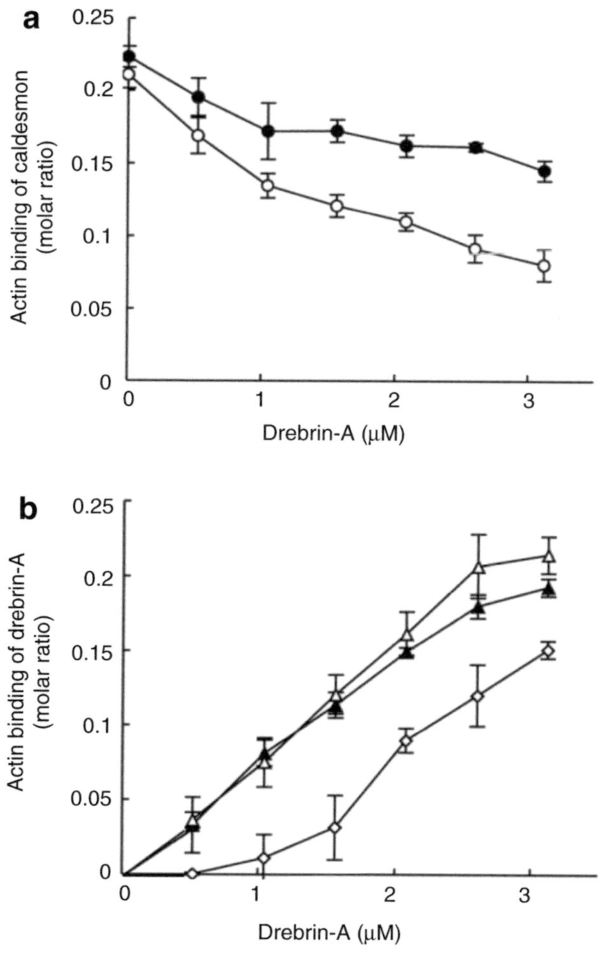

Fig. 3.2 Caldesmon and tropomyosin bind competitively with drebrin-A to F-actin. Different concentrations (0–3.1 μM) of drebrin-A were mixed with 7.1 μM F-actin in the presence of 2.3 μM smooth muscle caldesmon and/or 3.0 μM smooth muscle tropomyosin in 100 mM KCl and 20 mM Tris-Cl (pH 7.5) for 30 min, then centrifuged at 100,000 × *g* for 40 min. The supernatant and the sediment of each sample were separated and analyzed by SDS-PAGE. The amounts of each protein was determined by densitometry. (**a**) Actin binding of caldesmon was plotted against drebrin-A concentration. Open circles, in the presence of 2.3 μM caldesmon, closed circles, in the presence of 2.3 μM caldesmon and 3.0 μM tropomyosin. (**b**) Actin binding of drebrin-A was plotted against drebrin-A concentration. *Closed triangles*, in the presence of 2.3 μM caldesmon, *open triangles*, in the presence of 3.0 μM tropomyosin, *open diamonds*, in the presence of 2.3 μM caldesmon and 3.0 μM tropomyosin

(closed circles). Like tropomyosin, however, caldesmon had little effect on actin binding by drebrin-A when similar amounts of drebrin-A and caldesmon were present (compare Fig. 3.2b, closed triangles, and Fig. 3.1 in Ishikawa et al. 2007), whereas the presence of tropomyosin and caldesmon together inhibited drebrin binding to F-actin (Fig. 3.2b, open diamonds). These results could be explained by the following: (1) both tropomyosin and caldesmon compete with drebrin to bind to F-actin, (2) the affinity of tropomyosin or caldesmon alone for F-actin is too weak to inhibit drebrin binding to F-actin, and (3) the presence of tropomyosin and caldesmon together mutually stimulates their affinity for F-actin, and this enhanced affinity for F-actin is sufficient to inhibit drebrin binding to F-actin.

Fascin is a globular actin-binding protein with a molecular mass of 53–55 kDa that bundles F-actin tightly and localizes in microspikes and filopodia in cultured cells. As shown in Fig. 3.1, more fascin in the soluble fraction from E18 rat brain retained the ability to bind actin compared to fascin in the soluble fraction from adult brain. However, the concentration of fascin in the soluble fraction from E18 rat brains co-sedimented with F-actin was similar to that from the brains of 5-day-old rat, and brain weight of 5-day-old rat was much larger than that of E18 rat. We therefore purified brain fascin from the brains of 5-day-old rats and examined the effects of drebrin-E on the actin-binding and actin-bundling activity of fascin. An actin co-sedimentation assay revealed that actin binding of fascin was strongly inhibited by drebrin-E, while drebrin binding to F-actin was partially inhibited by fascin (Sasaki et al. 1996), suggesting that fascin and drebrin competitively bind to F-actin. We also found that the actin-bundling activity of fascin was inhibited by drebrin-E and confirmed this finding by a low-speed centrifugation assay and electron microscopy (Sasaki et al. 1996).

α-Actinin is a rod-shaped actin-cross-linking protein composed of two identical polypeptides aligned antiparallel to each other which localizes in the z-band in skeletal muscle, stress fibers, and adhesion plaques in non-muscle cells. We found that drebrin-E inhibited both the actin-binding and actin-cross-linking activity of α-actinin (Ishikawa et al. 1994).

Cofilin is an actin-binding protein with a molecular weight of ~19 kDa that has G-actin-binding activity, F-actin-binding activity, weak F-actin-severing activity, and F-actin-destabilizing activity. Drebrin and cofilin competitively bind to F-actin and inhibited the F-actin-severing activity of cofilin (Grinstevich and Reisler 2014).

3.4 Modulation of Myosin Activity by Drebrin

A variety of actin-binding proteins, such as tropomyosin (Yamaguchi et al. 1984), caldesmon (Ngai and Walsh 1984), and fodrin (Wagner 1984), were reported to modulate energy consumption through the actin-activated ATPase activity of myosin-II. We therefore examined whether drebrin also modulates myosin activity and found that the actin-activated ATPase activity of smooth muscle myosin-II (Hayashi et al. 1996) and skeletal muscle myosin-II (Fig. 3.3a) was inhibited by drebrin. An in vitro motility assay revealed that a low concentration (6 nM) of drebrin-E reduced the average sliding velocity of F-actin on smooth muscle myosin-II from 0.34 μm/s

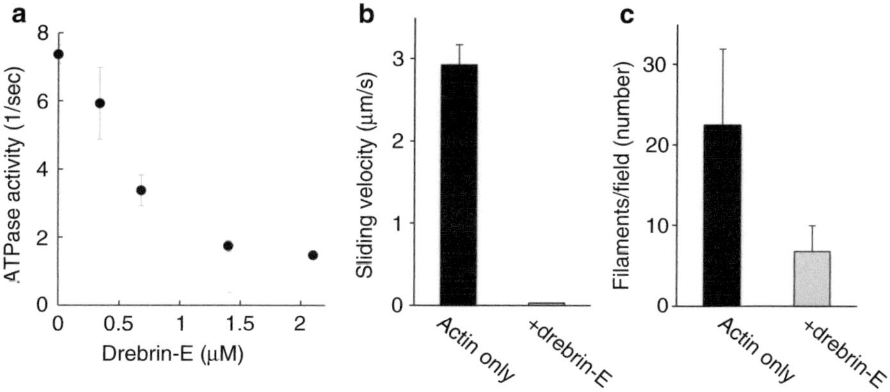

Fig. 3.3 Drebrin inhibits actin-activated ATPase activity (**a**), in vitro sliding on F-actin (**b**), and F-actin binding (**c**) of skeletal muscle myosin II. (**a**) 5.8 μM F-actin and different concentrations of drebrin-E were mixed in 25 mM KCl, 2.5 mM $MgCl_2$, 20 mM DTT, and 20 mM Tris-Cl (pH 7.5) for 30 min, and then 0.11 μM skeletal muscle myosin II and 1 mM ATP were added. ATPase activities were determined by a NADH-coupling ATP-regeneration system. (**b**, **c**) 5.8 μM F-actin (a mixture of rhodamine-phalloidin-labeled F-actin: unlabeled F-actin = 1:500) was incubated with 1.4 μM drebrin-E for 30 min at room temperature, then perfused into a flow chamber coated with 0.11 μM skeletal muscle myosin II in 25 mM KCl, 2.5 mM $MgCl_2$, 1 mM ATP, 20 mM DTT, and 20 mM Tris-Cl (pH 7.5) plus anti-bleach materials. Samples were observed with a total internal reflection fluorescence microscope. The average sliding velocity of the filaments (**b**) and the number of filaments attached to the glass surface 2–3 min after perfusion (**c**) were plotted

(without drebrin) to 0.10 μm/s (Hayashi et al. 1996) and that saturation levels of drebrin-E (F-actin concentration versus drebrin-E concentration of 5.8 μM:1.4 μM) completely inhibited the sliding of F-actin on skeletal muscle myosin-II (Fig.3.3b). One possible interpretation of these results is that drebrin may interfere with the power producing step of myosin-II, thus reducing the sliding velocity. We also found that the number of F-actin filaments attached to the myosin-II-coated glass surface decreased in the presence of drebrin-E (Fig. 3.3c), indicating that drebrin inhibits myosin-II binding to F-actin. We conclude that drebrin may modulate myosin-II activity not only by inhibiting the power producing step of myosin-II but also by inhibiting myosin-II-binding to F-actin, thus reducing the energy consumption of myosin-II.

Myosin-V is a key unconventional myosin in mammalian brain and play roles in membrane traffic in nerve cells (Vale 2003). We found that drebrin-A inhibited the actin-activated ATPase activity of myosin-V (Ishikawa et al. 2007). The number of F-actin filaments attached to a glass surface decorated with myosin-V decreased in the presence of drebrin-A, but once they attached to the surface, the average sliding velocities of F-actin filaments remained unchanged in the presence or absence of drebrin-A (Ishikawa et al. 2007). These results were obtained by an in vitro motility assay, in which multi-myosin-V molecules simultaneously attached to a single F-actin filament. Unlike myosin-II, however, myosin-V is a processive motor, so the movement of a single myosin-V molecule along F-actin filaments can be observed. The running velocities of a Q-dot-labeled myosin-V molecule on F-actin in the presence or absence of drebrin-E were the same, around 0.7 μm/s (Kubota et al. 2010), compatible with the results obtained by an in vitro motility assay. On the other hand, the average run-length of a single myosin-V "walk" decreased, from ~0.8 μm in the absence of drebrin to ~0.4 μm in the presence of 10 nM drebrin-E (Kubota et al. 2010). The binding frequency of myosin-V to F-actin was also decreased under these conditions (Fig. 3.2 of Kubota et al. 2010). How to evaluate these results?

Four states of single myosin-V head have been identified at different stages of the ATP hydrolysis cycle: ATP-bound form, ADP/Pi-bound form, ADP-bound form, and no-nucleotide form. Both the ATP-bound and ADP/Pi-bound forms are short-lived and have weak affinity for F-actin. The ADP-bound form is a rate-limiting step and is present for around 70% of the ATPase reaction. The no-nucleotide form is short-lived. The ATP-bound form and no-nucleotide form have strong binding affinity for F-actin. According to the hand over hand model, the leading head in the ADP/Pi-bound form binds to F-actin, while the trailing head in the ADP-bound form strongly attaches to F-actin (Fig. 3.4 step 2). Pi is released from the leading head, and the head binds strongly (Fig. 3.4 step 3). ADP is released from the trailing head (Fig. 3.4 step 4), then ATP binds to the trailing head, and the head detaches from F-actin (Fig. 3.4 step 5). The trailing head swings forward to become a leading head, while the former leading head binds tightly to remain in the same position, resulting in the myosin-V body moving one step forward (Fig. 3.4 step 6). During or after this process, ATP is dephosphorylated to ADP/Pi (Fig. 3.4 step 7). This cycle continues until the ATP binds to the trailing head before the leading head binds to F-actin. As long as one of the two feet strongly binds to F-actin, myosin-V can walk on F-actin, whereas when both feet are simultaneously weakly bound, myosin-V detaches from F-actin.

We postulate that drebrin inhibits step 2, namely, binding of the leading head to F-actin, but does not affect the other steps. If this is the case, the lifetime of the ADP/Pi-bound form should increase, but the speed of movement will be essentially unaffected because the rate-limiting step is not affected. Furthermore, because the average overlap time when both feet are tightly bound to F-actin becomes shorter, the probability that the trailing head detaches before the leading head binds increases, and so the run-length of myosin-V becomes shorter. Thus, the total run distance of myosin-V becomes shorter and the energy consumption (ATPase activity) decreases. Indeed, the amount of myosin-V that co-sedimented with F-actin decreased with increasing concentration of drebrin-E in the presence of ATP, whereas the amount of myosin-V co-sedimenting with F-actin remained unchanged in the presence of ADP (Kubota et al. 2010). This strongly suggests that binding of the ADP-bound form of myosin-V is not modulated by drebrin.

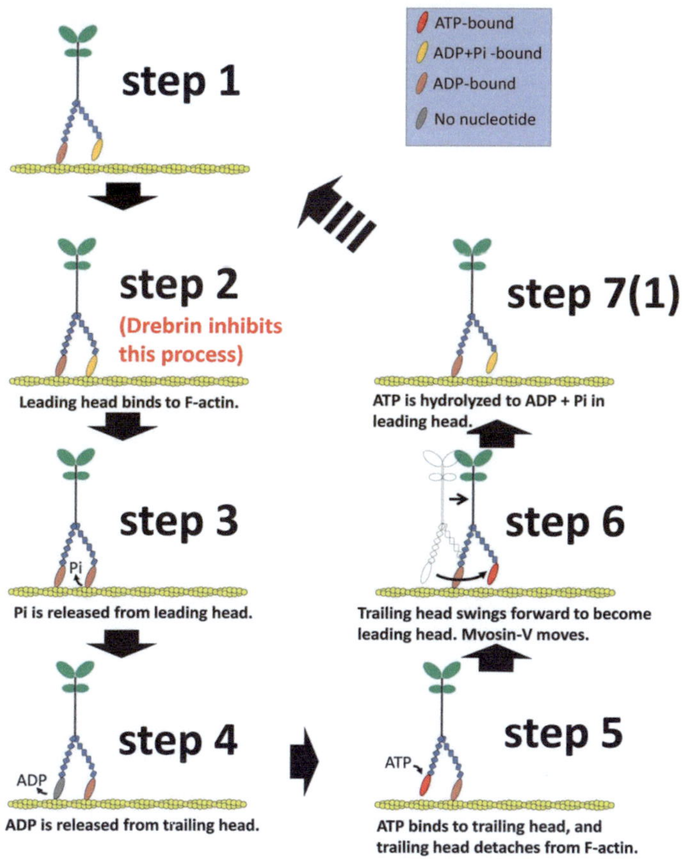

Fig. 3.4 Hand over hand motion of myosin-V

3.5 Conclusion

In this chapter, we showed that drebrin inhibits the actin-binding and actin-bundling activities of a wide variety of actin-binding proteins. Drebrin also inhibits the activity of myosin-II and myosin-V. Therefore, the expression of drebrin should cause drastic structural changes in actin organization.

In non-muscle cells such as neurons, a variety of actin-binding proteins differentially localize in and form specific actin structures such as stress fibers, filopodia, lamellipodia, and adhesion plaque. For example, drebrin-E concentrates in the basal region of filopodia, adhesion plaque, and actin arch, but not in the filopodia tip or lamellipodia in the growth cone of cultured neurons (Sasaki et al. 1996; Mizui et al. 2009). How does this "intracellular differentiation" in actin structures occur? We do not have a clear answer at this point, but the competitive binding between actin-binding proteins to F-actin should affect intracellular differentiation because they cannot simultaneously occupy the same position on F-actin filaments. Tropomyosin (Yang et al. 1979), cofilin (Hawkins et al. 1993; Hayden et al. 1993), and drebrin (Sharma et al. 2012) are known to bind to F-actin cooperatively, resulting in the formation of clusters on F-actin filaments (Schmidt et al. 2015; Ngo et al. 2015; Sharma et al. 2012). It is tempting to speculate that the initial point of intracellular differentiation in actin structures may be the spontaneous binding of such cooperative actin-binding proteins to F-actin. If a single molecule of drebrin, tropomyosin, or cofilin binds to F-actin, a cluster of these proteins will form and grow as long as the expression level of drebrin, tropomyosin, or cofilin is adequate and the F-actin filaments are bare. Spontaneous binding of competitors should slow or stop the growth of the cluster or cause the cluster on F-actin to break down, depending on the protein's expression level and affinity for F-actin. Rearrangements such as severing, polymerization, depolymerization, annealing, or cross-linking may accelerate the assembly and compartmentation of specific actin-binding proteins to form specific actin structures. Therefore, drebrin may be an important factor for the intracellular differentiation of actin structures.

Does the expression of drebrin cause the actin structure to be static or dynamic? At least in part, drebrin may cause the actin structure to be more dynamic because drebrin inhibits the binding of actin-stabilizers such as tropomyosin and caldesmon, both of which protect F-actin from gelsolin severing (Ishikawa et al. 1989a, b), in contrast to drebrin which does not protect F-actin from gelsolin severing (Ishikawa et al. 1994). However, drebrin also inhibits the binding of actin destabilizers such as cofilin (Grinstevich and Reisler 2014) or the highly mobile actin-bundler fascin (Sasaki et al. 1996). Furthermore, the inhibition of myosin-II and myosin-V activities by drebrin decreases the sliding speed of myosin-II (Hayashi et al. 1996) and the run length of myosin-V (Ishikawa et al. 2007; Kubota et al. 2010), possibly resulting in decreased cell motility compared to the absence of drebrin. The binding of drebrin to F-actin may also increase the stability of F-actin (Mikati et al. 2013). Therefore, it seems likely that the overall effect of drebrin expression on actin structure may be to make the cells more static.

References

Bernstein BW, Bamburg JR (1982) Tropomyosin binding to F-actin protects the F-actin from disassembly by brain actin-depolymerizing factor (ADF). Cell Motil 2:1–8

Grinstevich EE, Reisler E (2014) Drebrin inhibits cofilin-induced severing of F-actin. Cytoskeleton 71:472–483

Grinstevich EE, Galkin VE, Orlova A et al (2010) Mapping of drebrin binding site on F-actin. J Mol Biol 398:542–554

Hawkins M, Pope B, Maciver SK et al (1993) Human actin depolymerizing factor mediates a pH-sensitive destruction of actin filaments. Biochemistry 32:9985–9993

Hayashi K, Ishikawa R, Ye LH et al (1996) Modulatory role of drebrin on the cytoskeleton within dendritic spines in the rat cerebral cortex. J Neurosci 16:7161–7170

Hayden SM, Miller PS, Brauweiler A et al (1993) Analysis of the interaction of actin depolymerizing factor with G- and F-actin. Biochemistry 32:9994–10004

Ishikawa R, Yamashiro S, Matsumura F (1989a) Differential modulation of actin-severing activity of gelsolin by multiple isoforms of cultured rat cell tropomyosin: Potentiation of protective ability of tropomyosin by 83-kDa nonmuscle caldesmon. J Biol Chem 264:7490–7497

Ishikawa R, Yamashiro S, Matsumura F (1989b) Annealing of gelsolin-severed actin fragments by tropomyosin in the presence of Ca^{2+}: potentiation of the annealing process by caldesmon. J Biol Chem 264:16764–16770

Ishikawa R, Hayashi K, Shirao T et al (1994) Drebrin, a development-associated brain protein from rat embryo, causes the dissociation of tromomyosin from actin filaments. J Biol Chem 269:29928–29933

Ishikawa R, Yamashiro S. Kohama K et al (1998) Regulation of actin binding and actin bundling activities of fascin by caldesmon coupled with tropomyosin. J Biol Chem 273:26991–26997

Ishikawa R, Katoh K, Takahashi A et al (2007) Drebrin attenuates the interaction between actin and myosin-V. Biochem Biophys Res Commun 359:398–401

Kubota H, Ishikawa R, Ohki T et al (2010) Modulation of the mechano-chemical properties of myosin V by drebrin-E. Biochem Biophys Res Commun 400:643–648

Mikati MA, Grinstevich EE, Reisler E (2013) Drebrin-induced stabilization of actin filaments. J Biol Chem 288:19926–19938

Mizui T, Kojima N, Yamazaki H et al (2009) Drebrin E is involved in the regulation of axonal growth through actin-myosin interactions. J Neurochem 109:611–622

Ngai PK, Walsh MP (1984) Inhibition of smooth maucle actin-activated myosin Mg^{2+}-ATPase activity by caldesmon. J Biol Chem 259:13656–13659

Ngo KX, Kodera N, Katayama E et al (2015) Cofilin-induced unidirectional cooperative conformational changes in actin filaments revealed by high-speed atomic force microscopy. elife 4:e04806

Sasaki Y, Hayashi K, Shirao T et al (1996) Inhibition by drebrin of the actin-bundling activity of brain fascin, a protein localized in filopodia of growth cones. J Neurochem 66:980–988

Schmidt WM, Lehman W, Moore JR (2015) Direct observation of tropomyosin binding to actin filaments. Cytoskeleton 72:292–303

Sharma S, Grintsevich EE, Phillips ML et al (2011) Atomic force microscopy reveals drebrin induced remodeling of F-actin with subnanometer resolution. Nano Lett 11:825–827

Sharma S, Grintsevich EE, Hsueh C et al (2012) Molecular cooperativity of $drebrin_{1\text{-}300}$ binding and structural remodeling of F-actin. Biophys J 103:275–283

Shirao T, Obata K (1986) Immunochemical homology of 3 developmentally regulated brain proteins and their cevelopmental change in neuronal disribution. Dev. Brain Res 29:233–244

Shirao T, Hayashi K, Ishikawa R et al (1994) Formation of thick, curving bundles of actin by drebrin A expressed in fibroblast. Exp Cell Res 215:145–153

Vale RD (2003) The molecular motor toolbox for intracellular transport. Cell 112:467–480

Wagner PD (1984) Calcium-sensitive modulation of the actomyosin ATPase by fodrin. J Biol Chem 259:6306–6310

Worth DC, Daly C, Geraldo S et al (2013) Drebrin contains a cryptic F-actin-bundling activity regulated by Cdk5 phosphorylation. J Cell Biol 202:793–806

Yamaguchi M, Ver A, Carlos A et al (1984) Modulation of the actin-activated adenosinetriphosphatase activity of myosin by tropomyosin from vascular and gizzard smooth muscles. Biochemistry 23:774–779

Yang Y-Z, Korn ED, Eisenberg E (1979) Cooperative binding of tropomyosin to muscle and acanthamoeba actin. J Biol Chem 254:7137–7140

Chapter 4
Phosphorylation of Drebrin and Its Role in Neuritogenesis

Phillip R. Gordon-Weeks

Abstract Neuritogenesis is an early event in neuronal development in which newborn neurons first form growth cones, as a prerequisite for the formation of axons and dendrites. Growth cones emerge from segmented regions of the lamellipodium of embryonic neurons and grow away from the cell body leaving behind a neurite that will eventually polarise into an axon or dendrite. Growth cones also function to navigate precise routes through the embryo to locate an appropriate synaptic partner. Dynamic interactions between two components of the neuronal cytoskeleton, actin filaments and microtubules, are known to be essential for growth cone formation and hence neuritogenesis. The molecular mechanisms that coordinate interactions between actin filaments and dynamic microtubules during neuritogenesis are beginning to be understood. One candidate pathway coupling actin filaments to microtubules consists of the actin filament-binding protein drebrin and the microtubule-binding +TIP protein EB3. This pathway is regulated proximally by cyclin-dependent kinase 5 phosphorylation of drebrin but the upstream elements in the pathway have yet to be identified.

Keywords Neuritogenesis • Growth cone • Microtubule • Actin • Drebrin • EB3 Phosphorylation • Cdk5

4.1 Breaking Radial Symmetry: Neuritogenesis and the Birth of the Growth Cone

Newborn neurons are unremarkable, and often unpolarised, cells with a large nucleus and a thin rim of cytoplasm. To differentiate into the elaborate, polarised morphology of mature neurons, they must first break their spherical (radial) symmetry and grow axons and dendrites. This process, referred to as neuritogenesis,

P.R. Gordon-Weeks
MRC Centre for Developmental Neurobiology, King's College London,
New Hunt's House, Guy's Campus, London SE1 1UL, UK
e-mail: phillip.gordon-weeks@kcl.ac.uk

T. Shirao, Y. Sekino (eds.), *Drebrin*, Advances in Experimental Medicine and Biology 1006, DOI 10.1007/978-4-431-56550-5_4

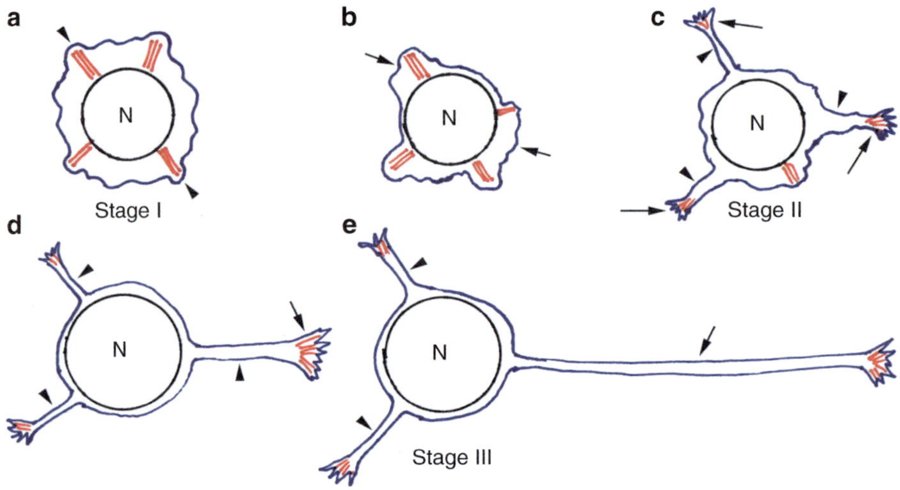

Fig. 4.1 Drawings depicting embryonic hippocampal neurons in culture undergoing neuritogenesis. (**a**) Within hours after plating the neuron has developed a circumferential lamellipodium supported by a dendritic actin filament network (*blue*) and radial filopodia containing bundled actin filaments (*red, arrowheads*). (**b**) The first sign of neuritogenesis is segmentation of the lamellipodium to form multiple individual lamellipodia (*small arrows*). (**c**) Subsequently, nascent growth cones (*arrows*) emerge from segmented lamellipodia and start to extend from the cell body leaving behind a neurite (*arrowheads*, stage II). (**d**) As these minor neurites elongate, they condense into thin shafts separating the cell body from the growth cone (*arrows*) and the growth cone of one minor neurite enlarges (*arrow*). (**e**) Subsequently, this neurite elongates rapidly to become an axon (*arrow*, stage III). The remaining neurites, after a delay, will differentiate into dendrites (*arrowheads*). N = nucleus

begins with the emergence of a growth cone at the cell surface, which subsequently grows away from the cell body leaving behind a neurite (Fig. 4.1). The growth cone has three important functions in neural development: it contributes to the assembly and growth of the neurite, it guides the growing neurite through the developing embryo to locate an appropriate synaptic partner, and it recognises and forms a synapse with an appropriate partner (Gordon-Weeks 2005). What determines the site(s) of growth cone initiation on the surface of a newborn neuron and the molecular mechanisms underlying growth cone formation are largely unknown (Flynn 2013). Neuritogenesis is a critical stage in neuronal development because unless growth cones emerge at the appropriate time and place they will not be in a position to respond to guidance cues that orchestrate the development of a correctly connected nervous system with appropriate neural circuits. Furthermore, understanding neuritogenesis has the potential to shed light on how to enhance growth cone formation by injured axons in the adult central nervous system and thereby improve regeneration (Bradke et al. 2012).

Much of our understanding of the cellular and molecular steps in neuritogenesis has come from studying neuritogenesis occurring spontaneously in dissociated embryonic neurons in two-dimensional cultures in vitro (Flynn 2013). This culture system was pioneered by Gary Banker (Dotti et al. 1988) and involves sparsely plating dissociated neurons on glass coverslips so that individual neuronal morphology can be observed without interference from neighbouring neurons. Although a system more divorced from in vivo conditions is difficult to imagine, this culture approach has, nonetheless, provided many insights into the cellular and molecular mechanisms underlying neuritogenesis and neuronal polarisation. It has particularly contributed to our understanding of the molecules involved in determining axonal polarity, although very few of these have been tested in vivo (Polleux and Snider 2010). In dissociated cell culture, primary embryonic neurons develop a circumferential lamellipodium soon after substrate attachment (Fig. 4.1). Subsequently, segmentation of the lamellipodium occurs, brought about by a localised collapse of the lamellipodium following localised cessation of protrusion and retraction of the lamellipodium. Growth cones form from the segmented lamellipodium where activity persists, and, once formed, growth cones migrate away from the cell body leaving behind a process, generally referred to as a neurite to reflect the ambiguity relating to whether it will become an axon or a dendrite. Most embryonic cortical neurons in culture produce multiple neurites that grow relatively slowly and are equi-potential in terms of whether they will become axons or dendrites. The growth cone of one neurite then enlarges while accumulating actin filaments, and this neurite advances more rapidly and becomes an axon. The growth of the remaining neurites slows down or stalls, and they later differentiate into dendrites. In culture, which neurite is selected to enlarge its growth cone and become an axon was originally thought to be random. There is evidence, however, that this might be predetermined since direct observation of cortical neurons polarising in culture has shown that the first neurite to emerge frequently becomes the axon (de Anda et al. 2005). This suggests that the cytoplasm of a newborn neuron might be polarised, and indeed one of the molecules found to be polarised in the plasma membrane of stage I neurons is the cell adhesion molecule N-cadherin (Gärtner et al. 2012; reviewed in: 2015). Some cortical neurons also go through a phase of bearing multiple, equi-potential neurites in vivo (Polleux and Snider 2010). The growth cone of one neurite contacts a radial glial fibre, and the neuron then becomes bipolar, as it migrates along the fibre, with a leading process that will become the dendrite and a trailing process that becomes the axon.

4.2 Filopodial Actin Filaments and Dynamic Microtubules Underlie Neuritogenesis

Early time-lapse microscopy studies in vitro revealed that dissociated embryonic neurons are initially surrounded by a lamellipodium supported by actin filaments that are organised distally into either dendritic meshworks or radial bundles that form the core cytoskeleton of filopodia (Smith 1994a, b) and proximally into tangential actin arcs (Fig. 4.2; Worth et al. 2013). Microtubules are mainly localised behind the actin arcs in circumferential bundles but occasionally extend individually

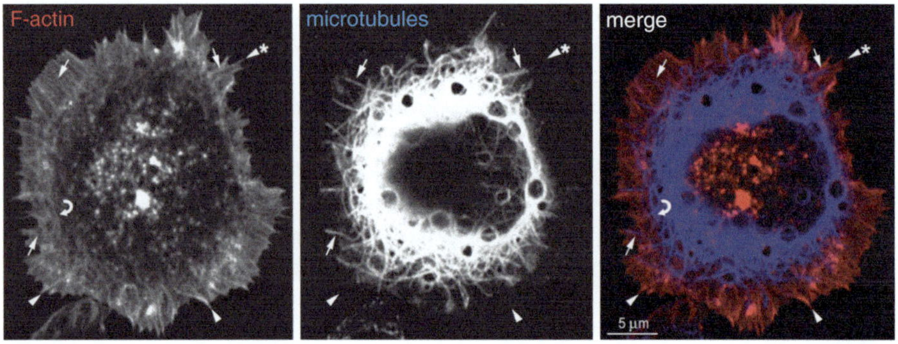

Fig. 4.2 Confocal images of a stage I embryonic cortical neuron labelled with phalloidin (*red*) to reveal actin filaments and with an antibody against tyrosinated α-tubulin (*blue*), a marker for dynamic microtubules. Actin filaments form the core of filopodia (*arrowheads*) and also actin arcs (*curved arrow*) in the lamellipodium. Most dynamic microtubules are located in a circumferential band between the nucleus and the actin arcs but individual microtubules (*arrows*) extend into the lamellipodium and occasionally invade filopodia from the circumferential band (*starred arrowhead*)

into the distal lamellipodium (Fig. 4.2). Local segmentation of the lamellipodium gives rise to a growth cone (Fig. 4.1; e.g. Dehmelt et al. 2003). Segmentation always occurs around the filopodia (Smith 1994a, b), which precede and are necessary for growth cone emergence (Dent et al. 2007). In heroic transgenic mouse experiments, Frank Gertler's group removed all three paralogues of the Ena/VASP proteins in embryonic cortical neurons and found that this led to defects in neuritogenesis, including failure to form actin filament bundles and therefore filopodia (Kwiatkowski et al. 2007). It had previously been shown that this family of actin-binding proteins regulate filopodia formation in developing neurons by preventing capping proteins from blocking actin filament polymerisation at the tips of filopodia thereby promoting filopodia formation. Embryonic cortical neurons cultured from these triple-knockout mice have circumferential lamellipodia but fail to develop filopodia and do not form growth cones (Dent et al. 2007). Filopodia giving rise to growth cones are invaded by dynamic microtubules which become bundled, by unidentified factors, in the emerging neurite (Flynn et al. 2012). One candidate for microtubule bundling during neuritogenesis is the microtubule-associated protein MAP2c (Dehmelt et al. 2003). In cortical neurons cultured from the triple Ena/VASP knockout mouse, microtubules still continue to invade the lamellipodia, bending parallel to the plasma membrane once they reach it, but fail to form bundles or to give rise to growth cones (Dent et al. 2007). The failure to produce filopodia can be rescued by culturing neurons on filopodia-inducing substrates such as laminin or expressing other anti-capping proteins such as mDia2 (Dent et al. 2007). These studies highlight the importance of filopodia in growth cone formation and confirm Smith's original observation that only those filopodia that are invaded by microtubules give rise to growth cones (Smith 1994a).

This localised reorganization of the actin filament and microtubule cytoskeleton is necessary for neuritogenesis since compounds that block microtubule dynamic instability or actin filament assembly inhibit neuritogenesis (Dent et al. 2007; Flynn et al. 2012). Consistent with a role for actin filament dynamics, knockout of the two actin depolymerising factor (ADF)/cofilin genes in mice cortical neurons compromises actin retrograde flow in lamellipodia and filopodia and inhibits neuritogenesis (Flynn et al. 2012). Any procedure that effectively uncouples microtubule and actin filament dynamics compromises neuritogenesis (Flynn 2013). An unanswered question is how are these cytoskeletal elements coordinated to drive neuritogenesis; what are the molecular pathways that couple actin filaments in filopodia to dynamic microtubules?

4.3 The Drebrin/EB3/Cdk5 Pathway and Neuritogenesis

We have shown that dynamic microtubules and actin filaments in growth cone filopodia are coupled during neuritogenesis by an interaction between the actin filament-binding protein drebrin and the +TIP protein EB3 (Geraldo et al. 2008; Worth et al. 2013). Drebrin, binding to parallel actin filaments at the base of filopodia, interacts with EB3 located at the plus ends of dynamic microtubules entering filopodia (Geraldo et al. 2008). The evidence for this includes the demonstration of direct binding between drebrin and EB3 in filopodia using FRET/FLIM (Geraldo et al. 2008) and the observation that a dominant negative construct named EB3M, that contains the drebrin binding domain of EB3 (Geraldo et al. 2008), blocks neuritogenesis. Conversely, overexpression of drebrin stimulates neuritogenesis leading to supernumerary neurites (unpublished observations). RNA interference methods have achieved only modest knockdown of drebrin in cultured neurons (~50%; Geraldo et al. 2008; Mizui et al. 2009), and at this level, neuritogenesis appears relatively normal, although neurites are shorter than controls (Geraldo et al. 2008). The drebrin/EB3 pathway has been independently confirmed in other cellular contexts where dynamic microtubules are interacting with actin filaments, suggesting that the pathway is canonical (Bazellieres et al. 2012; Merriam et al. 2013).

4.4 Drebrin Phosphorylation Regulates Its Interactions with Itself and Other Proteins

Drebrin is a phosphoprotein, as first recognised indirectly using mass spectrometry to identify proteins immunoprecipitated with phosphoserine/threonine antibodies from HeLa cells (Grönborg et al. 2002). Subsequently, tandem mass spectrometry screens directed at identifying phosphorylation sites on proteins from cell lines, whole tissue extracts or subcellular fractions, mapped multiple serine/threonine and tyrosine phosphorylation sites on drebrin (Ballif et al. 2004; Beausoleil et al. 2004;

Collins et al. 2005, 2008; Dephoure et al. 2008; Gauci et al. 2009; Guo et al. 2007; Mayya et al. 2009; Molina et al. 2007; Mukherji et al. 2006; Munton et al. 2007; Olsen et al. 2006; Rush et al. 2005; Sugiyama et al. 2007; Tao et al. 2005; Trinidad et al. 2006, 2007, 2008; Vosseller et al. 2005; Wang et al. 2007; Zahedi et al. 2008; Zheng et al. 2005). Curated public databases of phosphoproteins currently list 30 serine/threonine and 5 tyrosine phosphorylation sites on human drebrin (e.g. http://www.phosphosite.org/). Phospho-specific antibodies have been raised to some of the sites identified by mass spectrometry, and these have been used to further characterise drebrin phosphorylation (Kreis et al. 2013; Worth et al. 2013). A monoclonal antibody to drebrin phosphoserine 142 is commercially available (clone 3C14; http://www.merckmillipore.com/).

Phosphoserine 142 is the most frequently reported of the serine/threonine phosphorylation sites and is highly conserved across the vertebrates (Worth et al. 2013). It was first tentatively identified in affinity purified drebrin from MDCK cells using tandem mass spectrometry (Chew et al. 2005). More recently, the roles of this phosphorylation site in drebrin have been determined (Worth et al. 2013).

Drebrin has two actin filament-binding domains, a coiled coil (CC) domain and an adjacent helical (Hel) domain and can exist in two conformations: a closed one in which the C-terminus folds over to bind to the N-terminus occluding the actin filament-binding activity in the CC domain and an open conformation in which both actin filament-binding domains are exposed (Fig. 4.3; Worth et al. 2013). In the closed conformation, drebrin binds single actin filaments, whereas in the open conformation, it bundles actin filaments or binds to pre-existing actin filament bundles. We have recently shown that phosphorylation of drebrin by cyclin-dependent kinase 5 (Cdk5) at S142, which is in the CC domain, relieves the intramolecular inhibition of the CC domain binding to actin filaments, converting drebrin from a single actin filament binding protein, through the Hel domain, to an actin filament bundling protein co-operatively using both actin filament-binding domains (Worth et al. 2013). Phosphorylation is a common mechanism for relieving intramolecular autoinhibition, and examples are found in many cytoskeleton associated proteins, e.g. FAK (Chen et al. 2011), formins (Takeya et al. 2008) and WASP (Torres and Rosen 2006). Cdk5 phosphorylation of drebrin at S142 has been confirmed independently (Tanabe et al. 2014).

Since Cdk5 phosphorylated drebrin has two actin filament-binding domains available it can bundle, alternatively Cdk5 phosphorylation of drebrin might target it to pre-existing actin filament bundles such as those in filopodia. Consistent with this idea, pS142-drebrin accumulates, along with dynamic microtubules, at the base of filopodia in stage I neurons in culture where growth cones are forming (Fig. 4.4). Furthermore, phospho-mimetic and phospho-dead mutants of S142 enhance and inhibit neuritogenesis, respectively (Worth et al. 2013). Interestingly, although all stage I neurons express drebrin, only a sub-population, those in the early stages of neuritogenesis, have pS142 drebrin, suggesting that activation of Cdk5 and phosphorylation of S142 drebrin is an early event in growth cone formation (Worth et al. 2013). Since phosphorylation of S142 enables drebrin to bind to parallel actin filament bundles, such as those found in filopodia (Worth et al. 2013), Cdk5 phosphorylation of S142 drebrin might target it to filopodia in stage I neurons where

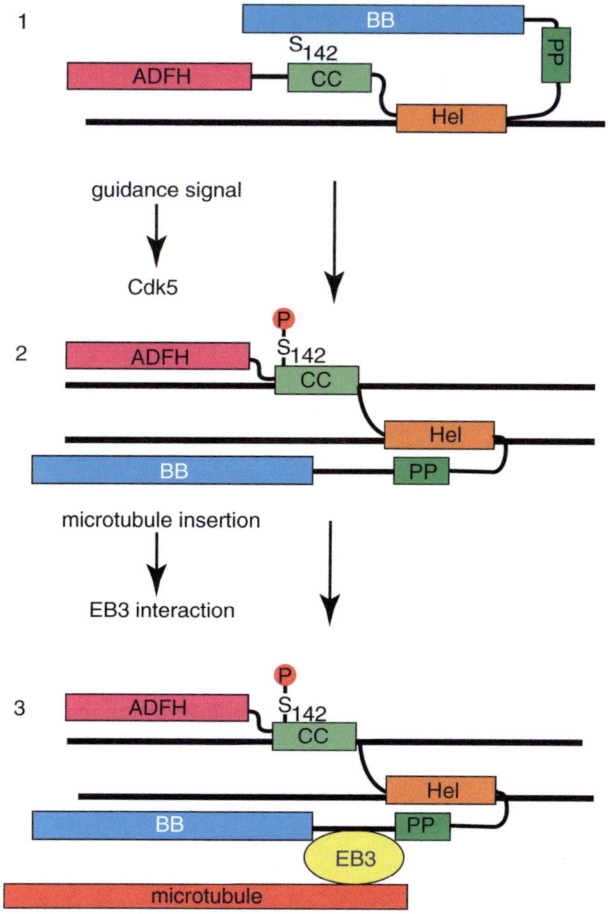

Fig. 4.3 Model of a molecular mechanism for coupling of actin filaments to microtubules via an interaction between drebrin, phosphorylated at S142, and the +TIP protein EB3. The diagram shows the domain structure of drebrin in the "closed" conformation when the BB domain binds at the border between the ADFH and CC domains where S142 is located (1). In this conformation, drebrin can bind to actin filaments but does not bundle or straddle pre-existing actin filament bundles. One filopodial actin filament, represented by a black line, is also shown. Activation of Cdk5 by intracellular signalling pathways driven by axon guidance cues leads to S142 phosphorylation (2) and the consequent relief of the intramolecular occlusion of one actin filament-binding domain by the BB domain. This produces an "open" conformation and drebrin now bundles actin filaments, or straddles existing actin filament bundles, by co-operative binding of the CC and Hel domains. Drebrin in the "open" conformation and bound to actin filaments at the base of filopodia is now in an appropriate state to interact with the +TIP protein EB3 located at the plus end of a microtubule (*red line*) entering the filopodium and thereby couples actin filaments to a microtubule (3). ©Originally published in Worth et al. (2013)

it is appropriately positioned to couple dynamic microtubules entering filopodia to the actin filaments in filopodia. After neuritogenesis, pS142 drebrin remains uniquely localised to all growth cones (Worth et al. 2013). How Cdk5 is activated in this context, and therefore what lies upstream of the drebrin/EB3/Cdk5 pathway, is

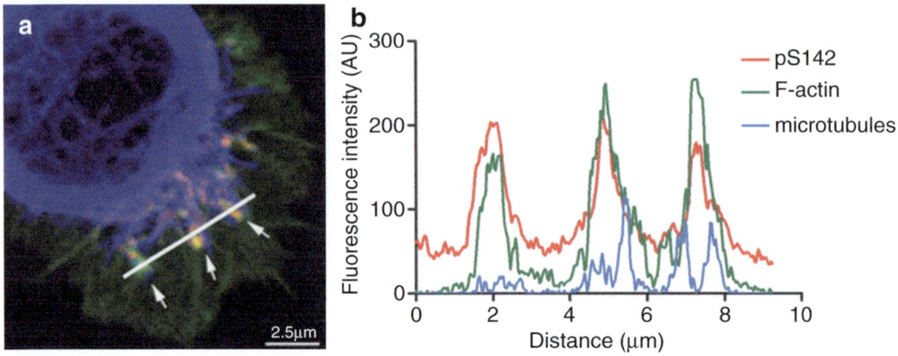

Fig. 4.4 (**a**) Immunofluorescence confocal image of a stage I cortical neuron commencing growth cone formation. The neuron was labelled with phalloidin for actin filaments (*green*) and an antibody against pS142 drebrin (*red*) and an antibody against tyrosinated α-tubulin which recognises dynamic microtubules (*blue*). In the nascent growth cone, drebrin, phosphorylated at S142, co-localises with bundles of actin filaments (*arrows*). Dynamic microtubules enter the nascent growth cone from a circumferential bundle of microtubules in the cell body and pass alongside the drebrin/actin filament bundles in the nascent growth cone (*arrows*). (**b**) Fluorescence intensity in arbitrary units (AU) of actin filaments (*green*), pS142 drebrin (*red*) and microtubules (*blue*) along the line plot shown in panel A. Actin filaments and pS142 drebrin co-localise in three peaks while three microtubules (*blue peaks*) are located at the sides. ©Originally published in Worth et al. (2013)

unknown. Several mechanisms have been identified by which Cdk5 can be activated, including by binding to its non-catalytic regulatory subunit, p35, or by direct phosphorylation on Y15 and S159 (Su and Tsai 2011), but these pathways have yet to be investigated in the context of neuritogenesis.

A second phosphorylation site on drebrin has also been characterised using phospho-specific antibodies to S647 in drebrin (Kreis et al. 2013). Although the kinase responsible has yet to be identified, there is evidence that S647 is dephosphorylated by the phosphatase and tensin homology protein (PTEN; Kreis et al. 2013). Drebrin is phosphorylated at S647 in the growth cone, but its role in neuritogenesis has not been examined. Interestingly, S647 phosphorylation can be regulated by neuronal activity.

4.5 In Vivo Experiments with Phospho-mimetic and Phospho-dead S142

Phosphorylation of drebrin S142 by Cdk5, first reported by Worth et al. (2013), has been confirmed and the role of this phosphorylation site explored in vivo (Tanabe et al. 2014). This group showed that in utero electroporation of rat embryos with phospho-mimetic (S142D) or phospho-dead (S142A) mutants of drebrin inhibited the migration of cerebral cortical neurons from the ventricular and sub-ventricular zones to the cortical plate. Unexpectedly, both mutants had the same effect on cortical migration. It is not yet known which steps in cortical migration are disrupted but these might include neuritogenesis, transition from a multipolar to a bipolar morphology or attachment and migration along radial glial fibres. One explanation for both mutants having the same phenotype is that there could be an optimal ratio or set point of phosphorylated to non-phosphorylated drebrin for cortical migration and disturbance of the ratio in either direction inhibits migration. In a study of the role of drebrin in the migration of olfactory bulb neuronal precursors in the rostral migratory stream, it was also found that S142D or S142A mutants of drebrin disrupt migration (Sonego et al. 2015).

4.6 Conclusions and Future Perspectives

It has been recognised for some time that co-operation between the microtubule and actin filament cytoskeletons is an essential step in the formation of a growth cone during neuritogenesis in embryonic neurons (Flynn 2013). The drebrin/EB3/Cdk5 pathway is a strong contender for coupling dynamic microtubules to actin filaments during neuritogenesis although it is unlikely that it is the only pathway, and, indeed, other candidates have been identified (Coles and Bradke 2015). In the future, we can expect to learn exactly what independent contribution these pathways make to neuritogenesis and, most importantly, how they are regulated in space and time in the embryo. Too much reliance has been placed on the results from in vitro experiments which may not reflect events in vivo. There is a pressing need to conduct these experiments in vivo, and emerging techniques facilitating in vivo imaging and genetic manipulation are likely to play key roles in this endeavour.

References

de Anda FC, Pollarolo G, Da Silva JS, Camoletto PG, Feiguin F, Dotti CG (2005) Centrosome localization determines neuronal polarity. Nature 436(7051):704–708. doi:10.1038/nature03811

Ballif BA, Villen J, Beausoleil SA, Schwartz D, Gygi SP (2004) Phosphoproteomic analysis of the developing mouse brain. Mol Cell Proteomics 3(11):1093–1101. doi:10.1074/mcp.M400085-MCP200

Bazellieres E, Massey-Harroche D, Barthelemy-Requin M, Richard F, Arsanto JP, Le Bivic A (2012) Apico-basal elongation requires a drebrin-E-EB3 complex in columnar human epithelial cells. J Cell Sci 125(Pt 4):919–931. doi:10.1242/jcs.092676

Beausoleil SA, Jedrychowski M, Schwartz D, Elias JE, Villen J, Li J, Cohn MA, Cantley LC, Gygi SP (2004) Large-scale characterization of HeLa cell nuclear phosphoproteins. Proc Natl Acad Sci USA 101(33):12130–12135. doi:10.1073/pnas.0404720101

Bradke F, Fawcett JW, Spira ME (2012) Assembly of a new growth cone after axotomy: the precursor to axon regeneration. Nat Rev Neurosci 13(3):183–193. doi:10.1038/nrn3176

Chen TH, Chan PC, Chen CL, Chen HC (2011) Phosphorylation of focal adhesion kinase on tyrosine 194 by Met leads to its activation through relief of autoinhibition. Oncogene 30(2):153–166. doi:10.1038/onc.2010.398

Chew CS, Okamoto CT, Chen X, Thomas R (2005) Drebrin E2 is differentially expressed and phosphorylated in parietal cells in the gastric mucosa. Am J Physiol Gastrointest Liver Physiol 289(2):G320–G331. doi:10.1152/ajpgi.00002.2005

Coles CH, Bradke F (2015) Coordinating neuronal actin-microtubule dynamics. Curr Biol 25(15):R677–R691. doi:10.1016/j.cub.2015.06.020

Collins MO, Yu L, Coba MP, Husi H, Campuzano I, Blackstock WP, Choudhary JS, Grant SG (2005) Proteomic analysis of in vivo phosphorylated synaptic proteins. J Biol Chem 280(7):5972–5982. doi:10.1074/jbc.M411220200

Collins MO, Yu L, Campuzano I, Grant SG, Choudhary JS (2008) Phosphoproteomic analysis of the mouse brain cytosol reveals a predominance of protein phosphorylation in regions of intrinsic sequence disorder. Mol Cell Proteomics 7(7):1331–1348. doi:10.1074/mcp.M700564-MCP200

Dehmelt L, Smart FM, Ozer RS, Halpain S (2003) The role of microtubule-associated protein 2c in the reorganization of microtubules and lamellipodia during neurite initiation. J Neurosci 23(29):9479–9490

Dent EW, Kwiatkowski AV, Mebane LM, Philippar U, Barzik M, Rubinson DA, Gupton S, Van Veen JE, Furman C, Zhang J, Alberts AS, Mori S, Gertler FB (2007) Filopodia are required for cortical neurite initiation. Nat Cell Biol 9(12):1347–1359. doi:10.1038/ncb1654

Dephoure N, Zhou C, Villen J, Beausoleil SA, Bakalarski CE, Elledge SJ, Gygi SP (2008) A quantitative atlas of mitotic phosphorylation. Proc Natl Acad Sci USA 105(31):10762–10767. doi:10.1073/pnas.0805139105

Dotti CG, Sullivan CA, Banker GA (1988) The establishment of polarity by hippocampal neurons in culture. J Neurosci 8(4):1454–1468

Flynn KC (2013) The cytoskeleton and neurite initiation. BioArchitecture 3(4):86–109

Flynn KC, Hellal F, Neukirchen D, Jacob S, Tahirovic S, Dupraz S, Stern S, Garvalov BK, Gurniak C, Shaw AE, Meyn L, Wedlich-Soldner R, Bamburg JR, Small JV, Witke W, Bradke F (2012) ADF/cofilin-mediated actin retrograde flow directs neurite formation in the developing brain. Neuron 76(6):1091–1107. doi:10.1016/j.neuron.2012.09.038

Gärtner A, Fornasiero EF, Munck S, Vennekens K, Seuntjens E, Huttner WB, Valtorta F, Dotti CG (2012) N-cadherin specifies first asymmetry in developing neurons. EMBO J 31(8):1893–1903. doi:10.1038/emboj.2012.41

Gärtner A, Fornasiero EF, Dotti CG (2015) Cadherins as regulators of neuronal polarity. Cell Adhes Migr 9(3):175–182. doi:10.4161/19336918.2014.983808

Gauci S, Helbig AO, Slijper M, Krijgsveld J, Heck AJR, Mohammed S (2009) Lys-N and trypsin cover complementary parts of the phosphoproteome in a refined SCX-based approach. Anal Chem 81(11):4493–4501. doi:10.1021/ac9004309

Geraldo S, Khanzada UK, Parsons M, Chilton JK, Gordon-Weeks PR (2008) Targeting of the F-actin-binding protein drebrin by the microtubule plus-tip protein EB3 is required for neuritogenesis. Nat Cell Biol 10(10):1181–1189. doi:10.1038/ncb1778

Gordon-Weeks PR (2005) Neuronal growth cones, Developmental and cell biology series, vol 37. Cambridge University Press, Cambridge

Grönborg M, Kristiansen TZ, Stensballe A, Andersen JS, Ohara O, Mann M, Jensen ON, Pandey A (2002) A mass spectrometry-based proteomic approach for identification of serine/threonine-phosphorylated proteins by enrichment with phospho-specific antibodies: identification of a novel protein, Frigg, as a protein kinase A substrate. Mol Cell Proteomics 1(7):517–527

Guo M, Galan J, Tao WA (2007) Soluble nanopolymer-based phosphoproteomics for studying protein phosphatase. Methods 42(3):289–297. doi:10.1016/j.ymeth.2007.02.019

Kreis P, Hendricusdottir R, Kay L, Papageorgiou IE, van Diepen M, Mack T, Ryves J, Harwood A, Leslie NR, Kann O, Parsons M, Eickholt BJ (2013) Phosphorylation of the actin binding protein Drebrin at S647 is regulated by neuronal activity and PTEN. PLoS One 8(8):e71957. doi:10.1371/journal.pone.0071957

Kwiatkowski AV, Rubinson DA, Dent EW, Edward van Veen J, Leslie JD, Zhang J, Mebane LM, Philippar U, Pinheiro EM, Burds AA, Bronson RT, Mori S, Fassler R, Gertler FB (2007) Ena/VASP Is Required for neuritogenesis in the developing cortex. Neuron 56(3):441–455. doi:10.1016/j.neuron.2007.09.008

Mayya V, Lundgren DH, Hwang SI, Rezaul K, Wu LF, Eng JK, Rodionov V, Han DK (2009) Quantitative phosphoproteomic analysis of T cell receptor signaling reveals system-wide modulation of protein-protein interactions. Sci Signal 2(84):ra46. doi:10.1126/scisignal.2000007

Merriam EB, Millette M, Lumbard DC, Saengsawang W, Fothergill T, Hu X, Ferhat L, Dent EW (2013) Synaptic regulation of microtubule dynamics in dendritic spines by calcium, F-actin, and drebrin. J Neurosci 33(42):16471–16482. doi:10.1523/JNEUROSCI.0661-13.2013

Mizui T, Kojima N, Yamazaki H, Katayama M, Hanamura K, Shirao T (2009) Drebrin E is involved in the regulation of axonal growth through actin-myosin interactions. J Neurochem 109(2):611–622. doi:10.1111/j.1471-4159.2009.05993.x

Molina H, Horn DM, Tang N, Mathivanan S, Pandey A (2007) Global proteomic profiling of phosphopeptides using electron transfer dissociation tandem mass spectrometry. Proc Natl Acad Sci USA 104(7):2199–2204. doi:10.1073/pnas.0611217104

Mukherji M, Brill LM, Ficarro SB, Hampton GM, Schultz PG (2006) A phosphoproteomic analysis of the ErbB2 receptor tyrosine kinase signaling pathways. Biochemistry 45(51):15529–15540. doi:10.1021/bi060971c

Munton RP, Tweedie-Cullen R, Livingstone-Zatchej M, Weinandy F, Waidelich M, Longo D, Gehrig P, Potthast F, Rutishauser D, Gerrits B, Panse C, Schlapbach R, Mansuy IM (2007) Qualitative and quantitative analyses of protein phosphorylation in naive and stimulated mouse synaptosomal preparations. Mol Cell Proteomics 6(2):283–293. doi:10.1074/mcp.M600046-MCP200

Olsen JV, Blagoev B, Gnad F, Macek B, Kumar C, Mortensen P, Mann M (2006) Global, in vivo, and site-specific phosphorylation dynamics in signaling networks. Cell 127(3):635–648. doi:10.1016/j.cell.2006.09.026

Polleux F, Snider W (2010) Initiating and growing an axon. Cold Spring Harb Perspect Biol 2(4):a001925. doi:10.1101/cshperspect.a001925

Rush J, Moritz A, Lee KA, Guo A, Goss VL, Spek EJ, Zhang H, Zha XM, Polakiewicz RD, Comb MJ (2005) Immunoaffinity profiling of tyrosine phosphorylation in cancer cells. Nat Biotechnol 23(1):94–101. doi:10.1038/nbt1046

Smith CL (1994a) Cytoskeletal movements and substrate interactions during initiation of neurite outgrowth by sympathetic neurons in vitro. J Neurosci 14(1):384–398

Smith CL (1994b) The initiation of neurite outgrowth by sympathetic neurons grown in vitro does not depend on assembly of microtubules. J Cell Biol 127(5):1407–1418

Sonego M, Oberoi M, Stoddart J, Gajendra S, Hendricusdottir R, Oozeer F, Worth DC, Hobbs C, Eickholt BJ, Gordon-Weeks PR, Doherty P, Lalli G (2015) Drebrin regulates neuroblast migration in the postnatal mammalian brain. PLoS One 10(5):e0126478. doi:10.1371/journal.pone.0126478

Su SC, Tsai LH (2011) Cyclin-dependent kinases in brain development and disease. Annu Rev Cell Dev Biol 27:465–491. doi:10.1146/annurev-cellbio-092910-154023

Sugiyama N, Masuda T, Shinoda K, Nakamura A, Tomita M, Ishihama Y (2007) Phosphopeptide enrichment by aliphatic hydroxy acid-modified metal oxide chromatography for nano-LC-MS/MS in proteomics applications. Mol Cell Proteomics 6(6):1103–1109. doi:10.1074/mcp.T600060-MCP200

Takeya R, Taniguchi K, Narumiya S, Sumimoto H (2008) The mammalian formin FHOD1 is activated through phosphorylation by ROCK and mediates thrombin-induced stress fibre formation in endothelial cells. EMBO J 27(4):618–628. doi:10.1038/emboj.2008.7

Tanabe K, Yamazaki H, Inaguma Y, Asada A, Kimura T, Takahashi J, Taoka M, Ohshima T, Furuichi T, Isobe T, Nagata K, Shirao T, Hisanaga S (2014) Phosphorylation of drebrin by cyclin-dependent kinase 5 and its role in neuronal migration. PLoS One 9(3):e92291. doi:10.1371/journal.pone.0092291

Tao WA, Wollscheid B, O'Brien R, Eng JK, Li XJ, Bodenmiller B, Watts JD, Hood L, Aebersold R (2005) Quantitative phosphoproteome analysis using a dendrimer conjugation chemistry and tandem mass spectrometry. Nat Methods 2(8):591–598. doi:10.1038/Nmeth776

Torres E, Rosen MK (2006) Protein-tyrosine kinase and GTPase signals cooperate to phosphorylate and activate Wiskott–Aldrich syndrome protein (WASP)/neuronal WASP. J Biol Chem 281(6):3513–3520. doi:10.1074/jbc.M509416200

Trinidad JC, Specht CG, Thalhammer A, Schoepfer R, Burlingame AL (2006) Comprehensive identification of phosphorylation sites in postsynaptic density preparations. Mol Cell Proteomics 5(5):914–922. doi:10.1074/mcp.T500041-MCP200

Trinidad JC, Thalhammer A, Specht CG, Baker PR, Lynn AJ, Schoepfer R, Burlingame AL (2007) Analysis of protein levels and phosphorylation stoichiometry from complex samples using the iTRAQ reagent. Mol Cell Proteomics 6(8):17–17

Trinidad JC, Thalhammer A, Specht CG, Lynn AJ, Baker PR, Schoepfer R, Burlingame AL (2008) Quantitative analysis of synaptic phosphorylation and protein expression. Mol Cell Proteomics 7(4):684–696. doi:10.1074/mcp.M700170-MCP200

Vosseller K, Hansen KC, Chalkley RJ, Trinidad JC, Wells L, Hart GW, Burlingame AL (2005) Quantitative analysis of both protein expression and serine/threonine post-translational modifications through stable isotope labeling with dithiothreitol. Proteomics 5(2):388–398. doi:10.1002/pmic.200401066

Wang Y, Ding SJ, Wang W, Jacobs JM, Qian WJ, Moore RJ, Yang F, Camp DG, 2nd, , Smith RD, Klemke RL (2007) Profiling signaling polarity in chemotactic cells. Proc Natl Acad Sci USA 104 (20):8328-8333. doi:10.1073/pnas.0701103104

Worth DC, Daly CN, Geraldo S, Oozeer F, Gordon-Weeks PR (2013) Drebrin contains a cryptic F-actin-bundling activity regulated by Cdk5 phosphorylation. J Cell Biol 202(5):793–806. doi:10.1083/jcb.201303005

Zahedi RP, Lewandrowski U, Wiesner J, Wortelkamp S, Moebius J, Schutz C, Walter U, Gambaryan S, Sickmann A (2008) Phosphoproteome of resting human platelets. J Proteome Res 7(2):526–534. doi:10.1021/pr0704130

Zheng HY, Hu P, Quinn DF, Wang YK (2005) Phosphotyrosine proteomic study of interferon alpha signaling pathway using a combination of immunoprecipitation and immobilized metal affinity chromatography. Mol Cell Proteomics 4(6):721–730. doi:10.1074/mcp.M400077-MCP200

Chapter 5
Remodeling of Actin Filaments by Drebrin A and Its Implications

Elena E. Grintsevich

Abstract Dendritic spines are small protrusions of dendrites that are critical for synaptic transmission. The plasticity and stability of dendritic spines is tightly linked to actin cytoskeleton. However, our understanding of specific properties and the fine-tuning of neuronal actin structures is incomplete. Drebrin A is highly enriched in dendritic spines, but its effects on actin morphology, dynamics, and interplay with other actin regulators are yet to be clarified. Here we review recent advances in understanding drebrin effects on actin morphology and dynamics.

Keywords Actin • Drebrin • Mapping • Cooperativity • Binding kinetics • Actin dynamics

5.1 Introduction and Significance

Drebrin E is expressed in many types of cells and tissues; however, drebrin A and s-drebrin A are exclusively neuronal isoforms (Jin et al. 2002; Aoki et al. 2005; Keon et al. 2000; Peitsch et al. 1999; Shirao et al. 1989). Drebrin A attracted initial attention due to its remarkable enrichment in brain and its co-localization with actin in dendritic spines—small dendritic protrusions participating in synaptic transmission (Aoki et al. 2005; Shirao and Obata 1986). What makes dendritic spines so important and separates them from other actin-based structures? From evolutionary perspective, dendritic spines are not an absolute requirement for synaptic transmission, which in some types of neurons can occur directly on dendritic shafts (Cowan et al. 2001). It is still debated why excitatory axons have preference for dendritic spines instead of forming synapses directly on dendritic shafts (Bellot et al. 2014). Interestingly, the presence of spines on the surface of dendrites is a common feature in vertebrates but is rarely seen in less evolved species (Sala et al. 2008). This suggests that dendritic

E.E. Grintsevich
Department of Chemistry and Biochemistry, University of California-Los Angeles, Los Angeles, CA 90095, USA
e-mail: egrin@ucla.edu

T. Shirao, Y. Sekino (eds.), *Drebrin*, Advances in Experimental Medicine and Biology 1006, DOI 10.1007/978-4-431-56550-5_5

spines evolved to implement higher-order brain functions such as advanced learning and memory (Holtmaat and Svoboda 2009; Kasai et al. 2010a). Over the years, many important discoveries related to dendritic spines were made. For instance, it was found that spine morphology and function are closely linked (reviewed in Kasai et al. 2010b). It was documented that spine morphology and functions are impaired in patients with complex neurodegenerative disorders such as Alzheimer's disease, Down syndrome, and epilepsy (Shim and Lubec 2002; Kojima and Shirao 2007; Sala et al. 2008; Lin and Koleske 2010). Moreover, it was shown that spine morphology, plasticity, and function are mediated through remodeling of actin cytoskeleton by its regulators (see (Frost et al. 2010) for review). Thus, understanding of actin remodeling and regulation in dendritic spines is important for our understanding of higher-order brain functions in normal and diseased states. On a molecular level, dendritic spines are very complex and contain many proteins, including drebrin—one of the key actin regulators in brain (Bellot et al. 2014; Hotulainen and Hoogenraad 2010; Sala et al. 2008; Aoki et al. 2005; Shirao and Obata 1986). Drebrin was shown to regulate spine plasticity and function (Biou et al. 2008; Jung et al. 2015; Ivanov et al. 2009a), with its level dramatically decreasing in patients with complex neurological disorders (Kojima and Shirao 2007; Shim and Lubec 2002). It was suggested that drebrin-decorated actin cytoskeleton in dendritic spines has unique properties separating it from other actin-based structures (Shirao and Sekino 2001); however, only recently, this was demonstrated directly (see below). Here we review some recent studies uncovering unique properties of drebrin-decorated F-actin and discuss their implications for the molecular cross-talk with other actin regulators.

5.2 Actin Structure: Basics and Terminology

Actin exists in two states: monomeric (G-actin) (Fig. 5.1a) and *filamentous*/polymeric (F-actin) (Fig. 5.1b). Structurally, actin monomer consists of four subdomains (*SD1–SD4*) coordinating adenine nucleotide (ATP, ADP) and the divalent cation associated with it (Mg^{2+} under physiological conditions) (Fig. 5.1a). Under polymerizing conditions, monomeric actin undergoes a transition to filamentous state, accompanied by ATP hydrolysis, and assembles into double-stranded helical polymer (Sheterline et al. 2002). Actin monomer incorporated into the polymer is called *protomer* or *actin subunit*. In actin filament, protomers are related to each other by an axial rise of 2.7 nm and a rotation of ~166–167° (Egelman et al. 1982). Images of the resulting double-stranded helical polymer show a characteristic crossover at average intervals of 36 nm (Fig. 5.1b). However, it has been shown that the length of helical crossover (also called *helical pitch* or *helical repeat*) in F-actin is not uniform. Such structural polymorphism in F-actin arises from the variable rotation and tilt of actin subunits within the filament, which translate into variable helical pitch (Galkin et al. 2002b, 2010). Thus, helical pitch of F-actin can be best described by distribution of lengths rather than a single number. It has been proposed that the structural heterogeneity of F-actin has important physiological role and affects recruitment and occupancy of actin regulators.

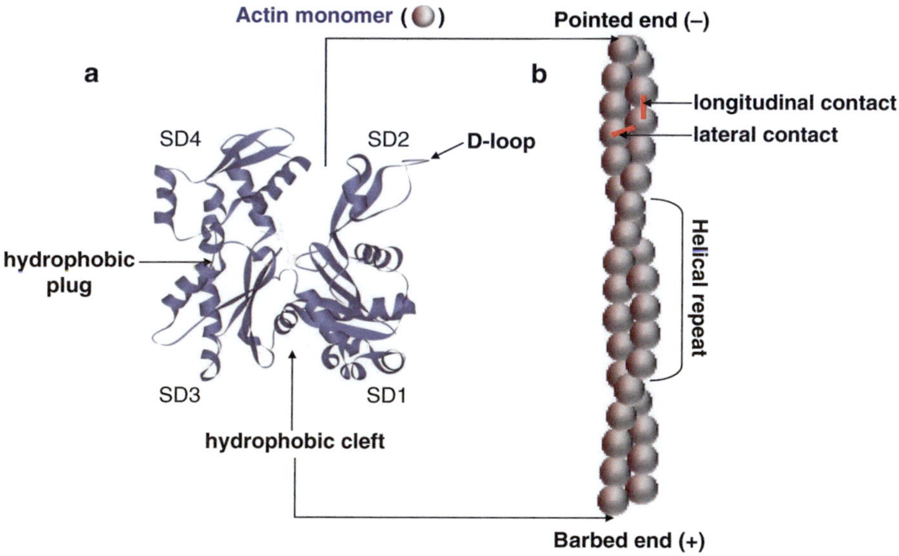

Fig. 5.1 Schematic representation of monomeric and filamentous actin. (**a**) Crystal structure of actin monomer (PDB: 1YAG) (Vorobiev et al. 2003). Actin subdomains are labeled SD1-SD4. Key structural elements involved in actin self-assembly (polymerization) are indicated. (**b**) Schematic representation of F-actin. Fast-growing (barbed end, B-end) and slow-growing (pointed end, P-end) ends are marked

Double-stranded helical structure of F-actin is stabilized through extensive protomer-protomer contacts. Two types of contacts can be distinguished: *longitudinal* and *lateral* (Fig. 5.1b, right panel, in red). Longitudinal contacts occur within the single strand of actin filament and are stabilized through electrostatic and hydrophobic interactions as well as geometric surface complementarity (von der Ecken et al. 2015). Lateral contacts are formed between protomers on the opposite strands. These contacts are formed by salt bridges, and their interface is less extensive compared to that of longitudinal contacts in F-actin (von der Ecken et al. 2015). The key regions on actin involved in its self-assembly are the DNase I-binding loop (D-loop, 41–50), hydrophobic cleft between SD1 and SD3, and a hydrophobic plug (residues 263–273) (Fig. 5.1a) (Fujii et al. 2010; Galkin et al. 2015; Oda et al. 2009; von der Ecken et al. 2015).

5.3 Mapping Drebrin-F-Actin Binding Interface

Drebrin's binding interface on F-actin is yet to be clarified. There are several challenges in determining high-resolution structure of the F-actin-drebrin complex. Traditional approaches often include X-ray crystallography, NMR, and electron microscopy (EM). However, determination of F-actin structure alone presents

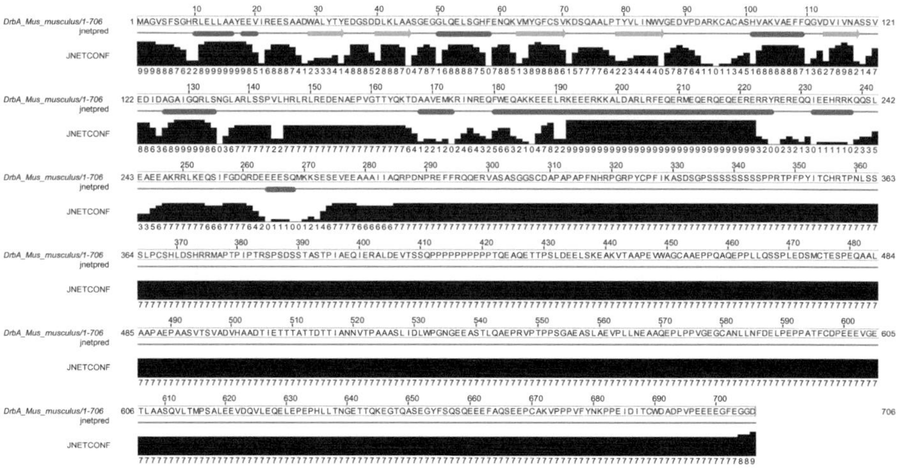

Fig. 5.2 Secondary structure prediction for drebrin A (*Mus musculus*). Prediction was made using Jpred4 protein secondary structure prediction server (Drozdetskiy et al. 2015). JNetPred line corresponds to the consensus prediction. Predicted α-helixes, β-sheets, and unstructured regions are shown in red, green, and gray, respectively. Confidence estimate for the prediction is given below (JNetCONF line, *black*). High values correspond to higher prediction confidence. Note that more than half of drebrin sequence are predicted to be unstructured

some challenges. The fact that actin filament's structure is double helical and the resulting polymers are heterogeneous in size makes F-actin and its complexes unsuitable for crystallization. Only recent technological advances in electron microscopy (such as use of direct electron detection cameras) allowed obtaining near-atomic resolution structure of F-actin (Egelman 2015). Drebrin is also a difficult object for structural work. As evident from secondary structure predictions (Fig. 5.2), the C-terminal part of drebrin's polypeptide chain is intrinsically disordered and could adopt a wide range of conformations, making it extremely challenging to crystallize this part of the protein. The N-terminal part of drebrin is predicted to be folded. However, examination of its sequence reveals that it is highly enriched in glutamic acid, which increases the surface entropy and dramatically decreases the chances of crystallization of these fragments (Goldschmidt et al. 2007).

Full-length drebrin binds to F-actin at 1:5 drebrin/actin stoichiometry (Ishikawa et al. 1994; Sharma et al. 2010). Its N-terminal sequence (residues 1–300, Drb1-300) can be defined as its actin-binding "core" since its binding affinity to F-actin is similar to that of the full-length construct (Grintsevich et al. 2010). Structurally, three actin-interacting domains can be identified in the N-terminal drebrin sequence: ADF-homology domain (drADF-hd), helical-charged motif (HCM), and its extension-actin binding domain (drABD). Together they result in 1:3 F-actin binding stoichiometry for the Drb1-300 construct (Grintsevich et al. 2010). Binding stoichiometry of the full-length drebrin (1:5) suggests the existence of additional unidentified actin-interacting sites in the C-terminal part of the molecule. Here, we review the information on drebrin-actin binding interface gathered from direct binding assays and from the competition data.

ADF-homology domain is a highly conserved motif and can be found in several classes of proteins (ADF/cofilins, twinfilins, Abp1/drebrins, glia maturation factors (GMFs), and coactosins) (for review see, Poukkula et al. 2011). This globular domain can interact with both monomeric and filamentous actins, which is true for conventional ADF/cofilins. However, due to variations in their sequence, different subclasses of ADFhd-s have clear preference for either G- (twinfilins) or F-actin (Abp1/drebrins, coactosins), with the exception of GMFs that do not bind actin at all. ADF/cofilins interact with two actin protomers within the same helical strand: with an upper protomer in its hydrophobic cleft between actin SD1 and SD2 (G-actin site) and with a lower protomer at the interface between SD2 and SD1 and also SD4 (F-actin site) (Galkin et al. 2011). Based on EM and crystallography, the G-actin-binding site on conventional cofilins includes their N-terminus, the tip of the long central helix 3 and loop 41–46 (hCofilin-2) (Galkin et al. 2011; Paavilainen et al. 2008). F-actin side on cofilin includes loops 94–98 and 154–158 (hCofilin-2) that interact with SD2 and SD4 of the lower actin protomer (Galkin et al. 2011). ADF homology domains of drebrin and Abp1 (SH3P7, HIP55) have highest sequence homology with each other and are often grouped together as a subclass (Hild et al. 2014; Poukkula et al. 2011). Atomic structures of ADFhd-s of Abp1/HIP55, combined with the mutational analysis, revealed that F-actin binding interfaces of these ADFhd-s are similar to that of conventional cofilins (Quintero-Monzon et al. 2005; Goroncy et al. 2009). Thus, by extrapolation, ADFhd of drebrin may interact with SD4 and SD2 of actin (residues 242–243 (SD4); 21–28; 90–96 (SD1)), which explains drebrin's competition with mammalian cofilins for F-actin binding (Table 5.1) (Zhao et al. 2006; Grintsevich and Reisler 2014). However, G-actin binding interface of ADFhd of Abp1/drebrins is less extensive, consistent with its lack of actin monomer binding (Quintero-Monzon et al. 2005). Structurally, truncation of the N-terminal region in ADFhd-s of Abp1/drebrins, and, possibly, unusual bending of β4 strand within ADFhd of HIP55 can be responsible for the weakening of G-actin site binding (Goroncy et al. 2009; Quintero-Monzon et al. 2005). It should be noted that binding of ADFhd to F-actin still requires interactions with both F- and G-binding sites in order to bridge the adjacent protomers. Intriguingly, when isolated from the rest of the sequence, ADFhd of drebrin has weak, if any, affinity for F-actin based on cosedimentation data (Grintsevich et al. 2010). At the same time, binding of homologous yeast Abp1 to F-actin is abolished upon genetic deletion of its ADFhd (Quintero-Monzon et al. 2005). Thus, it is tempting to speculate that other domains in drebrin's structure can assist the interaction of ADFhd with actin filaments by inducing allosteric changes in F-actin.

In drebrin and the homologous Abp1, ADFhd is followed by a helical-charged motif (HCM), which is unique only to these two proteins and has little homology to any known actin-binding motifs. There is no structural data on HCM of drebrin or Abp1. For drebrin, this domain is predicted to contain a coiled coil region which can potentially serve as a dimerization site. However, based on analytical centrifugation data, drebrin A seems to be monomeric in solution or, alternatively, might exist in a fast monomer-dimer equilibrium (Sharma et al. 2010).

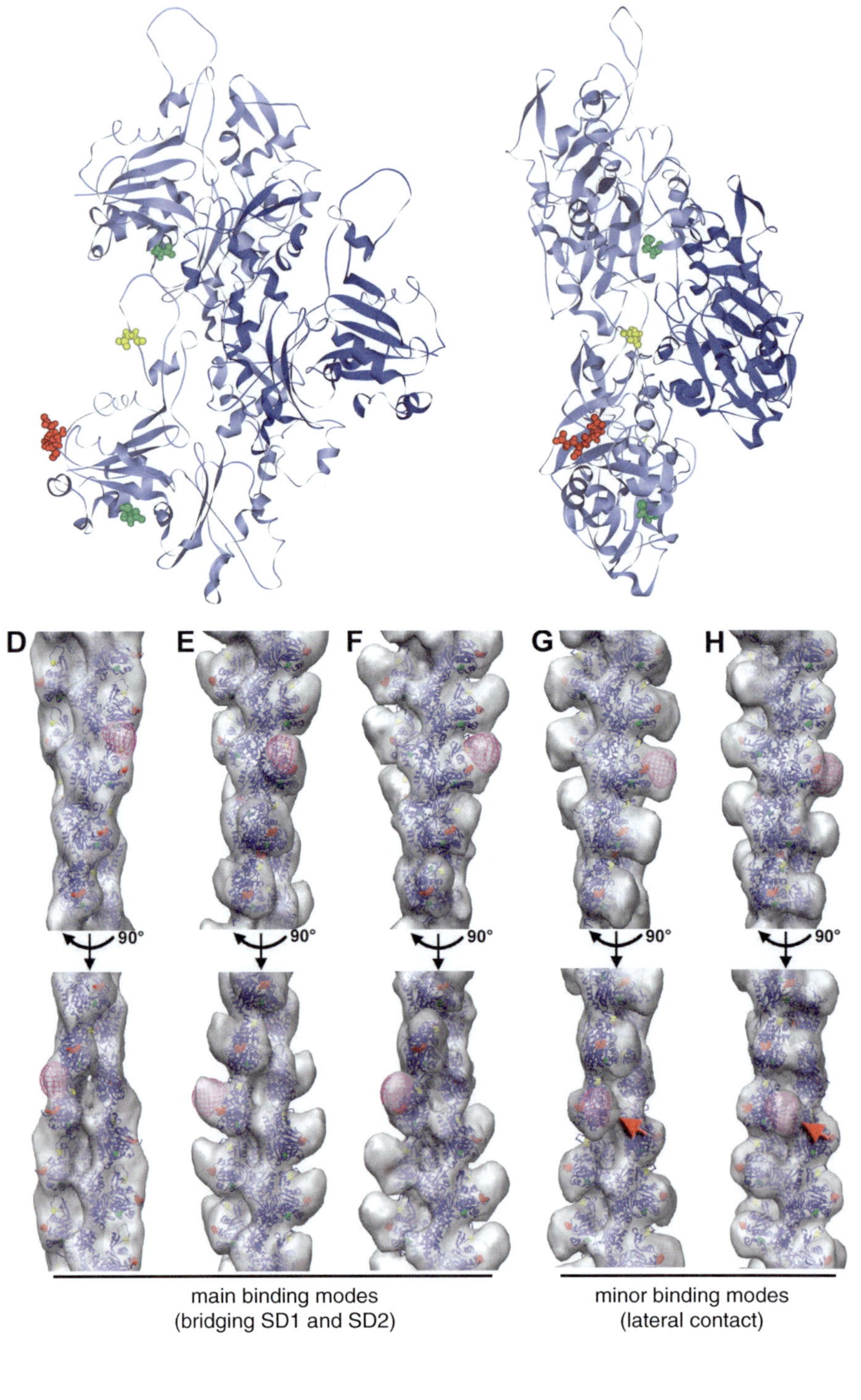
D
E
F
G
H
90°
90°
90°
90°
90°
main binding modes
(bridging SD1 and SD2)
minor binding modes
(lateral contact)

Actin-binding domain of drebrin (drABD, residues 233–300/317)—located in the C-terminal part of drebrin's HCM—was first defined on the basis of in vivo studies (Hayashi and Shirao 1999; Hayashi et al. 1999; Ivanov et al. 2009b; Biou et al. 2008). Later, its binding interface on F-actin was mapped directly, yielding a working model of the complex. These studies employed single-particle EM reconstruction and side-directed mutagenesis, combined with chemical cross-linking and mass spectrometry analysis (Grintsevich et al. 2010). It was found that drABD can simultaneously interact with two protomers in F-actin spanning between their C-terminal regions. It was documented that the N- and C-termini of drABD can interact with residues E99/E100 and C374 in SD1 of actin, respectively. Efficient cross-linking of drebrin residues 238, 248, 252, 270, and 271 to actin residue 51 indicates that drABD is centered on SD2 of actin (Fig. 5.3, top panel). EM reconstruction of F-actin-drABD complex was consistent with the cross-linking results but also revealed several surprising features. Specifically, several modes of drABD binding to F-actin have been observed which would support the existence of at least two binding modes for drABD on F-actin (Fig. 5.3, lower panel). In its main binding mode, drABD makes extensive contact with actin interface between SD1 and SD2, which includes residues 51, 99–100 identified in the cross-linking assays (Fig. 5.3). In its minor binding mode, drABD is also positioned close to the above residues on actin but makes cross-strand contact with another actin protomer (Fig. 5.3). EM reconstruction of the longer drebrin construct (Drb1-300) with F-actin also showed the same binding polymorphism, suggesting that it might be a functionally important feature of drebrin (Grintsevich et al. 2010). Moreover, such binding polymorphism is not uncommon among actin regulators and has been previously observed for cofilin (Galkin et al. 2011), utrophin (Galkin et al. 2002a), and tropomyosin (Bacchiocchi et al. 2004; Lehman et al. 1994; Xu et al. 1999). Taken together, binding of drABD to F-actin explains previously observed drebrin competition with α-actinin (Biou et al. 2008; McGough et al. 1994) and attenuation of actomyosin interaction (Ishikawa et al. 2007), as these two proteins interact also with the E99–E100 actin region (Table 5.1).

It was previously shown that drebrin displaces tropomyosin (Tm) from actin filaments. Such effect was observed with two Tm isoforms (brain (bovine) and smooth muscle (chicken)) (Ishikawa et al. 1994). Recent high-resolution cryoEM recon-

Fig. 5.3 Actin binding domain of drebrin (drABD) has several modes of F-actin binding. Actin trimer (*top panel*) corresponds to the highest-resolution EM reconstruction of F-actin available (von der Ecken et al. 2015). Actin protomers are shown in blue and drABD in magenta (*lower panel*). F-actin residues 99 and 100 (red), 51 (*yellow*), and 374 (*green*) are shown. These residues were mapped to be a part of drABD-F-actin interface based on cross-linking and mass spectrometry analysis. In its main binding modes, drABD bridges two adjacent actin protomers making extensive contacts with SD1 and SD2 (*lower panel*). This interaction affects longitudinal interface in F-actin. In minor binding modes, drABD makes a cross-strand (lateral) contact in F-actin. (Reprinted from Grintsevich et al. (2010) with permission from Elsevier; license number 3697321191127)

Table 5.1 Structural basis of drebrin competition with other F-actin regulators

Drebrin competitor	Competitor's interface on actin[a]	Competing domain on drebrin	References and notes
Tropomyosin	Arg147, Lys215, Lys291, Asp311, Lys326, Lys328 ("A-state", troponin-, myosin-, Ca^{2+}-free F-actin)	Unknown	(von der Ecken et al. 2015) (EM, 6.5 Å)
Myosin	Glu2, Glu3, Asp23, Asp24, Lys49, Glu55, Asp92, Glu98, Glu99, Lys325, Lys327, Ser349, Glu333, Thr350	drABD	(von der Ecken et al. 2016) (EM, 3.9 Å); human γ-actin)
Cofilin	143-147 (G), 343-346 (G) 349-354 (G), 242–243 (F), 21–28 (F), 90–96 (F)	ADFhd	(Galkin et al. 2011) (EM, 9 Å)

[a]Based on the highest-resolution EM reconstruction available to date

struction of F-actin-Tm complex clarified its interaction interface (Table 5.1) (von der Ecken et al. 2015). However, this information does not explain the competition between drebrin and Tm. Residue R147 on actin was found to be involved in F-actin-Tm interaction. The same region is a part of G-actin site of ADF/cofilin binding. However, as discussed above, G-actin-interacting interface in ADFhd of drebrin/Abp1 is weak. Drebrin competition with Tm might be providing valuable information about the residues that are involved in F-actin-drebrin interaction. These residues include K215, K291, D311, K326, and K328 (Table 5.1). Since these regions do not overlap with the binding interfaces of ADFhd and drABD, they might be involved in actin interaction with HCM of drebrin. Site-directed mutagenesis and high-resolution EM reconstruction of F-actin-drebrin complex will be required to clarify the role of the above residues in drebrin binding.

5.4 Drebrin Alters the Morphology and Mechanical Properties of Actin Filaments

It was suggested nearly two decades ago that drebrin "imparts a unique character to the actin cytoskeleton bound to it" (Shirao and Sekino 2001), but only recently, this has been demonstrated through direct imaging to be indeed the case (Sharma et al. 2010; Sharma et al. 2012). In two studies, individual drebrin-decorated filaments were examined by atomic force microscopy (AFM). The first important finding was that drebrin changes the length of helical repeat (helical pitch, Fig. 5.1) in F-actin to ~40 nm, compared to ~36 nm in bare F-actin (Sharma et al. 2010). Only few examples of actin regulatory proteins altering the morphology of F-actin have been reported (McGough et al. 1997; Schmid et al. 2004; Tsaturyan et al. 2005). The most striking example is actin-severing proteins ADF/cofilins that (in contrast to drebrin) dramatically shortens the length of helical repeat to 23–29 nm (Galkin et al. 2001; McGough et al. 1997; Ngo et al. 2015; Sharma et al. 2010). Interestingly, drebrin contains N-terminal ADFhd, but its effect on actin morphology appears to be opposite to that of cofilin. Another important finding was that full-length drebrin and its C-terminal truncated construct Drb1-300 cause the same morphological changes in F-actin (Sharma et al. 2012). This means that drebrin-induced morphological changes in F-actin are mediated through its N-terminal domains rather than its intrinsically disordered C-terminal part. This allows us to speculate that shorter drebrin isoforms (drebrin E and s-drebrin A) will induce the same changes in actin morphology as neuronal drebrin A.

Drebrin-decorated actin filaments were different not only in their morphology but also in mechanical properties. Drebrin-bound actin was 55% stiffer compared to the bare actin (Sharma et al. 2010). It is tempting to speculate that such a unique actin morphology and mechanical properties observed upon drebrin decoration may affect the recruitment and occupancy of other actin regulators in dendritic spines. Thus, drebrin alters neuronal F-actin networks not only by its direct binding but also allosterically, by inducing changes in F-actin morphology and structure.

5.5 Cooperativity of Drebrin Binding to F-Actin

Recent studies revealed another important feature of drebrin interaction with F-actin—its binding cooperativity (Sharma et al. 2010, 2012). Similar to Tm and cofilin, drebrin binding to actin filaments can't be described by a simple bimolecular binding reaction, and cooperativity coefficient should be taken into account (McGhee and von Hippel 1974; De La Cruz 2005). Cooperativity of the reaction means that binding of one molecule of actin-binding protein (ABP) to F-actin polymer influences the affinity for the next molecule. Interactions of Tm and cofilin with F-actin are both examples of positive binding cooperativity, meaning that the bound molecule increases the affinity for subsequent molecules (Fig. 5.4a). For cooperative binding, equilibrium dissociation constant, K_d, equals $1/\omega K_a$, where ω is a cooperativity coefficient and K_a is an association constant (McGhee and von Hippel 1974; De La Cruz 2005; Pollard 2010). In simple terms, cooperativity coefficient estimates a probability of a subsequent molecule binding next to another one as opposed to an isolated (no neighbors) binding site. For instance, cooperativity coefficient of 8 indicates that the molecule is eight times more likely to bind next to another one that is already bound to F-actin (a neighbor) as opposed to the isolated site on the filament (Fig. 5.4a). Cooperativity analysis is often complicated and can't be generalized because binding cooperativity may arise from different factors. For instance, cooperativity of Tm-F-actin binding arises mainly from the tail-to-head interactions between adjacent Tm dimers resulting in the formations of long Tm roads on actin filaments (for review, see (Gunning et al. 2008)). It should be noted that contribution of other factors was also suggested for Tm binding cooperativity (Willadsen et al. 1992). In contrast to Tm, the cooperativity of cofilin binding to F-actin is of different nature (Cao et al. 2006a; De La Cruz 2005; De La Cruz and Sept 2010). Based on the available data, we can conclude that neighboring cofilin molecules do not interact with each other. Instead, binding occurs in two

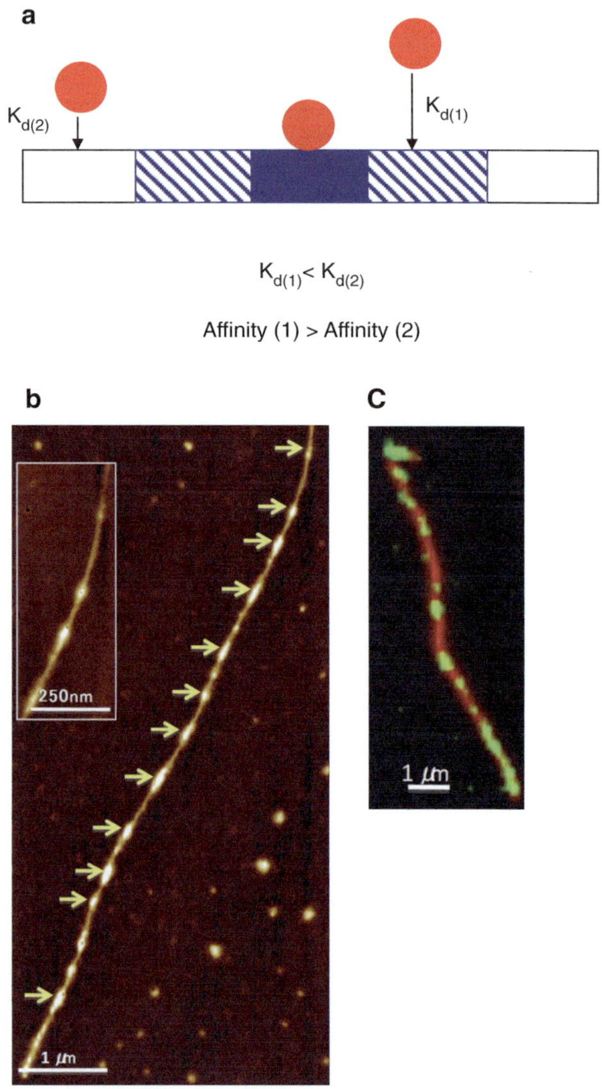

Fig. 5.4 Cooperativity of drebrin binding to actin filaments. (**a**) Schematic representation of cooperative binding. Rectangular shapes represent binding sites on hypothetical polymer for its binding partner (*red circles*). The site occupied by a molecule of binding protein on hypothetical polymer is shown in solid blue. Neighboring sites, adjacent to the occupied site, undergo allosteric changes upon protein binding (*blue stripes*). A case of positive cooperativity is demonstrated here: the first bound molecule increases the affinity for subsequent molecules. A site (*white*) removed from the bound site has lower affinity. (**b**, **c**) Drebrin binds in clusters on actin filaments as shown in AFM (**b**) and 2-color TIRFM images (**c**). (**b**) and (**c**) were reprinted from Sharma et al. (2012) with permission from Elsevier; license number 3697181433434)

phases. It was proposed that single cofilin molecules may be capturing and stabilizing infrequent dynamic states of F-actin that are possible due to the structural polymorphism in F-actin. Since suitable F-actin conformations are rarely occurring, initial binding of cofilin to actin filaments is slow (Cao et al. 2006a). This initial cofilin binding introduces allosteric change in the neighboring F-actin sites that facilitates the interaction of subsequent cofilin molecules with actin filaments. This results in cluster-like assembly of cofilin on F-actin (Ngo et al. 2015; Suarez et al. 2011). In cases of strong cooperativity of F-actin-ABP binding, clusters might not be observed. Instead, under subsaturating conditions in steady state, we would see two populations of filaments: decorated and undecorated. If such highly cooperative binding is monitored in real time, one may observe a formation of a continuous cluster spreading along actin filament as was recently shown for *Drosophila* Tm (Hsiao et al. 2015).

To determine binding cooperativity in bulk solution samples, one must find/develop a sensitive assay to measure the concentration of either free proteins or their complexes. Unfortunately, for drebrin, this is a challenging task. A contributing factor here is that its binding to F-actin is mediated through unique domains (rather than repeats, like in the case of Tm) which probably translates into nonredundant F-actin binding interfaces for each domain. This means that even if we identify a residue on actin that (when fluorescently labeled) is very sensitive to the binding of one of the drebrin's domains, the change in fluorescence signal will be most likely averaged over 3 or 5 actin protomers (for Drb1-300 and drebrin FL, respectively) due to drebrin's binding stoichiometry. For examination of the binding cooperativity of drebrin to F-actin in solution, fluorescence titration with acrylodan-labeled actin mutant A167C/C374A provided a rather sensitive assay (Sharma et al. 2012). Although binding isotherms of Drb1-300 to A167C/C374A F-actin showed persistent deviation from the hyperbolic curves (consistent with binding cooperativity) the choice of a suitable model to fit such data was nontrivial. First, several binding modes of drABD to F-actin indicated binding heterogeneity (Grintsevich et al. 2010). Also, abundant evidence suggests polymorphism in F-actin structure itself (Egelman et al. 1982; Galkin et al. 2010). Moreover, it was recently shown that bulk solution assays underestimate binding cooperativity in heterogeneous samples (Solomatin et al. 2011). Second, due to low drebrin/F-actin binding stoichiometry, potential overlap between drebrin binding sites has to be considered. This factor is usually dismissed in cases of stoichiometric (1:1) binding (e.g., cofilin) and/or if biophysics of the binding makes the possibility of such overlap unlikely (e.g., Tm). Third, potential propagation of drebrin-induced structural changes in F-actin further complicates the analysis of bulk solution data. Thus, probing the cooperativity of drebrin-F-actin interaction at single filament level was found to be the most informative option.

Uniform drebrin decoration observed by AFM gave a first indication of its cooperative binding to F-actin (Sharma et al. 2010). In the absence of drebrin preference for "neighbor" vs. "no-neighbor" binding sites on F-actin, we would expect to see a significant number of gaps in drebrin decoration that were not observed at least at ~30 nm resolution. Cooperativity was farther confirmed by AFM and total internal reflection fluorescence microscopy (TIRFM) in follow-up study employing Drb1-300 construct (Fig. 5.4b, c) (Sharma et al. 2012). Use of AFM for drebrin imaging allowed for cluster size analysis on the single filament level. Under subsaturating concentrations, formation of short (40–120 nm) clusters of Drb1-300 on actin filaments was documented by AFM (Fig. 5.4b). It should be noted that assembly of such a small clusters is difficult to monitor by fluorescence microscopy due to its limited spatial resolution (250–300 nm). Only large clusters and gaps in drebrin decoration of actin filaments can be easily resolved by TIRFM (Fig. 5.4c).

AFM imaging revealed also another unique feature of drebrin-decorated F-actin—a propagation of morphological changes from drebrin clusters to undecorated actin regions. Increased (40 nm) length of helical repeats was observed in undecorated actin regions adjacent to the drebrin-bound clusters (Sharma et al. 2012). It was shown that such conformational spread affects up to two helical repeats. Despite of >150 known ABPs (Dos Remedios et al. 2003), there are only a few examples of such a long-range effects on actin filaments (Ngo et al. 2015; Orlova et al. 1995; Ressad et al. 1998).

5.6 Kinetics of Drebrin Binding to F-Actin

Binding kinetics can provide valuable information about F-actin interactions with its binding partners (Pollard 2010). Several parameters contribute to the binding rates of macromolecules; among them are their diffusion rates and the size of their interacting surfaces. Drebrin molecule is relatively large (~80 KDa), and, based on analytical ultracentrifugation data, it adopts elongated conformation in solution. Thus, in solution, drebrin will diffuse slower than a small protein (such as cofilin), but it also has a larger F-actin interacting interface. It should be noted that interaction between macromolecules usually requires multiple collisions to occur. That is because most of these collisions are unproductive since they do not occur at the interacting interfaces. F-actin interactions have additional complexity. Due to the polymorphism of F-actin structure (Egelman et al. 1982; Galkin et al. 2010), some of the conformations suitable for its regulators' binding appear with low frequencies. This means that for some F-actin regulators, the number of productive collisions will be even lower than average. Initial binding of cofilin to F-actin is a good example of such phenomenon. For many macromolecules, association constants (K_a) are in the 10^6–10^7 $M^{-1}s^{-1}$ (Pollard 2010); however, for initial (noncooperative) cofilin binding $K_a = 8 \times 10^3$ $M^{-1}s^{-1}$ (Cao et al. 2006b). The slow cofilin binding to F-actin is due to its preference for "overtwisted" actin conformations that are infrequent.

Fluorescence resonance energy transfer (FRET) between F-actin-IADANSE and DABMI-labeled drebrin's actin-binding core (Drb1-300 construct) was employed to determine the "on" and "off" rates of drebrin-F-actin binding. It was found that drebrin association rate (k_{on}) with F-actin falls within 10^6–10^7 $M^{-1}s^{-1}$ range (Grintsevich and Reisler 2014). This suggests that drebrin binding does not depend on rarely occurring F-actin conformations.

Dissociation rate constants exhibit wider range of values compared to K_a and often determine affinities of macromolecular interactions (Pollard 2010). Using the same FRET system and competition with the unlabeled Drb1-300, it was shown that drebrin not only binds to but also dissociates from F-actin fast (k_{off} 0.9–1 s^{-1}) (Grintsevich and Reisler 2014). Intriguingly, drebrin's "off" rates were not affected significantly by the nucleotide state of actin filaments or F-actin stabilizing toxin phalloidin that is known to decrease the range of dynamic motion in filamentous actin (Grintsevich and Reisler 2014). The lack of phalloidin effect on drebrin dissociation kinetics suggests that in contrast to cofilin, a wide range of dynamic motions in F-actin is not required for drebrin dissociation from filaments as well as its binding to them. Together, this data suggests that drebrin binding interface on F-actin is exposed to allow for their fast interaction and dissociation. The affinity of Drb1-300 binding to F-actin estimated from kinetics experiments appears stronger (~40 nM) (Grintsevich and Reisler 2014) than that determined by co-sedimentation analysis performed with Drb1-300 and full-length protein (~100–200 nM) (Grintsevich et al. 2010; Sharma et al. 2010; Ishikawa et al. 1994). It is possible that drebrin co-sedimentation assays underestimate its affinity to F-actin, or the labeling of actin and drebrin somewhat improve their binding.

Several questions about F-actin-drebrin interaction are yet to be clarified. First, despite the fact that cooperativity of drebrin binding to F-actin was demonstrated in multiple assays, we are still lacking its quantitative analysis. In other words, we do not have an estimate of how much more likely drebrin will bind next to the drebrin cluster as opposed to isolated site on actin filament and what is the affinity of this interaction. Second, it would be important to investigate the kinetics of binding of different drebrin isoforms to F-actin using full-length constructs. Due to the low yields of full-length drebrin constructs, this probably would be best addressed in single-molecule TIRF experiments rather than by traditional stopped flow measurements.

How can our knowledge of drebrin cooperativity and binding improve the understanding of its cellular function? One of the important findings is that drebrin does not show strong preference for a particular nucleotide state of actin (Grintsevich and Reisler 2014). This suggests that both older (ADP-bound) and newer (ADP-Pi) F-actin regions can be decorated with drebrin. In contrast to that, it was recently shown that this is not the case for *Drosophila* Tm (Hsiao et al. 2015). It was demonstrated that its initial binding is biased toward the P-end (ADP-bound F-actin regions), which suggests that ADP-Pi regions will not be equally protected/stabilized by its binding. In line with that, Tm inhibits pointed (but not barbed) end depolymerization of actin filaments (Broschat et al. 1989). It should be noted that due to the large variety of Tm isoforms, this conclusion cannot be generalized. A comparison of drebrin with the same *Drosophila* Tm (which assembles on F-actin in one continuous cluster) suggests that the lower cooperativity of drebrin binding (assembles in small clusters) results in gaps in F-actin decoration by drebrin (Hsiao et al. 2015; Sharma et al. 2012). This may allow other actin regulators to bind in between drebrin clusters. For example, the important role of cofilin in dendritic spine morphology and maintenance is well established (Zhou et al. 2004; Kojima and Shirao 2007; Zhao et al. 2006). It was recently documented that cofilin-mediated severing is inhibited but not completely abolished in the presence of near-saturating concentrations of drebrin (Grintsevich and Reisler 2014), which may have physiological significance for dendritic spines. One of the possible explanations is that cofilin can bind in the gaps in F-actin decoration, i.e., between drebrin clusters.

It was reported previously that filaments saturated with the mixture of neuronal Tm isoforms are even more ADF/cofilin resistant than those decorated with skeletal muscle Tm (Kuhn and Bamburg 2008). It is possible that such a strong effect of some neuronal tropomyosins is too restrictive for dynamic actin structures present in dendritic spines. It is feasible that drebrin might be more suitable than some of Tm isoforms for the fast transition between stable and dynamic states of actin cytoskeleton in spines upon ADF/cofilin activation. Another question to be addressed is the effect of drebrin on Arp2/3-branched F-actin structures. It is feasible that drebrin can have an effect on branched actin networks that is distinct from that of neuronal Tm isoforms. It was recently shown that *Drosophila* Tm is excluded from Arp2/3 branches by an unknown mechanism (Hsiao et al. 2015). At the same time, it was shown that cofilin preferentially disassembles branched structures consistent with its role in lamellipodia function (Gressin et al. 2015). Moreover, it was shown that cofilin can cause quick dissociation of Arp2/3 branches presumably through the propagation of its morphological changes in actin filaments (Chan et al. 2009). It is possible that drebrin clusters do not allow for such propagation because they induce morphological changes in F-actin that are opposite to those of cofilin (McGough et al. 1997; Sharma et al. 2010).

5.7 Drebrin Effect on Actin Dynamics

Since actin can self-assemble only in one direction (head-to-tail), it forms polar double-helical polymers with two structurally different ends: barbed and pointed (Fig. 5.1). Actin monomers can associate and dissociate at either end, but both of these processes are more than tenfold faster at the barbed end (Kuhn and Pollard 2005; Pollard 1986). Thus, polymerization/depolymerization of uncapped actin filaments mostly reports on their barbed-end dynamics. In order to monitor depolymerization from only one end, capping proteins are often employed. For instance, gelsolin binds with high affinity to the barbed end of actin filaments (gelsolin-capped filaments) and restricts actin polymerization/depolymerization to the pointed end only.

Depolymerization of gelsolin-capped and uncapped actin filaments was examined in the presence of FL drebrin A (Fig. 5.5) and Drb1-300 construct (Mikati et al. 2013). Overall, with uncapped filaments, full-length drebrin showed stronger inhibition of F-actin depolymerization compared to the Drb1-300 construct (90% vs. 40%, respectively). However, both constructs caused similar extent of depolymerization inhibition of gelsolin-capped filaments (~40–50%). The simplest explanation would be that the weakly interacting C-terminal part of drebrin FL reduces protomers' off rate at the B-end, further contributing to the overall inhibition of actin depolymerization. Interestingly, drABD that reportedly mimics FL drebrin effect in vivo (Biou et al. 2008; Hayashi and Shirao 1999; Hayashi et al. 1999; Ivanov et al. 2009b) had negligible effect on the inhibition of F-actin depolymerization. Thus, P-end of actin is protected by drebrin to a lesser degree than the B-end. Results obtained with FL drebrin and Drb1-300 constructs suggest that the intrinsically disordered C-terminal part of drebrin molecule is responsible for the stronger inhibition of B-end than P-end depolymerization.

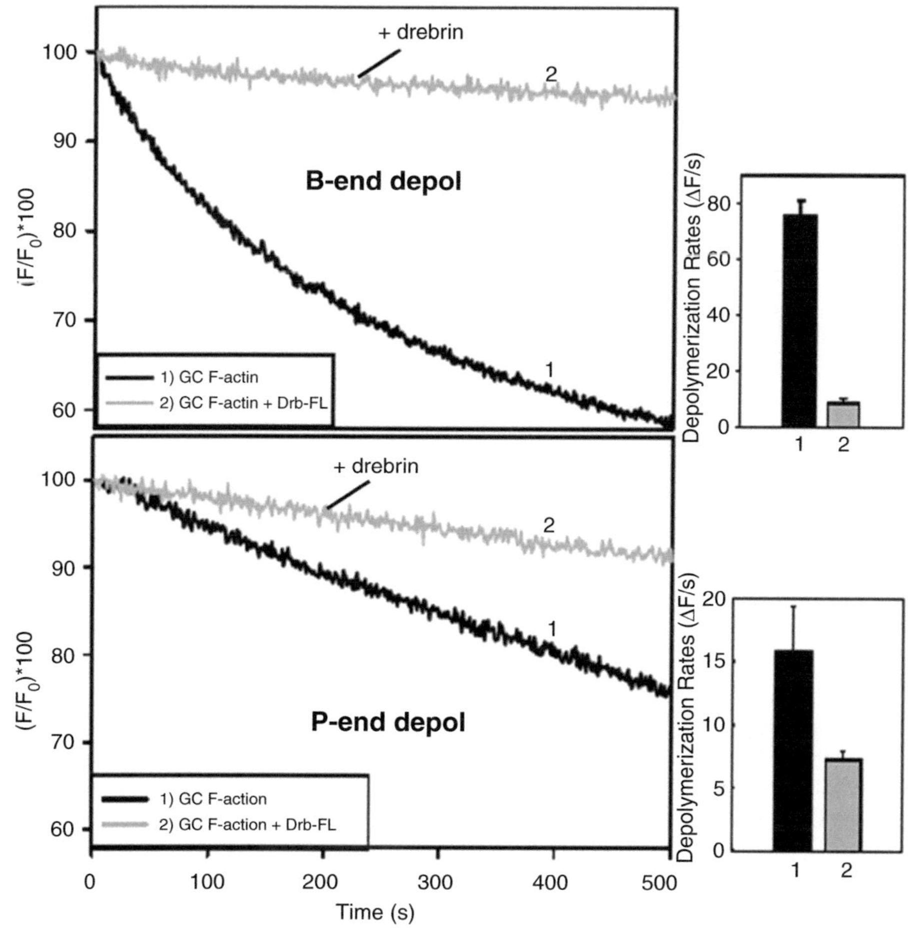

Fig. 5.5 Drebrin inhibits F-actin depolymerization from both ends. *Top panel, graph*: 2 μM 5% pyrenyl-labeled skeletal F-actin was depolymerized in the presence of Latrunculin A (Lat-A) (20 μM) only (*trace 1*) or with the additional presence of 0.8 μM full-length drebrin A (Drb-FL) (*trace* 2). The *bar plot* in the *top panel* shows the corresponding initial depolymerization rate of F-actin alone (*black bar*, 1) or in the presence of 0.8 μM Drb-FL (*gray bar*, 2). *Bottom panel, graph*: gelsolin-capped, 2 μM 5% pyrenyl-labeled skeletal F-actin depolymerized in the presence of Lat-A (20 μM) only (*trace* 1) or with Lat-A in the presence of 0.8 μM Drb-FL added (*trace* 2). The *bar plot* in the *bottom panel* shows the corresponding initial depolymerization rate of GC-F-actin alone (*black bar*) or in the presence of 0.8 μM Drb-FL (*gray bar*). For all conditions, the rates are the average of six separate measurements of the initial depolymerization over 60 s

What does this mean in cellular context? Dendritic spine heads contain branched F-actin structures nucleated by Arp2/3 complex. Consequentially, in these branched structures, pointed ends of daughter filaments are capped with Arp2/3 and therefore protected from depolymerization. At the same time, drebrin A dramatically inhibits depolymerization of the free B-ends which might be important for the maintenance of such structures (Mikati et al. 2013). Intrinsically disordered C-terminal part of drebrin contributes to its inhibition of actin depolymerization as judged from the effect of C-terminal truncation constructs. However, over half of the drebrin sequence is intrinsically disordered (Fig. 5.2), and it is so far unclear which site(s) in that sequence contribute to drebrin's inhibitory effect on F-actin depolymerization. It remains to be determined whether embryonic isoform—drebrin E—has the same effect on actin depolymerization as its neuronal isoform A. Drebrin interference with actin dynamics at B-ends suggests that drebrin may attenuate the function of B-end actin-binding factors such as cappers (Eps8, gelsolin), formins (mDia2, FMNL2), and small molecules (cytochalasin D). This speculation is in part supported by in vivo observations (Ikeda et al. 1996) showing that drebrin-expressing fibroblasts are resistant to cytochalasin D, which is known to depolymerize F-actin through a combination of B-end capping and monomer sequestration. Another interesting observation is that drebrin stabilized filaments show partial, concentration-dependent resistance to latrunculin A (they fail to depolymerize fully compared to uncomplexed control) (Mikati et al. 2013); however, the mechanism of this effect is unclear since latrunculin A is believed to exert its effect through actin monomer sequestration into non-polymerizable complex (Sheterline et al. 2002). Interestingly, a potential cross-talk between neuronal B-end capper Eps8 and embryonic isoform—drebrin E—has been also reported (Cheng et al. 2011). In a different study, it was shown that in bulk solution assays drebrin does not uncap actin filaments formed from gelsolin seeds and therefore capped by gelsolin (Mikati et al. 2013). This suggests that it probably can't actively displace gelsolin form B-ends. It is yet to be determined whether the same conclusion stands for Eps8 and/or any other neuronal capping proteins. Moreover, it is possible that similar to actin regulatory protein CARMIL, drebrin can affect the binding kinetics of B-ends cappers to actin filaments despite its inability to actively displace them (Uruno et al. 2006).

5.8 Drebrin Effect on the Polymerization-Deficient Actin Forms

Actin cytoskeleton is critical for eukaryotic cells. That is why this system should be very robust since it affects every single cell in the organism. Accumulation of some actin forms and certain mutations can result in pathologies (if not lethal), but some of it can be "mitigated" with the help of actin regulatory proteins. We already established in this chapter that drebrin has a stabilizing effect on F-actin. However, only recently, the effect of drebrin on polymerization rescue of the assembly deficient actin forms was systematically investigated (Mikati et al. 2013). Several C-terminal drebrin truncations were employed to investigate the domain requirements for such rescue as well as drebrin effects on longitudinal and lateral interface in F-actin. To this end, Cu^{2+}-catalyzed cross-linking, light scattering, and EM approaches were employed.

Cleaving skeletal actin with grimelysin (previously known as *E. coli* protease ECP) between residues 42 and 43 impairs its polymerization (Bozhokina et al.

2008; Klenchin et al. 2006). Based on the highest-resolution F-actin structure, residues 42–48 are part of its longitudinal interface and interact with the hydrophobic cleft (between SD1 and SD3) of an adjacent protomer (Fig. 5.1) (von der Ecken et al. 2015). It was found that FL drebrin A rescues the polymerization of grimelysin-cleaved actin. In terms of domain requirements for such a rescue, it was shown that C-terminal truncation constructs of drebrin—Drb1-300 and Drb2-252—also rescue grimelysin-cleaved actin (Mikati et al. 2013).

To probe lateral F-actin interface, V266G/L267G yeast actin mutant was employed (Wen et al. 2000). This mutant carries two substitutions in a hydrophobic loop of actin that result in decrease of hydrophobicity and weaken contacts between the two strands in F-actin. Similar to drebrin effect on impaired F-actin's longitudinal interface, N-terminal constructs Drb1-300, Drb2-252, and full-length protein rescued the polymerization of GG-actin (Mikati et al. 2013). Hydrophobic interactions are temperature sensitive, so changing the temperature of the reaction mixture will report on the extent and strength of the stabilizing effect. Lowering the temperature of GG-actin filaments to 4°C in the presence of drebrin constructs caused their partial depolymerization, but not as complete as previously reported for Tm. Re-polymerization of GG-actin in the presence of drebrin was not detected below 20°C, indicating similar temperature sensitivity to tropomyosins in stabilization of actin's lateral interface (Mikati et al. 2013).

Thus, the above documented results show that drebrin inhibits depolymerization of actin filaments from both barbed and pointed ends and stabilizes both lateral and longitudinal F-actin interfaces (Mikati et al. 2013). Drebrin's ability to rescue polymerization of actin forms with impaired lateral and longitudinal interfaces may contribute to the overall robustness of drebrin's stabilization of neuronal cytoskeleton.

A surprising result reported in Mikati et al. was that the actin-binding domain of drebrin (drABD) alone does not rescue filament formation of any of the polymerization-deficient actin forms used in that study (Mikati et al. 2013). Moreover, it did not inhibit F-actin depolymerization which disagrees with the in vivo observations (Ivanov et al. 2009b). It was shown that drABD inhibited Cu^{2+}-catalyzed disulfide cross-linking in yeast mutant Q41C (Mikati et al. 2013). Such inhibition suggests an increase in the mean distance between C374 and the DNase I binding loop on adjacent actin protomers in F-actin. Alternatively, it may be caused by steric hindrance due to drABD's proposed binding to actin's SD2 (Fig. 5.3) (Grintsevich et al. 2010). However, in vivo studies in different cell types indicate that drABD closely mimics the effects of full-length drebrin on the cellular actin organization (Biou et al. 2008; Hayashi and Shirao 1999; Hayashi et al. 1999; Ivanov et al. 2009b). In line with this, EM reconstructions of F-actin-drABD and F-actin-Drb1-300 complexes indicated that drABD is probably the strongest actin-binding module within drebrin sequence (Grintsevich et al. 2010). Intriguingly, in vitro results of Mikati et al. do not explain the similar in vivo effects of overexpression of drABD and full-length drebrin. It is tempting to speculate that the observed inhibition of Cys41-Cys374 cross-linking by drABD together with its inability to rescue the polymerization of actins with perturbed longitudinal interface can be explained by drABD-induced "under-twisting" of F-actin (helical pitch is 40 nm vs. 36 nm in bare actin). It is possible that this "stretched" conformation is stabilized by binding of N-terminal drebrin domains (ADF-H and predicted coil-coiled region, seq 2-252) to the filament. It is feasible that structural alterations in F-actin caused by drABD alone mimic those of the full-length protein. This may affect recruitment of other actin regulators in the studied cell types yielding the same phenotype as full-length drebrin. Structural studies employing drABD will be needed to address this possibility.

5.9 Summary and Future Directions

Recent biophysical and biochemical studies on drebrin A revealed some unique structural changes in drebrin-decorated F-actin. To summarize, we know now that drebrin binding to F-actin is fast, cooperative, and induces structural changes in actin filaments that translate into increased length of helical repeat. Moreover, these morphological changes propagate onto undecorated actin regions. Biochemical work also confirmed in vivo observations that drebrin has stabilizing effect on actin filaments and revealed that it inhibits F-actin dynamics at both (barbed and pointed) ends. However, it is yet to be clarified what are the functional consequences of such F-actin remodeling by drebrin. First, molecular cross-talk of drebrin with some key neuronal actin regulators (such as Arp2/3, formins, bundling proteins, etc.) is not fully understood. Second, the biochemical difference among drebrin isoforms remains unclear. Third, high-resolution structure of drebrin-F-actin complex is needed for better understanding of its function in dendritic spines and non-neuronal cells.

Acknowledgments I'm very grateful to Prof. Emil Reisler for his support and help with the preparation of this manuscript. I recognize that, due to space limitations, many important and relevant studies were not cited directly in this review, and I apologize for that to the authors of those publications. This work was supported by USPHS grant GM 077190.

References

Aoki C, Sekino Y, Hanamura K, Fujisawa S, Mahadomrongkul V, Ren Y, Shirao T (2005) Drebrin A is a postsynaptic protein that localizes in vivo to the submembranous surface of dendritic sites forming excitatory synapses. J Comp Neurol 483(4):383–402

Bacchiocchi C, Graceffa P, Lehrer SS (2004) Myosin-induced movement of (alpha)(alpha), (alpha)(beta), and (beta)(beta) smooth muscle tropomyosin on actin observed by multisite FRET. Biophys J 86(4):2295–2307

Bellot A, Guivernau B, Tajes M, Bosch-Morató M, Valls-Comamala V, Muñoz FJ (2014) The structure and function of actin cytoskeleton in mature glutamatergic dendritic spines. Brain Res 1573:1–16

Biou V, Brinkhaus H, Malenka RC, Matus A (2008) Interactions between drebrin and Ras regulate dendritic spine plasticity. Eur J Neurosci 27(11):2847–2859

Bozhokina E, Khaitlina S, Adam T (2008) Grimelysin, a novel metalloprotease from Serratia grimesii, is similar to ECP32. Biochem Biophys Res Commun 367(4):888–892
Broschat KO, Weber A, Burgess DR (1989) Tropomyosin stabilizes the pointed end of actin filaments by slowing depolymerization. Biochemistry 28(21):8501–8506
Cao W, Goodarzi JP, De La Cruz EM (2006a) Energetics and kinetics of cooperative cofilin-actin filament interactions. J Mol Biol 361(2):257–267
Cao W, Goodarzi JP, De La Cruz EM (2006b) Energetics and kinetics of cooperative cofilin-actin filament interactions. J Mol Biol 361(2):257–267
Chan C, Beltzner CC, Pollard TD (2009) Cofilin Dissociates Arp2/3 complex and branches from actin filaments. Curr Biol 19(7):537–545
Cheng CY, Lie PPY, Wong EWP, Mruk DD, Silvestrini B (2011) Adjudin disrupts spermatogenesis via the action of some unlikely partners: Eps8, Arp2/3 complex, drebrin E, PAR6 and 14-3-3. Spermatogenesis 1(4):291–297
Cowan WM, Südhof TC, Stevens CF (2001) Synapses. Johns Hopkins University Press, Baltimore, MD, pp 1–767
De La Cruz EM (2005) Cofilin binding to muscle and non-muscle actin filaments: isoform-dependent cooperative interactions. J Mol Biol 346(2):557–564
De La Cruz EM, Sept D (2010) The kinetics of cooperative cofilin binding reveals two states of the cofilin-actin filament. Biophys J 98(9):1893–1901
Dos Remedios CG, Chhabra D, Kekic M, Dedova IV, Tsubakihara M, Berry DA, Nosworthy NJ (2003) Actin binding proteins: regulation of cytoskeletal microfilaments. Physiol Rev 83(2):433–473
Drozdetskiy A, Cole C, Procter J, Barton GJ (2015) JPred4: a protein secondary structure prediction server. Nucleic Acids Res 43:W389–W394
von der Ecken J, Müller M, Lehman W, Manstein DJ, Penczek PA, Raunser S (2015) Structure of the F-actin-tropomyosin complex. Nature 519(7541):114–117
von der Ecken J, Heissler SM, Pathan-Chhatbar S, Manstein DJ, Raunser S. Cryo-EM structure of a human cytoplasmic actomyosin complex at near-atomic resolution. Nature (2016) V534 (7609), P724–728 pubmed ID: 27324845
Egelman EH (2015) Three-dimensional reconstruction of helical polymers. Arch Biochem Biophys 581:54–58
Egelman EH, Francis N, DeRosier DJ (1982) F-actin is a helix with a random variable twist. Nature 298(5870):131–135
Frost NA, Kerr JM, HE L, Blanpied TA (2010) A network of networks: cytoskeletal control of compartmentalized function within dendritic spines. Curr Opin Neurobiol 20(5):578–587
Fujii T, Iwane AH, Yanagida T, Namba K (2010) Direct visualization of secondary structures of F-actin by electron cryomicroscopy. Nature 467(7316):724–728
Galkin VE, Orlova A, Lukoyanova N, Wriggers W, Egelman EH (2001) Actin depolymerizing factor stabilizes an existing state of F-actin and can change the tilt of F-actin subunits. J Cell Biol 153(1):75–86
Galkin VE, Orlova A, VanLoock MS, Rybakova IN, Ervasti JM, Egelman EH (2002a) The utrophin actin-binding domain binds F-actin in two different modes: implications for the spectrin superfamily of proteins. J Cell Biol 157(2):243–251
Galkin VE, VanLoock MS, Orlova A, Egelman EH (2002b) A new internal mode in F-actin helps explain the remarkable evolutionary conservation of actin's sequence and structure. Curr Biol 12(7):570–575
Galkin VE, Orlova A, Schroder GF, Egelman EH (2010) Structural polymorphism in F-actin. Nat Struct Mol Biol 17(11):1318–1323
Galkin VE, Orlova A, Kudryashov DS, Solodukhin A, Reisler E, Schröder GF, Egelman EH (2011) Remodeling of actin filaments by ADF/cofilin proteins. Proc Natl Acad Sci USA 108(51):20568–20572
Galkin VE, Orlova A, Vos MR, Schröder GF, Egelman EH (2015) Near-atomic resolution for one state of F-actin. Structure 23(1):173–182

Goldschmidt L, Cooper DR, Derewenda ZS, Eisenberg D (2007) Toward rational protein crystallization: a web server for the design of crystallizable protein variants. Protein Sci 16(8):1569–1576

Goroncy AK, Koshiba S, Tochio N, Tomizawa T, Sato M, Inoue M, Watanabe S, Hayashizaki Y, Tanaka A, Kigawa T, Yokoyama S (2009) NMR solution structures of actin depolymerizing factor homology domains. Protein Sci 18(11):2384–2392

Gressin L, Guillotin A, Guérin C, Blanchoin L, Michelot A (2015) Architecture dependence of actin filament network disassembly. Curr Biol 25(11):1437–1447

Grintsevich EE, Reisler E (2014) Drebrin inhibits cofilin-induced severing of F-actin. Cytoskeleton (Hoboken) 71(8):472–483

Grintsevich EE, Galkin VE, Orlova A, Ytterberg AJ, Mikati MM, Kudryashov DS, Loo JA, Egelman EH, Reisler E (2010) Mapping of drebrin binding site on F-actin. J Mol Biol 398(4):542–554

Gunning P, O'Neill G, Hardeman E (2008) Tropomyosin-based regulation of the actin cytoskeleton in time and space. Physiol Rev 88(1):1–35

Hayashi K, Shirao T (1999) Change in the shape of dendritic spines caused by overexpression of drebrin in cultured cortical neurons. J Neurosci 19(10):3918–3925

Hayashi K, Ishikawa R, Kawai-Hirai R, Takagi T, Taketomi A, Shirao T (1999) Domain analysis of the actin-binding and actin-remodeling activities of drebrin. Exp Cell Res 253(2):673–680

Hild G, Kalmár L, Kardos R, Nyitrai M, Bugyi B (2014) The other side of the coin: functional and structural versatility of ADF/cofilins. Eur J Cell Biol 93(5-6):238–251

Holtmaat A, Svoboda K (2009) Experience-dependent structural synaptic plasticity in the mammalian brain. Nat Rev Neurosci 10(9):647–658

Hotulainen P, Hoogenraad CC (2010) Actin in dendritic spines: connecting dynamics to function. J Cell Biol 189(4):619–629

Hsiao JY, Goins LM, Petek NA, Mullins RD (2015) Arp2/3 Complex and cofilin modulate binding of tropomyosin to branched actin networks. Curr Biol 25(12):1573–1582

Ikeda K, Kaub PA, Asada H, Uyemura K, Toya S, Shirao T (1996) Stabilization of adhesion plaques by the expression of drebrin A in fibroblasts. Dev Brain Res 91(2):227–236

Ishikawa R, Hayashi K, Shirao T, Xue Y, Takagi T, Sasaki Y, Kohama K (1994) Drebrin, a development-associated brain protein from rat embryo, causes the dissociation of tropomyosin from actin filaments. J Biol Chem 269(47):29928–29933

Ishikawa R, Katoh K, Takahashi A, Xie C, Oseki K, Watanabe M, Igarashi M, Nakamura A, Kohama K (2007) Drebrin attenuates the interaction between actin and myosin-V. Biochem Biophys Res Commun 359(2):398–401

Ivanov A, Esclapez M, Ferhat L (2009a) Role of drebrin A in dendritic spine plasticity and synaptic function. Commun Integr Biol 2(3):268–270

Ivanov A, Esclapez M, Pellegrino C, Shirao T, Ferhat L (2009b) Drebrin A regulates dendritic spine plasticity and synaptic function in mature cultured hippocampal neurons. J Cell Sci 122(4):524–534

Jin M, Tanaka S, Sekino Y, Ren Y, Yamazaki H, Kawai-Hirai R, Kojima N, Shirao T (2002) A novel, brain-specific mouse drebrin: cDNA cloning, chromosomal mapping, genomic structure, expression, and functional characterization. Genomics 79(5):686–692

Jung G, Kim EJ, Cicvaric A, Sase S, Gröger M, Höger H, Sialana FJ, Berger J, Monje FJ, Lubec G (2015) Drebrin depletion alters neurotransmitter receptor levels in protein complexes, dendritic spine morphogenesis and memory-related synaptic plasticity in the mouse hippocampus. J Neurochem 134(2):327–339

Kasai H, Fukuda M, Watanabe S, Hayashi-Takagi A, Noguchi J (2010a) Structural dynamics of dendritic spines in memory and cognition. Trends Neurosci 33(3):121–129

Kasai H, Hayama T, Ishikawa M, Watanabe S, Yagishita S, Noguchi J (2010b) Learning rules and persistence of dendritic spines. Eur J Neurosci 32(2):241–249

Keon BH, Jedrzejewski PT, Paul DL, Goodenough DA (2000) Isoform specific expression of the neuronal F-actin binding protein, drebrin, in specialized cells of stomach and kidney epithelia. J Cell Sci 113(2):325–336

Klenchin VA, Khaitlina SY, Rayment I (2006) Crystal structure of polymerization-competent actin. J Mol Biol 362(1):140–150
Kojima N, Shirao T (2007) Synaptic dysfunction and disruption of postsynaptic drebrin-actin complex: a study of neurological disorders accompanied by cognitive deficits. Neurosci Res 58(1):1–5
Kuhn TB, Bamburg JR (2008) Tropomyosin and ADF/cofilin as collaborators and competitors. Adv Exp Med Biol 644:232–249
Kuhn JR, Pollard TD (2005) Real-time measurements of actin filament polymerization by total internal reflection fluorescence microscopy. Biophys J 88(2):1387–1402
Lehman W, Craig R, Vibert P (1994) Ca^{2+}-induced tropomyosin movement in Limulus thin filaments revealed by three-dimensional reconstruction. Nature 368(6466):65–67
Lin YC, Koleske AJ (2010) Mechanisms of synapse and dendrite maintenance and their disruption in psychiatric and neurodegenerative disorders. Annu Rev Neurosci 33:349–378
McGhee JD, von Hippel PH (1974) Theoretical aspects of DNA-protein interactions: co-operative and non-co-operative binding of large ligands to a one-dimensional homogeneous lattice. J Mol Biol 86(2):469–489
McGough A, Way M, DeRosier D (1994) Determination of the alpha-actinin-binding site on actin filaments by cryoelectron microscopy and image analysis. J Cell Biol 126(2):433–443
McGough A, Pope B, Chiu W, Weeds A (1997) Cofilin changes the twist of F-actin: implications for actin filament dynamics and cellular function. J Cell Biol 138(4):771–781
Mikati MA, Grintsevich EE, Reisler E (2013) Drebrin-induced stabilization of actin filaments. J Biol Chem 288:19926–19938
Ngo KX, Kodera N, Katayama E, Ando T, Uyeda TQP (2015) Cofilin-induced unidirectional cooperative conformational changes in actin filaments revealed by high-speed atomic force microscopy. Elife 4:e04806
Oda T, Iwasa M, Aihara T, Maeda Y, Narita A (2009) The nature of the globular- to fibrous-actin transition. Nature 457(7228):441–445
Orlova A, Prochniewicz E, Egelman EH (1995) Structural dynamics of F-actin: II. Cooperativity in structural transitions. J Mol Biol 245(5):598–607
Paavilainen VO, Oksanen E, Goldman A, Lappalainen P (2008) Structure of the actin-depolymerizing factor homology domain in complex with actin. J Cell Biol 182(1):51–59
Peitsch WK, Grund C, Kuhn C, Schnölzer M, Spring H, Schmelz M, Franke WW (1999) Drebrin is a widespread actin-associating protein enriched at junctional plaques, defining a specific microfilament anchorage system in polar epithelial cells. Eur J Cell Biol 78(11):767–778
Pollard TD (1986) Rate constants for the reactions of ATP- and ADP-actin with the ends of actin filaments. J Cell Biol 103(6):2747–2754
Pollard TD (2010) A guide to simple and informative binding assays. Mol Biol Cell 21(23):4061–4067
Poukkula M, Kremneva E, Serlachius M, Lappalainen P (2011) Actin-depolymerizing factor homology domain: a conserved fold performing diverse roles in cytoskeletal dynamics. Cytoskeleton 68(9):471–490
Quintero-Monzon O, Rodal AA, Strokopytov B, Almo SC, Goode BL (2005) Structural and functional dissection of the Abp1 ADFH actin-binding domain reveals versatile in vivo adapter functions. Mol Biol Cell 16(7):3128–3139
Ressad F, Didry D, Xia GX, Hong Y, Chua NH, Pantaloni D, Carlier MF (1998) Kinetic analysis of the interaction of actin-depolymerizing factor (ADF)/cofilin with G- and F-actins. J Biol Chem 273(33):20894–20902
Sala C, Cambianica I, Rossi F (2008) Molecular mechanisms of dendritic spine development and maintenance. Acta Neurobiol Exp 68(2):289–304
Schmid MF, Sherman MB, Matsudaira P, Chiu W (2004) Structure of the acrosomal bundle. Nature 431(7004):104–107
Sharma S, Grintsevich EE, Phillips ML, Reisler E, Gimzewski JK (2010) Atomic force microscopy reveals drebrin induced remodeling of F-actin with subnanometer resolution. Nano Lett 11(2):825–827

Sharma S, Grintsevich EE, Hsueh C, Reisler E, Gimzewski J (2012) Molecular cooperativity of drebrin1-300 binding and structural remodeling of F-actin. Biophys J 103(2):275–283

Sheterline P, Clayton J, Sparrow JC (2002) Actin, 4th edn. Protein profile. Oxford University Press, Oxford

Shim KS, Lubec G (2002) Drebrin, a dendritic spine protein, is manifold decreased in brains of patients with Alzheimer's disease and Down syndrome. Neurosci Lett 324(3):209–212

Shirao T, Obata K (1986) Immunochemical homology of 3 developmentally regulated brain proteins and their developmental change in neuronal distribution. Brain Res 394(2):233–244

Shirao T, Sekino Y (2001) Clustering and anchoring mechanisms of molecular constituents of postsynaptic scaffolds in dendritic spines. Neurosci Res 40(1):1–7

Shirao T, Kojima N, Nabeta Y, Obata K (1989) Two forms of drebrins, developmentally regulated brain proteins, in rat. Proceedings of the Japan Academy 65:169–172

Solomatin SV, Greenfeld M, Herschlag D (2011) Implications of molecular heterogeneity for the cooperativity of biological macromolecules. Nat Struct Mol Biol 18(6):732–734

Suarez C, Roland J, Boujemaa-Paterski R, Kang H, McCullough BR, Reymann AC, Guérin C, Martiel JL, De La Cruz EM, Blanchoin L (2011) Cofilin tunes the nucleotide state of actin filaments and severs at bare and decorated segment boundaries. Curr Biol 21(10):862–868

Tsaturyan AK, Koubassova N, Ferenczi MA, Narayanan T, Roessle M, Bershitsky SY (2005) Strong binding of myosin heads stretches and twists the actin helix. Biophys J 88(3):1902–1910

Uruno T, Remmert K, Hammer JA (2006) CARMIL is a potent capping protein antagonist. J Biol Chem 281(15):10635–10650

Vorobiev S, Strokopytov B, Drubin DG, Frieden C, Ono S, Condeelis J, Rubenstein PA, Almo SC (2003) The structure of nonvertebrate actin: Implications for the ATP hydrolytic mechanism. Proc Natl Acad Sci USA 100(10):5760–5765

Wen KK, Kuang B, Rubenstein PA (2000) Tropomyosin-dependent filament formation by a polymerization-defective mutant yeast actin (V266G, L267G). J Biol Chem 275(51):40594–40600

Willadsen KA, Butters CA, Hill LE, Tobacman LS (1992) Effects of the amino-terminal regions of tropomyosin and troponin T on thin filament assembly. J Biol Chem 267(33):23746–23752

Xu C, Craig R, Tobacman L, Horowitz R, Lehman W (1999) Tropomyosin positions in regulated thin filaments revealed by cryoelectron microscopy. Biophys J 77(2):985–992

Zhao L, Ma QL, Calon F, Harris-White ME, Yang F, Lim GP, Morihara T, Ubeda OJ, Ambegaokar S, Hansen JE, Weisbart RH, Teter B, Frautschy SA, Cole GM (2006) Role of p21-activated kinase pathway defects in the cognitive deficits of Alzheimer disease. Nat Neurosci 9(2):234–242

Zhou Q, Homma KJ, Mm P (2004) Shrinkage of dendritic spines associated with long-term depression of hippocampal synapses. Neuron 44(5):749–757

Chapter 6
Cell Shape Change by Drebrin

Kensuke Hayashi

Abstract Drebrin is localized in actin-rich regions of neuronal and non-neuronal cells. In mature neurons, its localization is strictly restricted to the postsynaptic sites. In order to understand the function of drebrin in cells, many studies have been performed to examine the effect of overexpression or knocking down of drebrin in various cell types, including neurons, myoblasts, kidney cells, and intestinal epithelial cells. In most cases alteration of cell shape and impairment or facilitation of actin-based activities of these cells were observed. Interestingly, overexpression of drebrin in matured neurons results in the alteration in dendritic spine morphology. Further studies have shown alteration in the localization of postsynaptic receptors and even changes in synaptic transmission caused by drebrin overexpression or depletion in neurons. These drebrin's effects are thought to come from drebrin's actin-cross-linking activity or competitive binding to actin against tropomyosin, fascin, and α-actinin. Furthermore, drebrin binds to various molecules, such as homer, EB3, and cell-cell junctional proteins, indicating that drebrin is a multifunctional cytoskeletal regulator.

Keywords Drebrin • Dendritic spine • Cell migration • Dendrite formation • Axon elongation

Since the earliest study of drebrin, it has already been known that transfection of drebrin cDNA into cultured cells induces dramatic change in cell shape. In the paper of 1992 by Dr. T. Shirao, in which he reported the first cloning of drebrin cDNA from the rat, he described formation of neurite-like processes in L cells transfected with rat drebrin cDNA (Shirao et al. 1992). Shortly later in 1994, he described abnormal structures of actin filaments, which he called "thick, curving bundle of actin," in these cells (Shirao et al. 1994). Changes in cell shape and actin bundles are very striking and meaningful, because they are direct and convincing evidence that drebrin has unknown but certain activities on cell shape.

K. Hayashi
Faculty of Science and Technology, Sophia University, 7-1, Kioicho, Chiyoda-ku, Tokyo 102-8554, Japan
e-mail: kensuk-h@sophia.ac.jp

T. Shirao, Y. Sekino (eds.), *Drebrin*, Advances in Experimental Medicine and Biology 1006, DOI 10.1007/978-4-431-56550-5_6

Drebrin was found to have a strong actin-binding activity by studies of Dr. Ishikawa (Ishikawa et al. 1994). His solid biochemical works revealed the nature of actin-binding activity of drebrin and opened a road to molecular study of drebrin function. Competitive binding of drebrin to actin with other actin-binding proteins unveiled the mechanisms for cell shape regulation by drebrin. However, the mechanisms have not yet fully understood. Recent works revealed that drebrin binds to various molecules other than actin filaments, such as homer, EB3, and cell-cell junctional proteins as well, indicating that drebrin is a multifunctional cytoskeletal regulator.

In this chapter, we review studies about what happens on actin filaments and cell shape when drebrin is overexpressed or knocked down in neuronal and non-neuronal cells. Possible mechanisms for those changes will also be discussed.

6.1 Drebrin Distribution in Neurons and in Non-neuronal Cells

6.1.1 Distribution in Neurons

In neural tissues of young rodents, high expression of drebrin is found in migrating young neurons. The most prominent example of drebrin-expressing migration neurons in vivo is the rostral migratory stream, where newly generated precursors of granule neurons of olfactory bulb, which are born in subventricular zone of lateral ventricle, are migrating toward the final destination. High expression of drebrin in the neurons in rostral migratory stream and in the core region of olfactory bulb was reported. As drebrin immunoreactivity decreases in postmigratory neuroblasts at the outer layers of olfactory bulb, the disappearance of drebrin from the cell might coincide with the cessation of migration (Song et al. 2008).

In these young neurons, drebrin is localized at submembranous region or in the cell processes. Such localization was firstly reported in differentiating neuroblastoma cell lines. In parallel with the neuronal differentiation of SH-SY5Y cells following retinoic acid treatment, drebrin accumulates at the submembranous cortical cytoplasm, accompanying with the accumulation of actin filaments in response to retinoic acid. Similar localization of drebrin was also described in primary cultured cerebellar granular cells, which have highly migratory activity (Asada et al. 1994).

Within neurons that are extending axons, drebrin is expressed at the growth cones. Immunohistochemical data have presented the enrichment of drebrin in the central region of growth cones of PC12 cells (Sasaki et al. 1996). Drebrin is not located at the distal part of filopodia, where fascin bundles actin filaments. This is consistent with the fact that drebrin and fascin bind to actin filaments competitively. Drebrin is found at the bottom regions of filopodia where drebrin is supposed to dissociate fascin and disrupt the bundles. In growth cones of primary cultured hippocampal neurons, triple staining for drebrin, actin, and tubulin revealed that drebrin is concentrated in interface region where actin filaments and microtubules overlap (Fig. 6.1) (Geraldo et al. 2008; Mizui et al. 2009).

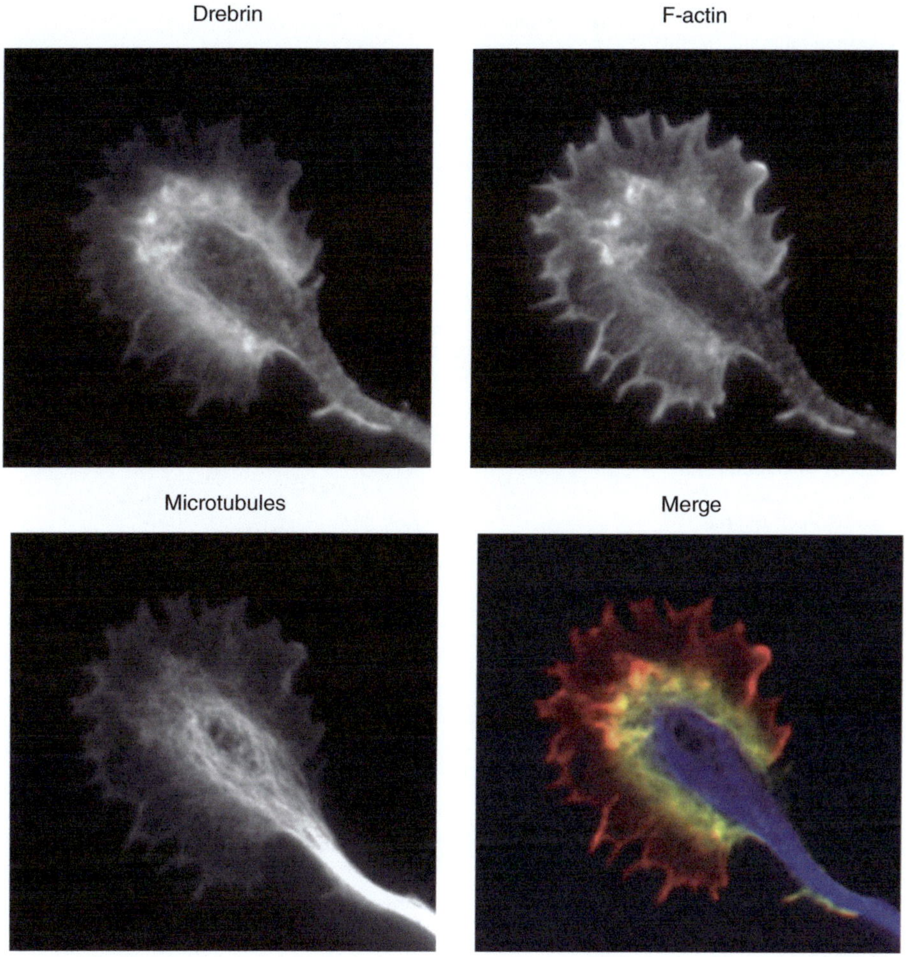

Fig. 6.1 Distribution of drebrin in growth cone. Growth cone was triple immunostained for drebrin, actin filaments, and microtubules. Drebrin is localized at the interphase between actin-rich region and microtubule-rich region. Adapted from (Mizui et al. 2009)

In adult, drebrin localization becomes more restricted to certain types of synaptic regions (Hayashi et al. 1996; Shirao and Obata 1986; Shirao et al. 1989). When sagittal section of adult rat is examined, drebrin immunoreactivity is intense in the gray matter of telencephalon (Fig. 6.2a, b). Gray matter of cerebellar cortex is not heavily stained, which is consistent with the absence of expression of adult-type drebrin as examined with Western blot (Fig. 6.2c). Immunostainings for drebrin in gray matter look as small grains scattered evenly over the gray matter. Observation using confocal laser microscope revealed that immunostaining of drebrin is always closely associated with immunostaining for synaptophysin, a presynaptic marker (Fig. 6.3a). Immuno-electron microscopic observation revealed that drebrin is

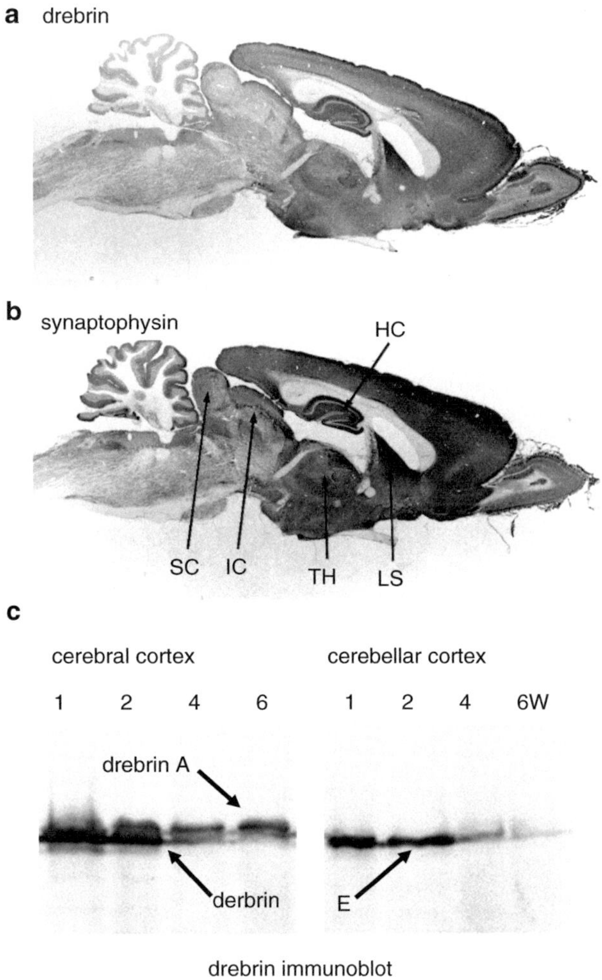

Fig. 6.2 Expression of drebrin in rat brain. (**a** and **b**) Sagittal section of adult rat brain stained with antibodies to drebrin (**a**) and synaptophysin (**b**). Drebrin immunoreactivity is intense in gray matter of forebrain but not in cerebellar cortex, whereas synaptophysin antibody stains both. *HC* hippocampus, *LS* lateral septum, *TH* thalamus, *IC* inferior colliculus, *SC* superior colliculus. (**c**) Immunoblot with drebrin antibody of cerebral and cerebellar cortices from 1- to 10-week-old rats. Expression of drebrin E decreases during the development, but expression of drebrin A increases in cerebral cortex. Note that drebrin A expression is much less in cerebellar cortex than in cerebral cortex. Adapted from (Hayashi et al. 1996)

strictly located at postsynaptic sites (Hayashi et al. 1996). Drebrin is not concentrated in postsynaptic sites of inhibitory synapses that do not form spines (Fig. 6.3b, c, unpublished data), while it exists at postsynaptic sites of excitatory synapses on GABAergic neurons (Hanamura et al. 2010), suggesting that drebrin expression is restricted to postsynapses with protruding morphology. Not all spines contain drebrin. Some spines are drebrin negative. Drebrin-positive spines were larger in head areas and have longer postsynaptic densities than drebrin-negative spines (Kobayashi et al. 2007).

In spines, drebrin likely binds to actin filaments of postsynaptic density. When synaptosomes from adult rat cerebral cortex were immunostained for drebrin and

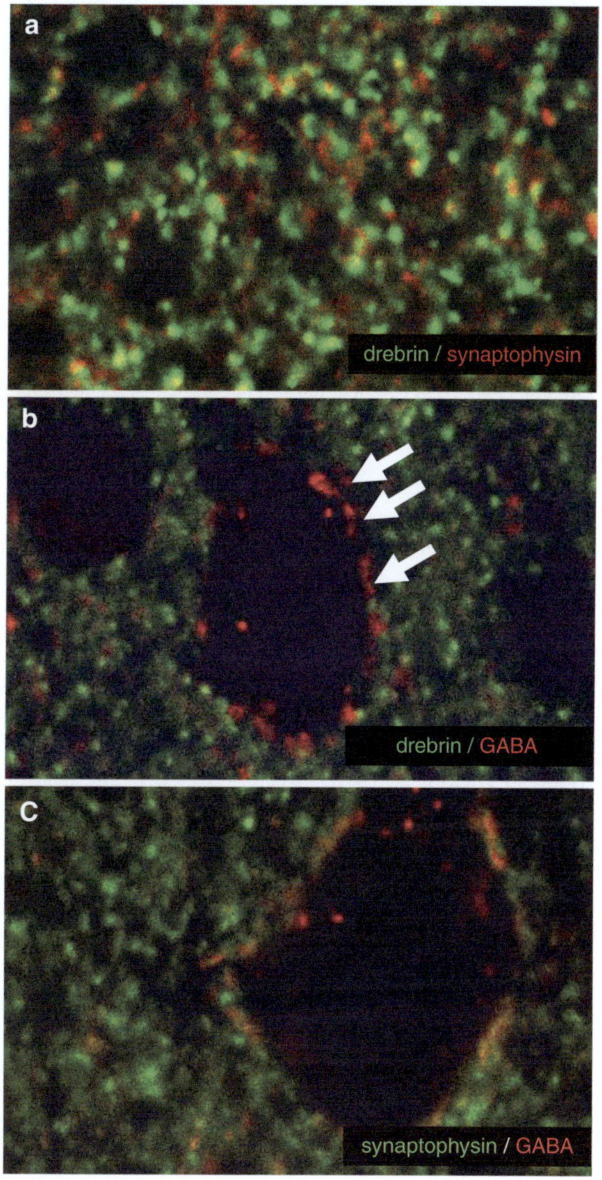

Fig. 6.3 Presence of drebrin at postsynaptic sites of most of synapses but not at inhibitory synapses attached to cell body of postsynaptic cells. (**a**) Confocal observation of gray matter of cerebral cortex of adult rat that were double stained with antibodies to drebrin and synaptophysin. Both immunoreactivities were seen as dots with close contact with each other. (**b**) Antibody to GABA-stained presynaptic sites associated to the cell body. Drebrin immunoreactivity was not found around the GABA-positive boutons. (**c**) Colocalization of synaptophysin and GABA confirmed the GABA-positive dots as presynaptic boutons

observed by EM, immunosignals were detected on fine filaments protruding from postsynaptic density (Fig. 6.4a). Drebrin-containing cytoskeletons can be isolated by immunoprecipitation using anti-drebrin monoclonal antibody. They included drebrin, actin, myosin, and gelsolin, suggesting that drebrin constitute multimolecular actin-based complex in spines (Fig. 6.4b) (Hayashi et al. 1996).

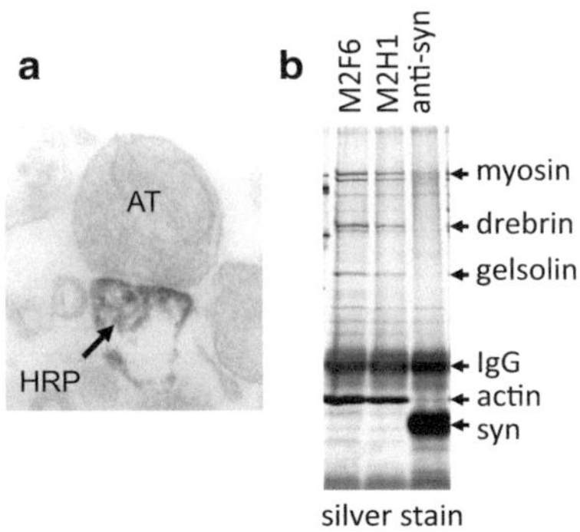

Fig. 6.4 Drebrin constitutes multimolecular complex that binds to postsynaptic actin filaments. (**a**) Synaptosomes were isolated from cerebral cortex of adult rat by sucrose density-gradient centrifugation and were labeled with an antibody to drebrin. Immunoreactivity was visualized with colloidal gold-conjugated (*right*) or with HRP-conjugated (*left*) second antibodies. Images show the presence of drebrin on fine filaments associated to postsynaptic densities. (**b**) NP-40 extracts from cerebral cortex of adult rat was immunoprecipitated with anti-drebrin-conjugated Sepharose. Two independent monoclonal antibodies (M2H1 and M2F6) were used. Both antibodies precipitated protein complex, in which myosin, drebrin gelsolin, and actin were detected. Adapted from (Hayashi et al. 1996)

6.1.2 Distribution in Non-neuronal Cells

By histological observation of non-neural tissues of adult rodents, drebrin was found in various cells of various tissues. It localizes at the intracellular organelles and cell processes that undergo robust morphological changes. For example in stomach, drebrin is specifically expressed in the acid-secreting parietal cells (Keon et al. 2000), where drebrin is phosphorylated (Chew et al. 2005). It is supposed that drebrin has a regulatory role on actin filament that triggers secretion of acids by robust reorganization of alveoli in parietal cells. In the kidney, drebrin is expressed in cells in distal parts of the nephron, where selective reabsorption and secretion occurs. In these cells, drebrin is localized at apical membranes which have extensive protrusions and invaginations with greatly amplified surface areas for efficient ion transport (Keon et al. 2000). Distribution of drebrin in the kidney was further studied by Peitsch (2003), who found the enrichment of drebrin in renal glomeruli (Peitsch et al. 2003). Intense drebrin immunosignals were found at the foot processes of developing podocytes, which form complicated protrusions of cell membrane. In a developing lung, temporal expression of drebrin is observed in cell protrusion of immature septal interstitial cells during alveolar maturation (Yamada et al. 2005). All these examples suggest the role of drebrin in regulation of membrane morphology of non-neuronal cells.

Further examples of high expression of drebrin in cells that undergo robust morphological change come from malignant cells. Drebrin is expressed in colorectal cancer cell lines and in malignant colorectal cancer tissue (Lin et al. 2014). Elevated drebrin expression is also found in basic cell carcinoma (Peitsch et al. 2005).

Keratinocytes of normal epidermis express almost no drebrin, but basal cell carcinomas are rich in drebrin. In these cells, drebrin is accumulated at adhering junctions. Close observation of cell line of this type of carcinoma revealed that drebrin was colocalized at tight junctions as well (Mizutani et al. 2014).

Precise subcellular localization of drebrin in non-neuronal cells has been investigated in cultured cells. Drebrin has been found in cultured cell lines including epithelial, endothelial, smooth muscle, and neural origins (Peitsch et al. 1999). Drebrin is localized at adherens junctions of these cells. In myoblasts, drebrin is enriched in cellular projections and at regions of cell-cell contact (Mancini et al. 2011). In immune cells, drebrin is found in the immune synapse, where drebrin colocalizes with the chemokine receptor (Perez-Martinez et al. 2010). In T cells infected with HIV viruses, drebrin is actively recruited toward cell-virus and virus-driven cell-cell contacts (Gordon-Alonso et al. 2013). At all of these sites, drebrin is colocalized with actin filaments.

Localization of drebrin at gap junctions was reported by Butkevich et al. (2004). Immunostaining of drebrin in astrocytes was seen throughout the cytosol, but relatively intense signals were detected at the cell periphery as fine spots, where it colocalized with gap junction protein (Cx43). The localization of drebrin at gap junctions in Vero cells (a cell line derived from the kidney of an African green monkey) were not dependent on its actin binding, as depolymerization of actin filaments with Latrunculin B did not disperse drebrin from plasma membrane. Direct molecular interaction between drebrin and Cx43 was suggested by FRET analysis in COS cells expressing CFP-drebrin and YFP-Cx43 (Butkevich et al. 2004).

Besides cell-cell contact sites, drebrin has been also found in terminal web of intestinal epithelial cells. The levels of drebrin expression increased in the terminal web both in vitro and in vivo during the formation of the apical domain of cell membrane (Bazellieres et al. 2012).

6.2 Drebrin-Dependent Remodeling of Actin Filaments Within Cells

Thick, curving bundles of actin filaments were described by Dr. T. Shirao in drebrin-cDNA transfected fibroblastic cells (Shirao et al. 1994). In cells expressing exogenous drebrin at low level, neither cytoskeletal organization nor cell shape did not change. Drebrin was found as patchy dot-like aggregates or sometimes it is bound to stress fibers. In cells expressing drebrin at high level, dramatic remodelings of cytoskeleton were induced (Fig. 6.5). Induced remodelings included disappearance of stress fibers, formation of thick, curving bundle of actin filaments, formation of ring structures of actin filaments, and formation of cell processes. The induced cell processes were sometimes highly branched especially when L cells were transfected (Shirao et al. 1994).

How drebrin remodels actin filaments into such thick and curving bundles? Hayashi et al. generated GFP-fusion constructs containing various subfragments of

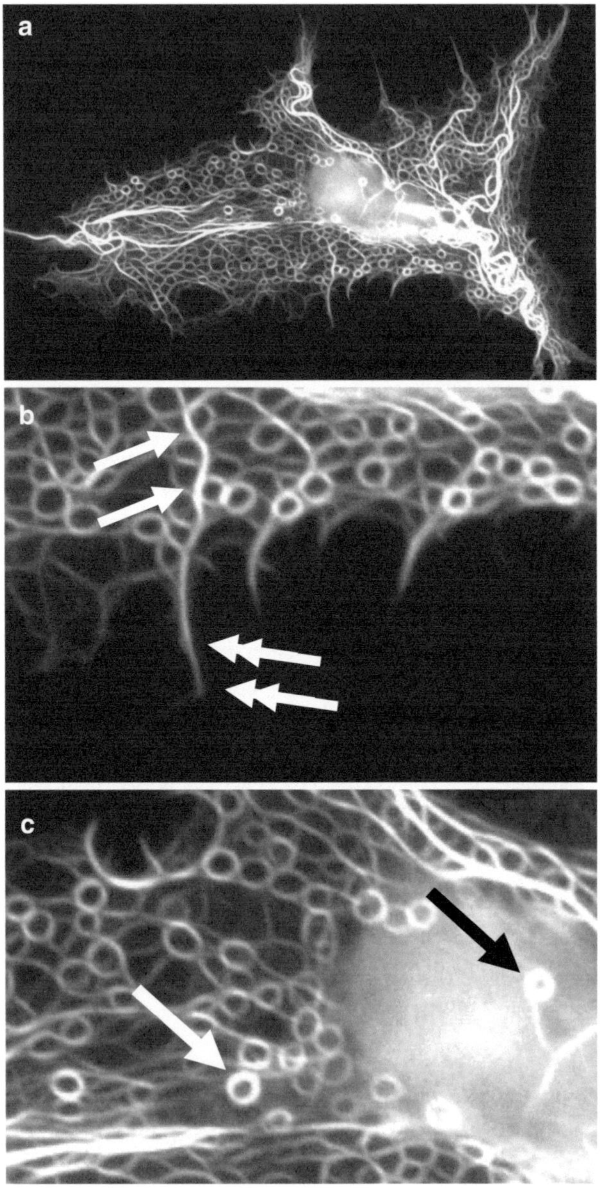

Fig. 6.5 Drebrin-induced thick and curving actin filaments of non-neuronal cells. (**a**) Wide view of single CHO cell transfected with N-terminal region of drebrin tagged by GFP. (**b**) Close observation showing a thick and curving actin filament (*arrows*) and a spike (*double arrows*). Note that actin bundles are continuous from intracellular curving bundles into the spike. (**c**) Thick curving bundles continuously make rings. Note that some of the rings are apparently isolated from other bundles (*arrows*)

drebrin and came to the conclusion that the actin-binding domain and actin-remodeling domain are identical and that this domain is located at the central region of drebrin (Hayashi et al. 1999) (see Fig. 1.3 in Chap. 1). By transfection with the fragment of actin-binding region, all types of cell and cytoskeletal change that are induced by transfection with full length drebrin were induced (Fig. 6.6a). In contrast, transfection of mutant drebrin that lacks actin-binding region did not bind to

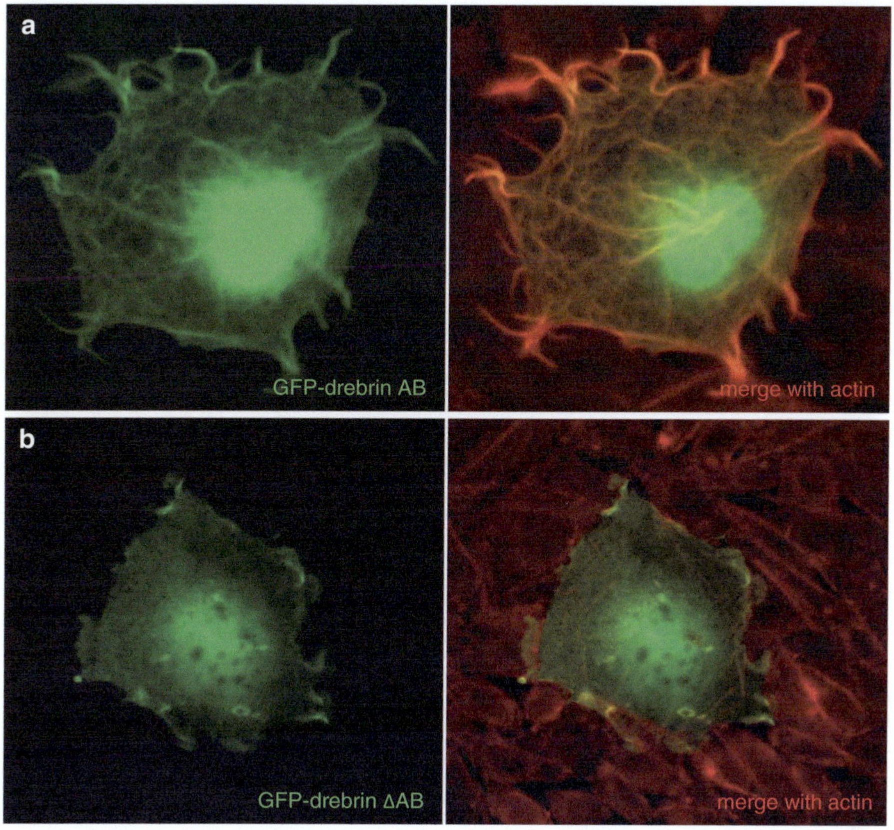

Fig. 6.6 Actin-binding domain of drebrin is necessary and sufficient to induce thick and curving actin filaments in non-neuronal cells. CHO cells were transfected with GFP fusion constructs that contain actin-binding domain of drebrin (**a**) or drebrin from which actin-binding domain is deleted (**b**). Actin-binding domain of drebrin induced all types of actin filament shape change including formation of thick curving bundles, ring structure, and cell spikes. Drebrin that lacks the central actin-binding domain did not bind to stress fibers nor induce any change in actin filaments

actin cytoskeleton nor showed any change in cytoskeleton and cell shape (Fig. 6.6b). Thus, actin binding is supposed to play the key role for remodeling of actin filaments. Since drebrin binds to actin filaments in competition with tropomyosin, Ishikawa et al. (1994) investigated the change in tropomyosin localization in drebrin-cDNA transfected cells (Ishikawa et al. 1994). They found that stress fibers which contained tropomyosin disappeared in these cells and the thick, curving bundles of actin filaments generated by exogenous drebrin were not stained for tropomyosin. These observations suggest that overexpression of drebrin might dissociate tropomyosin and probably α-actinin from stress fibers, and then actin filaments change their features. Interestingly, recent data indicated a possibility that drebrin cross-links actin filaments and bundles them (Worth et al. 2013). They discovered two domains in the N-terminal half of drebrin that independently bound to F-actin.

This activity is repressed by an intramolecular interaction regulated by phosphorylation. The existence of two actin-binding sites in drebrin molecules had been predicted previously by electron microscopic reconstruction (Grintsevich et al. 2010). Thus, the possible explanation of generation of thick, curving bundles of actin is that exogenous drebrin dissociate tropomyosin and α-actinin from stress fibers and cross-link actin filaments in place of them. The reason for formation of ring structure of actin filaments by drebrin, however, is completely unknown.

Besides the remodeling of stress fibers, remodeling of cortical actins and subsequent alteration of membrane functions have been observed in various types of cells after introduction of drebrin cDNA or depletion of drebrin.

In intestinal epithelial cells, actin filaments are condensed at the subapical region where they anchor microvilli constituting apical brush border. This actin-based network is called terminal web and the formation of terminal web is thought to be important for epithelial cells to be typical narrow, elongated shape. Drebrin is localized at terminal web. Drebrin depletion disrupted the normal morphology of terminal web and impaired cell compaction and elongation processes in the monolayer of human intestinal epithelial cell line (Bazellieres et al. 2012). Impairment in the cell elongation might involve interactions between drebrin and EB3, because decreasing the EB3 levels resulted in a similar elongation phenotype, without affecting the morphology of terminal webs. Impairment in the formation of apical membrane might involve drebrin function on membrane trafficking, because membrane components of drebrin knockdown cells are converted into lysosomal components (Vacca et al. 2014). Involvement of drebrin in membrane trafficking is also suggested by binding of drebrin to Golgi membrane isolated from rat liver (Xu and Stamnes 2006).

Another example of change in membrane function by drebrin overexpression or drebrin depletion comes from immune cells. Drebrin knockdown in T cells impairs the redistribution of virus receptors and inhibits actin polymerization at the immune synapse (Perez-Martinez et al. 2010).

6.3 Drebrin-Dependent Cell Adhesions

Cultured fibroblasts adhere to substratum with vinculin- and actin filaments-containing focal contacts. Treatment of these cells with cytochalasins disrupts the stress fibers and focal contacts, resulting in collapse of cells. Overexpression of drebrin in L cells made cells to be resistant to cytochalasin D treatment. The extent of cell spreading and cell attachment was greater in drebrin-expressing cells than that in control cells, suggesting that drebrin modulates cell-substratum adhesion (Ikeda et al. 1995). Adhesion plaques in drebrin-expressing cells were positive with vinculin-specific antibodies, and the expression level of vinculin was higher in drebrin-expressing cells. It was thought that expression of drebrin induced changes in the assembly of actin filaments of focal contacts and stabilized them (Ikeda et al. 1996).

As drebrin is found at adherens junctions and gap junctions of several cells, it is reasonable to suppose that overexpression or depletion of drebrin results in change in the structure and function of these junctions. In fact, drebrin knockdown in endothelial cells leads to weakening of cell-cell contacts, characterized by loss of nectin from adherens junctions (Rehm et al. 2013). Effect of drebrin knockdown on gap junctions was studied in Vero cells and astroglial cells. Decrease of drebrin expression by transfecting drebrin-siRNA resulted in a decrease in the endogenous levels of gap junctional proteins (Butkevich et al. 2004).

Cell-cell contacts between myogenic cells were also reported to be modulated by drebrin. Primary myoblasts and C2C12 cells depleted of drebrin by RNAi displayed reduced formation of multinucleated myotubes (Mancini et al. 2011). Modulation of cell-virus contacts by drebrin was also reported. Downregulation of drebrin expression in T cells decreased actin polymerization at the contact sites and promotes virus entry (Gordon-Alonso et al. 2013).

6.4 Drebrin-Induced Process Formation, Elongation, and Cell Migration

6.4.1 Drebrin-Induced Process Formation and Elongation

The first report of transfecting drebrin cDNA into mammalian cells described the formation of highly branched neurite-like cell processes in L cells (Shirao et al. 1992). The exogenously expressed drebrin was concentrated in submembraneous regions of these processes. Similar process formation was also found in drebrin-transfected CHO cells (Shirao et al. 1994). Within these processes, drebrin-bound thick actin bundle is found. These actin bundles are, in most cases, continuous to intracellular thick curving bundles of actin. Live cell imaging of these cell processes revealed that they slowly retracted and extended (Peitsch et al. 2006).

Although the induction of cell processes in cells that do not have any processes in nature does not by itself indicate any physiological function of drebrin in the formation of cell processes in neurons, many studies on neurons indicate physiological function of drebrin in neuritogenesis. As described above, drebrin is localized at the central part of growth cones of neurons. Overexpression of drebrin in cultured neurons led drebrin and actin filaments to accumulate throughout the growth cone and facilitated axonal growth. In contrast, knockdown of drebrin prevented axonal growth (Geraldo et al. 2008; Mizui et al. 2009). Similarly, introduction of antisense construct of drebrin into neuroblastoma cell line attenuated the neurite extension in response to serum deprivation and gangliosides treatment (Toda et al. 1999).

How does drebrin induce formation and elongation of cell processes? As for the growth cones, dissociation of fascin, which is a key actin-bundling protein important for formation of filopodia, from actin filaments at the basal region of filopodia might be one function of drebrin (Sasaki et al. 1996). Regulation of actomyosin interaction by drebrin at the growth cone is also suggested. Facilitating effect of drebrin overexpression on axonal growth in cultured hippocampal neurons was masked by inhibition of myosin II (Mizui et al. 2009).

Another interesting explanation for formation of cell process by drebrin came from the breakthrough discovery that drebrin binds to the plus-end capping protein EB3 (Geraldo et al. 2008). In growth cones, drebrin is located on actin filaments in the proximal region of filopodia, where it interacts with EB3 that is located at the tips of microtubules invading into peripheral region of growth cone. When this interaction was disrupted, the formation of growth cones and the extension of neurites were impaired. It is concluded that drebrin targets EB3 to coordinate actin filaments-microtubule interactions that underlie neuritogenesis (Geraldo et al. 2008). Expression of phosphomimetic and phospho-dead mutants of drebrin caused impairment of coupling of microtubules to actin filaments in growth cone and therefore interfered with neuritogenesis (Worth et al. 2013).

The same scenario can be applied to the formation of cell processes induced by drebrin in non-neuronal cells. Some processes induced in COS cells transfected with drebrin cDNA contained microtubules. Frequency of invasions of microtubules into the processes increased by co-transfection of drebrin and EB3 (Geraldo et al. 2008).

6.4.2 Drebrin-Induced Cell Migration

High expression of drebrin is found in migrating neurons such as rostral migratory stream of young brain and in malignant cells as reviewed above. Motility of drebrin-induced cell processes in non-neuronal cells also suggests the involvement of drebrin in controlling cell motility (Peitsch et al. 2006). In this context, various studies have been done to examine whether overexpression or depletion of drebrin alters the migration of neurons and non-neuronal cells.

In vivo assay of drebrin function on neuronal migration was first performed in chicken, in which in vivo electropolation technique is relatively easy. Dun et al. (2012) investigated the function of drebrin in the migration of oculomotor neuros of chick embryo. When drebrin was overexpressed, the leading processes of these neurons turned away from their target, and as a result, they moved along abnormal paths within the brain. Conversely, electropolation of drebrin shRNA blocked leading process formation and migration of these neurons (Dun et al. 2012). Rostral migratory stream of rodent brain was also used to investigate the role of drebrin in neuronal migration. Drebrin shRNA-transfected neuroblasts displayed reduced migrated distance and migration velocity in rostral migratory stream of mice (Sonego et al. 2015). Above, two examples of neuronal migration are classified into tangential migration in which neurons migrate relatively freely for a long distance. In contrast, migration of neurons along an existing scaffold of glial cells away from the neuroepithelium in developing cerebral cortex is called radial migration. Mechanisms for tangential migration and radial migration of neurons in development differ in some aspects. Drebrin was shown to be also involved in radial type of migration using electropolation of drebrin constructs into embryonic mouse cortex. Introduction of nonphosphorylatable or phosphorylation-mimicking mutants of drebrin decreased their migration, suggesting that the reversible phosphorylation/dephosphorylation reaction of drebrin is important (Tanabe et al. 2014).

As described earlier, high expression of drebrin is often observed in malignant cells. One such example is glioma cells. When invasiveness of glioma cell line was measured with Boyden chambers, stable overexpression of drebrin induced increased invasiveness, while knockdown of drebrin by small interfering RNA decreases invasion and migration (Terakawa et al. 2013). All these studies in neuronal and non-neuronal cells support the idea that drebrin is somehow involved in the motile machinery and possibly regulates the cell migration.

6.5 Drebrin-Induced Morphological Change of Dendritic Spines

Dendritic spine is an actin-rich structure that changes its shape during its maturation or in response to the synaptic stimuli. Shape change of spines exerts influence on the postsynaptic activity of synapses, because it alters the shape of postsynaptic active zones or distribution of receptors and because the volume of spines affects ionic concentration. Dynamism of actin filaments is thought to be responsible to the shape change of spines. Since drebrin is localized at dendritic spines, it is supposed to have a role in regulation of cytoskeletal organization and further in regulation of morphology and physiological function of spines.

Effect on spine shape by exogenously overexpressed drebrin cDNA was first examined in dissociate culture of cerebral cortex of rat (Hayashi and Shirao 1999). They made dense culture of dissociated neurons from fetal rat cerebral cortex and transfected them with drebrin-GFP fusion construct. Dense culture system was selected instead of low-density culture which is more common primary culture systems, because neurons are surrounded by other neurons and processes, where the environments are more like in vivo. Low efficiency of transfection made it possible to observe the morphology of transfected neurons. When GFP was transfected, fluorescence was seen intensely in cell body and dendritic shafts (Fig. 6.7a, b). In contrast, when GFP-drebrin was transfected, fluorescence were not observed intensely in cell body and dendritic shaft but observed as scattered dots around dendrites (Fig. 6.7c). Close observation revealed that these dots are spine heads (Fig. 6.7d). Using confocal microscope, morphology of spines labeled with GFP-drebrin were precisely analyzed. Measurement of spine length revealed that GFP-drebrin-expressing spines were significantly longer than GFP expressing or DiI labeled spines. Some of these long spines still attached to presynaptic boutons as examined by immunostaining with anti-synaptophysin antibody (Fig. 6.8b). Interestingly, GFP-drebrin that lacks the actin-binding domain did not localize to spines nor induce elongation of spines, indicating that the localization of drebrin

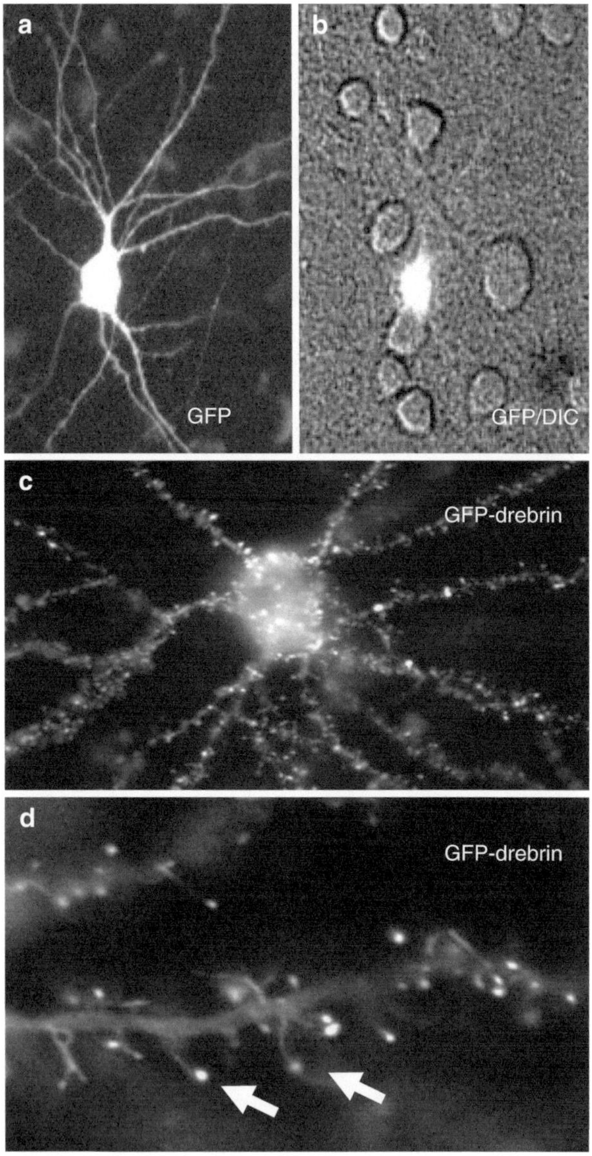

Fig. 6.7 Fluorescent microscopic view of neurons transfected with GFP or GFP-drebrin. (**a**) Dense culture of cortical neurons was transfected with GFP-cDNA. Fluorescence was observed at cell body and dendrites. (**b**) Image of (**a**) is overlapped with DIC image. The transfected neuron is surrounded by many untransfected neurons. (**c**) Neurons were transfected with drebrin cDNA. Fluorescence was weakly observed in cell body and thick shaft of dendrites, while fine dots scattered around the dendritic shafts were seen. (**d**) Close observation of C revealed that the fluorescent dots are tips of dendritic spines

at spines is due to its actin binding and that the elongation is due to the remodeling of actin filaments by drebrin. Similar alteration of spine morphology was reported when drebrin was transfected after 18 DIV, when spines already matured to form globular head, indicating that overexpression of drebrin causes shape change of already existing spines. Time lapse recording of drebrin-overexpressing spines revealed extreme motility that included loss or gain of entire protrusions (Biou et al. 2008).

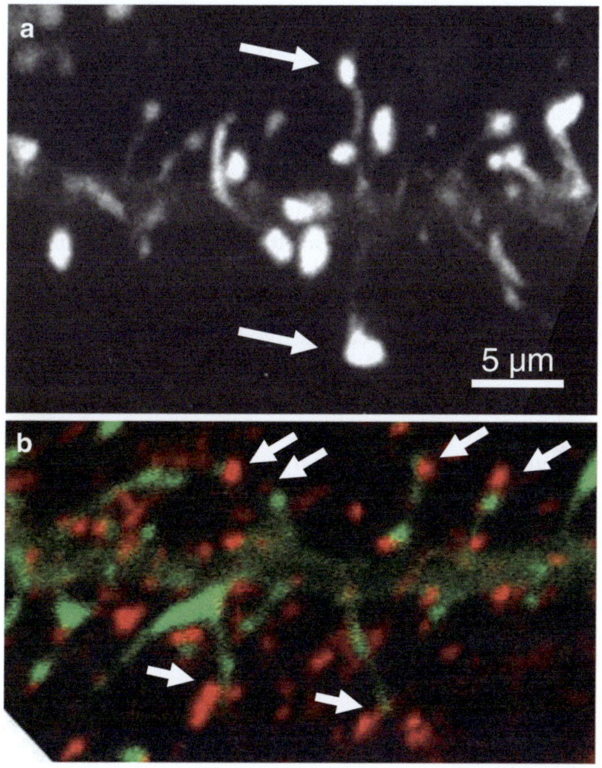

Fig. 6.8 Confocal microscopic view of spines of neurons transfected with GFP-drebrin. (**a**) 10–20 serial sections of 0.8 µm optical sections were projected onto one plane. Concentrated GFP-drebrin at spine heads was clearly observed. (**b**) Another neurons transfected with GFP-drebrin was stained with anti-synaptophysin antibody. GFP-drebrin-labeled spine heads were associated with synaptophysin labeled presynapses (*arrows*)

These observations indicate that drebrin is a regulator of spine shape, but they also led us hypothesize that drebrin induces de novo spines and molecular clustering of postsynaptic density. This possibility was examined by transfecting drebrin into immature neurons. It induced filopodia on neurites and induced accumulation of PSD-95 at its tip (Mizui et al. 2005). Conversely, suppression of drebrin in neurons by antisense oligonucleotides against it attenuated synaptic clustering of PSD-95, as well as clustering of actin filaments (Takahashi et al. 2003).

Regulation of neurotransmitter receptors by drebrin has been also examined. Drebrin depletion reduced levels of NMDA-R and dopamine receptor D1 in dendritic spines of hippocampal neurons (Takahashi et al. 2006; Jung et al. 2015). Overexpression of drebrin A augmented the glutamatergic synaptic transmission, probably through an increase of active synaptic site density. Downregulation of drebrin A led to a decrease of both glutamatergic and GABAergic synaptic activity (Ivanov 2009). These studies indicate that drebrin has activities to induce spines, to regulate their morphology, to assemble postsynaptic proteins, and to modulate synaptic transmission.

Mechanisms for these regulations by drebrin are poorly understood. A hint came from the finding that very similar alteration of spine morphology was obtained by transfecting constitutive active Ras mutants to hippocampal neurons. This alteration is always associated with growth of cell body and dendrites and is thought to be mediated by Ras-PI3K-Akt-mTOR pathway (Kumar et al. 2005). As far as the effect on spine morphology is concerned, the action of constitutive active Ras requires presence of drebrin. Alteration of spine morphology induced by Ras activation disappeared when drebrin shRNA was co-transfected. Alteration of spine morphology by overexpression of drebrin also disappeared when dominant negative Ras was co-expressed, suggesting interaction between these two proteins in shaping spines (Biou et al. 2008). Another hint is the finding of microtubule entry into spines. Microtubule invasion into spines is positively regulated by membrane depolarization and contributes to activity-dependent spine enlargement. Microtubule might deliver molecules or cargos into activated spines, thereby functioning as part of a synaptic tagging mechanism. Invasion of microtubules is guided by actin filaments, where drebrin might act as an important linker between them. In drebrin-overexpressing neurons, microtubule entry into spines occurred frequently while depleting drebrin expression with siRNA resulted in dramatically fewer spine invasions of microtubules (Merriam et al. 2013). Although these two hypotheses are focusing on independent mechanisms for drebrin-induced shape change of spines, they are not mutually exclusive. Because drebrin binds to many kinds of proteins including actin filament, microtubules, homer, etc., it can provide a scaffold for postsynaptic cytoskeleton and may be involved in multiple regulatory events for activity-dependent synaptic functions.

References

Asada H, Uyemura K, Shirao T (1994) Actin-binding protein, drebrin, accumulates in submembranous regions in parallel with neuronal differentiation. J Neurosci Res 38(2):149–159. doi:10.1002/jnr.490380205

Bazellieres E, Massey-Harroche D, Barthelemy-Requin M, Richard F, Arsanto JP, Le Bivic A (2012) Apico-basal elongation requires a drebrin-E-EB3 complex in columnar human epithelial cells. J Cell Sci 125(Pt 4):919–931. doi:10.1242/jcs.092676

Biou V, Brinkhaus H, Malenka RC, Matus A (2008) Interactions between drebrin and Ras regulate dendritic spine plasticity. Eur J Neurosci 27(11):2847–2859. doi:10.1111/j.1460-9568.2008.06269.x

Butkevich E, Hulsmann S, Wenzel D, Shirao T, Duden R, Majoul I (2004) Drebrin is a novel connexin-43 binding partner that links gap junctions to the submembrane cytoskeleton. Curr Biol 14(8):650–658. doi:10.1016/j.cub.2004.03.063

Chew CS, Okamoto CT, Chen X, Thomas R (2005) Drebrin E2 is differentially expressed and phosphorylated in parietal cells in the gastric mucosa. Am J Physiol Gastrointest Liver Physiol 289(2):G320–G331. doi:10.1152/ajpgi.00002.2005

Dun XP, Bandeira de Lima T, Allen J, Geraldo S, Gordon-Weeks P, Chilton JK (2012) Drebrin controls neuronal migration through the formation and alignment of the leading process. Mol Cell Neurosci 49(3):341–350. doi:10.1016/j.mcn.2012.01.006

Geraldo S, Khanzada UK, Parsons M, Chilton JK, Gordon-Weeks PR (2008) Targeting of the F-actin-binding protein drebrin by the microtubule plus-tip protein EB3 is required for neuritogenesis. Nat Cell Biol 10(10):1181–1189. doi:10.1038/ncb1778

Gordon-Alonso M, Rocha-Perugini V, Alvarez S, Ursa A, Izquierdo-Useros N, Martinez-Picado J, Munoz-Fernandez MA, Sanchez-Madrid F (2013) Actin-binding protein drebrin regulates HIV-1-triggered actin polymerization and viral infection. J Biol Chem 288(39):28382–28397. doi:10.1074/jbc.M113.494906

Grintsevich EE, Galkin VE, Orlova A, Ytterberg AJ, Mikati MM, Kudryashov DS, Loo JA, Egelman EH, Reisler E (2010) Mapping of drebrin binding site on F-actin. J Mol Biol 398(4):542–554. doi:10.1016/j.jmb.2010.03.039

Hanamura K, Mizui T, Kakizaki T, Roppongi RT, Yamazaki H, Yanagawa Y, Shirao T (2010) Low accumulation of drebrin at glutamatergic postsynaptic sites on GABAergic neurons. Neuroscience 169(4):1489–1500. doi:10.1016/j.neuroscience.2010.06.043

Hayashi K, Shirao T (1999) Change in the shape of dendritic spines caused by overexpression of drebrin in cultured cortical neurons. J Neurosci 19(10):3918–3925

Hayashi K, Ishikawa R, Ye LH, He XL, Takata K, Kohama K, Shirao T (1996) Modulatory role of drebrin on the cytoskeleton within dendritic spines in the rat cerebral cortex. J Neurosci 16(22):7161–7170

Hayashi K, Ishikawa R, Kawai-Hirai R, Takagi T, Taketomi A, Shirao T (1999) Domain analysis of the actin-binding and actin-remodeling activities of drebrin. Exp Cell Res 253(2):673–680. doi:10.1006/excr.1999.4663

Ikeda K, Shirao T, Toda M, Asada H, Toya S, Uyemura K (1995) Effect of a neuron-specific actin-binding protein, drebrin A, on cell-substratum adhesion. Neurosci Lett 194(3):197–200

Ikeda K, Kaub PA, Asada H, Uyemura K, Toya S, Shirao T (1996) Stabilization of adhesion plaques by the expression of drebrin A in fibroblasts. Brain Res Dev Brain Res 91(2):227–236

Ishikawa R, Hayashi K, Shirao T, Xue Y, Takagi T, Sasaki Y, Kohama K (1994) Drebrin, a development-associated brain protein from rat embryo, causes the dissociation of tropomyosin from actin filaments. J Biol Chem 269(47):29928–29933

Ivanov A, Esclapez M, Pellegrino C, Shirao T, Ferhat L (2009) Drebrin A regulates dendritic spine plasticity and synaptic function in mature cultured hippocampal neurons. J Cell Sci 122(4):524–534

Jung G, Kim EJ, Cicvaric A, Sase S, Gröger M, Höger H, Sialana FJ, Berger J, Monje FJ, Lubec G (2015) Drebrin depletion alters neurotransmitter receptor levels in protein complexes, dendritic spine morphogenesis and memory-related synaptic plasticity in the mouse hippocampus. J Neurochem 134(2):327–339

Keon BH, Jedrzejewski PT, Paul DL, Goodenough DA (2000) Isoform specific expression of the neuronal F-actin binding protein, drebrin, in specialized cells of stomach and kidney epithelia. J Cell Sci 113(Pt 2):325–336

Kobayashi C, Aoki C, Kojima N, Yamazaki H, Shirao T (2007) Drebrin a content correlates with spine head size in the adult mouse cerebral cortex. J Comp Neurol 503(5):618–626. doi:10.1002/cne.21408

Kumar V, Zhang MX, Swank MW, Kunz J, GY W (2005) Regulation of dendritic morphogenesis by Ras-PI3K-Akt-mTOR and Ras-MAPK signaling pathways. J Neurosci 25(49):11288–11299. doi:10.1523/JNEUROSCI.2284-05.2005

Lin Q, Tan HT, Lim TK, Khoo A, Lim KH, Chung MC (2014) iTRAQ analysis of colorectal cancer cell lines suggests Drebrin (DBN1) is overexpressed during liver metastasis. Proteomics 14(11):1434–1443. doi:10.1002/pmic.201300462

Mancini A, Sirabella D, Zhang W, Yamazaki H, Shirao T, Krauss RS (2011) Regulation of myotube formation by the actin-binding factor drebrin. Skelet Muscle 1(1):36. doi:10.1186/2044-5040-1-36

Merriam EB, Millette M, Lumbard DC, Saengsawang W, Fothergill T, Hu X, Ferhat L, Dent EW (2013) Synaptic regulation of microtubule dynamics in dendritic spines by calcium, F-actin, and drebrin. J Neurosci 33(42):16471–16482. doi:10.1523/jneurosci.0661-13.2013

Mizui T, Takahashi H, Sekino Y, Shirao T (2005) Overexpression of drebrin A in immature neurons induces the accumulation of F-actin and PSD-95 into dendritic filopodia, and the formation of large abnormal protrusions. Mol Cell Neurosci 30(4):630–638

Mizui T, Kojima N, Yamazaki H, Katayama M, Hanamura K, Shirao T (2009) Drebrin E is involved in the regulation of axonal growth through actin-myosin interactions. J Neurochem 109(2):611–622. doi:10.1111/j.1471-4159.2009.05993.x

Mizutani Y, Iwamoto I, Kanoh H, Seishima M, Nagata K (2014) Expression of drebrin, an actin binding protein, in basal cell carcinoma, trichoblastoma and trichoepithelioma. Histol Histopathol 29(6):757–766

Peitsch WK, Grund C, Kuhn C, Schnolzer M, Spring H, Schmelz M, Franke WW (1999) Drebrin is a widespread actin-associating protein enriched at junctional plaques, defining a specific microfilament anchorage system in polar epithelial cells. Eur J Cell Biol 78(11):767–778. doi:10.1016/s0171-9335(99)80027-2

Peitsch WK, Hofmann I, Endlich N, Pratzel S, Kuhn C, Spring H, Grone HJ, Kriz W, Franke WW (2003) Cell biological and biochemical characterization of drebrin complexes in mesangial cells and podocytes of renal glomeruli. J Am Soc Nephrol 14(6):1452–1463

Peitsch WK, Hofmann I, Bulkescher J, Hergt M, Spring H, Bleyl U, Goerdt S, Franke WW (2005) Drebrin, an actin-binding, cell-type characteristic protein: induction and localization in epithelial skin tumors and cultured keratinocytes. J Invest Dermatol 125(4):761–774. doi:10.1111/j.0022-202X.2005.23793.x

Peitsch WK, Bulkescher J, Spring H, Hofmann I, Goerdt S, Franke WW (2006) Dynamics of the actin-binding protein drebrin in motile cells and definition of a juxtanuclear drebrin-enriched zone. Exp Cell Res 312(13):2605–2618. doi:10.1016/j.yexcr.2006.04.017

Perez-Martinez M, Gordon-Alonso M, Cabrero JR, Barrero-Villar M, Rey M, Mittelbrunn M, Lamana A, Morlino G, Calabia C, Yamazaki H, Shirao T, Vazquez J, Gonzalez-Amaro R, Veiga E, Sanchez-Madrid F (2010) F-actin-binding protein drebrin regulates CXCR4 recruitment to the immune synapse. J Cell Sci 123(Pt 7):1160–1170. doi:10.1242/jcs.064238

Rehm K, Panzer L, van Vliet V, Genot E, Linder S (2013) Drebrin preserves endothelial integrity by stabilizing nectin at adherens junctions. J Cell Sci 126(Pt 16):3756–3769. doi:10.1242/jcs.129437

Sasaki Y, Hayashi K, Shirao T, Ishikawa R, Kohama K (1996) Inhibition by drebrin of the actin-bundling activity of brain fascin, a protein localized in filopodia of growth cones. J Neurochem 66(3):980–988

Shirao T, Obata K (1986) Immunochemical homology of 3 developmentally regulated brain proteins and their developmental change in neuronal distribution. Brain Res 394(2):233–244

Shirao T, Kojima N, Nabeta Y, Obata K (1989) Two forms of drebrins, developmentally regulated brain proteins, in rat. Proc Jpn Acad Ser B 65(7):169–172

Shirao T, Kojima N, Obata K (1992) Cloning of drebrin A and induction of neurite-like processes in drebrin-transfected cells. Neuroreport 3(1):109–112

Shirao T, Hayashi K, Ishikawa R, Isa K, Asada H, Ikeda K, Uyemura K (1994) Formation of thick, curving bundles of actin by drebrin A expressed in fibroblasts. Exp Cell Res 215(1):145–153. doi:10.1006/excr.1994.1326

Sonego M, Oberoi M, Stoddart J, Gajendra S, Hendricusdottir R, Oozeer F, Worth DC, Hobbs C, Eickholt BJ, Gordon-Weeks PR, Doherty P, Lalli G (2015) Drebrin regulates neuroblast migration in the postnatal mammalian brain. PLoS One 10(5):e0126478. doi:10.1371/journal.pone.0126478

Song M, Kojima N, Hanamura K, Sekino Y, Inoue HK, Mikuni M, Shirao T (2008) Expression of drebrin E in migrating neuroblasts in adult rat brain: coincidence between drebrin E disappearance from cell body and cessation of migration. Neuroscience 152(3):670–682. doi:10.1016/j.neuroscience.2007.10.068

Takahashi H, Sekino Y, Tanaka S, Mizui T, Kishi S, Shirao T (2003) Drebrin-dependent actin clustering in dendritic filopodia governs synaptic targeting of postsynaptic density-95 and dendritic spine morphogenesis. J Neurosci 23(16):6586–6595

Takahashi H, Mizui T, Shirao T (2006) Down-regulation of drebrin A expression suppresses synaptic targeting of NMDA receptors in developing hippocampal neurones. J Neurochem 97:110–115

Tanabe K, Yamazaki H, Inaguma Y, Asada A, Kimura T, Takahashi J, Taoka M, Ohshima T, Furuichi T, Isobe T, Nagata K, Shirao T, Hisanaga S (2014) Phosphorylation of drebrin by cyclin-dependent kinase 5 and its role in neuronal migration. PLoS One 9(3):e92291. doi:10.1371/journal.pone.0092291

Terakawa Y, Agnihotri S, Golbourn B, Nadi M, Sabha N, Smith CA, Croul SE, Rutka JT (2013) The role of drebrin in glioma migration and invasion. Exp Cell Res 319(4):517–528. doi:10.1016/j.yexcr.2012.11.008

Toda M, Shirao T, Uyemura K (1999) Suppression of an actin-binding protein, drebrin, by antisense transfection attenuates neurite outgrowth in neuroblastoma B104 cells. Brain Res Dev Brain Res 114(2):193–200

Vacca B, Bazellieres E, Nouar R, Harada A, Massey-Harroche D, Le Bivic A (2014) Drebrin E depletion in human intestinal epithelial cells mimics Rab8a loss of function. Hum Mol Genet 23(11):2834–2846. doi:10.1093/hmg/ddt670

Worth DC, Daly CN, Geraldo S, Oozeer F, Gordon-Weeks PR (2013) Drebrin contains a cryptic F-actin-bundling activity regulated by Cdk5 phosphorylation. J Cell Biol 202(5):793–806. doi:10.1083/jcb.201303005

Xu W, Stamnes M (2006) The actin-depolymerizing factor homology and charged/helical domains of drebrin and mAbp1 direct membrane binding and localization via distinct interactions with actin. J Biol Chem 281(17):11826–11833. doi:10.1074/jbc.M510141200

Yamada M, Kurihara H, Kinoshita K, Sakai T (2005) Temporal expression of alpha-smooth muscle actin and drebrin in septal interstitial cells during alveolar maturation. J Histochem Cytochem 53(6):735–744. doi:10.1369/jhc.4A6483.2005

Part III
Drebrin in Nervous System

Chapter 7
Localization of Drebrin: Light Microscopy Study

Tomoaki Shirao, Noriko Koganezawa, Hiroyuki Yamazaki, Kenji Hanamura, and Kazuyuki Imamura

Abstract Developmental changes in the expression and localization of drebrin has been mainly analyzed in chick embryo and young rat by various anti-drebrin polyclonal and monoclonal antibodies. Immunoblot analysis demonstrated that the adult drebrin isoform (drebrin A) is restricted to neural tissues, while the embryonic drebrin isoforms (drebrin E1 and E2 in chicken and drebrin E in mammals) are found in a wide variety of tissues. In the developing brain, drebrin E (including chicken drebrin E2) is expressed in newly generated neurons. During neuronal migration, drebrin E is distributed ubiquitously within the neurons. Once drebrin A is expressed in the developing neuron, drebrin E is no longer present within the cell soma and accumulates in the growth cone of growing processes, resulting in the cessation of neuronal migration. The limited subcellular localization of drebrin A, which is possibly regulated by a drebrin A-specific mechanism, is likely to affect the localization of drebrin E. In the adult brain, drebrin is mainly localized in dendritic spines, but in some nuclei, drebrin can be detected in neuronal somata as well as dendritic spines. The fact that the developmental changes in drebrin expression highly correlate in time with the sensitive period of visual cortical plasticity in kittens suggests that synaptic plasticity depends on drebrin.

Keywords Immunohistochemistry • Monoclonal antibody • Neuronal migration Axonal growth cone • Drebrin A-specific antibody • Image subtraction method Isoform conversion • Optic tectum • Cerebellum • Visual cortical plasticity

---

T. Shirao (✉) • N. Koganezawa • H. Yamazaki • K. Hanamura
Department of Neurobiology and Behavior, Gunma University Graduate School of Medicine, Maebashi, Gunma 371-8511, Japan
e-mail: tshirao@gunma-u.ac.jp

K. Imamura
Department of Systems Life Engineering, Maebashi Institute of Technology, Maebashi, Japan

T. Shirao, Y. Sekino (eds.), *Drebrin*, Advances in Experimental Medicine and Biology 1006, DOI 10.1007/978-4-431-56550-5_7

7.1 Introduction

Immunological analysis combined with biochemical and anatomical techniques can be used to elucidate the physiological role of molecules of interest. Therefore, soon after the drebrin discovery (Shirao and Obata 1985), we produced various anti-drebrin antibodies, demonstrated the relationships between drebrin isoforms, and visualized their cellular and subcellular distribution. In this chapter, we review the production and characterization of drebrin antibodies and then the analysis of drebrin isoforms by immunoblotting and immunohistochemistry using anti-drebrin antibodies. The immunoelectron microscopy of drebrin will be reviewed in detail in the next chapter.

7.2 Production and Characterization of Drebrin Antibodies

Partially purified drebrin E1 and drebrin E2 from the optic tectum of embryonic day (ED) 11 chick brains were subjected to sodium dodecyl sulfate-polyacrylamide gel electrophoresis, and after the gel was stained with Coomassie Blue, about 0.1 mg of drebrin E1 or drebrin E2 was eluted from each band. Then, using either of the purified isoforms, we raised two independent rabbit polyclonal antibodies. Interestingly, each of the resultant polyclonal antibodies recognized both drebrin E1 and E2 by immunoblot analysis. Moreover, they recognized drebrin A. This indicates that the protein structures of drebrin E1, E2, and A are similar, which is consistent with the similar peptide-mapping pattern of each purified protein (Fig. 7.1). As expected, immunohistochemistry using these polyclonal antibodies demonstrated an indistinguishable staining pattern, which is columnar staining of the tectobulbar fiber layer as well as the marginal zone of ED7 chick optic tectum. On the other hand, the ventricular zone was hardly stained (Shirao and Obata 1986).

To obtain a specific antibody against each drebrin, we tried to produce monoclonal antibodies. We immunized a female BALB/c mouse with 0.1 mg of purified drebrin E2 for 6 months. After the intravenous booster injection of 0.1 mg drebrin E2, splenocytes were fused with X63-Ag8-653 myeloma cells. The positive clones were first selected immunohistochemically using the aforementioned unique drebrin immunostaining pattern of ED7 tectum. Then, we confirmed their specificity using immunoblot analysis of extracts from ED11 optic tectum. Consequently, we produced five independent monoclonal antibodies (mAbs) M2A6, M2D8, M2E7, M2F6, and M2H1(Shirao and Obata 1986). Unfortunately, we could not raise any isoform-specific antibodies, and each mAb recognized all of the drebrin isoforms (E1, E2, and A). Three of them including mAb M2F6 can recognize mammalian drebrin in addition to chicken drebrin. mAb M2F6 recognizes drebrin in paraffin and frozen sections of formalin-fixed tissues and can be used for immunoelectron microscopy as well as immunohistochemistry. Thus, mAb M2F6 has been used as the standard antibody in the drebrin studies. mAb M2F6 is now commercially available from many vendors.

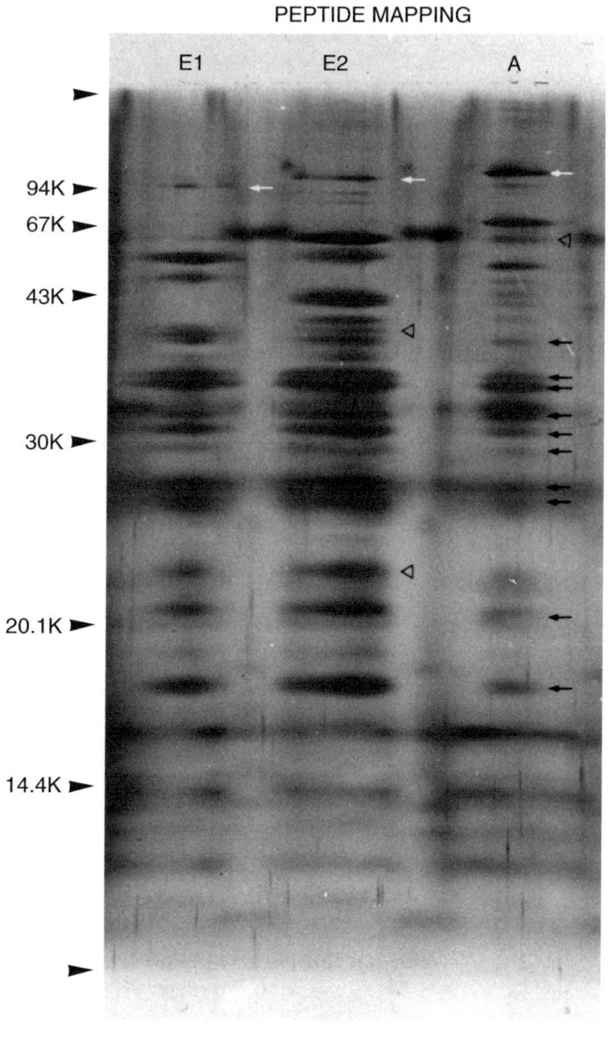

Fig. 7.1 Peptide mapping of chicken drebrin E1, E2, and A. Each protein was treated with *S. aureus* V8 protease, and the peptides were separated by SDS-PAGE. Intact proteins, *white arrows*; common peptides, *black arrows*; distinct peptides. *open triangles*. Silver stain

One of the characteristic features of mAb M2F6 is that it hardly recognizes degraded drebrin fragments, which can be recognized by other polyclonal and monoclonal antibodies. This suggests that the epitope is near the cleavage sites of a proteinase. In 1991, Arai and his colleagues reported that drebrin degradation is calcium dependent (Arai et al. 1991). Furthermore, Chimura and his colleagues have recently reported that calpain cleaves drebrin near the mAb M2F6 epitope (Chimura et al. 2015). We have elucidated that the mAb M2F6 epitope is a conformational epitope composed of discontinuous sections of the C-terminal amino acid sequence (AA 580–707 in rat drebrin A) that includes both of the Homer-binding motifs (Fig. 7.2).

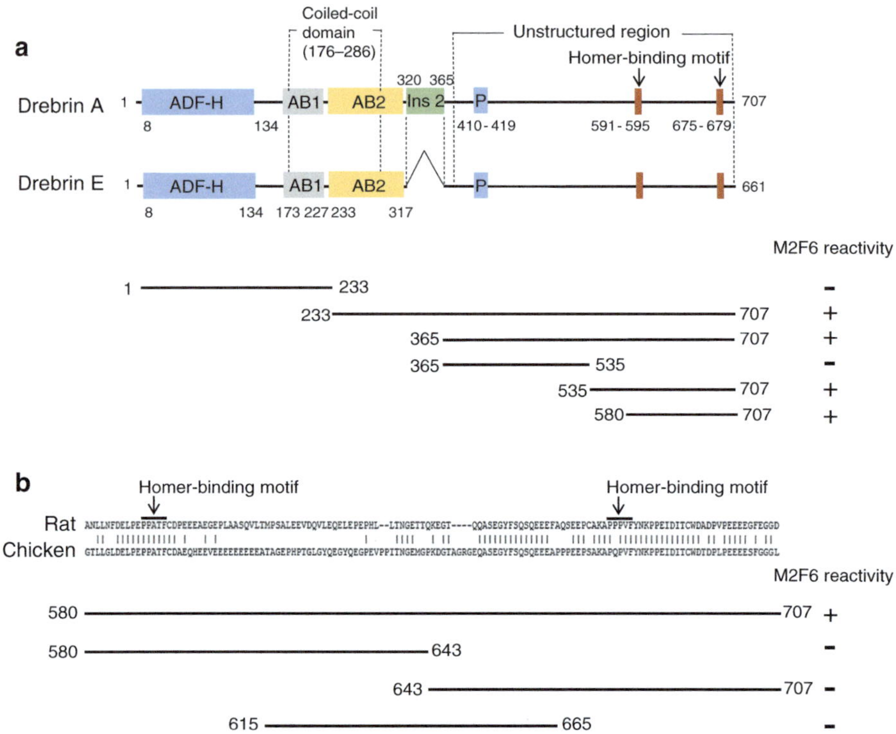

Fig. 7.2 Epitope mapping of monoclonal antibody M2F6. (**a**) Diagram of full-length drebrin and putative M2F6 epitope. The drebrin fragments were expressed in HEK293 cells and analyzed with M2F6 by Western blotting. (**b**) Comparison between rat and chicken in M2F6 recognition region. The conserved homologous region is predicted to be M2F6 epitope; however, these three fragments (580–643, 643–707, and 615–665) did not react to M2F6. It suggests that M2F6 recognizes discontinuous sequence of drebrin

7.3 Drebrin Expression During Development

Immunoblot analysis demonstrated that drebrins are highly expressed in various tissues in ED11 chick embryo. Drebrin is present in the skeletal and intestinal muscles but absent from the liver. On the other hand, high drebrin expression in the adult is mostly limited to neural tissues (Shirao and Obata 1986). A ribonuclease protection assay using the antisense cRNA probe of a drebrin A cDNA fragment showed three protected bands corresponding to drebrin E1, E2, and A in the chicken brain RNA. Using the ribonuclease assay, we showed that drebrin E1 and E2 mRNAs were detected in the whole body of the 5-day chick embryo, and drebrin E2 mRNA was detected in the kidney of postnatal day-1 chick (Kojima et al. 1993). Together, the results indicate that drebrin A is restricted to neural tissues, while drebrin E1 and E2 are found in a wide variety of tissues.

As shown below, most of the developmental studies about drebrin localization have been done using chicken brain. Chickens have two embryonic isoforms, drebrin E1 and E2, but mammals have only a single embryonic isoform, drebrin E. As discussed in Chap. 2, mammalian drebrin E is comparable to chicken drebrin E2. In addition, in situ hybridization analysis of the chick cerebellum showed that the drebrin E2 distribution included the drebrin E1 distribution. Thus, the immunostaining of chick embryos using mAb M2F6 can be considered to show the distribution of drebrin E2. The cellular localization of drebrin E2 in chickens was similar to that of drebrin E in rats.

We immunohistochemically stained frozen sections of chick embryos at various developmental stages using mAb M2F6. Drebrin first appeared on ED2 in the myotome. On ED11, drebrin was observed in the longitudinal muscular layer of the intestine but not in the intestinal epithelium. Pectoral muscles and mesonephros were also immunostained (Shirao and Obata 1986).

In the brain, we first examined drebrin localization in the optic tectum, not only because drebrin was first identified in the optic tectum but also because the morphological development of the optic tectum had been well studied (LaVail and Cowan 1971a; LaVail and Cowan 1971b). In the ED4 optic tectum, drebrin was observed in the neuronal somata and processes in the superficial layer just under the pia mater (Fig. 7.3a, b). On ED7, mAb M2F6 showed a characteristic staining pattern similar to the aforementioned polyclonal antibodies, namely, growing cell processes in the tectobulbar fiber layer and the marginal zone. Undifferentiated cells in the ventricular zone do not express drebrin (Fig. 7.3c). On ED11 drebrin is widely distributed in the optic tectum, except for the ventricular zone. Interestingly, the drebrin content in the tectobulbar fiber layer was decreased, but in the stratum opticum, which contains a lot of growing retinal fibers, it was increased (Fig. 7.3d). On ED16, a high drebrin content was observed in the superficial layer of the stratum griseum et fibrosum superficiale. On that day, the retinal fibers of the stratum opticum change their direction and grow into and terminate in this layer. On the other hand, the drebrin content was decreased in the stratum opticum as well as the tectobulbar layer (Fig. 7.3e) (Shirao and Obata 1986). These observations raised the hypothesis that drebrin is ubiquitously distributed within newly generated neurons but later highly accumulates in the growth cone of cell processes (Shirao et al. 2017).

Next, we examined the drebrin localization in the cerebellar cortex, because the cerebellar cortex consists of four well-organized layers, namely, the external granular layer (EGL), the molecular layer, the Purkinje cell layer, and the internal granular layer (IGL). In ED16 chick cerebellum, newly generated granule cells migrate from the EGL to the IGL through the molecular layer and grow their dendrites in the IGL (Altman 1972). In the EGL, drebrin was less expressed in the proliferative zone than in the premigratory zone. Interestingly, the intracellular distribution of drebrin in the IGL was different from that in the EGL. In the IGL, drebrin was observed in cell processes, but hardly in cell somata, although drebrin was ubiquitously observed within the cell in the EGL (Shirao et al. 1990; Shirao and Obata 1986). A similar subcellular localization change in drebrin was observed in postnatal

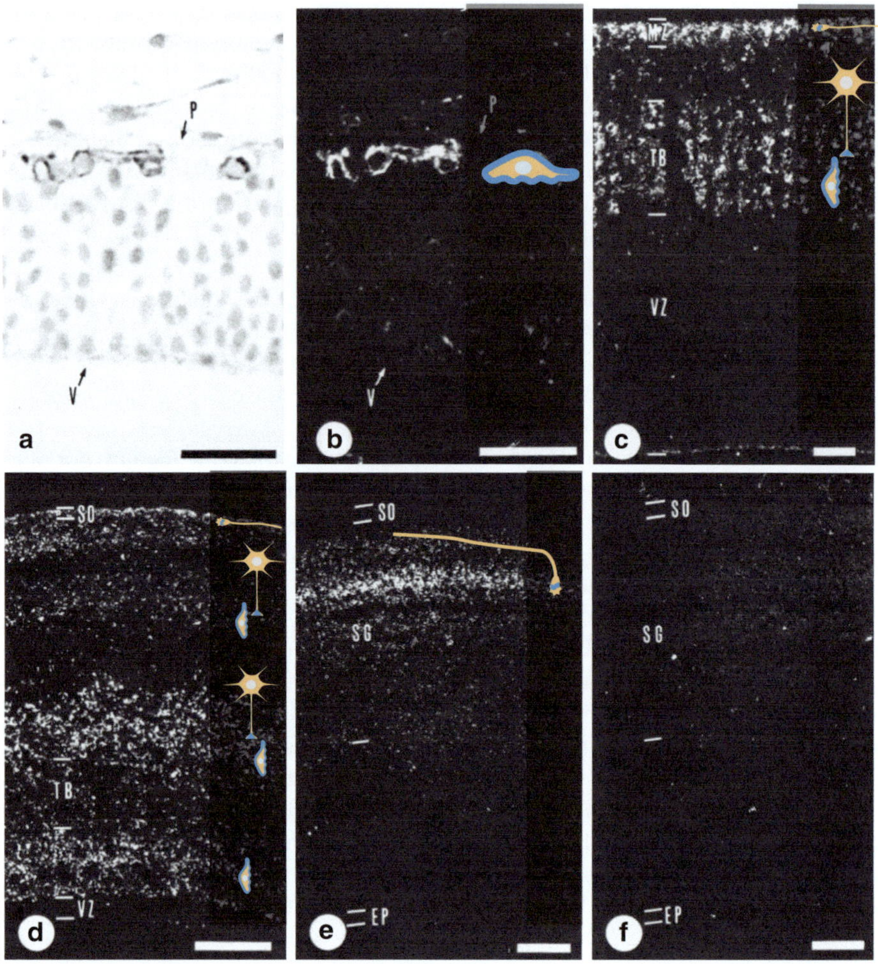

Fig. 7.3 Layer-specific drebrin distribution in chick optic tectum at various developmental stages. (**a**)–(**e**) were immunostained with anti-pan-drebrin antibody (mAb M2F6). (**a**) was counterstained with hematoxylin. (**f**) is a control section immunostained with myeloma culture supernatant. (**a**, **b**) are ED4; (**c**) is ED7; (**d**) is ED11; and (**e**, **f**) are ED16. Intense immunostaining with anti-pan-drebrin antibody is in the layers which contains newly generated migrating neurons or axonal growth cones. Note that neural epithelium in VZ are hardly immunostained. Schematic neurons are drawn in (**b**–**e**). Representative positions of migrating neurons, immature neurons, and axonal growth cones are indicated. High drebrin-content regions are shown in *cyan blue*. *P* pial surface, *V* ventral surface, *MZ* the marginal zone, *VZ* the ventricular zone, *SO* the stratum opticum, *TB* tecto-bulbar fiber layer, *SG* the stratum griseum et fibrosum superficiale. Scale bars are 20 μm in (**a**–**c**) and 100 μm in (**d**–f). Adapted from Shirao and Obata (1986) with permission

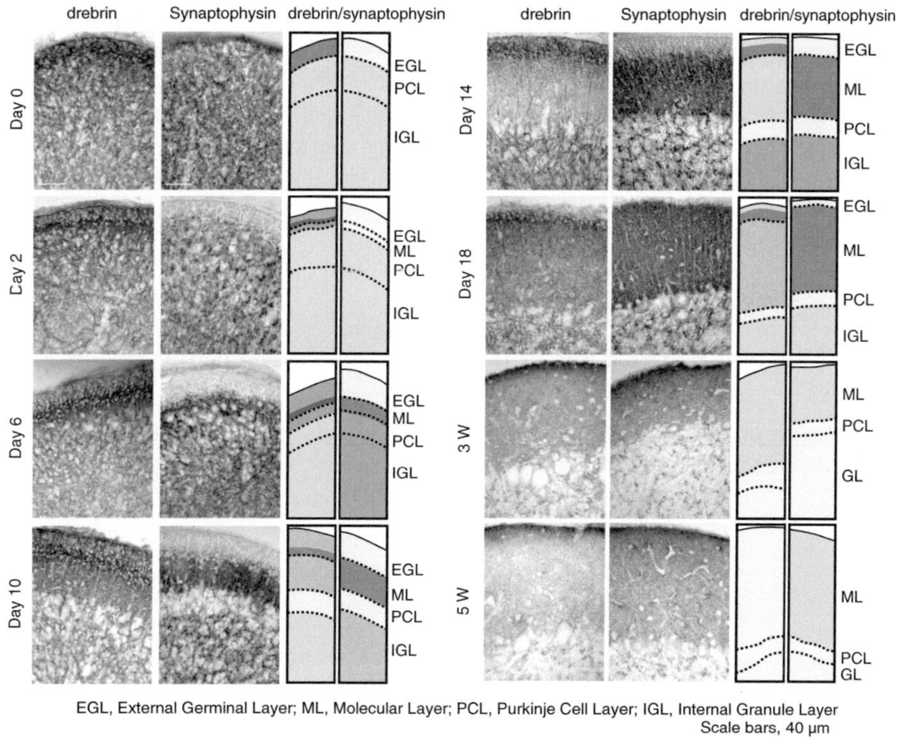

Fig. 7.4 Expression pattern of drebrin and synaptophysin during development in rat cerebellum. Anti-pan-drebrin antibody (mAb M2F6) or anti-synaptophysin antibody (mAb 171B5) was used as a first antibody, and immunoreaction was visualized with diaminobenzidine. Schematic drawings indicate the strength of immunoreactivity (*left*, drebrin; *right*, synaptophysin; *darker gray* indicates stronger reaction). *EGL* external germinal layer, *ML* molecular layer, *PCL* Purkinje cell layer, *IGL* internal granule layer; scale bars, 40 μm

rat cerebellum (Fig. 7.4). These observations support the above hypothesis proposed by the optic tectum study.

Later, the localization of drebrin E in the growth cone was shown in PC12 cells (Sasaki et al. 1996), and its physiological role in the axonal and dendritic growth cones was elucidated using cultured hippocampal neurons (Mizui et al. 2009). In the hippocampal neurons, drebrin E specifically bound to the actin arc between the lamellipodia and the microtubule-rich regions within the growth cones. Aggregation cell cultures of cerebellar granule cells demonstrated that drebrin was localized in the migrating neurons along radially outgrowing neurites. After they changed their direction perpendicular to the neurites, drebrin disappeared from the cell somata and localized on the leading trail. During adult neurogenesis following rostral migration in the rat brain, similar subcellular localization changes in drebrin were observed (Song et al. 2008).

Taken together, the results indicate that drebrin E (including chicken drebrin E2) is expressed in newly generated neurons. During neuronal migration, drebrin E is distributed in the cell somata and the leading and trailing processes within the neurons but then accumulates in the growth cone of growing processes after the neurons stop migrating.

7.4 Drebrin Expression in the Adult Nervous Tissue

In the adult chicken optic tectum, drebrin was widely distributed, except for the ependyma and the layer of the retinal fibers, namely, the stratum opticum (SO) (Fig. 7.5). The dot-like staining pattern enriched in the neuropil suggests that

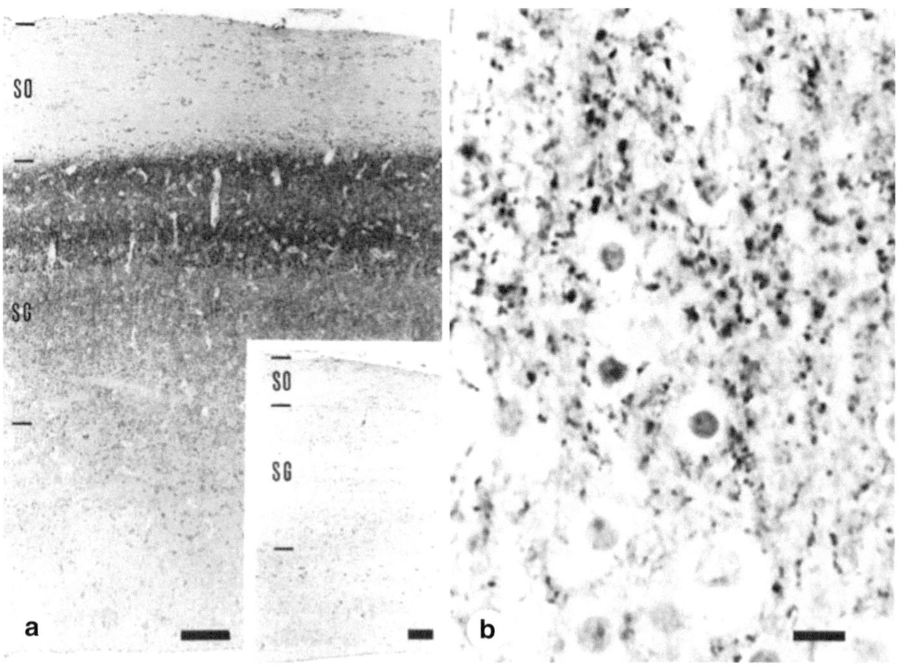

Fig. 7.5 Drebrin distribution in neuropil region of adult optic tectum. The sagittal sections of adult chicken optic tectum were immunostained with anti-pan-drebrin antibody (mAb M2F6) and counterstained with hematoxylin. An intense immunostaining is visible in the stratum griseum et fibrosum superficiale (SG). Drebrin is not observed in mature fiber tracts such as the stratum opticum (SO). The inset is a control. (**b**) High magnification of the SG. Immunoreactive dots were not observed in the cytoplasm of cell somata. Scale bars are 100 μm in (**a**) and 5 μm in (**b**). Adapted from Shirao and Obata (1986) with permission

drebrin is associated with synapses (Shirao and Obata 1986). Immunoelectron microscopy using mAb M2F6 clarified that drebrin is specifically enriched in dendritic spines (Shirao et al. 1987).

In the cerebellar cortex, drebrin was intensely expressed in the granular layer. This layer consists of granule cell somata and so-called cerebellar glomeruli, where many granule cell dendrites make synapses on a single mossy fiber terminal, and drebrin is specifically localized in cerebellar glomeruli, but not in cell somata. Interestingly, drebrin was hardly distributed in the molecular layer where synapses between parallel fibers (granule cell axons) and Purkinje cell dendrites are densely packed. This suggests that drebrin is localized at some synapses, but not all of them (Shirao et al. 1987; Shirao and Obata 1986).

Drebrin is not detected in most neuronal somata of the adult brain. However, drebrin was detected in neuronal somata as well as synapses in some nuclei including pars parvocellularis of the subtectal nucleus isthmi and deep cerebellar nuclei. Moreover, in the dorsal root ganglion (DRG) of the peripheral nervous system, drebrin was predominantly localized in the cytosol of the cell somata (Shirao and Obata 1986). In addition, the expression level of drebrin varied among the neurons in the DRG. It has not been clarified yet how the drebrin distribution in the cell somata is regulated, although it is known that the drebrin E isoform is distributed in the cell somata of migrating neurons as discussed above.

7.5 Drebrin and Visuocortical Plasticity

During the sensitive period of early postnatal development, the functional architecture of the visual cortex is highly susceptible to visual experience. The sensitivity of the visual cortex to 1 week of monocular deprivation is highest around postnatal week 4 (PW4), and after PW13, it is lower and barely detected around PW16. Since the discovery of this plasticity in the cat visual cortex, lots of studies have searched the factors restricting the phase of plasticity to a few months of early postnatal development (Daw et al. 1992). In 1992, we reported that the developmental changes in drebrin expression highly correlated in time with the sensitive period of visual cortical plasticity in kittens (Imamura et al. 1992). The expression level of drebrin A was highest around PW4. Drebrin A expression decreased in the visual cortex of PW13 kittens (Fig. 7.6) and was barely detectable at PW21 and in the adult. On the other hand, the localization of the presynaptic marker protein, synaptophysin, did not change after PW5, indicating that the synapse number remains rather constant. Therefore, we proposed that synaptic plasticity depends on drebrin and the disappearance of drebrin is closely associated with the decrease in the sensitivity of the visual cortex. This explanation was further supported by the finding that in brain regions sustaining synaptic plasticity even in the adult, such as the hippocampus, drebrin A expression was comparable between juvenile and adult cats. This suggests that drebrin is involved in the mechanism of cortical plasticity.

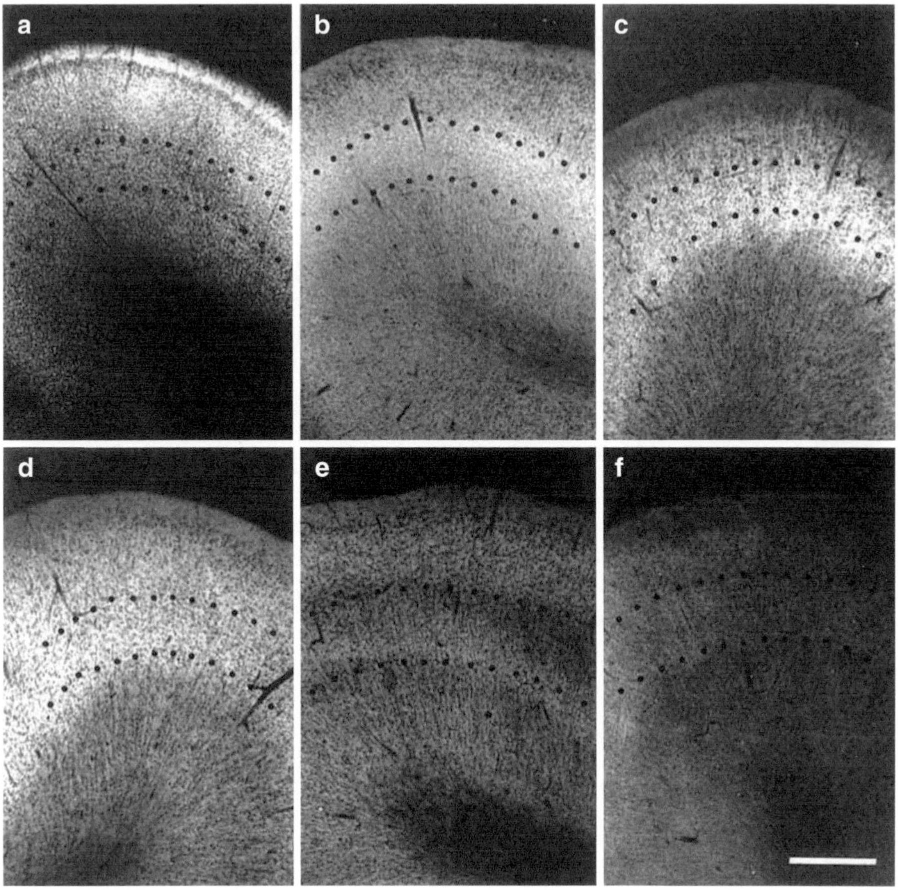

Fig. 7.6 Developmental changes of drebrin distribution in the kitten visual cortex (medial bank of the postlateral gyrus, area 17). Coronal section was obtained from kittens aged 1 week (**a**), 3 weeks (**b**), 5 weeks (**c**), 7 weeks (**d**), 13 weeks (**e**), and 21 weeks (**f**), respectively, and immunostained with anti-pan-drebrin antibody (mAb M2F6). *Dolled lines* in each panel indicate the border of layer IV. Scale bar is 500 μm. Adapted from Imamura et al. (1992) with permission

7.6 Production of a Drebrin A-Specific Antibody

The lack of an isoform-specific antibody compatible with immunohistochemistry made it difficult to study the physiological role of drebrin isoform conversion in detail. In 1992, we started to produce a drebrin A-specific antibody using drebrin A-specific amino acid sequences deduced from the rat Ins2 nucleotide sequence (Shirao et al. 1994). In the first trial, we immunized rabbits with a mixture of three peptides: Gly-Arg-Pro-Tyr-Cys-Pro-Phe-Ile-Lys-Ala-Ser-Asp-Ser-Gly-Pro-Ser-Ser (residues 319–335), Ser-Pro-Pro-Arg-Thr-Pro-Phe-Pro-Tyr-Ile-Thr-Cys (residues 342–353), and His-Arg-Thr-Pro-Asn-Leu-Ser-Ser-Ser-Leu (residues 354–363) of rat drebrin A (Shirao et al. 1992). These sequences are completely conserved between mouse, rat, and human. The resultant antibody specific to drebrin A (DAS1) could be used for immunoblot analysis. Using DAS1, we showed that the Chinese hamster ovary (CHO) cell line expresses drebrin E but not drebrin A. Unfortunately, however, DAS1 is not compatible with immunohistochemistry.

Therefore, we then immunized rabbits with peptide Phe-Ile-Lys-Ala-Ser-Asp-Ser-Gly-Pro-Ser-Ser-Ser (residues 325–336). The resultant antibody (DAS2) is compatible with both immunoblot analysis and immunohistochemistry. DAS2 is now commercially available. Immunoblotting showed that drebrin A emerges at postnatal day 6 in the cerebral cortex. Light microscopy using DAS2 showed that drebrin A is highly concentrated in the synaptic layers of the hippocampus and cortex (Aoki et al. 2005). Immunoblot analysis after biochemical fractionation indicated that most of the drebrin A binds to F-actin, but more than half of drebrin E does not bind to F-actin in vivo (Aoki et al. 2005). Electron microscopy analysis elucidated that drebrin A localizes in dendritic spines, which is reviewed in more detail in the next chapter.

7.7 Isoform-Dependent and Isoform-Independent Distribution of Drebrin Within a Neuron

As we mentioned previously, immunoblot analysis showed that the major isoform of drebrin changes from drebrin E to drebrin A in the brain during development. In parallel with the isoform conversion, the drebrin distribution became limited to the dendritic spines. Using mAb M2F6, we showed that in the early developmental stages before the isoform conversion, drebrin is distributed throughout the entire cell body and processes within the cell (Shirao and Obata 1986). On the other hand, in the adult neurons, we clarified using DAS2 that drebrin A is localized in dendritic spines within the neuron.

Double immunostaining with M2F6 and DAS2 antibodies demonstrated that the ratio of drebrin A to total drebrin varied between neighboring dendritic spines of cultured neurons, suggesting the balance between drebrin E and A varies among spines. In addition, we have recently found that the drebrin dynamics in the dendritic spine are different between drebrin E and A (Hanamura et al. 2017). A drebrin A-specific mechanism may regulate the accumulation of drebrin in dendritic spines. Furthermore, the drebrin A-specific mechanism may depend on the insertion of the Ins2 sequence in the middle of the drebrin molecule, because drebrin E, which does not have the Ins2 insertion, tends to be ubiquitously distributed within the cell.

On the other hand, in immature neurons, exogenously expressed GFP-drebrin A can accumulate in the growth cone as well as endogenous drebrin E (Fig. 7.7), suggesting that the drebrin distribution does not entirely depend on the drebrin isoform. Thus, it is possible that the spine localization of drebrin depends on a mature neuron-specific mechanism.

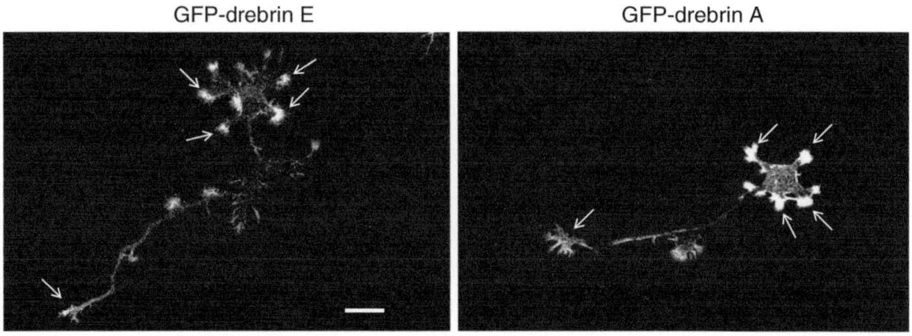

Fig. 7.7 Expression of exogenous drebrin in the growth cone. DIV 1 primary cultured hippocampal neurons expressing GFP-drebrin E (*left*) and GFP-drebrin A (*right*). Note that not only GFP-drebrin E but also GFP drebrin A are accumulated at growth cones (*arrows*). Scale bar, 20 μm

A drebrin E-specific antibody may be useful for settling the above issue, but unfortunately production of a drebrin E-specific antibody has not been achieved. Therefore, instead of a drebrin-E specific antibody, we used an image subtraction method with mAb M2F6 and DAS2 to identify the subcellular distribution of drebrin E. We acquired images of mAb M2F6 and DAS2 staining separately and subtracted the image of DAS2 from that of mAb M2F6 to visualize the drebrin E-specific signal. Using this method, we analyzed the drebrin expression in the rostral migratory stream from the lateral ventricle to the olfactory bulb in the adult rat brain and found that drebrin E, but not drebrin A, was distributed within the cell soma of migrating neurons (Song et al. 2008). Once drebrin A is expressed in a developing neuron, drebrin E is no more present within the cell soma. This suggests that the limited subcellular localization of drebrin A, which is possibly regulated by a drebrin A-specific mechanism, affects the localization of drebrin E.

As mentioned above, most of drebrin A binds to F-actin, but more than half of drebrin E does not bind to F-actin in vivo. Although there is no difference in the binding affinity to F-actin between drebrin E and drebrin A in vitro, drebrin E seems to have a lower preference for F-actin than drebrin A in vivo. It is known that drebrin binding to F-actin induces structural changes in F-actin. The resultant characteristic F-actin, with a long crossover that is normally infrequent, is more preferable for both drebrin isoforms (Sharma et al. 2012, 2011). Therefore, we speculate that the appearance of drebrin A-decorated F-actin can facilitate the binding of drebrin E to F-actin, resulting in the disappearance of drebrin E from the cell soma and the simultaneous cessation of cell migration.

Taken together, the results suggest that drebrin E is highly expressed in newly generated neurons and is distributed in the cell soma in addition to the actin arc in the growth cone within the cell. When drebrin A is expressed in later developmental stages, drebrin A-decorated F-actin localizes in limited subcellular regions, possibly via a drebrin A-specific mechanism, and induces the disappearance of drebrin E from the cell soma, because of the preference of drebrin E to drebrin A-decorated

F-actin. In parallel to further maturation, drebrin A highly accumulates in dendritic spines by a combination of a drebrin A-specific mechanism and a mature neuron-specific mechanism.

References

Altman J (1972) Postnatal development of the cerebellar cortex in the rat. I. The external germinal layer and the transitional molecular layer. J Comp Neurol 145:353–397

Aoki C, Sekino Y, Hanamura K, Fujisawa S, Mahadomrongkul V, Ren Y, Shirao T (2005) Drebrin A is a postsynaptic protein that localizes in vivo to the submembranous surface of dendritic sites forming excitatory synapses. J Comp Neurol 483:383–402

Arai H, Sato K, Uto A, Yasumoto Y (1991) Effect of transient cerebral ischemia in mongolian gerbils on synaptic vesicle protein (SVP-38) and developmentally regulated brain protein (drebrin). Neurosci Res Commun 9:143–150

Chimura T, Launey T, Yoshida N (2015) Calpain-mediated degradation of drebrin by excitotoxicity in vitro and in vivo. PLoS One 10:e0125119

Daw NW, Fox K, Sato H, Czepita D (1992) Critical period for monocular deprivation in the cat visual cortex. J Neurophysiol 67:197–202

Hanamura K, Kamata Y, Yamazaki H, Kojima N, Shirao T (2017) Isoform-dependent regulation of drebrin dynamics in dendritic spines. Neuroscience

Imamura K, Shirao T, Mori K, Obata K (1992) Changes of drebrin expression in the visual cortex of the cat during development. Neurosci Res 13:33–41

Kojima N, Shirao T, Obata K (1993) Molecular cloning of a developmentally regulated brain protein, chicken drebrin A and its expression by alternative splicing of the drebrin gene. Brain Res Mol Brain Res 19:101–114

LaVail JH, Cowan WM (1971a) The development of the chick optic tectum. I. Normal morphology and cytoarchitectonic development. Brain Res 28:391–419

LaVail JH, Cowan WM (1971b) The development of the chick optic tectum. II. Autoradiographic studies. Brain Res 28:421–441

Mizui T, Kojima N, Yamazaki H, Katayama M, Hanamura K, Shirao T (2009) Drebrin E is involved in the regulation of axonal growth through actin-myosin interactions. J Neurochem 109:611–622

Sasaki Y, Hayashi K, Shirao T, Ishikawa R, Kohama K (1996) Inhibition by drebrin of the actin-bundling activity of brain fascin, a protein localized in filopodia of growth cones. J Neurochem 66:980–988

Sharma S, Grintsevich EE, Phillips ML, Reisler E, Gimzewski JK (2011) Atomic force microscopy reveals drebrin induced remodeling of f-actin with subnanometer resolution. Nano Lett 11:825–827

Sharma S, Grintsevich EE, Hsueh C, Reisler E, Gimzewski JK (2012) Molecular cooperativity of drebrin1-300 binding and structural remodeling of F-actin. Biophys J 103:275–283

Shirao T, Obata K (1985) Two acidic proteins associated with brain development in chick embryo. J Neurochem 44:1210–1216

Shirao T, Obata K (1986) Immunochemical homology of 3 developmentally regulated brain proteins and their developmental change in neuronal distribution. Brain Res 394:233–244

Shirao T, Inoue HK, Kano Y, Obata K (1987) Localization of a developmentally regulated neuron-specific protein S54 in dendrites as revealed by immunoelectron microscopy. Brain Res 413:374–378

Shirao T, Kojima N, Terada S, Obata K (1990) Expression of three drebrin isoforms in the developing nervous system. Neurosci Res Suppl 13:S106–S111

Shirao T, Kojima N, Obata K (1992) Cloning of drebrin A and induction of neurite-like processes in drebrin-transfected cells. Neuroreport 3:109–112

Shirao T, Hayashi K, Ishikawa R, Isa K, Asada H, Ikeda K, Uyemura K (1994) Formation of thick, curving bundles of actin by drebrin A expressed in fibroblasts. Exp Cell Res 215:145–153

Song M, Kojima N, Hanamura K, Sekino Y, Inoue HK, Mikuni M, Shirao T (2008) Expression of drebrin E in migrating neuroblasts in adult rat brain: coincidence between drebrin E disappearance from cell body and cessation of migration. Neuroscience 152:670–682

Chapter 8
Making of a Synapse: Recurrent Roles of Drebrin A at Excitatory Synapses Throughout Life

Chiye Aoki and Ang D. Sherpa

Abstract Mature excitatory synapses are composed of more than 1500 proteins postsynaptically and hundreds more that operate presynaptically. Among them, drebrin is an F-actin-binding protein that increases noticeably during juvenile synaptogenesis. Electron microscopic analysis reveals that drebrin is highly enriched specifically on the postsynaptic side of excitatory synapses. Since dendritic spines are structures specialized for excitatory synaptic transmission, the function of drebrin was probed by analyzing the ultrastructural characteristics of dendritic spines of animals with genetic deletion of drebrin A (DAKO), the adult isoform of drebrin. Electron microscopic analyses revealed that these brains are surprisingly intact, in that axo-spinous synaptic junctions are well-formed and not significantly altered in number. This normal ultrastructure may be because drebrin E, the alternate embryonic isoform, compensates for the genetic deletion of drebrin A. However, DAKO results in the loss of homeostatic plasticity of *N*-methyl-D-aspartate receptors (NMDARs). The NMDAR activation-dependent trafficking of the NR2A subunit-containing NMDARs from dendritic shafts into spine head cytoplasm is greatly diminished within brains of DAKO. Conversely, within brains of wild-type rodents, spines respond to NMDAR blockade with influx of F-actin, drebrin A, and NR2A subunits of NMDARs. These observations indicate that drebrin A facilitates the trafficking of NMDAR cargos in an F-actin-dependent manner to mediate homeostatic plasticity. Analysis of the brains of transgenic mice used as models of Alzheimer's disease (AD) reveals that the loss of drebrin from dendritic spines predates the emergence of synaptic dysfunction and cognitive impairment, suggesting that this form of homeostatic plasticity contributes toward cognition. Two studies suggest that the nature of drebrin's interaction with NMDARs is dependent on the receptor's subunit composition. Drebrin A can be found co-clustering with NR2B-containing NMDARs at the plasma membrane,

C. Aoki (✉) • A.D. Sherpa
Center for Neural Science, New York University, 4 Washington Place,
Room 809, New York, NY 10003, USA
e-mail: ca3@nyu.edu

T. Shirao, Y. Sekino (eds.), *Drebrin*, Advances in Experimental Medicine and Biology 1006, DOI 10.1007/978-4-431-56550-5_8

while NR2A-containing NMDARs co-traffic into the spine cytoplasm but do not co-cluster at the plasma membrane. Most recently, we encountered a physiological condition that supports this idea. When adolescent female rats are reared under a condition of restricted food access and ad libitum wheel access, they paradoxically become excessive runners, choosing to run, even during the limited hours of food availability. This behavioral pattern is termed activity-based anorexia (ABA) and has served as an animal model for anorexia nervosa. Those animals that exhibit the greatest ABA vulnerability, in that they lose the most amount of body weight and run with greatest exuberance to the point of risking their lives, exhibit the highest levels of NR2B-NMDARs and drebrin at the postsynaptic membrane of hippocampal pyramidal neurons. Those animals that exhibit the greatest resilience to ABA, in that they run minimally under such condition, thereby losing minimal amount of weight, exhibit the highest level of NR2A-NMDARs in the spine cytoplasm and lowest levels of drebrin at the postsynaptic membrane. This pattern suggests that drebrin has dual roles: retention of NR2A-NMDARs in the reserve pool and trafficking of NR2B-NMDARs to the postsynaptic membrane, ultimately contributing to an individual's reactivity to stress. Altogether, these observations indicate that drebrin is a protein that is important for synaptic plasticity and deserves the attention of neuroscientists studying the neurobiological basis of cognition and stress reactivity.

Keywords Homeostatic plasticity • NMDA receptor • Reserve pool • PSD • Trafficking • Juvenile • Adolescent • Hippocampus • Electron microscopic immunocytochemistry

8.1 The Critical Period for Developmental Plasticity Overlaps with the Period of Robust Synaptogenesis

So much of who we are is a result of the sensory, emotional, and cognitive experiences that we encounter during the first phase of life—the juvenile period. The unique environmental circumstances that we encounter during the juvenile period stimulate the afferents entering the neocortex, thereby driving the formation of new synapses—synaptogenesis—within the neocortex. For rodents and cats, almost all of the synaptogenesis in the neocortex begins postnatally. Synapse number increases by more than 15-fold, attaining adult-like levels toward the end of the juvenile stage of development (Blue and Parnavelas 1983; Cragg 1975; De Felipe et al. 1997; Kageyama and Robertson 1993; Micheva and Beaulieu 1997; Ramaswamy and Markram 2015). The critical period for developmental plasticity is defined as the time during which sensory experience sculpts receptive field properties of neocortical neurons to last a lifetime. This critical period coincides with the latter half of the juvenile phase for exuberant synaptogenesis (Gordon and Stryker 1996;

Hubel and Wiesel 1970; Wiesel 1982). Synaptogenesis in the primate neocortex follows a slightly different time course, since synaptogenesis begins prenatally: however, at least a twofold increase in synapse density occurs postnatally, again coinciding with the critical period for the formation of the receptive field properties (Bourgeois 1997).

The critical period for developmental plasticity is constant within any one species, but varies across species, indicating that most of the molecular events dictating the beginning and the end of the critical period are genetically defined. There are also slight differences in the timing of developmental plasticity across cortical regions within any one species. One cortical region with a significantly prolonged phase of plasticity and synaptogenesis is the hippocampus. Here, the dendritic growth phase is prolonged to the end of adolescence, rather than ending with the juvenile stage. The formation of dendritic spines keeps pace with dendritic branch growths, resulting in a roughly constant density of spines (Chowdhury et al. 2014b). However, a closer look at the spines of adolescent hippocampus indicates that as many as half of the spines are filopodia-like: immature, elongated, and thin (Chowdhury et al. 2014b). In accordance with this prolonged phase of synaptogenesis, mature place fields do not emerge until late adolescence (Martin and Berthoz 2002). The reason for this prolonged phase of synaptogenesis in the hippocampus and other neocortical regions that are primarily cortico-cortical (e.g., area V4) may be that they must wait for their afferents to attain maturity, before they initiate the activity-dependent synaptogenesis.

Mature excitatory synapses are composed of more than 1500 proteins postsynaptically and hundreds more that operate presynaptically (Ryan and Grant 2009). One among them that increases in level during synaptogenesis is drebrin, an F-actin-binding protein that occurs highly enriched, specifically on the postsynaptic side. In this chapter, I will review findings indicative of the function of the adult isoform, drebrin A, that could be gleaned from electron microscopic analyses of intact brain tissue. One idea that emerges from those studies is that drebrin A participates in the activity-dependent trafficking of NMDARs (*N*-methyl-D-aspartate receptors) via interaction with F-actin.

8.2 Drebrin A Is Involved in Excitatory Synaptic Function

Drebrin A is a soluble neuronal protein with a distinctive property of binding to F-actin (Hayashi and Shirao 1999; Shirao and Obata 1986). Although F-actin occurs abundantly in presynaptic axon terminals and dendritic spines, drebrin A occurs only on the postsynaptic side (Aoki et al. 2005), indicating that its preferential trafficking to dendrites and spines is somehow specified. Its putative involvement in synaptic function was first surmised, based on the dramatic increase of its level during the first postnatal week within chicken brains (Shirao et al. 1988). It was later shown that one of the two isoforms of drebrin, called drebrin A, emerges after around

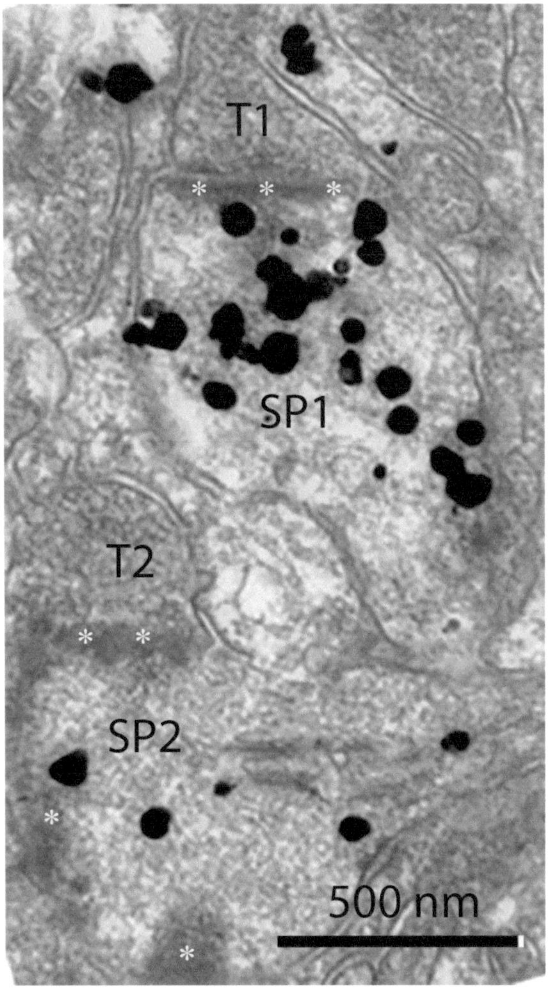

Fig. 8.1 Drebrin A immunoreactivity within dendritic spines is revealed by using silver-intensified gold (SIG) as the immunolabel for electron microscopic immunocytochemistry. SP1 and SP2 are both dendritic spines. Five SIG particles are evident within SP2, while SP1 contains a greater number of SIG particles, indicating higher level of drebrin A within. The presynaptic axon terminal for SP1 and SP2 are labeled as T1 and T2, respectively. The synaptic junctions associated with T1 and T2 are asymmetric and presumed to be excitatory, based on the presence of thick PSDs (*white asterisks*). This figure was adapted from Fig. 2 of a published paper (Mahadomrongkul et al. 2005)

postnatal day (PNd) 10 in the cortex and the hippocampus of rat brains to replace an embryonic isoform, drebrin E (Aoki et al. 2005). Using an antibody that is specific to drebrin A, electron microscopic immunocytochemistry (EM ICC) was performed to determine the subcellular distribution pattern of drebrin A. In adulthood, it was shown that drebrin A is neuronal, detectable on the postsynaptic side of about 70% of the asymmetric (presumably excitatory) axo-synapses (Fig. 8.1). These asymmetric synapses belong to spines of pyramidal (i.e., glutamatergic) neurons and dendritic shafts of GABAergic interneurons. In contrast, none of the symmetric synapses are immunoreactive to drebrin A, whether on dendritic shafts of pyramidal neurons or on GABAergic interneurons. Thus, at least in the hippocampus and cortex, drebrin A is involved in excitatory synaptic function of all cell types and is not involved in inhibitory postsynaptic function, regardless of the cell types (Fig. 8.2).

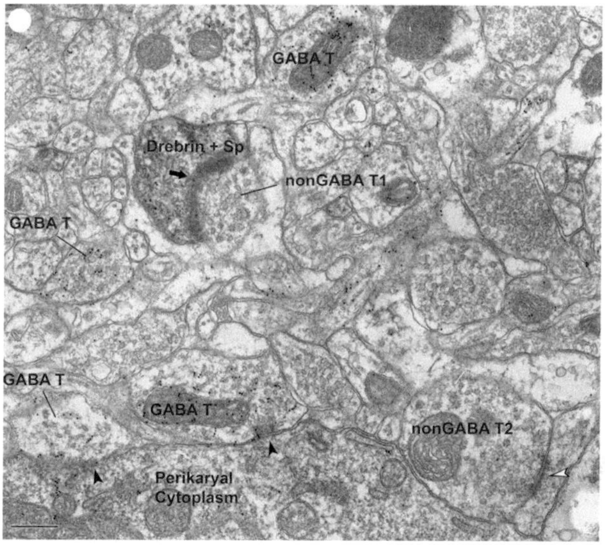

Fig. 8.2 Drebrin A immunoreactivity occurs postsynaptic to excitatory synapses but are absent at GABAergic inhibitory synapses. Adult hippocampal tissue was immunolabeled dually—by DAB—to immunolabel drebrin A and by post-embed gold (PEG) to immunolabel GABA. The spine immunoreactive to drebrin A (Drebrin + Sp) exhibits a thick PSD (*arrow*), indicating that it is forming an excitatory synapse. The presynaptic terminal lacks PEG particles, indicating that it is not GABAergic. There are three GABAergic terminals shown in this field, all of which contain numerous (an excess of ten) PEG particles that are 10 nm in diameter. Arrowheads point to the postsynaptic side of the inhibitory synaptic junctions, which lack the thick PSDs. This figure was modified from Fig. 5 of a published paper (Aoki et al. 2005)

8.3 Is Drebrin A for Morphogenesis or Trafficking of Synaptic Molecules?

Soon after its discovery, drebrin (i.e., drebrin A or E) was recognized to have multiple properties conducive of orchestrating the interaction of cytoskeletal protein complexes within spines. For one, drebrin competes with tropomyosin for binding to F-actin (Shirao 1995). Drebrin also inhibits the actin-activated myosin ATPase and reduces the sliding velocity of actin filaments along immobilized myosin (Hayashi et al. 1996). These were clues that drebrin could be involved in one or both types of cellular processes: (1) neuronal morphogenesis and (2) actomyosin trafficking of cargos.

In support of the idea that drebrin A is involved in spine morphogenesis, it was shown that when drebrin is overexpressed in fibroblasts or CHO (Chinese hamster ovary) cells, F-actin filaments bind preferentially to drebrin rather than tropomyosin, the consequence of which is that thick, curving F-actin bundles form, leading to the appearance of neurite-like processes (Hayashi and Shirao 1999; Shirao 1995).

EM-ICC revealed that, unlike the adult tissue, which showed drebrin A to be predominantly cytoplasmic within dendritic spines, drebrin A in neonatal tissue (rat cortex at PNd7) occurs more frequently associated with the dendritic plasma membrane (Fig. 8.3). These dendritic plasma membranes are part of the flat dendritic shafts, since most dendrites at this stage are immature and have yet to form spines (Fig. 8.3, panel A) or are spinous membranes lacking well-defined PSDs (Fig. 8.3, panel B) and contain only low levels of NMDARs (Fig. 8.4) (Aoki et al. 2005). Through elegant molecular techniques to increase or decrease the expression of drebrin A, it has been shown that dendritic spines of cultured hippocampal neurons contain drebrin A-F-actin clusters before the arrival of PSD-95. When PSD-95 arrives to the spines, they co-cluster precisely where drebrin A-F-actin clusters are located. If drebrin A-F-actin clusters are experimentally depleted, then PSD-95 is no longer able to enter and be retained within spines (Takahashi et al. 2003). Conversely, if the F-actin-binding domain of drebrin A is deleted, drebrin A does not cluster in spines, even if overexpressed: instead, it remains diffusely distributed in both shafts and spines (Hayashi and Shirao 1999). These results suggest that the drebrin A-F-actin complex arrives to newly forming synaptic junctions and enables their maturation.

Within mature tissue, there appears to be a limited capacity for drebrin A to enter spines. This idea is based on an observation that overexpression of drebrin A by as much as 20% has no effect of increasing the proportion or level of drebrin A within spines, although the number of spines per unit volume is increased (Kobayashi et al. 2005). This indicates that drebrin A does not diffuse passively into spines. Indeed, there is evidence to indicate that the spine's capacity to house drebrin A is activity dependent. An in vivo study has shown that blockade of NMDARs within intact neocortex by superfusing the pial surface with the NMDAR antagonist, D-APV (D-2-aminopho-5-phosphovalerate), increases the proportion of spines immunoreactive to drebrin A as well as the levels of drebrin A within individual spines within 30 min (Fujisawa et al. 2006) (Fig. 8.5). The same tissue also increases the level of F-actin within spines (Fujisawa et al. 2006), together with the NR2A-containing NMDARs (Aoki et al. 2003), which are followed by a rise of PSD-95 within another 30 min (unpublished observations). The D-APV blockade-evoked upregulation of NMDARs can be considered to be a homeostatic plasticity mechanism that compensates for the pharmacologically induced reduction of NMDAR activity. A converse experiment, conducted using hippocampal neurons grown in vitro, confirms the link between NMDAR activity and drebrin-F-actin complexes in spines. Specifically, it has been shown that activation of NMDARs evokes the exodus of F-actin-bound drebrin A via activation of myosin II ATPase but not by the myosin light chain kinase or the Rho-associated kinase (Mizui et al. 2014). A scenario compatible with these observations is that drebrin A enters spine heads as a complex with F-actin from dendritic shafts. The net influx of drebrin A-F-actin complexes into spines could contribute toward trapping cargos containing NMDARs and PSD-95 within spine heads, once they have entered through the spine necks from dendritic shafts

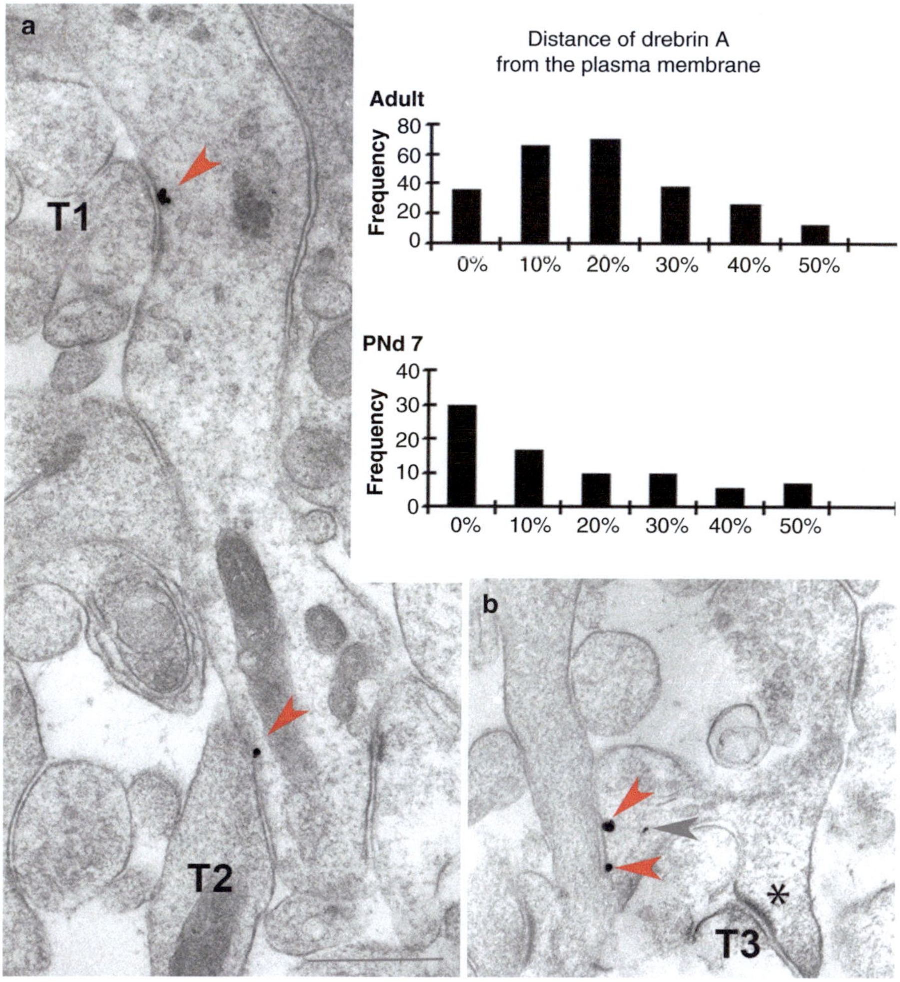

Fig. 8.3 Drebrin A immunoreactivity occurs more often over the plasma membrane of dendrites at PNd 7, when compared to adulthood. Silver-intensified gold particles (SIG) reflecting drebrin A immunoreactivity were quantified across multiple dendritic profiles in the hippocampus and tallied, based on their position as being exactly on the plasma membrane (0% distance from the plasma membrane in the bar graph), 50% from the plasma membrane, if equal distant from the two opposing plasma membranes sides of the dendrite, or at an intermediate distance between the midpoint and on the plasma membrane (10–40%). The frequency of SIG particles on the plasma membrane was higher at PNd7 than in adulthood (histogram on *top right*). Panels A and B show examples of such plasmalemmal labeling of drebrin A (*red arrowheads*) at PNd 7. Both immunolabels in panel A occur along smooth aspiny portions of the dendritic plasma membrane. T1 and T2 are presynaptic axon terminals, indicating that drebrin A is associated with postsynaptic membranes. In panel B, two drebrin A immunolabels occur along the plasma membrane of a spine head that has yet to acquire PSDs, such as the one that is postsynaptic to T3 (*asterisk*). The gray arrowhead is pointing to an example of an SIG that resides at a position removed from the plasma membrane. The calibration bar in panel A = 500 nm and also applies to panel B. This figure was modified from Figs. 10 and 11 of a published paper (Aoki et al. 2005)

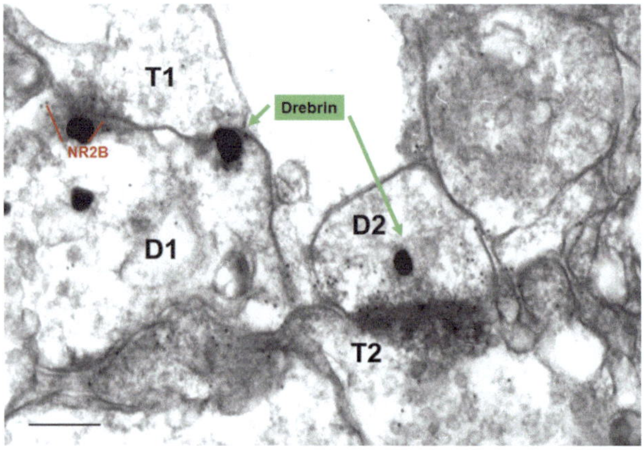

Fig. 8.4 Drebrin A immunoreactivity of PNd7 hippocampal tissue occurs at the plasma membrane patches of dendrites expressing low levels of immunoreactivity for the NR2B subunit of NMDA receptors and no PSD but occur removed from the plasma membrane of synapses with PSDs and higher levels of NR2B. Two dendrites, D1 and D2, with contrasting morphological features and immunoreactivity for the NR2B subunit of NMDA receptors are shown. Immunoreactivity of synapses to the NR2B subunit of NMDA receptors was visualized using 10 nm colloidal gold (PEG, post-embed gold). Red lines point to two examples of PEG particles reflecting NR2B immunoreactivity that are associated with the synaptic junction between dendrite D1 and axon terminal T1. PEG particles are more numerous for D2 than for D1, indicating higher levels of NR2B immunoreactivity in D2. T2 that is presynaptic to dendrite D2 contains more vesicles than in T1. Another difference between the two synapses is that D2 exhibits a PSD, while D1 does not. These three features together indicate that the synapse T2–D2 is relatively more mature than synapse T1–D1. Drebrin A immunoreactivity was detected using silver-intensified gold (SIG, *green arrows*) as immuno-labels. SIG particles reside over the plasma membrane of D1 but is removed from the plasma membrane of D2. These contrasting patterns, together with the quantification shown in Fig. 8.3, indicate that drebrin A tends to be associated with the plasma membrane of immature axo-dendritic synapses but becomes displaced from the plasma membrane once synapse stabilization is attained. Calibration bar = 500 nm. This figure was modified from Fig. 8 of a published paper (Aoki et al. 2005)

via inhibition of the actomyosin ATPase (Hayashi et al. 1996). Such a sequence of events explains the net rise of NMDARs and PSD-95 that we have observed at spines following NMDAR-blockade with D-APV. However, one important point to make here is that although the D-APV treatment increases the influx of NR2A immunoreactivity in the spine cytoplasm, we have not observed a rise of NR2A-NMDARs at the postsynaptic plasma membrane. This may be because the final step involving the exocytosis of cargos carrying NR2A-NMDARs requires NMDAR activity (Barria and Malinow 2002). The signal that promotes the entry of the drebrin A-F-actin complex into spines could be the lowered level of intracellular calcium: NMDARs permeate calcium, when activated, so blockade of NMDARs by D-APV would have the consequence of lowering intracellular levels of calcium.

A critical test of this working hypothesis was to determine the consequence of removing drebrin A from cortical spines. Indeed, when the D-APV treatment of

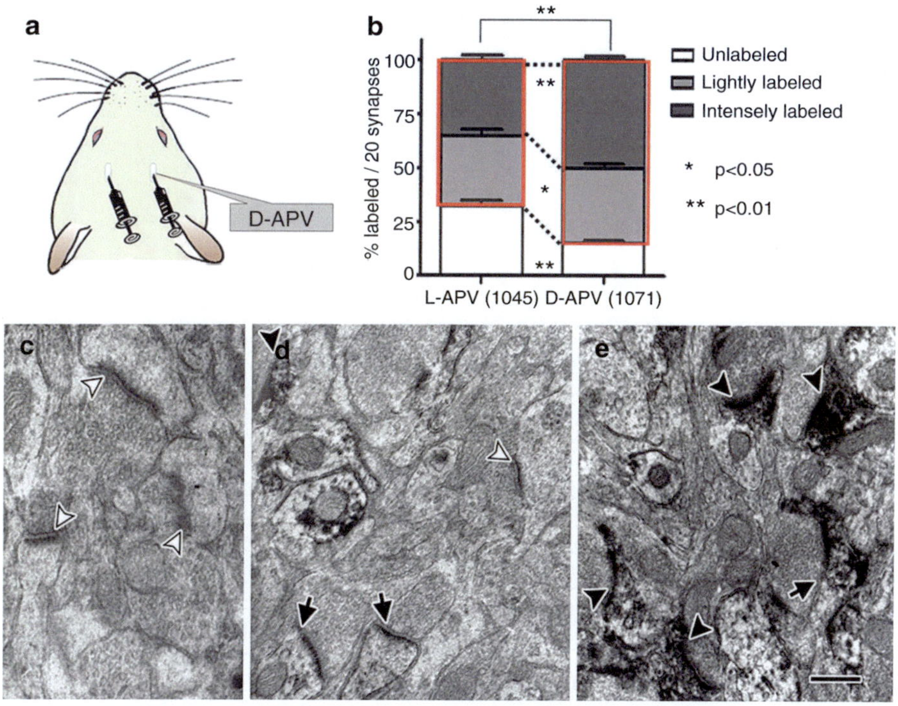

Fig. 8.5 D-APV blockade of NMDARs elicits trafficking of drebrin A into dendritic spines. Panel *A* illustrates the procedure followed for in vivo blockade of NMDARs. Small holes overlying the somatosensory cortex were drilled, and D-APV, an NMDAR blocker, was infused over the pia matter of one hemisphere while maintaining the animal anesthetized with isoflurane. The contralateral hemisphere received L-APV, an inactive enantiomer. 30 min to 2 h later, animals were transcardially perfused with a fixative to localize drebrin A and F-actin in place. Panel *C* shows an example of an electron micrograph of a tissue that was incubated with all immunoreagents minus the primary antibody directed against drebrin A. Panel *D* shows an example from the hemisphere that received L-APV, while panel *E* shows an example of the neuropil from the hemisphere that received D-APV. *Black arrowheads* point to spines labeled intensely for drebrin A, using DAB as the immunolabel. White arrowheads point to unlabeled spines. *Arrows* point to less completely immunolabeled spines. Panel *B* shows the results of the quantification, indicating that the proportion of spines intensely immunolabeled, lightly labeled, and unlabeled were all significantly different between the two hemispheres. D-APV significantly increased the proportion of spines that were immunolabeled for drebrin A. Each bar of the graph represents mean + SEM. Horizontal calibration bar in panel $E = 500$ nm and applies to all three micrographs. Adapted from Fig. 4 of a published article (Fujisawa et al. 2006)

cortical tissue was repeated upon mice with genetic deletion (knockout) of drebrin A (DAKO), these cortical spines no longer responded with increased levels of NR2A-NMDARs in the cytoplasm (Aoki et al. 2009) (Fig. 8.6). While these results indicate that trafficking of NR2A-NMDARs into spines is dependent on the expression of drebrin A, we have not tested the interdependence of drebrin A expression

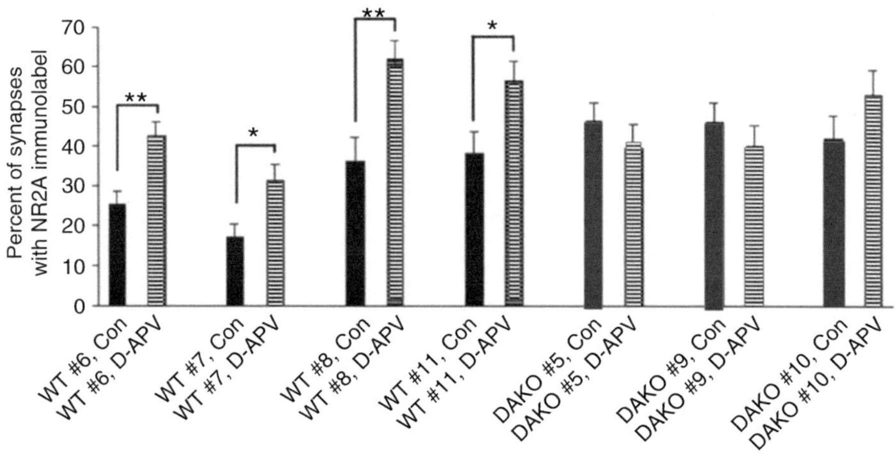

Fig. 8.6 Drebrin A knockout (DAKO) eliminates the D-APV blockade-evoked increase of NR2A immunoreactivity within dendritic spines. Each pair of bars depicts the proportion of dendritic spines in layer 1 of somatosensory cortex with immunoreactivity to the NR2A subunit of NMDARs within the two hemispheres of a single animal (mean + SEM). One hemisphere received D-APV, the NMDAR blocker (*striped bars*), while the contralateral hemisphere received the inactive enantiomer, L-APV ("Con," *solid bars*). Black pair of bars depict the results obtained from wild-type adult mice, while the purple pair of bars depict results from DAKO mice. In four out of four wild-type cases, the interhemispheric differences in immunolabeling for the NR2A subunit of NMDARs were significant. In three out of the three DAKO cases, the interhemispheric differences in immunolabeling were not significant. Modified from Fig. 5 of a published article (Aoki et al. 2009)

with the activity-dependent trafficking of NR2B-NMDARs using DAKO tissue. This point is relevant, because the trafficking of NR2A-NMDAR and NR2B-NMDAR is in opposite directions (NR2B-NMDAR declines, while the NR2A-NMDAR levels rise) following D-APV blockade of wild-type neocortex (Fujisawa and Aoki 2003).

8.4 Additional Phenotypes of Drebrin A KO: Beyond the Loss of Homeostatic Plasticity?

As was summarized above, a number of works involving cultured hippocampal neurons have indicated that drebrin A facilitates spine morphogenesis. Moreover, since our previous EM analysis had revealed that drebrin A level within spines correlates positively with spine size within adult cortex (Kobayashi et al. 2007), we expected DAKO to cause a reduction in the number and size of dendritic spines. Contrary to

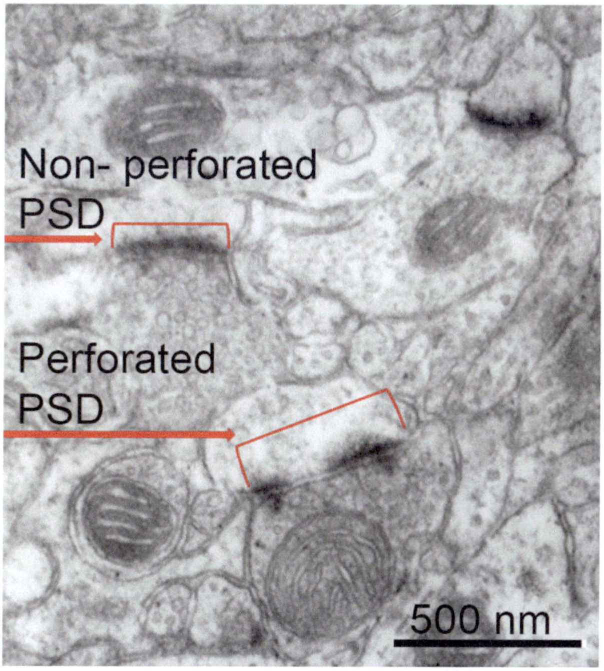

Fig. 8.7 Perforated and non-perforated PSDs. Some of the PSDs of excitatory synapses occur perforated. Typically, perforated PSDs are associated with the larger mushroom head spines. This micrograph was taken from layer 1 of a young adult mouse cortex

this expectation, DAKO does not induce any obvious alteration in spine density or spine morphology. The parameters we measured were the number of synapses per unit volume, the width of spines, and the width of PSDs. We also could not detect any difference in the synaptic or cytoplasmic localization of NR2A-containing NMDARs within spines at basal levels. Such lack of phenotypic difference suggests that when drebrin A is genetically deleted, drebrin E persists and may provide the compensatory biochemical pathway necessary for spine morphogenesis, even though they cannot compensate for the biochemical pathway necessary for the activity-dependent trafficking of NMDARs.

The one phenotypic feature that was different within brains of DAKO mice was the frequency of synapses with perforated PSDs, which were increased by 40% within brains of DAKO cortex (Fig. 8.7). The perforated PSDs are about 60% larger than the non-perforated PSDs and are usually associated with mushroom spines. Such well-developed larger spines are hypothesized to be sites for storage of memory, while the smaller spines are hypothesized to be the sites for acquiring new memory (Matsuzaki et al. 2004). If this association between perforated PSDs and memory storage holds for the DAKO brains, then DAKO animals might be predicted to exhibit deficits with extinction of memory and LTD, due to the excess number of mushroom spines.

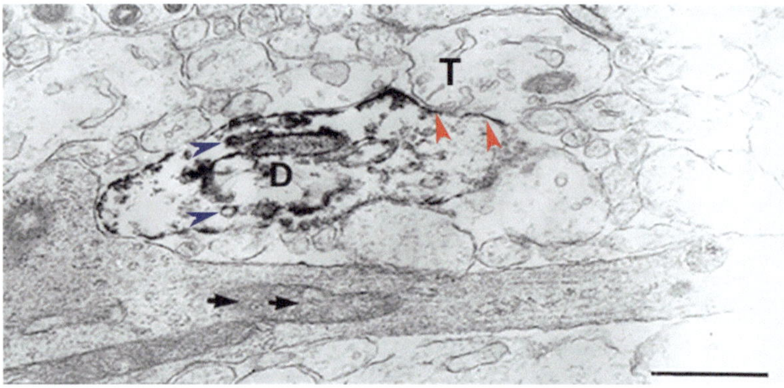

Fig. 8.8 NMDAR immunoreactivity appears along dendrites prior to the formation of spines. This micrograph was taken from the visual cortex of a PNd 4 rat. Immunoreactivity for the NR1 subunit of NMDAR was visualized using DAB as the immunolabel within a dendrite, D. A terminal with few vesicles (T) is directly contacting the dendrite at the portion highlighted (*small red arrowheads*). Immunoreactivity is also associated with membranous organelles in the cytoplasm, two of which are highlighted (*blue arrowheads*). The two black arrows highlight a glial process coursing below the dendrite. Calibration bar = 500 nm. This figure was taken from a published paper (Aoki et al. 1994)

8.5 The Putative Role of Drebrin in Transporting NMDARs to the Plasma Membrane of Newly Forming Spines

As was described earlier, drebrin A is found at the plasma membrane of dendritic shafts at early developmental stages, when dendritic spines are still scarce and PSDs are not yet evident (Aoki et al. 2005). By EM-ICC, NR1 subunits of NMDARs within neonatal cortices have also been visualized along smooth portions of the plasma membrane of dendritic shafts, prior to the formation of dendritic spines (Aoki et al. 1994, 1997) (Fig. 8.8). Synaptogenesis proceeds through fusion of membranous packets containing newly formed NR2B-NMDARs (Washbourne et al. 2002). An intriguing possibility is that these NR2B-NMDAR cargos are transported along actomyosin lattices toward the plasma membrane and become trapped at the membrane surface, where drebrin presides, remaining stabilized at the plasma membrane while "waiting" for the eventual arrival of other anchoring proteins, such as PSD-95 and the NR2A-NMDARs (Fig. 8.9). In support of this idea, EM-ICC has revealed that those plasma membrane patches expressing drebrin A overlap with patches expressing NR2B-NMDARs (Fig. 8.4).

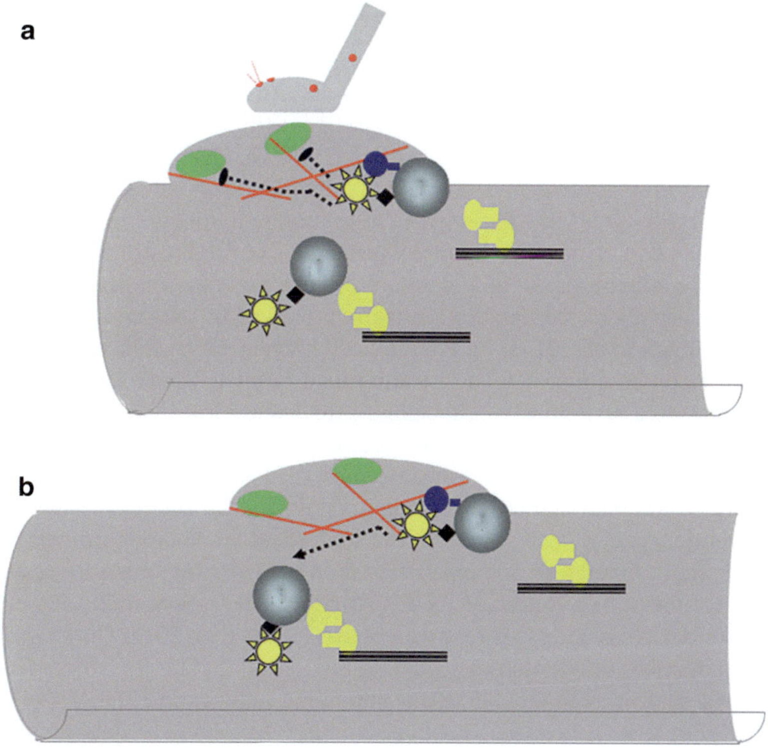

Fig. 8.9 A working model of drebrin A's involvement in the trafficking of NMDAR cargos to the plasma membrane of a newly forming spine. *Green ball* = drebrin A; *yellow sun* = actomyosin ATPase motor; *red lines* = F-actin bundles; *cobalt blue spheres* = membranous cargo carrying PSD-95 and NMDARs; *small blue clubs* = α-actinin; *yellow clubs* = motor proteins associated with microtubules; *double lines* = microtubules in the dendritic shaft microtubules; *red dots* = vesicles within a nascent axon terminal. Panel *A*: a cargo containing NMDAR and PSD-95 moves along F-actin bundles via the actomyosin ATPase motor. The same cargo can move into and toward the plasma membrane until it encounters drebrin A at the plasma membrane, as indicated by the black dotted lines. Once the cargo encounters drebrin A, the cargo becomes stationary, due to drebrin A's inhibitory influence upon actomyosin ATPase. In this way, NMDAR cargos accumulate wherever drebrin A is located. If drebrin A is located at the plasma membrane, NMDAR cargo will accumulate by the plasma membrane. These events lead to a buildup of receptors into membranous clusters at newly forming spines, becoming increasingly sensitive to the neurotransmitter that is released from the nearby axon terminal (located above the spinous protrusion). Panel *B*: the same cargo can be transported elsewhere within spines and out of spines by following the portion of F-actin lattice that is not associated with drebrin A

8.6 Drebrin's Role During Adolescence

Recently, we have begun to explore the role of drebrin A in the hippocampus during adolescence, after realizing that synaptogenesis is still robust during this phase of development for this brain region (summarized above). Our recently collected data indicate that during adolescence, not only is synaptogenesis robust but also that the level of expression of NMDARs at spines in stratum radiatum is strongly influenced by the animal's experience. For example, just 4 days of food restriction or 8 days of wheel running activity or the combination of these two environmental factors can each evoke changes to the dendritic branching pattern in stratum radiatum of the CA1 (Chowdhury et al. 2014a, b). Paradoxically, when adolescent rodents are subjected to the combination of food restriction and running wheel access, these animals become excessive runners, choosing to run even during the limited period of food availability. Those animals that become obsessive wheel runners must be removed from the environment of wheel+food restriction within 4 days, or else they will die from self-starvation. This phenomenon, called activity-based anorexia (ABA), can be induced in about half of the adolescent female rodents and has been used as an animal model of anorexia nervosa. ABA captures four core features of this mental illness: (1) voluntary food restriction, (2) hyperactivity, (3) elevated anxiety-like behavior, and (4) severe weight loss (Aoki et al. 2012; Gutierrez 2013). There are a number of hypotheses regarding the reason that food-restricted animals become hyperactive, with one being that starvation induces an innate foraging-like behavior, which is converted to wheel running for animals held in captivity (Guisinger 2003; Gutierrez 2013). Individuals respond to the wheel+food restriction environment with different degrees of excessive running, which, in turn, is strongly correlated to the extent of weight loss and anxiety-like behavior (the more that the animals run, the stronger is their anxiety-like behavior, as tested on the elevated plus maze (Wable et al. 2015). We examined whether spine density or dendritic branching in the hippocampus might be correlated with these behavioral measurements and found none. However, it is possible that what is altered is not simply the number of excitatory synaptic junctions formed upon dendrites of the hippocampus but, instead, the strength of these synapses and modifiability of their strengths. Indeed, we have found that dendritic spines of the animals that underwent this wheel+food restriction treatment exhibit elevated levels of both the NR2B- and NR2A-NMDARs at the plasma membrane (Klingensmith et al. 2015; Chen et al. 2016). Closer examination of these changes reveals that NR2B-NMDARs are increased precisely at the postsynaptic plasma membrane and especially among individuals that respond to food restriction with greatly excessive wheel running—i.e., the ABA-vulnerable individuals (ABA vulnerable, Fig. 8.10). Conversely, although the NR2A-NMDARs also rise, the pattern is the opposite from that observed for the NR2B-NMDARs. NR2A-NMDARs increase at the cytoplasm, and this rise in the cytoplasm correlates negatively with activity: the more that NR2A-NMDARs are at the cytoplasm, the less that they are hyperactive (ABA resilient, Fig. 8.10), because they are better able to suppress the innate response of foraging. Membranous NR2B-NMDAR levels are inversely correlated with cytoplasmic

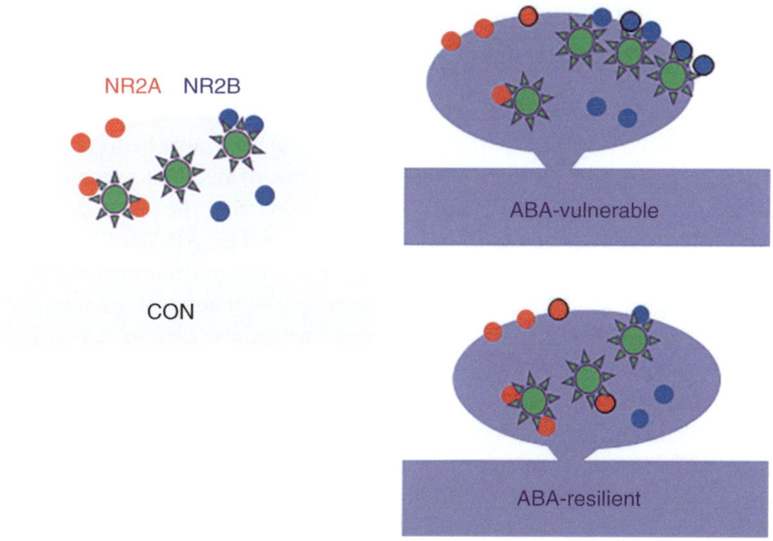

Fig. 8.10 The proposed role of drebrin A in the trafficking of the NR2A- and the NR2B-NMDAR content within spines of the hippocampus after activity-based anorexia (ABA) induction. ABA vulnerability is defined by the degree of hyperactivity evoked by food restriction, leading to exacerbated body weight loss. This behavior correlates with anxiety-like behavior. We propose that hyperexcitability of hippocampal pyramidal neurons may contribute to this. *Red dots* = NR2A-NMDAR immunoreactivity; *blue dots* = NR2B-NMDAR immunoreactivity; *green stars* = drebrin A interacting with the actomyosin ATPase. *Dots with black outlines* reflect those NMDARs that are hypothesized to have been added as a consequence of ABA induction. Within spines of ABA-vulnerable adolescent rats, both NR2A and NR2B-NMDAR immunoreactivity have increased at the postsynaptic plasma membrane, as has the proportion of spines with drebrin A immunoreactivity. Those animals with heightened expression of drebrin A also exhibit heightened levels of NR2B-NMDARs at the postsynaptic plasma membrane. Within the spines of animals exhibiting ABA resilience, defined as animals that have suppressed the food restriction-evoked hyperactivity, levels of drebrin A and postsynaptic membranous NR2B-NMDAR are lower, while the level of the non-synaptic (including cytoplasmic) NR2A-NMDAR is high. Correlations of the level of NMDARs at specific locations within spines with drebrin A are shown by the close association of the symbols. The correlation analysis suggests that drebrin A interacts directly with both the membranous NR2B-NMDARs but only with the cytoplasmic NR2A-NMDARs. It is assumed that the interaction between drebrin A and NMDARs remains the same before and after ABA induction. Modified from Chen et al. 2016

NR2A-NMDARs. This tight and inverse relationship suggests the existence of an exchange mechanism between membranous NR2B-NMDARs and cytoplasmic NR2A-NMDARs. In light of the topic of this chapter, drebrin A, we speculate that drebrin A may be involved in this exchange mechanism. First, it is interesting to note that ABA induction during adolescence also increases drebrin A levels within spines and that this, too, is related to the animal's behavior. The proportion of spines of the ABA hippocampus immunolabeled for drebrin A ranged from 30% to 67%, and this proportion strongly correlated positively with NR2B-NMDAR labelings at the membrane and negatively with NR2A-NMDAR labelings within spine cytoplasm (Fig. 8.10) (Chen et al. 2016). The rise of drebrin A within spines of ABA-induced animals may enable the delivery of ABA-evoked newly synthesized NR2B-NMDARs (blue circles with black outlines in Fig. 8.10) to the spine plasma membrane, as we have already hypothesized for newly forming synapses of neonatal tissue (Fig. 8.4). Conversely, the role of drebrin A may also be to favor the location of NR2A-NMDARs to the cytoplasm of spines of resilient animals, such that some of the newly synthesized NR2A-NMDARs entering the spine head of ABA-induced animals (red circles with black outlines in the "ABA-resilient" spine of Fig. 8.10) are prevented from reaching the postsynaptic plasma membrane. In support of this idea, we have observed that a pharmacological treatment that enhances the rise of cytoplasmic NR2A-NMDAR immunoreactivity also increases cytoplasmic drebrin A and F-actin (Fig. 8.5 and Fujisawa et al. 2006). Although highly speculative, this working model makes specific predictions that are testable: for this model to work, drebrin A's affinity to cytoplasmic NR2A-NMDARs would have to be higher than drebrin A's affinity to cytoplasmic NR2B-NMDARs or plasmalemmal NR2A-NMDARs. Moreover, drebrin A's affinity for the membranous NR2B-NMDARs would have to be higher than drebrin A's affinity for the membranous NR2A-NMDARs or cytoplasmic NR2B-NMDARs.

The contrasting pattern of regulation for the NR2A and NR2B-NMDARs is a recurrent theme. During development, the expression of NR2B-NMDAR is earlier than that of the NR2A-NMDAR [rev. in Haberny et al. (2002)], with the NR2A-NMDAR expression becoming relatively more dominant sometime between P10 and P35 in the rat hippocampal CA1 (Petralia et al. 2005). In this review article, we described results demonstrating that D-APV treatment to intact adult cortical neuropil evoked a rise of NR2A-NMDARs and a decline of NR2B-NMDARs in the spine cytoplasm (Aoki et al. 2003; Fujisawa and Aoki 2003).

What might be the consequence of the rise of membranous NR2B-NMDARs at spines of the adolescent hippocampus? One simplest scenario is that the rise of membranous NR2B-NMDARs increases excitability of hippocampal pyramidal neurons. Since increased excitability of the hippocampus is linked to anxiogenesis (Shen et al. 2007) and elevated anxiety is a strong contributor to hyperactivity that leads to self-starvation (Wable et al. 2015), the rise of NR2B-NMDARs could be the culprit contributing toward self-starvation. However, an opposite scenario is also possible, involving NR2B-NMDAR-mediated LTD (long-term depression). Due to the difference in the NMDAR channel kinetics, the frequency used to induce long-term depression (LTD; 1 Hz) generates a larger charge transfer of calcium influx via the NR2B-NMDARs than the NR2A-NMDARs (Erreger et al. 2005). Therefore, LTD will be enhanced within a network receiving low-frequency stimulation and heightened NR2B-NMDAR expression. Conversely, the frequency used to induce long-term potentiation (LTP; 100 Hz) is more effective in evoking calcium charge transfer via the NR2A-NMDARs than of the NR2B-NMDARs (Erreger et al. 2005). In the perirhinal cortex at ages 7–12 weeks (adults) and in the juvenile hippocampus (3–4 weeks), NR2B-NMDAR activation has been shown to be required for the long-term depression (LTD), while NR2A-NMDAR activation leads to long-term potentiation (LTP) or depotentiation (Liu et al. 2004; Massey et al. 2004). If the same rule applies to the adolescent hippocampus, then both the rise of NR2B-NMDARs at the synaptic membrane (for the ABA vulnerable) and the trafficking of NR2A-NMDARs into the cytoplasm (for the ABA resilient), which we speculated to involve drebrin A, may dampen excitability of the hippocampus. If so, then drebrin A could ultimately be serving to protect individuals from self-starvation by dampening the anxiety-driven hyperactivity through NR2B-NMDAR-mediated LTD and reduction of the NR2A-NMDAR-mediated LTP.

For many brain regions, adolescence is the final developmental stage for brain development. This stage coincides with the rampant rise of the first incidences of mental illness, risky behaviors, and environmental stresses that are associated with "leaving the nest" [rev. in Aoki et al. (2016)]. Given that dendritic branches in the hippocampus are still undergoing growth and retraction during adolescence, it is important that we strive to determine whether a number of cell biological principles that have been identified for adult and juvenile brains also apply toward adolescent brains. There is a possibility that at least some of those rules are turned upside-down during adolescence, as the dendrites replay the cellular events of synaptogenesis that have been characterized for neonatal tissue. I believe we have made one contribution toward this endeavor by identifying a role for drebrin A in the activity-dependent trafficking of NMDARs to spines of adolescent hippocampus.

8.7 Drebrin A's Role in the Aging Brain

One of the greatest challenges for neuroscientists of this century is to understand the etiology of Alzheimer's disease (AD). Brain samples from individuals diagnosed with Alzheimer's disease are reported to contain lowered levels of drebrin A than those of age-matched healthy adults (Counts et al. 2006; Harigaya et al. 1996; Hatanpaa et al. 1999; Shim and Lubec 2002). By the time patients are diagnosed with the pre-AD stage of mild cognitive impairment (MCI), the temporal cortex begins to exhibit the loss of drebrin A. In contrast, the frontal cortex is reported to exhibit an increase of drebrin A among those diagnosed with MCI, but this converts to drebrin A becoming lower than normal by the time the disease has progressed to AD (Counts et al. 2006). The loss of this protein cannot be accounted for simply by cell loss or neuronal loss, because these assays are normalized to F-actin and neuron-specific enolase levels (Shim and Lubec 2002). Is this a phenomenon associated with the loss of excitatory synapses and spines or the loss more specifically of drebrin A? We have addressed this question through the use of an animal model of familial AD (FAD) that was generated by targeted knock-in of two genes (2xKI)—one encoding the mutated amyloid precursor protein (APP: K670N/M671L, Swedish mutation with humanized amyloid beta sequence) and another encoding the mutated presenilin-1 (PS1, P264L). In this animal model, the expression of the wild-type gene is completely eliminated, and the two humanized genes are controlled by endogenous promotors, so as to avoid their overexpression (Flood et al. 2002). Histological studies had established that brains of these animals are not distinguishable from those of the wild-type littermates at 6 months of age. However, by the ninth postnatal month, plaque load and amyloid beta42 emerge and rise monotonically thereafter. Hippocampal LTP deficiency emerges around the 7–9 months of age, and deficiency in AMPAR currents is apparent by the 9–12 months of age (Chang et al. 2006). In order to determine whether synaptic changes that are detectable by EM-ICC might precede even the biophysical and the histological changes, we analyzed drebrin A levels within brains of these animals at a time point preceding the emergence of the biophysical or histological changes—i.e., at 6 months of age. We analyzed drebrin A immunoreactivity in spines across three regions of the 2xKI mouse brains—the hippocampal CA1, the entorhinal cortex, and the somatosensory cortex (Aoki et al. 2007; Mahadomrongkul et al. 2005)—to determine whether the limbic areas (hippocampus and entorhinal cortex) exhibit greater and earlier signs of neurodegeneration than the neocortex, as has been reported to be for the brains of individuals with AD (Braak and Braak 1991). In the hippocampal CA1, we detected a 56% decrease in the density of drebrin A-containing spines, which contributed toward a 25% decrease in overall synapse density (i.e., counting both the drebrin A-containing and drebrin A-lacking spines). In the entorhinal cortex of the same age, we did not detect a change in spine density or in the proportion of spines containing drebrin A. However, those spines that were drebrin A positive contained only 30% of the normal levels. This pattern in the entorhinal cortex suggest to us that, with time, these would convert to becoming drebrin A negative, i.e., containing levels of drebrin A that are below the level of detectability. Finally, in the somatosensory cortex of the same 2xKI animals, we also detected a more modest (15%) decrease in the proportion of drebrin A-containing spines, but this was due mostly to the increase in the number of spines lacking drebrin A, rather than to a decrease in the number of drebrin A-containing spines (Mahadomrongkul et al. 2005). This pattern in the somatosensory cortex indicates to us that the sparing of the neocortex from degeneration (Braak and Braak 1991) may be related to their retention of synaptogenic properties. Conversely, the enhanced vulnerability of the hippocampus may be related to its high level of expression of drebrin A (Aoki et al. 2005). In support of this idea, we observed that the distribution of drebrin A within spines was more frequent at the synaptic plasma membrane than at non-synaptic portions of spines for the 2xKI aged brain (greater than 18 months old) and that this enhancement of synaptic localization was already evident by 6 months postnatal (Fig. 8.11). If, as was proposed for the plasmalemmal drebrin A during neonatal synaptogenesis (Fig. 8.9), drebrin A within brains of the 2xKI is triggered to become localized more to the plasma membrane, then drebrin A's actomyosin ATPase inhibiting property could cause NMDARs to become trapped more at the plasma membrane. If, as was proposed for the adolescent hippocampus, the synaptic drebrin A is enabling the localization of NR2B-NMDARs to the plasma membrane and retention of the NR2A-NMDARs in the cytoplasm, this could favor LTD over LTP in the somatosensory cortex, where modest levels of drebrin A was found, thereby protecting the somatosensory cortex from excitotoxicity and eventual death of neurons. It has been proposed that Alzheimer's disease is a manifestation of synaptic failure (Selkoe 2002). The loss of drebrin A, especially in brain regions such as the

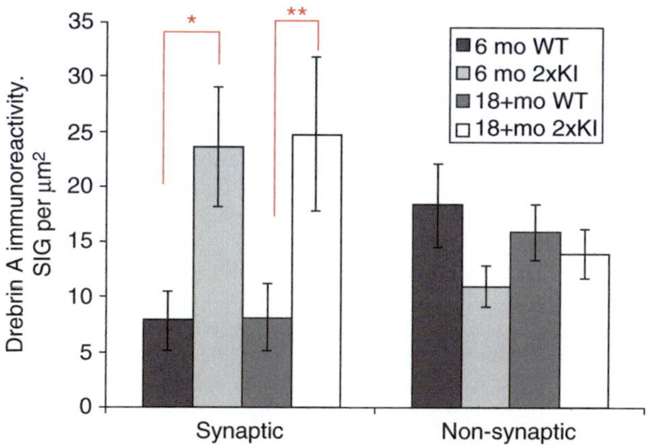

Fig. 8.11 Drebrin A immunoreactivity within spines of the somatosensory cortex of FAD (familial-type Alzheimer's disease) model mice (2xKI) occurs more often at the synaptic membrane, compared to spines from age-matched wild-type littermates. Drebrin A immunoreactivity was detected using silver-intensified gold (SIG) as the immunolabel and determined to be clustered over the postsynaptic region or removed from the synaptic membrane by analyzing electron micrographs of images acquired from layer 1. Adapted from Fig. 4 of a published paper (Mahadomrongkul et al. 2005)

hippocampus, where the loss is profound, may be one of the contributors to synaptic dysfunction, leading to excessive excitation of those spines via NR2A-NMDAR-mediated LTP, excitotoxicity of those neurons, and eventual cell death within brains of patients with FAD.

References

Aoki C et al (1994) Cellular and subcellular localization of NMDA-R1 subunit immunoreactivity in the visual cortex of adult and neonatal rats. J Neurosci 14:5202–5222

Aoki C et al (1997) NMDA-R1 subunit of the cerebral cortex co-localizes with neuronal nitric oxide synthase at pre- and postsynaptic sites and in spines. Brain Res 750:25–40

Aoki C et al (2003) NMDA receptor blockade in intact adult cortex increases trafficking of NR2A subunits into spines, postsynaptic densities, and axon terminals. Brain Res 963:139–149

Aoki C et al (2005) Drebrin A is a postsynaptic protein that localizes in vivo to the submembranous surface of dendritic sites forming excitatory synapses. J Comp Neurol 483:383–402

Aoki C et al (2007) Chemical and morphological alterations of spines within the hippocampus and entorhinal cortex precede the onset of Alzheimer's disease pathology in double knock-in mice. J Comp Neurol 505:352–362

Aoki C et al (2009) Drebrin a knockout eliminates the rapid form of homeostatic synaptic plasticity at excitatory synapses of intact adult cerebral cortex. J Comp Neurol 517:105–121

Aoki C et al (2012) Adolescent female rats exhibiting activity-based anorexia express elevated levels of GABA(A) receptor alpha4 and delta subunits at the plasma membrane of hippocampal CA1 spines. Synapse 66:391–407

Aoki C et al (2016) Synaptic changes in the hippocampus of adolescent female rodents associated with resilience to anxiety and suppression of food restriction-evoked hyperactivity in an animal model for anorexia nervosa. Brain Res 1654:102–115

Barria A, Malinow R (2002) Subunit-specific NMDA receptor trafficking to synapses. Neuron 35:345–353

Blue ME, Parnavelas JG (1983) The formation and maturation of synapses in the visual cortex of the rat. II. Quantitative analysis. J Neurocytol 12:697–712

Bourgeois JP (1997) Synaptogenesis, heterochrony and epigenesis in the mammalian neocortex. Acta Paediatr Suppl 422:27–33

Braak H, Braak E (1991) Neuropathological stageing of Alzheimer-related changes. Acta Neuropathol 82:239–259

Chang EH et al (2006) AMPA receptor downscaling at the onset of Alzheimer's disease pathology in double knockin mice. Proc Natl Acad Sci U S A 103:3410–3415

Chowdhury TG et al (2014a) Activity-based anorexia has differential effects on apical dendritic branching in dorsal and ventral hippocampal CA1. Brain Struct Funct 219:1935–1945

Chowdhury TG et al (2014b) Activity-based anorexia during adolescence disrupts normal development of the CA1 pyramidal cells in the ventral hippocampus of female rats. Hippocampus 24:1421–1429

Counts SE et al (2006) Differential expression of synaptic proteins in the frontal and temporal cortex of elderly subjects with mild cognitive impairment. J Neuropathol Exp Neurol 65:592–601

Chen Y-W et al (2016) Hunger-evoked excessive exercise correlates with levels of post-synaptic NR2B-NMDA receptors and drebrin at hippocampal dendritic spines, while the NR2A-NMDA receptors in the reserve pool of spines contribute to suppression of excessive exercise. Brain Struct Funct. Dec 3 2016 [Epub ahead of print]

Cragg BG (1975) The development of synapses in kitten visual cortex during visual deprivation. Exp Neurol 46:445–451

De Felipe J et al (1997) Inhibitory synaptogenesis in mouse somatosensory cortex. Cereb Cortex 7:619–634

Erreger K et al (2005) Subunit-specific gating controls rat NR1/NR2A and NR1/NR2B NMDA channel kinetics and synaptic signalling profiles. J Physiol 563:345–358

Flood DG et al (2002) FAD mutant PS-1 gene-targeted mice: increased A beta 42 and A beta deposition without APP overproduction. Neurobiol Aging 23:335–348

Fujisawa S, Aoki C (2003) In vivo blockade of N-methyl-D-aspartate receptors induces rapid trafficking of NR2B subunits away from synapses and out of spines and terminals in adult cortex. Neuroscience 121:51–63

Fujisawa S, Shirao T, Aoki C (2006) In vivo, competitive blockade of N-methyl-D-aspartate receptors induces rapid changes in filamentous actin and drebrin A distributions within dendritic spines of adult rat cortex. Neuroscience 140:1177–1187

Gordon JA, Stryker MP (1996) Experience-dependent plasticity of binocular responses in the primary visual cortex of the mouse. J Neurosci 16:3274–3286

Guisinger S (2003) Adapted to flee famine: adding an evolutionary perspective on anorexia nervosa. Psychol Rev 110:745–761

Gutierrez E (2013) A rat in the labyrinth of anorexia nervosa: contributions of the activity-based anorexia rodent model to the understanding of anorexia nervosa. Int J Eat Disord 46:289–301

Haberny KA et al (2002) Ontogeny of the N-methyl-D-aspartate (NMDA) receptor system and susceptibility to neurotoxicity. Toxicol Sci 68:9–17

Harigaya Y et al (1996) Disappearance of actin-binding protein, drebrin, from hippocampal synapses in Alzheimer's disease. J Neurosci Res 43:87–92

Hatanpaa K et al (1999) Loss of proteins regulating synaptic plasticity in normal aging of the human brain and in Alzheimer disease. J Neuropathol Exp Neurol 58:637–643

Hayashi K et al (1996) Modulatory role of drebrin on the cytoskeleton within dendritic spines in the rat cerebral cortex. J Neurosci 16:7161–7170

Hayashi K, Shirao T (1999) Change in the shape of dendritic spines caused by overexpression of drebrin in cultured cortical neurons. J Neurosci 19:3918–3925

Hubel DH, Wiesel TN (1970) The period of susceptibility to the physiological effects of unilateral eye closure in kittens. J Physiol 206:419–436
Kageyama GH, Robertson RT (1993) Development of geniculocortical projections to visual cortex in rat: evidence early ingrowth and synaptogenesis. J Comp Neurol 335:123–148
Klingensmith LM et al (2015) Immunoreactivity for the NR2A and NR2B subunits of NMDA receptors is increased at axo-spinous synapses of hippocampal CA1 of adolescent female rats exhibiting food restriction-evoked hyperactivity, an animal model of anorexia nervosa. In: Annual meeting of the society for neuroscience. Society for Neuroscience, Chicago, IL
Kobayashi C et al (2007) Drebrin a content correlates with spine head size in the adult mouse cerebral cortex. J Comp Neurol 503:618–626
Kobayashi C, Aoki C, Hanamura K, Shirao T (2005) Ultrastructural analysis of spine density and distribution of drebrin A within spines of drebrin A-transgenic mice. In: Annual meeting abstract of the society for neuroscience. Society for Neuroscience, Washington, DC
Liu L et al (2004) Role of NMDA receptor subtypes in governing the direction of hippocampal synaptic plasticity. Science 304:1021–1024
Mahadomrongkul V et al (2005) Stability of the distribution of spines containing drebrin A in the sensory cortex layer I of mice expressing mutated APP and PS1 genes. Brain Res 1064:66–74
Martin PD, Berthoz A (2002) Development of spatial firing in the hippocampus of young rats. Hippocampus 12:465–480
Massey PV et al (2004) Differential roles of NR2A and NR2B-containing NMDA receptors in cortical long-term potentiation and long-term depression. J Neurosci 24:7821–7828
Matsuzaki M et al (2004) Structural basis of long-term potentiation in single dendritic spines. Nature 429:761–766
Micheva KD, Beaulieu C (1997) Development and plasticity of the inhibitory neocortical circuitry with an emphasis on the rodent barrel field cortex: a review. Can J Physiol Pharmacol 75:470–478
Mizui T et al (2014) Myosin II ATPase activity mediates the long-term potentiation-induced exodus of stable F-actin bound by drebrin A from dendritic spines. PLoS One 9:e85367
Petralia RS et al (2005) Ontogeny of postsynaptic density proteins at glutamatergic synapses. Mol Cell Neurosci 29:436–452
Ramaswamy S, Markram H (2015) Anatomy and physiology of the thick-tufted layer 5 pyramidal neuron. Front Cell Neurosci 9:233
Ryan TJ, Grant SG (2009) The origin and evolution of synapses. Nat Rev Neurosci 10:701–712
Selkoe DJ (2002) Alzheimer's disease is a synaptic failure. Science 298:789–791
Shen H et al (2007) Reversal of neurosteroid effects at alpha4beta2delta GABAA receptors triggers anxiety at puberty. Nat Neurosci 10:469–477
Shim KS, Lubec G (2002) Drebrin, a dendritic spine protein, is manifold decreased in brains of patients with Alzheimer's disease and Down syndrome. Neurosci Lett 324:209–212
Shirao T, Obata K (1986) Immunochemical homology of 3 developmentally regulated brain proteins and their developmental change in neuronal distribution. Brain Res 394:233–244
Shirao T et al (1988) Molecular cloning of a cDNA for the developmentally regulated brain protein, drebrin. Brain Res 464:71–74
Shirao T (1995) The roles of microfilament-associated proteins, drebrins, in brain morphogenesis: a review. J Biochem (Tokyo) 117:231–236
Takahashi H et al (2003) Drebrin-dependent actin clustering in dendritic filopodia governs synaptic targeting of postsynaptic density-95 and dendritic spine morphogenesis. J Neurosci 23:6586–6595
Wable GS et al (2015) Anxiety is correlated with running in adolescent female mice undergoing activity-based anorexia. Behav Neurosci 129:170–182
Washbourne P, Bennett JE, McAllister AK (2002) Rapid recruitment of NMDA receptor transport packets to nascent synapses. Nat Neurosci 5:751–759
Wiesel TN (1982) The postnatal development of the visual cortex and the influence of environment. Nature 299(5884):583–591. www.nobelprize.org

Chapter 9
Drebrin in Neuronal Migration and Axonal Growth

Kenji Hanamura

Abstract During development, production of neurons from neural stem cells, migration of neurons from their birthplace to their final location, and extension of neurites, axons, and dendrites are important for the formation of functional neuronal circuits. The actin cytoskeleton has major roles in the morphological development of neurons. In this chapter, we focused on the distribution and function of the actin-binding protein, drebrin, to elucidate the importance of drebrin-bound F-actin in neurons during early developmental stages of neurons in embryonic, postnatal, and adult brains. There are three major isoforms of drebrin in the chicken brain (E1, E2, and A) and two major isoforms in the mammalian brain (E and A). Among these drebrin isoforms, drebrin E1 and E2 in chicken and drebrin E in the mammalian brain are involved in these neuronal stages. In migrating neurons of the developing and adult brain, drebrin is localized at the base of filopodia of leading processes, to regulate neuronal migration. In axonal growth cones, drebrin is localized in the transitional zone to regulate axonal growth by inhibiting actomyosin interactions and mediating the interactions between F-actin and microtubules. For axonal collateral branching, drebrin is localized at axonal actin patches and the base of filopodia, to accelerate the transition from actin patches to filopodia and stabilize the filopodia.

Keywords Drebrin • Neuronal migration • Axonal growth • F-actin • Leading process • Growth cone • Filopodia • Axonal collateral branching

9.1 Introduction

Neurogenesis from neural stem cells, migration from their birthplace to their final destination, and extension and branching of neurites, axons, and dendrites are important for the formation of functional neuronal circuits. In the morphological

K. Hanamura
Department of Neurobiology and Behavior, Gunma University Graduate School of Medicine, Maebashi, Gunma 371-8511, Japan
e-mail: kenji_hanamura@gunma-u.ac.jp

T. Shirao, Y. Sekino (eds.), *Drebrin*, Advances in Experimental Medicine and Biology 1006, DOI 10.1007/978-4-431-56550-5_9

development of neurons, the actin cytoskeleton has pivotal roles. Although there are many actin-binding proteins involved in these processes, we focused on drebrin, because drebrin-bound F-actin has distinct characteristics among F-actin bound with other actin-binding proteins.

9.2 Localization of Drebrin E Isoforms in Early Developmental Stages of Neurons

Drebrin is a side-binding protein of F-actin (Asada et al. 1994; Shirao et al. 1994, b; Ishikawa et al. 1994) that was originally identified in the developing brain (Shirao and Obata 1985) with a unique expression pattern during development. Drebrin has three major isoforms in chicken (drebrin E1, E2, and A; Kojima et al. 1988, 1993) and two major isoforms (drebrin E and A; Shirao et al. 1989a, b) and one minor isoform (s-drebrin A; Jin et al. 2002) in mammals. Among these isoforms, drebrin Es are involved in the early development of neurons.

The distribution of drebrin during development has been studied in chick brains (Shirao and Obata 1986). In the developing cerebellum on day 16 of incubation, drebrin is localized in the premigratory zone but not in the proliferative zone. This indicates that the expression of drebrin becomes prominent just after neuronal differentiation for formation and extension of neuronal processes during migration. Consistent with this observation, in the developing tectum on day 4 of incubation, drebrin is localized in cell bodies and processes of differentiated neurons in the superficial layer but not in undifferentiated cells. On day 7 of incubation, drebrin is concentrated in the axonal pathway, where axonal growth cones are enriched. However, in the late stages of the developing brain, drebrin does not appear to be expressed in the axonal pathway, where mature axonal fibers pass (Shirao and Obata 1986). The decrease in drebrin expression in the axonal pathway during development is also observed in kitten visual cortex (Imamura et al. 1992). In 1-week-old kittens, drebrin staining was occasionally observed in the white matter, but there was no drebrin staining in the white matter of 3-week-old kittens. These observations indicate that drebrin is localized in growing axons.

Detailed observations of drebrin localization in the early developmental stages of neurons have been done in hippocampal neuronal cultures (Mizui et al. 2009). In these observations, the developmental stages were classified according to Dotti's scheme (Dotti et al. 1988). After plating dissociated neurons, they were surrounded by lamellipodia, followed by extension of short processes and polarization to have longer neurites than the others. F-actin was distributed throughout the entire lamellipodia and the tips of processes and neurites, but drebrin was only localized at the base of the lamellipodia and filopodia (Mizui et al. 2009). Closer observation of the growth cones showed that drebrin was localized in the transitional zone, where F-actin and microtubules partially overlapped (Fig. 9.1a). This suggests that drebrin is involved in the interactions between microtubules and F-actin. We have also demonstrated that the growth cones of human-induced pluripotent stem cell (iPSC)-derived neurons have

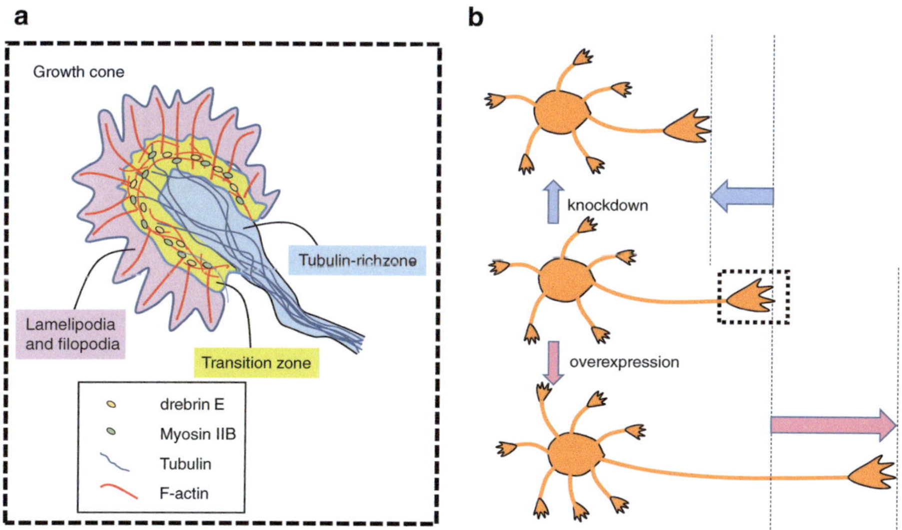

Fig. 9.1 Localization of drebrin in the axonal growth cone and its function in axonal growth. (**a**) Schematic drawing of the localization of F-actin, drebrin, myosin IIB, and tubulin in the growth cone of cultured neurons. (**b**) Schematic drawing of the effects of drebrin levels on axonal growth of cultured hippocampal neurons

distribution patterns of drebrin, F-actin, and microtubules that are similar to rat neurons, although the axonal growth is slower (Ohara et al. 2015).

9.3 Drebrin Function in Axonal Growth and Branching

A recent study has demonstrated the function of drebrin in the interaction between F-actin and microtubules during the early developmental stages of neurons (Geraldo et al. 2008). Pull-down assays and fluorescence energy transfer (FRET) using fluorescence-lifetime imaging microscopy (FLIM) demonstrated that drebrin directly binds to microtubule-binding protein EB3. In addition, inhibition of the EB3-drebrin interaction and knockdown of drebrin reduced neurite length. The drebrin's function of mediating the interaction between microtubules and F-actin has also been shown in dendritic spines (Merriam et al. 2013).

In addition to regulating the interaction between microtubules and F-actin, drebrin promotes axonal growth by regulating actomyosin interactions (Mizui et al. 2009; Fig. 9.2). Previous studies have demonstrated that inhibition of actomyosin interactions by the myosin II ATPase inhibitor, blebbistatin, promoted axonal growth (Ketschek et al. 2007; Rösner et al. 2007). Overexpression of drebrin E also increased the axonal length (Fig. 9.1b), but inhibition of myosin II ATPase activity masked the promoting effects of drebrin E overexpression on axonal growth. The association between drebrin and the actomyosin interactions has been studied in

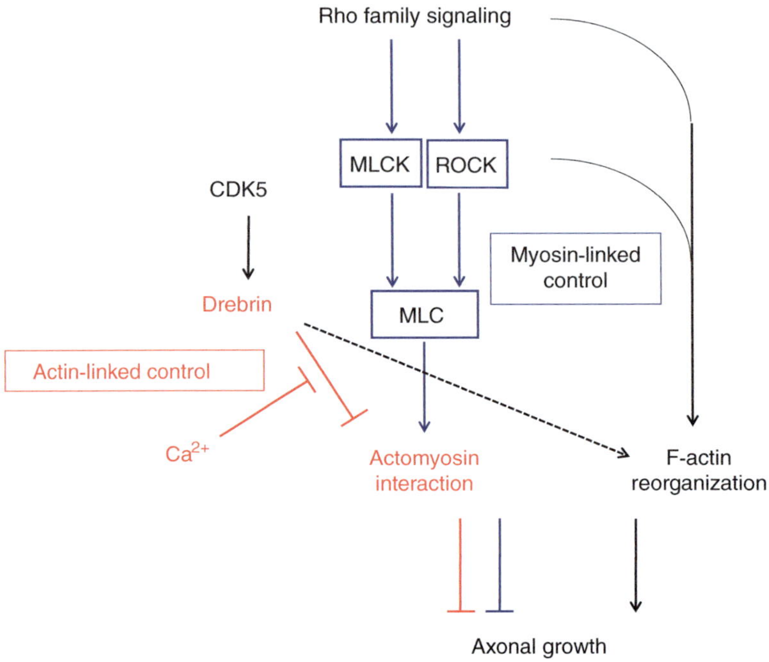

Fig. 9.2 Signaling pathways regulating axonal growth. Actomyosin interaction negatively regulates axonal growth. For this interaction phosphorylation of myosin light chain (MLC) of myosin II needs to be induced by myosin light chain kinase (MLCK) and Rho-associated kinase (ROCK) (myosin-linked control). On the other hand, this actomyosin interaction is inhibited by binding of drebrin to F-actin for accelerating the axonal growth (actin-linked control). Biochemical experiments have demonstrated that the binding of drebrin to F-actin reduced the sliding velocity of F-actin on myosin II and inhibited the actin-dependent ATPase activity of myosin II by up to 40% (Hayashi et al. 1996). In the skeletal muscle, the elevation of intracellular Ca^{2+} disinhibits the inhibition by tropomyosin. By analogy, the function of drebrin may be inhibited by Ca^{2+} in axonal growth cones of developing neurons (Sutherland et al. 2014). Consistent with this idea, accumulation level of drebrin is negatively regulated by intracellular Ca^{2+} concentration in dendritic spines of mature neurons (Mizui et al. 2014). Furthermore, F-actin reorganization induced by drebrin and Rho family signaling is also involved in the regulation of axonal growth. Phosphorylation of drebrin by CDK5 is associated with this F-actin reorganization

in vitro experiments. Experiments with purified drebrin indicated that drebrin can reduce the sliding velocity of F-actin on myosin and inhibit up to 40% of myosin II ATPase activity (Hayashi et al. 1996). These data support the idea that drebrin accelerates axonal growth by inhibiting the actomyosin interactions. Inhibition of myosin II ATPase activity enables microtubules to penetrate the peripheral domain. This suggests that drebrin's function of mediating the interaction between microtubules and F-actin is closely associated with the inhibitory effects of drebrin on myosin II ATPase activity in the axonal growth cone.

Previous studies have demonstrated that in neuronal growth cones, the speed of the movement of F-actin bundles slows down when they enter the transitional zone

(Medeiros et al. 2006). Our time-lapse imaging of GFP-drebrin E showed that the speed of the retrograde transport of drebrin in the transitional zone was comparable to that of severed F-actin in the transitional zone and slower than that of F-actin in the peripheral domain of growth cones (Mizui et al. 2009). These observations indicate that drebrin and F-actin are transported simultaneously in the transitional zone of axonal growth cones. In the transitional zone, F-actin bundles exhibit kinking or buckling and are terminated by a severing process, although in the peripheral domain F-actin bundles form filopodia (Medeiros et al. 2006). These morphological changes of F-actin may be regulated by drebrin. In the border between the peripheral region and the transitional zone, the actin cytoskeleton composition is altered from fascin-bound F-actin to drebrin-bound F-actin. Previous biochemical experiments have demonstrated that the binding of fascin to F-actin and the fascin-induced bundling of F-actin are inhibited by drebrin (Sasaki et al. 1996). In addition, the binding of alpha-actinin and tropomyosin to F-actin, but not filamin or caldesmon, is inhibited by drebrin (Ishikawa et al. 1994). These data suggest that competitive binding of these proteins to F-actin may be involved in the morphological changes of F-actin in the transitional zone of growth cones.

Furthermore, drebrin inhibits up to 40% of myosin II ATPase activity (Hayashi et al. 1996), thus promoting axonal growth, as described above. Because the inhibition of myosin II ATPase activity induces elongation of F-actin bundles in growth cones without newly formed kinking or buckling (Medeiros et al. 2006), kinking and buckling in the transitional zone may be regulated by the balance between myosin II and drebrin. Moreover, previous biochemical experiments have demonstrated that drebrin-bound F-actin forms a complex with gelsolin in the brain (Hayashi et al. 1996) and is sensitive to gelsolin activity (Ishikawa et al. 1994). In gelsolin-knockout neurons, a delayed retraction of filopodia has been observed (Lu et al. 1997a, b). These data suggest that severing of F-actin bundles in the transitional zone may be regulated by drebrin-induced changes in the sensitivity of F-actin to gelsolin, to control growth cone behavior. Further studies are required to understand how drebrin-induced morphological changes of F-actin are involved in the regulation of axonal growth (Fig. 9.2).

The function of drebrin in axonal growth has also been studied in the developing brain (Dun et al. 2012). Overexpression of drebrin E in the medial longitudinal fasciculus increased the number of filopodia per growth cone and the number of growth cones per axon. Interestingly, drebrin E without its C-terminal region (1–315) had more profound effects on these morphological changes than the full-length drebrin. In contrast, drebrin without its N-terminal region reduced the number of filopodia per growth cone and produced clubbed growth cones. In the oculomotor nerve, overexpression of drebrin constructs seemed to produce the same morphological changes in the axonal growth cones as in the longitudinal fasciculus, but neither overexpression nor shRNA knockdown affected the overall trajectory to their targets, the extraocular muscles.

In addition to axonal growth, drebrin has promoting effects on the initiation of axon collateral branches in embryonic dorsal root ganglion (DRG) sensory neurons (Ketschek et al. 2016). Overexpression of drebrin E1 increased the number of filopodia and branches of their axons in the spinal cord. In cultured DRG neurons, *drebrin* shRNA reduced the number of collateral branches and the length of the filopodia. Axonal collateral branches are transformed from axonal filopodia. These axonal filopodia emerge after the formation of actin patches in axons (Loudon et al. 2006). Therefore, the spontaneously formed actin patches have been thought to be the precursors of filopodia. Although neither the frequency of the appearance of new patches nor the life span of the patches increased by overexpression of drebrin E1, the transition from actin patches to filopodia was accelerated by the overexpression (Ketschek et al. 2016). In contrast, *drebrin* shRNA reduced both the frequency of actin patch emergence and the transition from actin patches to filopodia but increased the life span of actin patches. These data indicate that drebrin E1 has a role in the formation of filopodia by accelerating the transition from actin patches to filopodia and stabilizing filopodia. In dendrites, drebrin A has a pivotal role in the accumulation of F-actin in dendritic filopodia (Takahashi et al. 2003). The function of drebrin-bound F-actin in the stabilization of filopodia may be a common mechanism for the formation of axonal collateral branches and dendritic filopodia of developing neurons.

9.4 Drebrin Localization in Migrating Neurons

The localization of drebrin in migrating neurons has also been studied in neuronal cultures. In migrating cerebellar granule cells, drebrin is localized near submembranous regions of the cell bodies and processes (Asada et al. 1994; Tanaka et al. 2000). In migrating oculomotor neurons of the developing brain, drebrin is localized in the distal portion of the leading processes (Dun et al. 2012). This distribution of drebrin in the leading processes is different from the distribution of GAP-43 along their length and from that of neurofilament heavy chains in the base of the leading processes. In migrating neuroblasts of the mouse rostral migratory stream (RMS), drebrin is localized in the cell bodies and leading processes (Sonego et al. 2015). Consistent with the localization of drebrin in growth cones, drebrin is localized at the base of filopodia of leading processes and is excluded from the peripheral region of these filopodia, where Lifeact and fascin are localized. The differential localization of drebrin and fascin is consistent with the observation in the growth cones of PC12 cells, where fascin was localized in the filopodia of growth cones, but drebrin was localized in the basal region of growth cones (Sasaki et al. 1996).

9.5 Drebrin Function in Migrating Neurons

The function of drebrin in migrating neurons has been analyzed in the RMS of mouse pups (Sonego et al. 2015). *Drebrin* shRNA reduced the migration distance and velocity of migrating neurons. In these neurons, *drebrin* shRNA increased the number of secondary branches of leading processes. In contrast, drebrin in migrating oculomotor neurons of the developing brain used a different manner for regulating the morphology of the leading processes (Dun et al. 2012), although *drebrin* shRNA inhibited the migration of both neuronal populations. In oculomotor neurons, *drebrin* shRNA inhibited the formation of leading processes. The differential effects of *drebrin* shRNA on the morphology of leading processes may be caused by the surrounding substratum. Drebrin increases the axonal growth by inhibiting myosin II ATPase activity (Mizui et al. 2009), as described above, and the effects of myosin II ATPase activity on axonal growth depend on the substratum (Ketschek et al. 2007). Inhibition of myosin II ATPase activity decreased the speed of axonal growth on laminin but increased the rate of axonal growth on polylysine. The environmental conditions of migrating neurons in the RMS of mouse pups may be different from those of oculomotor neurons in the developing brain.

In addition to the morphology of leading processes and the migration distance and velocity, drebrin can affect the orientation of the migrating neurons. Overexpression of drebrin E1 and E2 reduced the size of the column of migrating neurons and induced the caudal elongation of the oculomotor nucleus (Dun et al. 2012). Interestingly, drebrin E without the C-terminal region (1–315) had more profound effects on the navigation of migrating oculomotor neurons than full-length drebrin E1 or E2 by inducing abnormal migration into the hindbrain. Hence, reduction in the size of the oculomotor nucleus overexpressing drebrin E without the C-terminal region was more prominent than overexpression of the full-length drebrin E1 or E2. These data indicate the importance of the C-terminal region in the navigation of migrating neurons.

9.6 Drebrin E in Migrating Neurons of the Adult Brain

The localization of drebrin E in migrating neurons was observed in the adult brain (Song et al. 2008; Figs. 9.3 and 9.4) as well as in the developing brain. In the adult brain, neurogenesis continuously occurs in some restricted regions. Neurons produced in the subventricular zone (SVZ) of the lateral ventricles of the forebrain migrate through the RMS to the olfactory bulb. Drebrin is localized in migrating neurons that have polysialylated-neural cell adhesion molecule (PSA-NCAM) and doublecortin. In the adult brain, drebrin A is the major isoform, but newly generated migratory neurons express drebrin E in their cell bodies and processes. This indicates that the isoform conversion of drebrin from E to A depends on the developmental stage of the individual neurons and not on the developmental stage of the animal.

After newly generated neurons from the SVZ enter the olfactory bulb, the RMS is localized in the core region of the olfactory bulb, where a strong drebrin signal is still observed in the cell bodies (Song et al. 2008). In the olfactory bulb, migrating neurons change their leading processes to a radial direction. In the granule cell layer of the olfactory bulb, the number of newly generated neurons that have drebrin in their cell body is reduced, and the amount of drebrin in their cell bodies seems to

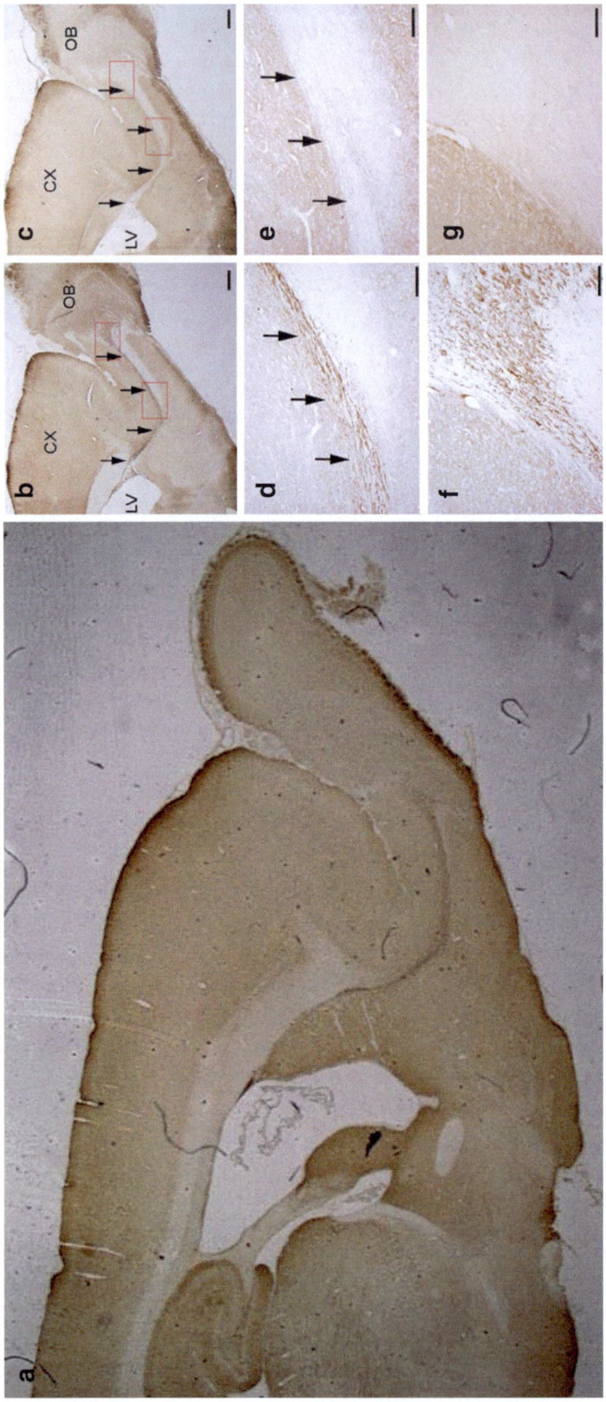

Fig. 9.3 Drebrin E is strongly expressed in the RMS of adult rat forebrain. Sagittal sections of adult rat forebrain stained with anti-pan-drebrin (**a**, **b**, **d**, and **f**) and anti-drebrin A (**c**, **e**, and **g**) antibodies. Intense immunostaining in the RMS was observed in (**a**), (**b**), (**d**), and (**f**) but not in (**c**), (**e**), and (**g**). (**d**) and (**e**) are magnified images of the lower-left red boxes in (**b**) and (**c**), respectively. (**f**) and (**g**) are magnified images of the upper-right red boxes in (**b**) and (**c**), respectively. Note that the stream spreads in the olfactory bulb (OB). *LV* lateral ventricles, *CX* cerebral cortex. Scale bars: 500 μm for (**b**) and (**c**), 100 μm for (**d**–**g**)

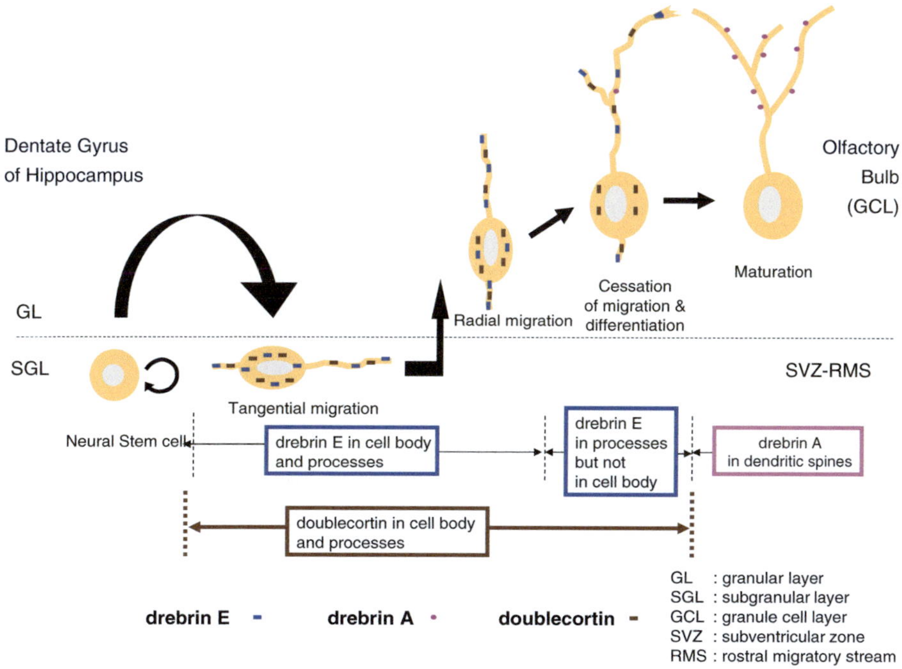

Fig. 9.4 Localization of drebrin E in newly generated neurons of the adult brain. Schematic drawing of the localization of drebrin E, drebrin A, and doublecortin in newly generated neurons of the adult brain. Cells in the lower-left part are neural stem cell and migrating neuron in the GL of the dentate gyrus and in the SVZ-RMS of the adult brain. Neurons in the upper-right portion are around or in their final destination (SGL in the dentate gyrus or GCL in olfactory bulb) to finish their migration, extend their dendrites and make synapses

decrease. The coincidence of drebrin reduction and disappearance from the cell bodies and the cessation of migration suggest that the level of drebrin in the cell body controls neuronal migration. Consistent with this idea, in vivo experiments demonstrated that the amount of drebrin E affected the distribution of newly generated neurons in the olfactory bulb of mouse pups (Sonego et al. 2015). Overexpression of drebrin increased the proportion of neurons around the core region of the olfactory bulb (inner olfactory bulb) and reduced the newly generated neurons around the granule cell layer of the olfactory bulb (outer olfactory bulb). This suggests that the reduction in drebrin in cell bodies and processes induces the transition from tangential migration to radial migration.

Although the amount of drebrin in the cell bodies is lower in migrating neurons in the granule cell layer of the olfactory bulb during radial migration, drebrin seems to also be important for radial migration of glutamatergic neurons in the developing cerebral cortex (Tanabe et al. 2014). Overexpression of phosphomimetic and non-phosphorylatable double mutants of drebrin A at Ser142 and Ser342 decreased the migration of cortical neurons differentiated at the ventricular zone. Because overexpression of wild-type (WT) drebrin A does not affect neuronal migration, these results suggest that the reversible phosphorylation and dephosphorylation are important for radial migration.

In addition to the SVZ, the subgranular layer of the dentate gyrus of hippocampus produces newly generated neurons in the adult brain (Song et al. 2008; Fig. 9.4). Smaller doublecortin-positive cells contain drebrin E in their cell bodies, but larger doublecortin-positive cells with apical dendrites do not have drebrin in their cell bodies (Song et al. 2008). This supports the idea that drebrin E is localized in migrating neurons in the dentate gyrus. Furthermore, the forebrain of songbirds produces newly generated neurons at adulthood. Drebrin is localized in the cell body and processes of these newly generated neurons migrating to song control nuclei (Nakatsubo et al. 1995). These data indicate that drebrin in the neurons' cell body may be a good marker of migrating neurons in the adult brain.

9.7 Drebrin in Neurons of the Injured Brain and Spinal Cord

Drebrin can also be useful as a marker of migrating neurons in injured adult brains. The number of drebrin-positive migrating neurons increased after a brain incision in infant rats (Inoue et al. 2003) or an olfactory bulbectomy in adult rats (Song et al. 2008). In addition to newly generated migrating neurons, drebrin is localized in the cell bodies of a small population of neurons (Shirao and Obata 1986). In the spinal ganglion, deep cerebellar nuclei and the subtectal nucleus isthmi pars parvocellularis, a small amount of drebrin is localized in the cytoplasm of some cell bodies. Interestingly, the amount of drebrin in the cell bodies of motor neurons increased after axotomy in the adult rat spinal cord (Kobayashi et al. 2001). Because drebrin can increase the axonal length of developing neurons, as described above, the increased level of drebrin in axotomized motor neurons may affect the regeneration of axons by promoting axonal growth.

9.8 Additional Molecular Mechanisms Associated with Drebrin in Early Developmental Stages of Neurons

The molecular mechanisms regulating the amount of drebrin and its phosphorylation status have been studied in developing neurons. Immunocytochemistry of cultured DRG neurons showed that nerve growth factor increases the level of drebrin in developing axons (Ketschek et al. 2016). Interestingly, new protein synthesis in axons is not required for this increase, because the protein synthesis inhibitors, anisomycin and cycloheximide, did not block this increase. In contrast, oxytocin increases the neurite length in parallel with enhancing the level of *drebrin* mRNA in neuroblastoma (Lestanova et al. 2016).

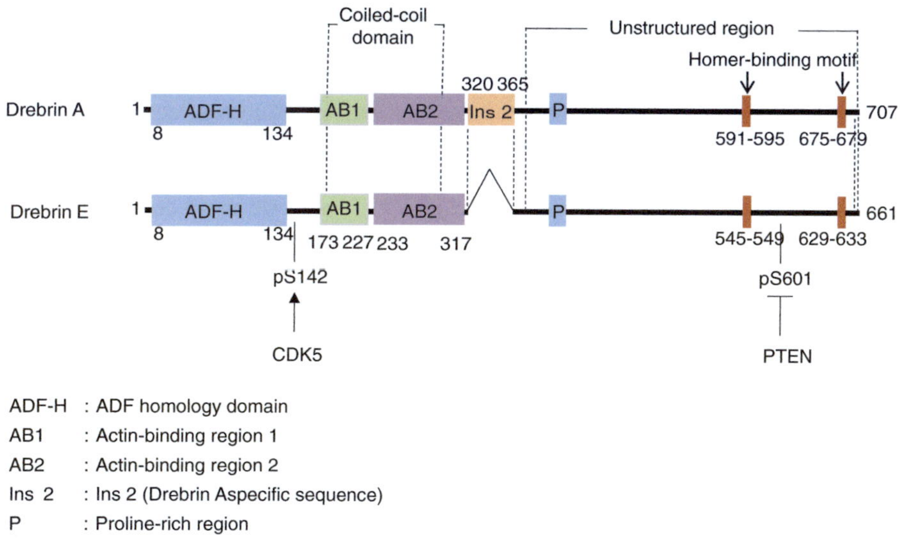

Fig. 9.5 Drebrin phosphorylation sites. Diagrammatic representation of drebrin showing the ADF homology (ADF-H) domain, actin-binding region 1 and actin-binding region 2 (AB1, AB2), drebrin A-specific sequence, ins2 (Ins 2), and the proline-rich region (P). Phosphorylation sites are shown

Posttranslational modifications of drebrin have been studied to understand the molecular mechanisms underlying drebrin-dependent regulation in the early stages of neuronal development. The Ser/Thr kinase, Cdk5, is known to phosphorylate Ser142 of drebrin E and A and Ser342 in the drebrin A-specific region (Worth et al. 2013; Tanabe et al. 2014; Figs. 9.2 and 9.5). This is associated with neuronal migration (Tanabe et al. 2014) and neurite outgrowth (Worth et al. 2013). In vitro experiments suggested that phosphorylation of drebrin at Ser142 regulates F-actin bundling by relieving intramolecular interactions (Worth et al. 2013). In addition to phosphorylation, a recent study has shown that the protein phosphatase, phosphatase and tensin homolog (PTEN), can dephosphorylate drebrin at a serine residue located between two homer-binding sites in the C-terminal region (Kreis et al. 2013; Fig. 9.5). PTEN is known to regulate neuronal migration and axonal growth (Chadborn et al. 2006; Sakakibara and Horwitz 2006). Therefore, the absence of regulation of drebrin phosphorylation in the C-terminal region may be the cause of the deterioration in abnormal migration of drebrin-overexpressing neurons induced by deletion of the C-terminal region (Dun et al. 2012). Because Cdk5 and PTEN mediate signaling from extracellular molecules, drebrin may mediate navigational signals during neuronal migration and axonal growth.

Chemokines are extracellular molecules that may be involved in the drebrin-dependent regulation of neuronal migration, because drebrin can interact with the chemokine receptor, CXCR4, via the C-terminal region to recruit CXCR4 to immune synapses (Pérez-Martínez et al. 2010). Previous studies have demonstrated that CXCR4 is essential for neuronal migration in various brain regions (Zou et al. 1998; Bagri et al. 2002; Stumm et al. 2003; Zhu et al. 2009). Thus, the drebrin-dependent CXCR4 function may also be involved in these processes. Drebrin is also known to regulate the localization of the cell-surface molecule, connexin 43, to maintain GAP junctions in astrocytes (Butkevich et al. 2004). Because abnormal neuronal migration has been observed in the cerebral cortex of connexin 43-knockout mice (Fushiki et al. 2003), drebrin may be involved in connexin 43-dependent regulation of neuronal migration. These results suggest that drebrin regulates neuronal migration by targeting cell-surface molecules in appropriate subcellular regions.

9.9 Summary and Future Perspectives

During the early developmental stages of neurons, drebrin E has pivotal roles in neuronal migration and axonal growth. Drebrin E is localized at the base of F-actin both in the leading processes of migrating neurons and in axonal growth cones, suggesting the presence of common regulatory machineries for regulating drebrin-dependent neuronal migration and axonal growth. Drebrin E is potentially involved in several molecular mechanisms that regulate neuronal migration and axonal growth. More detailed analysis of the actin cytoskeletal network with live cell imaging in combination with FRET and/or super resolution techniques will shed light on the molecular and cellular mechanisms underlying neuronal migration and axonal growth. Differential molecular mechanisms underlying drebrin-dependent neuronal migration and axonal growth are probably associated with different stages and/or populations of developing neurons. Drebrin appears to mediate the extracellular guidance signals around their pathways and from their targets, to find appropriate pathways and targets. The molecular mechanisms underlying drebrin-dependent neuronal migration may be different between tangential migration and radial migration. In addition, attractive and repulsive signals, diffusible and contact-mediated regulation, and/or terminal and collateral branching may have differential dependence on drebrin-dependent machineries. Drebrin mutant mice have been produced and analyzed in several laboratories including our group and The Jackson Laboratory (Aoki et al. 2009; Kojima et al. 2010; Jung et al. 2015; Law et al. 2015; Kojima et al. 2016; Stiber et al. 2016). Detailed analysis of the cytoarchitecture of the nervous system and the neuronal connections in these knockout mice will be evaluated in future studies.

References

Aoki C, Kojima N, Sabaliauskas N, Shah L, Ahmed TH, Oakford J, Ahmed T, Yamazaki H, Hanamura K, Shirao T (2009) Drebrin a knockout eliminates the rapid form of homeostatic synaptic plasticity at excitatory synapses of intact adult cerebral cortex. J Comp Neurol 517:105–121

Asada H, Uyemura K, Shirao T (1994) Actin-binding protein, drebrin, accumulates in submembranous regions in parallel with neuronal differentiation. J Neurosci Res 38:149–159

Bagri A, Gurney T, He X, Zou YR, Littman DR, Tessier-Lavigne M, Pleasure SJ (2002) The chemokine SDF1 regulates migration of dentate granule cells. Development 129:4249–4260

Butkevich E, Hülsmann S, Wenzel D, Shirao T, Duden R, Majoul I (2004) Drebrin is a novel connexin-43 binding partner that links gap junctions to the submembrane cytoskeleton. Curr Biol 14:650–658

Chadborn NH, Ahmed AI, Holt MR, Prinjha R, Dunn GA, Jones GE, Eickholt BJ (2006) PTEN couples Sema 3A signalling to growth cone collapse. J Cell Sci 119:951–957

Dotti CG, Sullivan CA, Banker GA (1988) The establishment of polarity by hippocampal neurons in culture. J Neurosci 8:1454–1468

Dun XP, Bandeira de Lima T, Allen J, Geraldo S, Gordon-Weeks P, Chilton JK (2012) Drebrin controls neuronal migration through the formation and alignment of the leading process. Mol Cell Neurosci 49:341–350

Fushiki S, Perez Velazquez JL, Zhang L, Bechberger JF, Carlen PL, Naus CC (2003) Changes in neuronal migration in neocortex of connexin 43 null mutant mice. J Neuropathol Exp Neurol 62:304–314

Geraldo S, Khanzada UK, Parsons M, Chilton JK, Gordon-Weeks PR (2008) Targeting of the F-actin-binding protein drebrin by the microtubule plus-tip protein EB3 is required for neuritogenesis. Nat Cell Biol 10:1181–1189

Hayashi K, Ishikawa R, Ye LH, He XL, Takata K, Kohama K, Shirao T (1996) Modulatory role of drebrin on the cytoskeleton within dendritic spines in the rat cerebral cortex. J Neurosci 16:7161–7170

Imamura K, Shirao T, Mori K, Obata K (1992) Changes of drebrin expression in the visual cortex of the cat during development. Neurosci Res 13:33–41

Inoue HK, Tezuka M, Oda O, Nishi H, Nagaseki Y, Kobahyashi S, Shirao T (2003) The role of subventricular zone (SVZ) cells in the repair of brain damage in infant rats -Immunohistochemical analysis compared with adult rats. Nerv Syst Child 28:331–335

Ishikawa R, Hayashi K, Shirao T, Xue Y, Takagi T, Sasaki Y, Kohama K (1994) Drebrin, a development-associated brain protein from rat embryo, causes the dissociation of tropomyosin from actin filaments. J Biol Chem 269:29928–29933

Jin M, Tanaka S, Sekino Y, Ren Y, Yamazaki H, Kawai-Hirai R, Kojima N, Shirao T (2002) A novel, brain-specific mouse drebrin: cDNA cloning, chromosomal mapping, genomic structure, expression, and functional characterization. Genomics 79:686–692

Jung G, Kim EJ, Cicvaric A, Sase S, Gröger M, Höger H, Sialana FJ, Berger J, Monje FJ, Lubec G (2015) Drebrin depletion alters neurotransmitter receptor levels in protein complexes, dendritic spine morphogenesis and memory-related synaptic plasticity in the mouse hippocampus. J Neurochem 134:327–339

Ketschek AR, Jones SL, Gallo G (2007) Axon extension in the fast and slow lanes: substratum-dependent engagement of myosin II functions. Dev Neurobiol 67:1305–1320

Ketschek A, Spillane M, Dun XP, Hardy H, Chilton J, Gallo G (2016) Drebrin coordinates the actin and microtubule cytoskeleton during the initiation of axon collateral branches. Dev Neurobiol 76:1092–1110

Kobayashi S, Shirao T, Sasaki T (2001) Drebrin expression is increased in spinal motoneurons of rats after axotomy. Neurosci Lett 311:165–168

Kojima N, Kato Y, Shirao T, Obata K (1988) Nucleotide sequences of two embryonic drebrins, developmentally regulated brain proteins, and developmental change in their mRNAs. Brain Res 464:207–215

Kojima N, Shirao T, Obata K (1993) Molecular cloning of a developmentally regulated brain protein, chicken drebrin a and its expression by alternative splicing of the drebrin gene. Brain Res Mol Brain Res 19:101–114

Kojima N, Hanamura K, Yamazaki H, Ikeda T, Itohara S, Shirao T (2010) Genetic disruption of the alternative splicing of drebrin gene impairs context-dependent fear learning in adulthood. Neuroscience 165:138–150

Kojima N, Yasuda H, Hanamura K, Ishizuka Y, Sekino Y, Shirao T (2016) Drebrin a regulates hippocampal LTP and hippocampus-dependent fear learning in adult mice. Neuroscience 324:218–226

Kreis P, Hendricusdottir R, Kay L, Papageorgiou IE, van Diepen M, Mack T, Ryves J, Harwood A, Leslie NR, Kann O, Parsons M, Eickholt BJ (2013) Phosphorylation of the actin binding protein Drebrin at S647 is regulated by neuronal activity and PTEN. PLoS One 8:e71957

Law M, Lee Y, Morales JL, Ning G, Huang W, Pabon J, Kannan AK, Jeong AR, Wood A, Carter C, Mohinta S, Song J, August A (2015) Cutting edge: drebrin-regulated actin dynamics regulate IgE-dependent mast cell activation and allergic responses. J Immunol 195:426–430

Lestanova Z, Bacova Z, Kiss A, Havranek T, Strbak V, Bakos J (2016) Oxytocin increases Neurite length and expression of cytoskeletal proteins associated with neuronal growth. J Mol Neurosci 59:184–192

Loudon RP, Silver LD, Yee HF, Gallo G (2006) RhoA-kinase and myosin II are required for the maintenance of growth cone polarity and guidance by nerve growth factor. J Neurobiol 66:847–867

Lu M, Witke W, Kwiatkowski DJ, Kosik KS (1997) Delayed retraction of filopodia in gelsolin null mice. J Cell Biol 138(6):1279–1287

Medeiros NA, Burnette DT, Forscher P (2006) Myosin II functions in actin-bundle turnover in neuronal growth cones. Nat Cell Biol 8:215–226

Merriam EB, Millette M, Lumbard DC, Saengsawang W, Fothergill T, Hu X, Ferhat L, Dent EW (2013) Synaptic regulation of microtubule dynamics in dendritic spines by calcium, F-actin, and drebrin. J Neurosci 33:16471–16482

Mizui T, Kojima N, Yamazaki H, Katayama M, Hanamura K, Shirao T (2009) Drebrin E is involved in the regulation of axonal growth through actin-myosin interactions. J Neurochem 109:611–622

Mizui T, Sekino Y, Yamazaki H, Ishizuka Y, Takahashi H, Kojima N, Kojima M, Shirao T (2014) Myosin II ATPase activity mediates the long-term potentiation-induced exodus of stable F-actin bound by drebrin a from dendritic spines. PLoS One 9:e85367

Nakatsubo N, Arai O, Shirao T (1995) Identification of the newly generated neurons of canaries during adulthood. Dokkyo J Med Sci 22:65–72

Ohara Y, Koganezawa N, Yamazaki H, Roppongi RT, Sato K, Sekino Y, Shirao T (2015) Early-stage development of human induced pluripotent stem cell-derived neurons. J Neurosci Res 93:1804–1813

Pérez-Martínez M, Gordón-Alonso M, Cabrero JR, Barrero-Villar M, Rey M, Mittelbrunn M, Lamana A, Morlino G, Calabia C, Yamazaki H, Shirao T, Vázquez J, González-Amaro R, Veiga E, Sánchez-Madrid F (2010) F-actin-binding protein drebrin regulates CXCR4 recruitment to the immune synapse. J Cell Sci 123:1160–1170

Rösner H, Möller W, Wassermann T, Mihatsch J, Blum M (2007) Attenuation of actinomyosin II contractile activity in growth cones accelerates filopodia-guided and microtubule-based neurite elongation. Brain Res 1176:1–10

Sakakibara A, Horwitz AF (2006) Mechanism of polarized protrusion formation on neuronal precursors migrating in the developing chicken cerebellum. J Cell Sci 119:3583–3592

Sasaki Y, Hayashi K, Shirao T, Ishikawa R, Kohama K (1996) Inhibition by drebrin of the actin-bundling activity of brain fascin, a protein localized in filopodia of growth cones. J Neurochem 66:980–988

Shirao T, Obata K (1985) Two acidic proteins associated with brain development in chick embryo. J Neurochem 44:1210–1216

Shirao T, Obata K (1986) Immunochemical homology of 3 developmentally regulated brain proteins and their developmental change in neuronal distribution. Brain Res 394:233–244

Shirao T, Kojima N, Nabeta Y, Obata K (1989a) Two forms of drebrins, developmentally regulated brain proteins, in rat. Proc Jpn Acad Ser B Phys Biol Sci 65:169–172

Shirao T, Kojima N, Nabeta Y, Obata K (1989b) Two forms of drebrins, developmentally regulated brain proteins, in rat. Proc Jpn Acad Ser B Phys Biol Sci 65(7):169–172

Shirao T, Hayashi K, Ishikawa R, Isa K, Asada H, Ikeda K, Uyemura K (1994a) Formation of thick, curving bundles of actin by drebrin A expressed in fibroblasts. Exp Cell Res 215:145–153

Shirao T, Hayashi K, Ishikawa R, Isa K, Asada H, Ikeda K, Uyemura K (1994b) Formation of thick, curving bundles of actin by drebrin A expressed in fibroblasts. Exp Cell Res 215(1):145–153

Sonego M, Oberoi M, Stoddart J, Gajendra S, Hendricusdottir R, Oozeer F, Worth DC, Hobbs C, Eickholt BJ, Gordon-Weeks PR, Doherty P, Lalli G (2015) Drebrin regulates neuroblast migration in the postnatal mammalian brain. PLoS One 10:e0126478

Song M, Kojima N, Hanamura K, Sekino Y, Inoue HK, Mikuni M, Shirao T (2008) Expression of drebrin E in migrating neuroblasts in adult rat brain: coincidence between drebrin E disappearance from cell body and cessation of migration. Neuroscience 152:670–682

Stiber JA, Wu JH, Zhang L, Nepliouev I, Zhang ZS, Bryson VG, Brian L, Bentley RC, Gordon-Weeks PR, Rosenberg PB, Freedman NJ (2016) The actin-binding protein Drebrin inhibits Neointimal hyperplasia. Arterioscler Thromb Vasc Biol 36:984–993

Stumm RK, Zhou C, Ara T, Lazarini F, Dubois-Dalcq M, Nagasawa T, Höllt V, Schulz S (2003) CXCR4 regulates interneuron migration in the developing neocortex. J Neurosci 23:5123–5130

Sutherland DJ, Pujic Z, Goodhill GJ (2014) Calcium signaling in axon guidance. Trends Neurosci 37:424–432

Takahashi H, Sekino Y, Tanaka S, Mizui T, Kishi S, Shirao T (2003) Drebrin-dependent actin clustering in dendritic filopodia governs synaptic targeting of postsynaptic density-95 and dendritic spine morphogenesis. J Neurosci 23:6586–6595

Tanabe K, Yamazaki H, Inaguma Y, Asada A, Kimura T, Takahashi J, Taoka M, Ohshima T, Furuichi T, Isobe T, Nagata K, Shirao T, Hisanaga S (2014) Phosphorylation of drebrin by cyclin-dependent kinase 5 and its role in neuronal migration. PLoS One 9:e92291

Tanaka S, Sekino Y, Shirao T (2000) The effects of neurotrophin-3 and brain-derived neurotrophic factor on cerebellar granule cell movement and neurite extension in vitro. Neuroscience 97:727–734

Worth DC, Daly CN, Geraldo S, Oozeer F, Gordon-Weeks PR (2013) Drebrin contains a cryptic F-actin-bundling activity regulated by Cdk5 phosphorylation. J Cell Biol 202:793–806

Zhu Y, Matsumoto T, Mikami S, Nagasawa T, Murakami F (2009) SDF1/CXCR4 signalling regulates two distinct processes of precerebellar neuronal migration and its depletion leads to abnormal pontine nuclei formation. Development 136:1919–1928

Zou YR, Kottmann AH, Kuroda M, Taniuchi I, Littman DR (1998) Function of the chemokine receptor CXCR4 in haematopoiesis and in cerebellar development. Nature 393:595–599

Chapter 10
Drebrin and Spine Formation

Hideto Takahashi and Yusuke Naito

Abstract Dendritic spines are the postsynaptic receptive regions of most excitatory synapses in the central nervous system. Thus, spines are supposed to act as a fundamental unit for information processing of the brains. Previous studies have demonstrated the roles of drebrin in the formation of dendritic spines and in the recruitment of synaptic proteins to postsynaptic sites. Further, a live imaging study has revealed the unique dynamics of drebrin in dendritic spines, which help to understand how drebrin is involved in dendritic spine formation. This review will provide a basic knowledge about dendritic spine and overview recent progresses in understanding of the roles of drebrin in dendritic spine morphogenesis and synaptogenesis.

Keywords Dendritic spine • Dendritic filopodia • Actin cytoskeleton • Postsynaptic density (PSD) • Activity dependent • Protein trafficking • FRAP

10.1 The Biology of Dendritic Spine

10.1.1 Dendritic Spine

In the neurons of the adult brains, the most glutamatergic presynaptic terminals innervate dendritic spines, tiny protoplasmic protrusions emerging from dendritic shafts of the target neurons (Harris and Kater 1994; Rochefort and Konnerth 2012; DeFelipe 2015). Dendritic spines constitute fundamental neuronal contact sites for excitatory synapses. Spines are composed of three major components: a base connected to a dendritic shaft, a neck, and a head which connects with an axon. Spines are typically less than 2 μm in length and possess a spine head approximately 1 μm in width. The morphology of dendritic spines varies largely among spines and within one spine over time. Based on their morphology in fixed brain tissues and fixed cultured neurons, dendritic spines are generally classified into three

H. Takahashi (✉) • Y. Naito
Institut de Recherches Cliniques de Montréal (IRCM), Montréal, QC H2W 1R7, Canada
e-mail: Hideto.Takahashi@ircm.qc.ca

T. Shirao, Y. Sekino (eds.), *Drebrin*, Advances in Experimental Medicine and Biology 1006, DOI 10.1007/978-4-431-56550-5_10

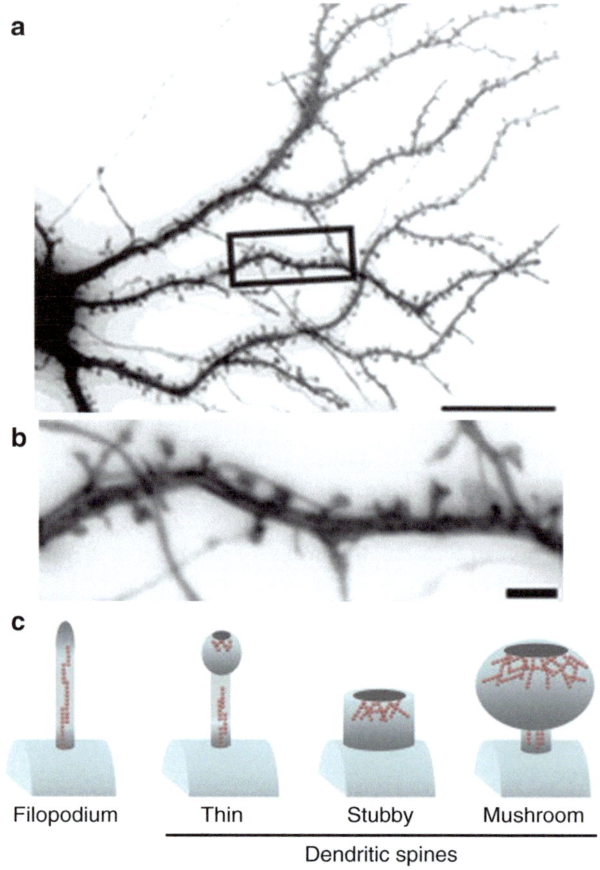

Fig. 10.1 (**a**, **b**) A microscope image of a hippocampal neuron in low-density cultures at 21 days in vitro by DiI labeling. Many mushroom-shaped spines emerging from dendritic shaft are observed in (**b**). Scale bars: 20 μm in (**a**) and 2 μm in (**b**). Adapted from Shirao and González-Billault (2013) with permission. (**c**) Schematic representation of morphologies of dendritic protrusions. There are two types of dendritic protrusions, dendritic filopodium, a long thin headless protrusion, and dendritic spine, which has bulbous head. Dendritic spines also exhibit three different types: thin, stubby, and mushroom type. *Gray disks* represent the PSD structure and chains of *red circles* represent F-actin. Adapted from Sekino et al. (2007) with permission

morphological types: (1) thin spines, which have a small head and narrow neck; (2) stubby spines, which don't have neck; and (3) mushroom-shape spines, which have a relatively large head and short neck (Fig. 10.1). Time-lapse imaging studies have shown that one spine is not static and takes transition among these morphological states, suggesting that observed spine shapes are snapshots of morphologically dynamic spines (Dailey and Smith 1996; Ziv and Smith 1996; Dunaevsky et al. 1999). Importantly, the size and shape of dendritic spines are changed in response to synaptic activity and experiences, reflecting the nature of synaptic plasticity. For example, long-term potentiation (LTP) is associated with the increase in spine head volume (Matsuzaki et al. 2004; Maletic-Savatic et al. 1999) and the length of spine neck (Araya et al. 2006). In contrast, long-term depression is accompanied by the shrink of dendritic spines (Zhou et al. 2004). Thus, the morphological plasticity of dendritic spines is supposed to be the cellular basis of synaptic plasticity underlying learning and memory. This idea also suggests the structure-function relationship of dendritic spines, which will be more discussed later.

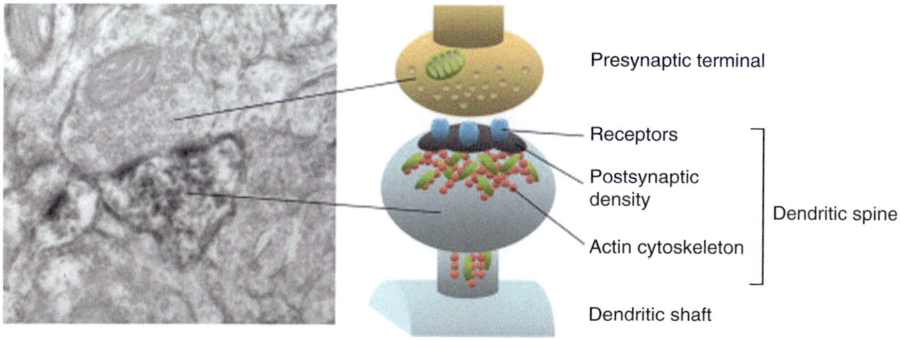

Fig. 10.2 Ultrastructure of dendritic spine synapse. The *left* panel shows an immunno-electron microscopic image for drebrin at an asymmetric synapse. Drebrin is enriched in dendritic spine but not presynaptic terminal. The *right* schematic image shows basic components of an asymmetric synapse. Dendritic spine consists of an actin cytoskeleton with actin-binding proteins and a PSD structure scaffolding neurotransmitter receptors. Adapted from Sekino et al. (2007) with permission

10.1.2 Structural Elements of Dendritic Spines: Actin Cytoskeleton and PSD

Dendritic spines are composed of two major structural elements: actin cytoskeleton and postsynaptic density (PSD) (Kennedy 1997; Sekino et al. 2007) (Fig. 10.2). Actin is a highly abundant protein (typically 5–10% of total proteins) in eukaryotic cells including neurons. Individual actin molecules have two states, a monomeric state as a globular actin (G-actin) and a polymeric state (Kabsch and Vandekerckhove 1992). Polymerization of monomeric actins results in the formation of filamentous actin (F-actin). G-actin molecules reversibly bind to the end of F-actin. Thereby, actin has dynamic equilibrium between G-actin and F-actin, which is one of key factors that determine actin cytoskeleton in dendritic spines (Okamoto et al. 2004). F-actin generally forms two types of structures, F-actin bundles and F-actin networks (Svitkina 2013). Cells including neurons express many F-actin binding proteins, which determine and/or modify F-actin bundles and networks. In general, F-actin regulates cell structures by modulating its bundle and network structures. For example, cellular projections including filopodia and microvilli contain F-actin bundles, whereas the peripheral submembranous region of cells mainly contains F-actin networks. In neuron, F-actin is enriched in dendritic spines (Matus et al. 1982; Cohen et al. 1985). Electron microscopic studies have shown that a spine head contains the network (lattice structure) of F-actin, whereas a spine neck contains F-actin bundles like filopodia (Landis and Reese 1983). In addition, F-actin generally regulates the motile force of cells through the modulation of polymerization and depolymerization states of F-actin and the activity of myosin ATPase (Pollard and Borisy 2003). Time-lapse imaging studies using green fluorescent protein (GFP)-tagged actin have revealed that dendritic spines undergo rapid continuous motility based on actin dynamics (Star et al. 2002). The reagents that change actin polymerization states such as cytochalasin D and latrunculin A alter spine motility. Interestingly, spine motility is also regulated by glutamate receptor activations (Fischer et al. 2000), suggesting that spine motility may contribute to the plasticity of excitatory synapses. In dendritic spines, many F-actin binding proteins including drebrin exist and regulate the shapes, size, and dynamics of dendritic spines.

The PSD is an electron-dense highly organized protein structure attached to the postsynaptic membrane of dendritic spines (Kennedy 1997) (Fig. 10.2). PSD is composed of various scaffold proteins such as membrane-associated guanylate kinases (MAGUKs) and many PSD-95/Dlg/ZO-1 (PDZ)-containing proteins (Kim et al. 1995; Kim and Sheng 2004). PSD-95 is a member of MAGUK family and a typical PDZ protein abundant in PSD (Cho et al. 1992; Kistner et al. 1993). PSD-95 directly binds to NMDA-type glutamate receptor subunit GluN2 (Niethammer et al. 1996) and indirectly binds to AMPA-type glutamate receptor via transmembrane AMPA receptor regulatory proteins (TARPs) such as stargazin (Tomita et al. 2005, 2004; Chen et al. 2000), suggesting that PSD-95 plays roles in regulating the trafficking and anchoring of glutamate receptors. PSD-95 is further involved in dendritic spine morphology. Overexpression of PSD-95 in cultured neurons enlarges dendritic spine (El-Husseini et al. 2000). Overexpression studies have also shown that other PSD proteins such as Homer and Shank regulate spine morphology (Sala et al. 2001). Thus, PSD proteins are also a key element that controls dendritic spine structure. Interestingly, many PSD proteins are physically and functionally associated with F-actin. Thus, it has been suggested that actin cytoskeleton and PSD cooperatively, rather than independently, regulate dendritic spine morphology.

10.1.3 Structure-Function Relationship of Dendritic Spines

Many evidences about the morphological changes of dendritic spines in synaptic plasticity have suggested structure-function relationship of dendritic spines (Kasai et al. 2003). Several previous studies have revealed that the size of dendritic spines is closely correlated with their synaptic function. An electromicroscopic study with postembedding immunogold labeling has shown that smaller spines with less than about 180 nm of PSD lack AMPA receptors (Takumi et al. 1999). These smaller spines are called "silent synapses," which do not generate excitatory postsynaptic potentials under basal stimulation. In addition, the size of PSD in dendritic spines is correlated with the number of AMPA receptors, supporting that larger spines are more functional in terms of AMPA receptor-mediated excitatory postsynaptic potentials. Another study using a glutamate uncaging technique has further revealed that spine head volume is correlated with the expression of functional AMPA receptors (Matsuzaki et al. 2001). Such structure-function relationship of dendritic spines suggests that the molecular and cellular mechanisms that control spine structure are important for our understanding of how functional neuronal circuits in the brains develop and are modified by experiences for higher brain functions such as cognition, learning, and memory (Kasai et al. 2003, 2010).

10.1.4 Dendritic Spine Morphogenesis: The Role of Dendritic Filopodia

How do dendritic spines form during neuronal development? Immature neurons, especially before synaptogenesis period, do not have dendritic spines. Instead, they possess dendritic filopodia, thin headless protrusions emerging from dendritic shaft (Yoshihara et al. 2009; Ebrahimi and Okabe 2014). The major difference of dendritic filopodia from spines is the lack of a bulbous head. As neurons mature, dendritic filopodia disappear, and dendritic spines emerge. The developmental transition from dendritic filopodia to spines is supposed to be crucial for the development of functional excitatory synapses on target neurons. The developmental transition further suggests that dendritic filopodia may act as a precursor of dendritic spines. This possibility is supported by several time-lapse imaging studies (Dailey and Smith 1996; Ziv and Smith 1996; Dunaevsky et al. 1999). A majority of dendritic filopodia rapidly extends and retracts with short-term lifetime (around 10 min) in an early developmental stage. As innervated axons increase along dendrites during development, some filopodia can make physical contact with axons and are stabilized by the association of functional presynaptic terminals. This observation suggests that dendritic filopodia can be transformed into dendritic spines after the physical contact with axons. What molecular mechanism governs the transformation of dendritic filopodia to spines? A previous time-lapse imaging has shown that the initiation of spine morphogenesis precedes synaptic accumulation of PSD-95 (Okabe et al. 2001). Further, PSD-95 mutant mice exhibit normal dendritic spine morphogenesis (Migaud et al. 1998). Thus, PSD-95 would be not required for spine morphogenesis during development. On the other hand, the pharmacological experiment has shown that depolymerization of F-actin by latrunculin A causes the disassembly of some synaptic structural elements (Allison et al. 2000). Further, the important evidence is that dendritic filopodia are actin-rich structure like spines (Fischer et al. 1998; Dunaevsky et al. 1999). Therefore, actin-based mechanisms would be crucial for the developmental transition from dendritic filopodia to dendritic spines.

10.2 The Roles of Drebrin in Dendritic Spine Morphogenesis

As described in the previous chapters (see Chaps. 3 and 7), drebrin is an F-actin-binding protein and enriched in dendritic spines (Figs. 10.2 and 10.3). Drebrin has two major isoforms, drebrin E and drebrin A based on alternative splicing of exon 12a. The expression of drebrin A isoform is upregulated at synaptogenesis period. Further, the amount of drebrin positively correlates with spine head size in the adult mouse cerebral cortex (Kobayashi et al. 2007). Together, drebrin would be one of the best candidates as an actin-based molecular mechanism that regulates dendritic spine morphogenesis during development. A cell-biological study on drebrin has

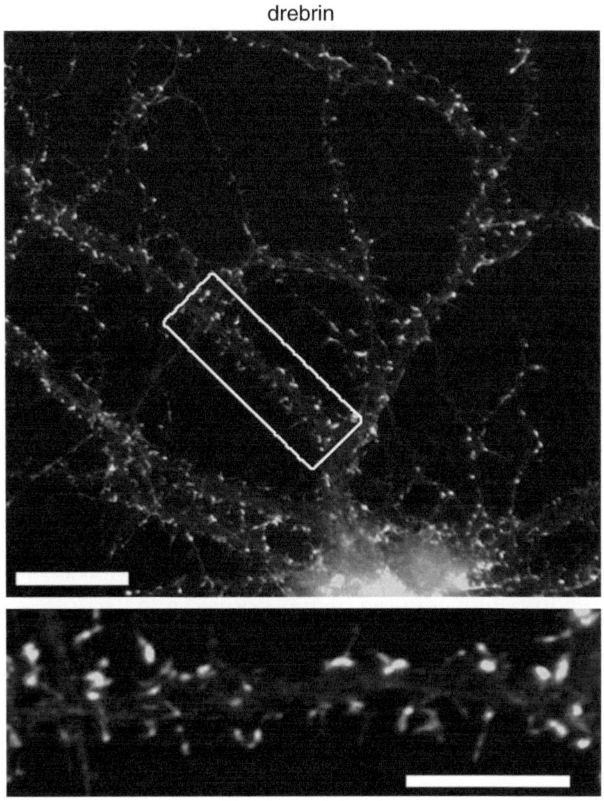

Fig. 10.3 Drebrin immunolabeling of a cultured hippocampal neuron at 15 days in vitro by mouse monoclonal anti-drebrin antibody M2F6. Drebrin clusters are observed at dendritic spines. The lower panel presents the boxed region in the upper panel. Scale bars: 20 μm (*upper*), 10 μm (*lower*). Adapted from Takahashi et al. (2009) with permission

demonstrated that the cluster formation of drebrin occurs in dendritic filopodia, especially at the contact site between dendritic filopodia and axons (Takahashi et al. 2003). In addition, drebrin A knockdown experiments have revealed that drebrin clustering in dendritic filopodia is required for dendritic spine morphogenesis during development and synaptic accumulation of PSD-95 (Takahashi et al. 2003, 2006). The following paragraphs will focus on describing the developmental changes of drebrin localization in dendritic filopodia and the roles of drebrin in dendritic spine morphogenesis during development.

10.2.1 Drebrin Clustering in Dendritic Filopodia

As dendritic filopodia and spines both are actin-rich structure, developmental transition from dendritic filopodia to spines can be visualized by F-actin labeling of hippocampal neurons in low-density culture using fluorescent-conjugated phalloidin

(Takahashi et al. 2003). At 7 days in vitro (DIV), almost all of F-actin-positive dendritic protrusions are observed as long, thin, and headless (filopodia-like) protrusions. At 14 DIV, the dendritic protrusions show the mixed population of dendritic filopodia and spines, including long and thin protrusions, with or without a small head, and mushroom-type or stubby spine like protrusions. At 21 DIV, almost all of the dendritic protrusions exhibit mushroomlike or stubby shapes with enriched F-actin levels. Thus, F-actin labeling of cultured hippocampal neurons is useful for observing the developmental shift from dendritic filopodia to spines. This observation is also consistent with the previous studies using fluorescent dyes (Papa et al. 1995; Ziv and Smith 1996).

How is drebrin distribution changed in association with the developmental changes of F-actin-positive dendritic protrusions? Double labeling for drebrin and F-actin has shown interesting changes in drebrin distribution in dendritic shaft and protrusions. At 7 DIV, the intense immunoreactivity of drebrin with F-actin staining is detected at the submembranous regions along dendrites. The distributions of drebrin and F-actin in dendritic filopodia exhibit the fibrous staining pattern. At 21 DIV, most of the drebrin immunoreactivity is detected as clusters colocalized with F-actin in dendritic spines. At dendritic shaft, drebrin immunoreactivity is not detectable although F-actin staining remained there. These data support that drebrin is enriched in dendritic spines and also indicate that drebrin distribution is not completely overlapped with F-actin distribution although drebrin binds to F-actin. At 14 DIV, transitional distribution of drebrin can be detected. The detailed analysis of drebrin-F-actin labeling at 14 DIV has revealed that there are two types of drebrin distribution in dendritic filopodia, diffuse pattern and cluster pattern. Thus, dendritic filopodia can be classified into either (1) cluster-type filopodia or (2) diffuse-type filopodia in terms of the existence of drebrin clusters (Takahashi et al. 2003). According to the quantitative analysis, at 7 DIV, approximately 90% of dendritic filopodia is classified as diffuse-type filopodia, and the rest is classified as cluster-type filopodia. From 7 to 14 DIV, the population of diffuse-type and cluster-type filopodia is drastically changed. At 14 DIV, the number of cluster-type filopodia doubles that at 7 DIV and accounts for 50% of total filopodia. In contrast, the number of diffuse-type filopodia at 14 DIV drops by half compared to that at 7 DIV. Thus, drebrin-F-actin labeling reveals the developmental transition from diffuse-type filopodia to cluster-type ones. The transition of diffuse-type filopodia to cluster-type ones was also observed by the time-lapse imaging of GFP-tagged drebrin A (Fig. 10.4). Previous time-lapse studies have shown that some filopodia are very dynamic, whereas other filopodia are converted to stable spines (Dailey and Smith 1996; Ziv and Smith 1996; Dunaevsky et al. 1999). In addition, the dynamics of dendritic filopodia and spines are stabilized during development (Ziv and Smith 1996). The diffuse-type filopodia may represent the dynamic filopodia, whereas the cluster-type filopodia may represent the transitional stage to stable spines.

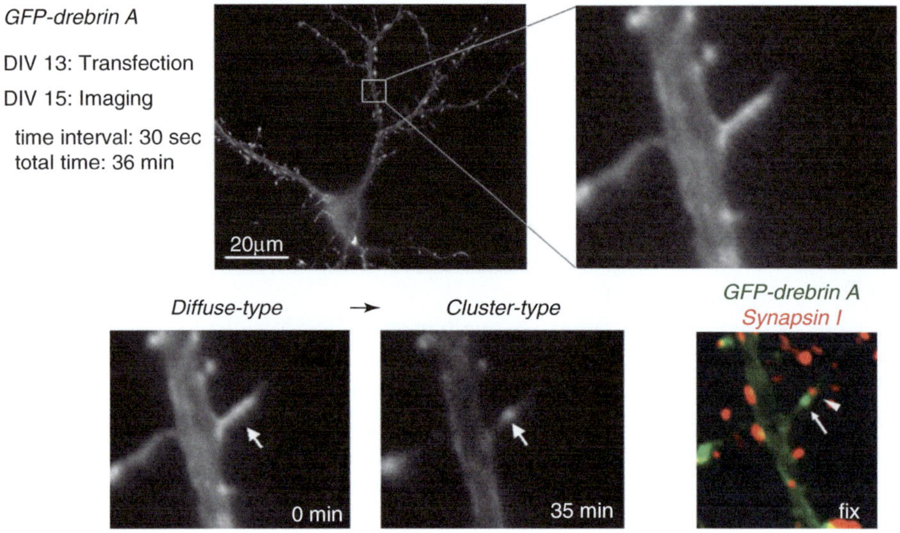

Fig. 10.4 The time-lapse imaging of GFP-tagged drebrin A. GFP-tagged drebrin A was transfected into hippocampal neurons at 13 days in vitro. Two days after the transfection, GFP signal was imaged every 30 s for 35 min by confocal microscopy. The bottom panel shows that drebrin A was diffusely distributed throughout a filopodium (an *arrow* in the *left bottom* panel) in the beginning of the imaging. 35 min after starting the imaging, the diffuse pattern of drebrin A distribution shifted to a cluster pattern (an *arrow* in the *middle bottom* panel). This drebrin A cluster was juxtaposed with presynaptic maker synapsin I (*arrowhead* in the *right bottom* panel)

10.2.2 Drebrin Clustering at Synaptic Contact of Dendritic Filopodia

Drebrin clustering in dendritic filopodia is detected from synaptogenesis period, suggesting that drebrin clustering is associated with synaptic contact of dendritic filopodia (Takahashi et al. 2003). The quantitative analysis of triple labeling of cultured hippocampal neurons for drebrin, F-actin, and synapsin I, a presynaptic vesicle protein, has shown that a majority of cluster-type filopodia is associated with synapsin I clusters, indicating that cluster-type filopodia have synaptic contact. In contrast, most of diffuse-type filopodia is not associated with synapsin I clusters. The important observation is that drebrin clusters in the synaptic cluster-type filopodia are always juxtaposed with synapsin I clusters, demonstrating the tight association of drebrin clustering in dendritic filopodia with presynaptic contact. These data suggest that drebrin clusters in filopodia are precursors of postsynaptic structures of spine synapses. Further, it has been suggested that drebrin clustering would be triggered by synaptic contact. The molecular and cellular mechanisms that trigger drebrin clustering will be discussed later.

10.2.3 *Drebrin Clustering vs PSD-95 Clustering in Dendritic Filopodia*

The major structural components of dendritic spines are actin cytoskeleton and PSD (Kennedy 1997; Sekino et al. 2007). Given drebrin clustering at the contact site between dendritic filopodia and presynaptic terminals, it is interesting to address whether the formation of PSD at the contact sites also occurs together with drebrin clustering. Triple labeling for F-actin, synapsin I, and PSD-95 or drebrin has revealed that almost all of dendritic spines with presynaptic contacts (synaptic spines) contain both PSD-95 and drebrin clusters (Takahashi et al. 2003). In contrast, only about half (57%) of dendritic filopodia with presynaptic contacts (synaptic filopodia) contains PSD-95 clusters, but almost all (86%) of synaptic filopodia contains drebrin clusters (Takahashi et al. 2003). These quantitative data indicate that postsynaptic clustering of drebrin in dendritic filopodia may be followed by that of PSD-95. After dendritic filopodia contact with axons, actin reorganization tends to precede PSD formation in dendritic filopodia, supporting that dendritic filopodia with drebrin clusters (cluster-type filopodia) would be a precursor of dendritic spines (protospines). According to a previous electromicroscopical study (Fiala et al. 1998), some filopodia possess nascent synapses containing PSD structure. A time-lapse imaging study has also shown that filopodia with PSD-95 clusters directly transform into mature spines during development (Marrs et al. 2001). Therefore, in developing neurons, most of filopodia containing both drebrin and PSD-95 clusters may exhibit the transitional stage into mature spines although it is also possible that some filopodia with drebrin and PSD-95 clusters represent temporal snapshots of mature spines.

10.2.4 *Drebrin Clustering Is Required for PSD-95 Clustering*

The evidence that drebrin clustering precedes PSD-95 clustering in dendritic filopodia raises an interesting question whether drebrin clustering is required for PSD-95 clustering. As described in Chap. 2, drebrin has two major alternative splicing isoforms, drebrin E and drebrin A. Notably, the upregulation of drebrin A expression occurs in synaptogenesis period, in which drebrin clustering is also observed (Hayashi et al. 1998). These data suggest that developmental upregulation of drebrin A is essential for drebrin clustering and the initial step of synaptogenesis. To suppress the drebrin A upregulation, the antisense oligonucleotide that targets drebrin A-specific nucleotide sequence has been developed (Takahashi et al. 2003). The drebrin A antisense oligonucleotides selectively suppress the expression of drebrin A but not that of drebrin E or actin. As expected, drebrin A knockdown by the antisense oligonucleotides indeed inhibits drebrin clustering in dendritic protrusions

(Takahashi et al. 2003). Drebrin A knockdown significantly reduces the number of cluster-type filopodia but increases that of diffuse-type filopodia, indicating that drebrin A knockdown inhibits the developmental transition from diffuse-type filopodia to cluster-type ones (Takahashi et al. 2003). How about PSD-95 clustering in the case of drebrin A knockdown? Drebrin A knockdown significantly suppresses synaptic clustering of PSD-95. In addition, the rescue of drebrin A expression into the knocked-down neurons restores synaptic clustering of PSD-95 (Takahashi et al. 2003). Together drebrin clustering triggered by drebrin A upregulation is crucial for synaptic clustering of PSD-95 during development.

10.2.5 Drebrin Clustering Is Required for Dendritic Spine Morphogenesis

Drebrin A knockdown in cultured neurons also impairs dendritic spine morphogenesis. Drebrin A knockdown significantly reduces the number and the width of dendritic spines (Takahashi et al. 2006). In consequence, drebrin A knockdown causes filopodia-like thin dendritic protrusions. Given that drebrin A is necessary for drebrin clustering and PSD-95 clustering, it has been suggested that drebrin clustering is crucial for the morphogenesis of mature dendritic spines with PSD structure. Together, the following model of how drebrin is involved in dendritic spine morphogenesis during development has been proposed (Fig. 10.5). At the early stage of synaptogenesis, dendrites are bristled with many dendritic filopodia in which drebrin is diffusely distributed (diffuse-type filopodia). Diffuse-type filopodia are supposed to be very dynamic with short-term life span. After diffuse-type filopodia

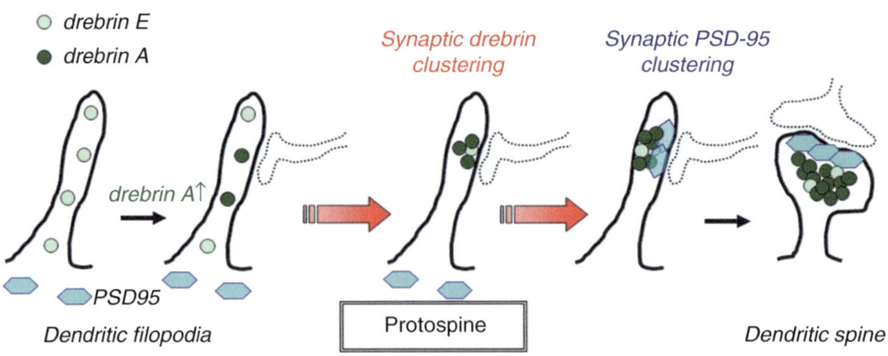

Fig. 10.5 Schematic summery of dendritic spine morphogenesis from dendritic filopodia. At early of synaptogenesis, drebrin E is gradually replaced by drebrin A, which is developmentally upregulated, in filopodia. Once a presynaptic terminal contacts with a filopodium, drebrin A is accumulated at nascent contact sites. At this stage, the filopodia with drebrin clusters (cluster-type filopodia) don't have PSD-95 clusters. Afterward, PSD-95 is further accumulated depending on drebrin clustering. Consequently, the filopodia with drebrin clusters transform into mature spines. Therefore, the filopodia that have drebrin clusters but not PSD-95 ones are suggested to be "protospines"

contact with an axon, drebrin is accumulated with F-actin at a postsynaptic site in the filopodia. As a result, the diffuse-type filopodia are transformed to be cluster type as a protospine. The drebrin clustering with F-action requires developmental upregulation of drebrin A expression. In the cluster-type filopodia, PSD-95 is then accumulated depending on the cluster formation of drebrin with F-actin. Finally, the cluster-type filopodia are transformed to mature dendritic spines, which contain both drebrin-based actin components and PSD structure (Fig. 10.5). Thus, the drebrin-F-actin complex anchors the postsynaptic components and governs the morphological maturation of dendritic spines.

10.2.6 Can Drebrin A Promote Dendritic Spine Morphogenesis?

The developmental upregulation of drebrin A is required for synaptic clustering of PSD-95 and dendritic spine morphogenesis during development. These findings suggest that forced expression of drebrin A in immature neurons might promote dendritic spine morphogenesis and/or synaptic clustering of PSD-95. However, the experiments of drebrin A overexpression do not support these possibilities (Mizui et al. 2005). The overexpression of drebrin A in immature hippocampal neurons from 7 to 9 DIV does not induce the morphologically mature types of dendritic spines. Instead, drebrin A overexpression induces extremely large dendritic

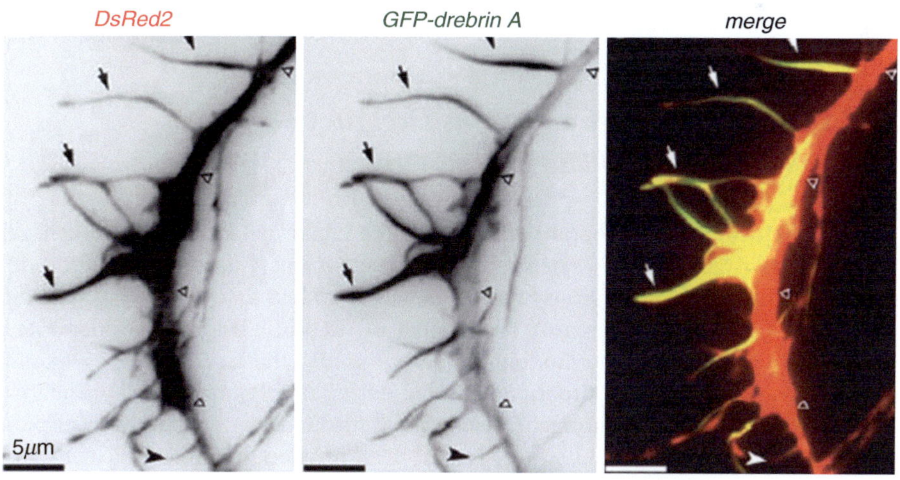

Fig. 10.6 "Megapodia" induced by drebrin A overexpression. GFP-tagged drebrin A vector was transfected to cultured hippocampal neuron at 7 days in vitro together with DsRed2 vector to visualize dendritic morphology. Neurons expressing GFP-drebrin A and DsRed2 were imaged 2 days after transfection. GFP-drebrin A (*green*) was accumulated in dendritic protrusions and formed large abnormal headless protrusions named "megapodia" (*arrow*). Scale bars: 5 μm. Adapted from Mizui et al. (2005) with permission

protrusions without head (Fig. 10.6). The large headless protrusions contain the large amount of drebrin A, F-actin, and PSD-95. However, drebrin A and PSD-95 are diffusely distributed and not clustered in the large protrusions. Further, the large protrusions induced by drebrin A do not contain MAP2, indicating that they do not correspond to the branch of dendrites. Thus, the large headless and F-actin-rich dendritic protrusions are called as "megapodia" meaning large filopodia (Fig. 10.6). The induction of megapodia by drebrin A overexpression suggests that drebrin A upregulation is necessary but not sufficient for spine morphogenesis. Given that drebrin clustering occurs locally at the axon-dendritic filopodia contact sites, it is possible that drebrin clustering is regulated by not only drebrin A upregulation but also by local signaling from presynaptic terminals such as the release of neurotransmitters and/or growth factors from presynaptic terminals and/or transsynaptic adhesion signals.

10.3 Activity-Dependent Modulation of Drebrin Dynamics at Dendritic Spines

Many previous studies have shown that synaptic activity regulates the formation, maintenance, and plasticity of dendritic spines (Kasai et al. 2010; Saneyoshi et al. 2010; Papa and Segal 1996; Nimchinsky et al. 2002), suggesting that drebrin-based actin cytoskeleton in dendritic spines is highly regulated by synaptic activity. In the following paragraphs, activity-dependent clustering and dynamics of drebrin will be discussed.

10.3.1 Synaptic Activity Controls Drebrin Clustering During Development

Pharmacological perturbation of neuronal activity has revealed the involvement of synaptic activity in drebrin clustering during development. Chronic treatment (8 to 15 DIV) of hippocampal neurons in cultures with tetrodotoxin, which blocks action potentials, significantly decreases the number of drebrin clusters along dendrites (Takahashi et al. 2009). This finding supports the idea that spontaneous synaptic activity positively regulates drebrin clustering during development. What if neuronal activity is increased? Chronic treatment with the $GABA_A$-receptor blocker picrotoxin, which enhances excitatory synaptic transmission, significantly increases the number of drebrin clusters (Takahashi et al. 2009). Thus, spontaneous synaptic activity, especially its excitatory components, promotes drebrin clustering. Given the phenotypes of drebrin A overexpression in immature neurons, these results also suggest that not only developmental upregulation of drebrin A but also spontaneous synaptic activity is necessary for drebrin clustering during development and consequently for dendritic spine morphogenesis.

10.3.2 AMPA Receptor Activity Regulates Synaptic Clustering of Drebrin

In the brains, excitatory synaptic transmission is generally mediated by presynaptically released glutamate and postsynaptic glutamate receptors. Pharmacological treatments with subtype-specific blockers of glutamate receptors have demonstrated that AMPA receptor activity is involved in drebrin clustering in spine morphogenesis (Takahashi et al. 2009) (Fig. 10.7). Cultured hippocampal neurons were chronically treated with subtype-specific blockers of glutamate receptors

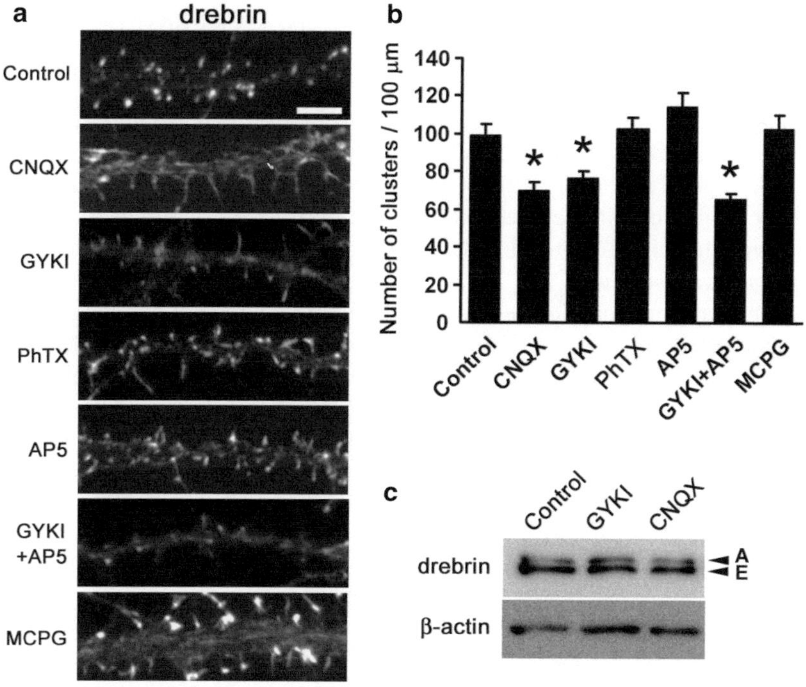

Fig. 10.7 The involvement of AMPA-type glutamate receptor (AMPAR) activity in drebrin clustering on dendritic spines. (**a**) Hippocampal neurons were treated with the following pharmacological agents from 8 days in vitro for 7 days: the AMPAR and kainite-receptor blocker CNQX (40 μM), the specific AMPAR blocker GYKI (10 μM), the Ca^{2+}-permeable (GluA2-subunit-lacking) AMPAR blocker philanthotoxin (PhTX, 10 μM), the NMDA-type glutamate receptor blocker AP5 (50 μM), GYKI (10 μM) plus AP5 (50 μM), and metabotropic glutamate receptor blocker MCPG (100 μM). Scale bar: 5 μm. (**b**) Quantitative analysis of drebrin clusters after these treatments. CNQX, GYKI, and GYKI plus AP5 treatments significantly suppressed drebrin clustering (*$P < 0.01$ versus control), indicating that Ca^{2+}-impermeable AMPAR regulates drebrin clustering. (**c**) Western blot data show that GYKI and CNQX treatments, blocking AMPAR or kainite receptor, do not affect the expression level of drebrin protein. Adapted from Takahashi et al. (2009) with permission

including (1) CNQX, a competitive inhibitor for AMPA receptors and kainite receptors; (2) GYKI, a noncompetitive inhibitor specific for AMPA receptors; (3) philanthotoxin, a blocker for only Ca^{2+}-permeable AMPA receptors; (4) AP5, an NMDA-type glutamate receptor blocker; and (5) MCPG, a selective blocker of metabotropic glutamate receptors. Interestingly, chronic treatment with CNQX or GYKI, which can block AMPA receptor activation, significantly diminishes drebrin clustering on dendrites. Philanthotoxin does not affect drebrin clustering, indicating that Ca^{2+}-impermeable AMPA receptors regulate drebrin clustering. Drebrin clustering is not affected by AP5 or MCPG. Thus, the activation of NMDA receptors and metabotropic glutamate receptors is dispensable for drebrin clustering. The suppression of drebrin clustering by AMPA receptor inhibition is not due to the alteration of drebrin expression level because the Western blot analysis has shown that the chronic treatment with CNQX and GYKI does not affect total expression level of drebrin protein (Fig. 10.7c). Further, according to double immunolabeling with drebrin and synapsin I, the chronic treatment with GYKI does not affect the clustering of synapsin I despite the decrease in drebrin cluster number (Takahashi et al. 2009). The selective diminishment of drebrin clustering is thus not due to secondary effects of impaired formation of presynaptic contact sites. According to more detailed analysis of drebrin clustering in dendritic filopodia by triple labeling for drebrin, synapsin I, and F-actin, the inhibition of AMPA receptor activation with GYKI suppresses drebrin clustering on dendritic protrusions, especially at synaptic sites, without affecting it on dendritic shafts. Notably, the enhancement of AMPA receptor activity with the AMPA receptor desensitization blocker, cyclothiazide (CTZ), increases total drebrin cluster density on dendritic protrusions in the presence of NMDA receptor blocker AP5 without affecting synapsin I cluster density (Takahashi et al. 2009). Together, the spontaneous activation of Ca^{2+}-impermeable (GluA2-containing) AMPA receptors specifically facilitates drebrin clustering on dendritic protrusions without affecting the formation of presynaptic contact sites.

10.3.3 Chronic Inhibition of AMPA Receptors Impairs Dendritic Spine Morphogenesis During Development

Given that AMPA receptors regulate drebrin clustering in dendritic protrusions and that drebrin clustering governs dendritic spine morphogenesis, it has been suggested that AMPA receptor activity also regulates spine morphogenesis during development through drebrin. As expected, chronic treatment of hippocampal neurons with AMPA receptor blocker GYKI indeed impairs spine morphogenesis and instead induces filopodia-like thin spines, some of which are malformed with Y-shaped or multibranched structures (Takahashi et al. 2009). In contrast, chronic treatment with other blockers such NMDA receptor blocker AP5 or metabotropic glutamate receptor blocker MCPG does not significantly affect spine morphogenesis (Takahashi et al. 2009). Thus, AMPA receptor activity conducts normal spine morphogenesis presumably through regulating drebrin clustering in dendritic filopodia and spines. These results are consistent with those by using hippocampal slices in culture showing that chronic blockage of AMPA receptor activation reduces the density and length of dendritic spines in hippocampal CA1 pyramidal neurons (McKinney et al. 1999; Mateos et al. 2007). Further, chronic treatment with GYKI diminishes synaptic clustering of PSD-95 in dendritic spines but not that on dendritic shaft (Takahashi et al. 2009). These data also support a model by which spontaneous AMPA receptor activity induces drebrin clustering, which consequently promotes the formation of PSD-95 clusters in spines, thereby enabling the normal development of dendritic spines.

10.3.4 AMPA Receptor Activity Stabilizes Drebrin in Dendritic Spines

Activity-dependent clustering of drebrin would be due to the changes of drebrin dynamics in dendritic filopodia and spines. Fluorescent recovery after photobleaching (FRAP) analysis is a useful and widely used approach to examine the turnover of synaptic proteins within individual spines in real time. FRAP analysis using cultured hippocampal neurons transfected with GFP-tagged drebrin A (GFP-drebrin A) has revealed an intriguing activity-dependent turnover of drebrin in dendritic spines (Takahashi et al. 2009) (Fig. 10.8). Under normal physiological condition that allows spontaneous synaptic activity, GFP-drebrin A shows relatively dynamic turnover in dendritic spines. In spines at hippocampal neurons at 2 weeks in vitro, about 23% of drebrin is immobile as a stable fraction, whereas the rest of drebrin (about 75%) is mobile. The exchange rate of drebrin mobile fraction follows a time constant of approximately 6 min (Takahashi et al. 2009). Thus, about a quarter of total drebrin within a single spine is stabilized under spontaneous neuronal activity. In addition, FRAP analysis has demonstrated the interesting roles of AMPA receptor activity in drebrin turnover in dendritic spines. The treatment with CNQX decreases the stable fraction of GFP-drebrin A to approximately 9% without affecting the exchange rate. Conversely, the treatment with AMPA or CTZ in the presence of AP5, which enhances only AMPA receptor activity, increases the fraction of stable drebrin without changing its exchange rate (Takahashi et al. 2009). AP5 treatment itself has no effect on the stable drebrin fraction but decreases the time constant for drebrin turnover to about 4 min. These results indicate that the activity of AMPA receptors, but not that of NMDA receptors, contributes to stabilize drebrin in spines.

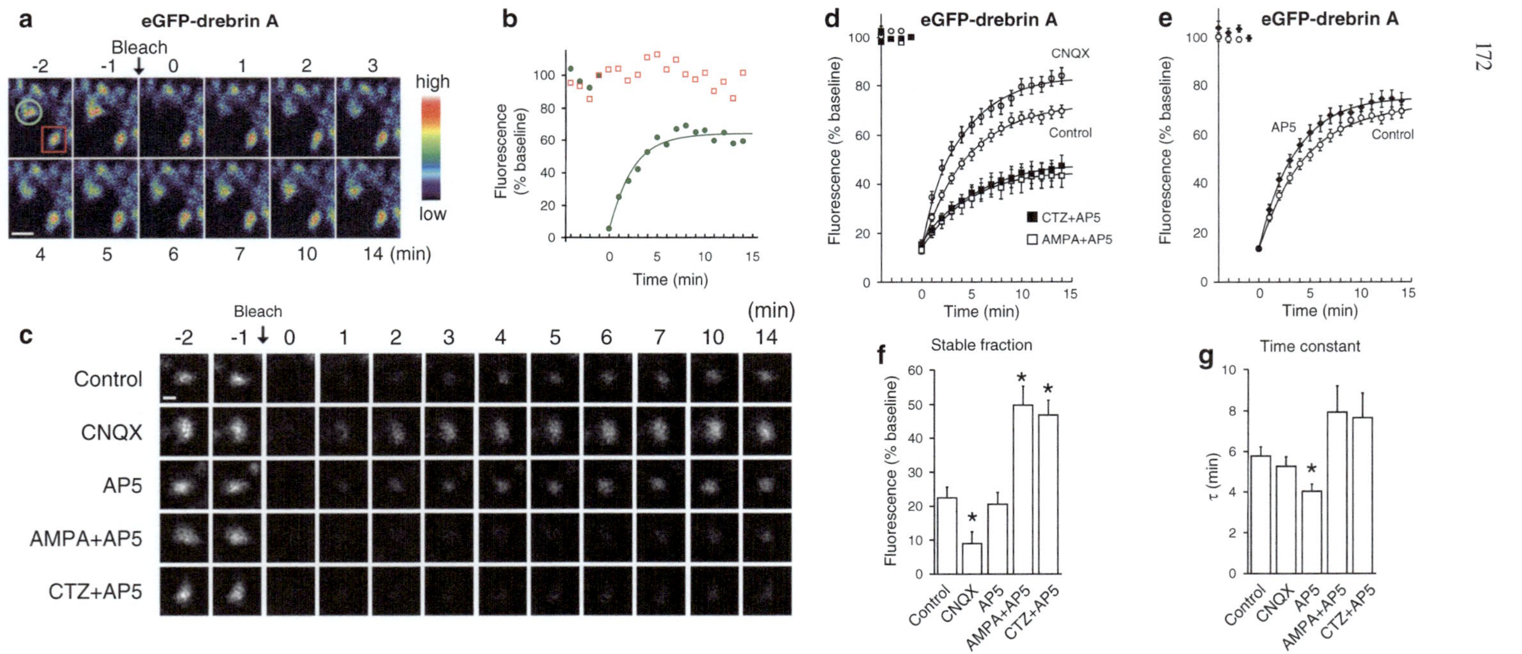

Fig.10.8 AMPAR activity increases stable fraction of drebrin in dendritic spines. (**a**) A GFP-drebrin A transfected neuron was analyzed by time-lapse imaging during fluorescence recovery after photobleaching (FRAP). GFP-drebrin A in a single spine (*green circle*) was selectively and completely photobleached by a small fast light pulse, whereas a neighboring cluster (*red square*) was unaffected. The subsequent images show gradual recovery of the GFP-drebrin A signal. Scale bar: 3 μm. (**b**) Normalized fluorescent intensity in the photobleached spine and the neighboring spine. The green curve was derived from the equation $F(t) = 1 - fs - fm*\exp(-t/\tau)$, where fs, fm, t, and τ are the stable fraction, dynamic fraction, time (minutes), and time constant, respectively. (**c**) The representative time-lapse sequences during FRAP of GFP-drebrin A of the vehicle control neurons and those in the presence of CNQX (40 μM), AP5 (100 μM), AMPA (1 μM) plus AP5 (100 μM), or CTZ (100 μM), a positive allosteric modulator of the AMPARs and kainate receptors (100 μM) plus AP5 (100 μM). Scale bar: 1 μm. (**d**, **e**) FRAP curves of drebrin in the indicated conditions. (**f**, **g**) The comparison of stable fraction (**f**) and time constant (**g**) of drebrin under each condition. AMPAR blockade by CNQX decreases the size of the stable drebrin fraction. By contrast, AMPAR activation by AMPA, or CTZ plus AP5, raises the stable drebrin fraction (**f**). Thus, AMPAR activity positively regulates the stable fraction of drebrin. NMDA receptor blockade by AP5 shortens the time constant of dynamic drebrin without affecting the stable drebrin fraction, indicating spontaneous NMDA receptor activity regulates the transport efficiency of mobile drebrin into spines (**g**). Adapted from Takahashi et al. (2009) with permission

10.3.5 An Increase in Stable Drebrin Is Accompanied by Increased Drebrin Clustering

An interesting question is whether the increase in the stable drebrin fraction corresponds with the increase in drebrin clustering. FRAP analysis of GFP-tagged actin has shown that short-term (15- to 30-min) treatment with cytochalasin D, an F-actin capping reagent, increases the stable fraction of actin in dendritic spines (Star et al. 2002). Short-term treatment with cytochalasin D to nonneuronal cells also induces the aggregation of drebrin-associated F-actin in the cytoplasm (Asada et al. 1994), presumably by changing the turnover of drebrin-F-actin complexes. These data suggest that cytochalasin D would be useful to examine what happens to drebrin clustering when the stable drebrin fraction is changed. Notably, FRAP analysis has shown that a 1-h cytochalasin D treatment increases stable drebrin fraction threefold (Takahashi et al. 2009). The short-term cytochalasin D treatment also promotes the clustering of drebrin with F-actin on dendrites (Takahashi et al. 2009). Thus, the increased fraction of stable drebrin is tightly accompanied by the enhancement of drebrin clusters. Given that AMPA receptor activity increases the stable fraction of drebrin in dendritic spines, AMPA receptor activity enhances drebrin stability in spines and consequently promotes drebrin clustering in spines. Taken together, AMPA receptor-dependent stabilization of drebrin in dendritic spines is an activity-dependent cellular mechanism underlying spine morphogenesis during development.

10.3.6 How Does AMPA Receptor-Mediated Drebrin Stabilization Contribute to Spine Morphogenesis and Plasticity?

According to previous studies on drebrin in nonneuronal cells (see also Chap. 6), drebrin-bound F-actin is resistant to cytochalasin D. Interestingly, F-actin in dendritic spines is also resistant to cytochalasin D. Given that dendritic spines contain co-clusters of drebrin and F-actin, these evidences suggest that drebrin stabilizes actin cytoskeleton in spines by providing their F-actin with cytochalasin D resistance. In addition, time-lapse imaging studies have shown that dendritic filopodia have short lifetime, whereas dendritic spines are maintained for a long term (Dailey and Smith 1996; Marrs et al. 2001; Grutzendler et al. 2002; Portera-Cailliau et al. 2003). Spine head volume is correlated with the amount of functional AMPA receptors (Matsuzaki et al. 2001). As drebrin contents are also correlated with spine head volume (Kobayashi et al. 2007), it has been further suggested that the rapid turnover of filopodia is due to an insufficient amount of stable drebrin, whereas spine persistence is due to an AMPA receptor-regulated abundance of stable drebrin. Thus, AMPA receptor-mediated drebrin stabilization may regulate the turnover of dendritic filopodia and spines and determine structure-function relationship of spines by stabilizing their actin cytoskeleton. Importantly, AMPA receptor-mediated drebrin stabilization would also contribute to synaptic plasticity. The induction of long-term potentiation (LTP) is accompanied by the enlargement of spine head (Matsuzaki et al. 2004; Okamoto et al. 2004; Honkura et al. 2008) and synaptic accumulation of drebrin and F-actin in input-specific manner (Fukazawa et al. 2003), suggesting that AMPA receptor-mediated drebrin stabilization could be involved in the enlargement of stimulated spines. How do AMPA receptors increase stable drebrin? Channel activation of AMPA receptors not only induces ion influx but also activates intracellular signaling pathways such as mitogen-activated protein kinase (MAPK), protein tyrosine kinase Lyn, and Rho GTPase (Wang and Durkin 1995; Wang et al. 1997; Hayashi et al. 1999; Kim et al. 2004). Interestingly, a previous study has revealed that drebrin physically and functionally interacts with Ras GTPase in spine plasticity (Biou et al. 2008). Therefore, it would be interesting to address how AMPA receptor-induced signaling pathways mediate drebrin-based actin reorganization for spine formation, maintenance, and/or plasticity.

10.3.7 Impairment of Activity-Dependent Drebrin Clustering and Dendritic Spine Pathology

Abnormal spine morphology is accompanied by many neurodevelopmental disorders such as Fragile X syndrome and neurodegenerative diseases such as Alzheimer's disease (Hering and Sheng 2001; Fiala et al. 2002). It has been well established that drebrin downregulation is a pathognomonic feature of Down syndrome and Alzheimer's disease (Harigaya et al. 1996; Shim and Lubec 2002; Calon et al. 2004; Counts et al. 2006). Previous studies involving a mouse model of Alzheimer's disease have shown that a decrease in AMPA receptors at the synapse is followed by a decrease in the proportion of drebrin-positive spines (Chang et al. 2006; Mahadomrongkul et al. 2005). Normal spine morphogenesis is impaired by declustering of drebrin using drebrin A antisense oligonucleotides (Takahashi et al. 2003) and also by drebrin knockdown using RNA interference (Biou et al. 2008). Further AMPA receptor blockade results in malformed spines like filopodia-like thin spines with Y-shape and multibranched protrusions (Takahashi et al. 2009). It has been thus suggested that the impairment of AMPA receptor-mediated drebrin clustering in spines would be one of the pathogenesis of spine abnormality in the neurological conditions.

10.4 Drebrin Involvement in Activity-Dependent Synaptic Clustering of NMDA Receptors

It has been well known that NMDA receptor blockade induces synaptic accumulation of NMDA receptors as homeostatic scaling of NMDA receptors (Rao and Craig 1997; Perez-Otano and Ehlers 2005). In general, homeostatic synaptic plasticity is supposed to be crucial for maintaining the excitability of neurons within the dynamic range (Turrigiano 2008). As described above, FRAP analysis has revealed that NMDA receptor blockade decreases the time constant of dynamic drebrin without affecting the stable drebrin fraction (Takahashi et al. 2009), indicating that spontaneous NMDA receptor activity regulates the transport efficiency of drebrin into spines. In the following, the involvement of drebrin in activity-dependent trafficking of NMDA receptors will be discussed.

10.4.1 Drebrin A Is Required for Activity-Dependent Synaptic Accumulation of NMDA Receptors

PSDs contain high concentrations of NMDA receptors as well as PSD-95 (Kennedy 2000). PSD-95 binds to the NMDA receptor subunits GluN2A and GluN2B (Niethammer et al. 1996). As described above, drebrin A upregulation is required for synaptic clustering of PSD-95. Further, the inhibition of NMDA receptors changes the exchange rate of mobile drebrin fraction (Takahashi et al. 2009). Therefore, it is possible that drebrin A regulates synaptic targeting of NMDA receptors via PSD-95. Originally, NMDA receptors are not highly clustered at synapses under the condition that allow spontaneous neuronal activity. Drebrin A knockdown by antisense oligonucleotides does not affect synaptic clustering of NMDA receptors in cultured hippocampal neurons under such condition. On the other hand, NMDA receptor blockage induces significant accumulation of NMDA receptors at synaptic sites (Rao and Craig 1997). Interestingly, drebrin A knockdown suppresses synaptic clustering of NMDA receptors induced by NMDA receptor blocker AP5 (Takahashi et al. 2006). Thus, drebrin A knockdown selectively blocks activity-dependent synaptic clustering of NMDA receptors (Fig. 10.9). Curiously, drebrin A knockdown increases PSD-95 clustering on dendrites when neurons are treated with AP5 (Takahashi et al. 2006), suggesting that synaptic clustering of NMDA receptors accelerated by AP5 treatment is independent of the molecular interaction between PSD-95 and NMDA receptors. Previous studies have demonstrated the involvement of cAMP-dependent protein kinase (PKA) in the activity-dependent synaptic clustering of NMDA receptors (Crump et al. 2001). Notably, PKA activation by IBMX, which inhibits cAMP phosphodiesterase and consequently increases cAMP levels to activate PKA, is sufficient to induce synaptic clustering of NMDA receptors (Crump et al. 2001). Thus, it would be interesting to examine whether and how PKA regulates drebrin clustering and dynamics, e.g., through drebrin phosphorylation. As described above, FRAP analysis has shown that NMDA receptor activity controls the exchange rate of mobile drebrin fraction without affecting stable drebrin fraction. Therefore, it has been suggested that mobile drebrin fraction contributes to the transport of NMDA receptors into spines.

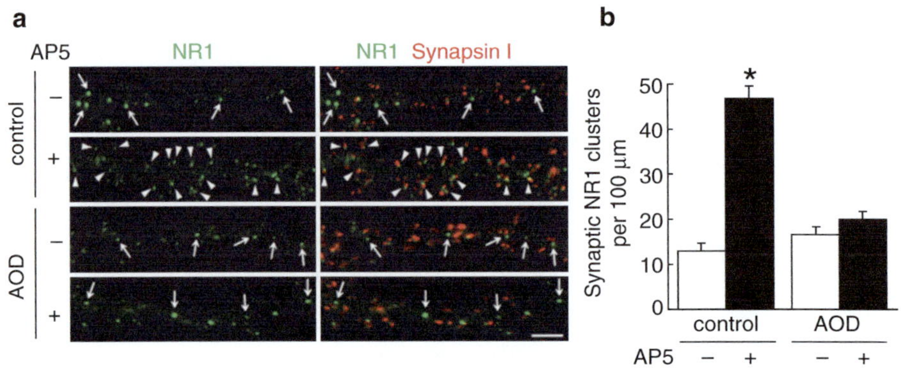

Fig. 10.9 The suppression of drebrin A expression inhibits activity-dependent synaptic targeting of NMDA-type glutamate receptor (NMDAR). (**a**) Cultured hippocampal neurons were treated with antisense oligonucleotides against drebrin A-specific exon (AOD) at 12 days in vitro for 2 days and further incubated in the medium with or without 50 μM NMDAR blocker AP5 for additional 2 days. The neurons were then immunostained for common NMDAR subunit GluN1 (*green*) and synapsin I (*red*). In control neurons, AP5 treatment increased synaptic clusters of GluN1 (NR1) associated with synapsin I clusters (*second top*, *arrowheads*). In contrast, in AOD-treated neurons, AP5 treatment had no significant effect on GluN1 clustering (*bottom*). *Arrows* in all panels indicate non-synaptic GluN1 clusters. Scale bar, 5 μm. (**b**) Quantitative analysis of total and synaptic NMDAR clusters. Note that AP5 enhances NMDAR clustering in the control neurons, whereas AP5 does not enhance it in drebrin A knockdown neurons. Adapted from Takahashi et al. (2009) with permission

10.4.2 In Vivo Role of Drebrin A in Activity-Dependent Synaptic Targeting of NMDA Receptors

The involvement of drebrin A in activity-dependent synaptic targeting of NMDA receptors has been also demonstrated by the analysis of drebrin A knockout (DAKO) mice. As only unique exon (exon 12a) of drebrin gene is deleted in DAKO mice, DAKO mice lose drebrin A but retain drebrin E, thus maintaining same expression level of total drebrin to wild-type mice (Aoki et al. 2009). In wild-type mice, the in vivo application of AP5 to cortical surface induces rapid accumulation of NMDA receptors in dendritic spines within 30 min. In contrast, the AP5 treatment of DAKO mice does not induce NMDA receptor accumulation in spines (Aoki et al. 2009). These data indicate that drebrin A is involved in the rapid form of activity-dependent homeostatic accumulation (synaptic scaling) of NMDA receptors in dendritic spines of adult cortex in vivo. Similar to in vitro results, the basal level of NMDA receptors in spines in DAKO mice is not significantly different from that in wild-type mice (Aoki et al. 2009), supporting that drebrin A is dispensable for basal targeting of NMDA receptors in dendritic spines.

10.5 Conclusion

Drebrin clustering is crucial for dendritic spine morphogenesis and synaptic clustering of PSD-95 during development. Drebrin clustering requires not only the developmental upregulation of drebrin A but also excitatory synaptic activity mediated by AMPA receptors. AMPA receptor activity promotes drebrin clustering by increasing the stable drebrin faction in dendritic spines. In contrast, NMDA receptor activity controls the turnover speed of mobile drebrin. Further, drebrin A is required for activity-dependent synaptic targeting of NMDA receptors but not for their basal targeting. These findings highlight the pivotal roles of drebrin on dendritic spine formation and synaptic plasticity.

References

Allison DW, Chervin AS, Gelfand VI, Craig AM (2000) Postsynaptic scaffolds of excitatory and inhibitory synapses in hippocampal neurons: maintenance of core components independent of actin filaments and microtubules. J Neurosci 20(12):4545–4554

Aoki C, Kojima N, Sabaliauskas N, Shah L, Ahmed TH, Oakford J, Ahmed T, Yamazaki H, Hanamura K, Shirao T (2009) Drebrin a knockout eliminates the rapid form of homeostatic synaptic plasticity at excitatory synapses of intact adult cerebral cortex. J Comp Neurol 517(1):105–121. doi:10.1002/cne.22137

Araya R, Jiang J, Eisenthal KB, Yuste R (2006) The spine neck filters membrane potentials. Proc Natl Acad Sci U S A 103(47):17961–17966. doi:10.1073/pnas.0608755103

Asada H, Uyemura K, Shirao T (1994) Actin-binding protein, drebrin, accumulates in submembranous regions in parallel with neuronal differentiation. J Neurosci Res 38(2):149–159. doi:10.1002/jnr.490380205

Biou V, Brinkhaus H, Malenka RC, Matus A (2008) Interactions between drebrin and Ras regulate dendritic spine plasticity. Eur J Neurosci 27(11):2847–2859. doi:10.1111/j.1460-9568.2008.06269.x

Calon F, Lim GP, Yang F, Morihara T, Teter B, Ubeda O, Rostaing P, Triller A, Salem N Jr, Ashe KH, Frautschy SA, Cole GM (2004) Docosahexaenoic acid protects from dendritic pathology in an Alzheimer's disease mouse model. Neuron 43(5):633–645. doi:10.1016/j.neuron.2004.08.013

Chang EH, Savage MJ, Flood DG, Thomas JM, Levy RB, Mahadomrongkul V, Shirao T, Aoki C, Huerta PT (2006) AMPA receptor downscaling at the onset of Alzheimer's disease pathology in double knockin mice. Proc Natl Acad Sci U S A 103(9):3410–3415. doi:10.1073/pnas.0507313103

Chen L, Chetkovich DM, Petralia RS, Sweeney NT, Kawasaki Y, Wenthold RJ, Bredt DS, Nicoll RA (2000) Stargazin regulates synaptic targeting of AMPA receptors by two distinct mechanisms. Nature 408(6815):936–943. doi:10.1038/35050030

Cho KO, Hunt CA, Kennedy MB (1992) The rat brain postsynaptic density fraction contains a homolog of the Drosophila discs-large tumor suppressor protein. Neuron 9(5):929–942

Cohen RS, Chung SK, Pfaff DW (1985) Immunocytochemical localization of actin in dendritic spines of the cerebral cortex using colloidal gold as a probe. Cell Mol Neurobiol 5(3):271–284

Counts SE, Nadeem M, Lad SP, Wuu J, Mufson EJ (2006) Differential expression of synaptic proteins in the frontal and temporal cortex of elderly subjects with mild cognitive impairment. J Neuropathol Exp Neurol 65(6):592–601

Crump FT, Dillman KS, Craig AM (2001) cAMP-dependent protein kinase mediates activity-regulated synaptic targeting of NMDA receptors. J Neurosci 21(14):5079–5088

Dailey ME, Smith SJ (1996) The dynamics of dendritic structure in developing hippocampal slices. J Neurosci 16(9):2983–2994

DeFelipe J (2015) The dendritic spine story: an intriguing process of discovery. Front Neuroanat 9:14. doi:10.3389/fnana.2015.00014

Dunaevsky A, Tashiro A, Majewska A, Mason C, Yuste R (1999) Developmental regulation of spine motility in the mammalian central nervous system. Proc Natl Acad Sci U S A 96(23):13438–13443

Ebrahimi S, Okabe S (2014) Structural dynamics of dendritic spines: molecular composition, geometry and functional regulation. Biochim Biophys Acta 1838(10):2391–2398. doi:10.1016/j.bbamem.2014.06.002

El-Husseini AE, Schnell E, Chetkovich DM, Nicoll RA, Bredt DS (2000) PSD-95 involvement in maturation of excitatory synapses. Science 290(5495):1364–1368

Fiala JC, Feinberg M, Popov V, Harris KM (1998) Synaptogenesis via dendritic filopodia in developing hippocampal area CA1. J Neurosci 18(21):8900–8911

Fiala JC, Spacek J, Harris KM (2002) Dendritic spine pathology: cause or consequence of neurological disorders? Brain Res Brain Res Rev 39(1):29–54

Fischer M, Kaech S, Knutti D, Matus A (1998) Rapid actin-based plasticity in dendritic spines. Neuron 20(5):847–854

Fischer M, Kaech S, Wagner U, Brinkhaus H, Matus A (2000) Glutamate receptors regulate actin-based plasticity in dendritic spines. Nat Neurosci 3(9):887–894. doi:10.1038/78791

Fukazawa Y, Saitoh Y, Ozawa F, Ohta Y, Mizuno K, Inokuchi K (2003) Hippocampal LTP is accompanied by enhanced F-actin content within the dendritic spine that is essential for late LTP maintenance in vivo. Neuron 38(3):447–460

Grutzendler J, Kasthuri N, Gan WB (2002) Long-term dendritic spine stability in the adult cortex. Nature 420(6917):812–816. doi:10.1038/nature01276

Harigaya Y, Shoji M, Shirao T, Hirai S (1996) Disappearance of actin-binding protein, drebrin, from hippocampal synapses in Alzheimer's disease. J Neurosci Res 43(1):87–92. doi:10.1002/jnr.490430111

Harris KM, Kater SB (1994) Dendritic spines: cellular specializations imparting both stability and flexibility to synaptic function. Annu Rev Neurosci 17:341–371. doi:10.1146/annurev.ne.17.030194.002013

Hayashi K, Suzuki K, Shirao T (1998) Rapid conversion of drebrin isoforms during synapse formation in primary culture of cortical neurons. Brain Res Dev Brain Res 111(1):137–141

Hayashi T, Umemori H, Mishina M, Yamamoto T (1999) The AMPA receptor interacts with and signals through the protein tyrosine kinase Lyn. Nature 397(6714):72–76. doi:10.1038/16269

Hering H, Sheng M (2001) Dendritic spines: structure, dynamics and regulation. Nat Rev Neurosci 2(12):880–888. doi:10.1038/35104061

Honkura N, Matsuzaki M, Noguchi J, Ellis-Davies GC, Kasai H (2008) The subspine organization of actin fibers regulates the structure and plasticity of dendritic spines. Neuron 57(5):719–729. doi:10.1016/j.neuron.2008.01.013

Kabsch W, Vandekerckhove J (1992) Structure and function of actin. Annu Rev Biophys Biomol Struct 21:49–76. doi:10.1146/annurev.bb.21.060192.000405

Kasai H, Matsuzaki M, Noguchi J, Yasumatsu N, Nakahara H (2003) Structure-stability-function relationships of dendritic spines. Trends Neurosci 26(7):360–368. doi:10.1016/S0166-2236(03)00162-0

Kasai H, Fukuda M, Watanabe S, Hayashi-Takagi A, Noguchi J (2010) Structural dynamics of dendritic spines in memory and cognition. Trends Neurosci 33(3):121–129. doi:10.1016/j.tins.2010.01.001

Kennedy MB (1997) The postsynaptic density at glutamatergic synapses. Trends Neurosci 20(6):264–268

Kennedy MB (2000) Signal-processing machines at the postsynaptic density. Science 290(5492):750–754

Kim E, Sheng M (2004) PDZ domain proteins of synapses. Nat Rev Neurosci 5(10):771–781. doi:10.1038/nrn1517

Kim E, Niethammer M, Rothschild A, Jan YN, Sheng M (1995) Clustering of Shaker-type K+ channels by interaction with a family of membrane-associated guanylate kinases. Nature 378(6552):85–88. doi:10.1038/378085a0

Kim SJ, Jeon S, Shin EY, Kim EG, Park J, Bae CD (2004) AMPA, not NMDA, activates RhoA GTPases and subsequently phosphorylates moesin. Exp Mol Med 36(1):98–102. doi:10.1038/emm.2004.14

Kistner U, Wenzel BM, Veh RW, Cases-Langhoff C, Garner AM, Appeltauer U, Voss B, Gundelfinger ED, Garner CC (1993) SAP90, a rat presynaptic protein related to the product of the Drosophila tumor suppressor gene dlg-A. J Biol Chem 268(7):4580–4583

Kobayashi C, Aoki C, Kojima N, Yamazaki H, Shirao T (2007) Drebrin a content correlates with spine head size in the adult mouse cerebral cortex. J Comp Neurol 503(5):618–626. doi:10.1002/cne.21408

Landis DM, Reese TS (1983) Cytoplasmic organization in cerebellar dendritic spines. J Cell Biol 97(4):1169–1178

Mahadomrongkul V, Huerta PT, Shirao T, Aoki C (2005) Stability of the distribution of spines containing drebrin A in the sensory cortex layer I of mice expressing mutated APP and PS1 genes. Brain Res 1064(1-2):66–74. doi:10.1016/j.brainres.2005.10.012

Maletic-Savatic M, Malinow R, Svoboda K (1999) Rapid dendritic morphogenesis in CA1 hippocampal dendrites induced by synaptic activity. Science 283(5409):1923–1927

Marrs GS, Green SH, Dailey ME (2001) Rapid formation and remodeling of postsynaptic densities in developing dendrites. Nat Neurosci 4(10):1006–1013. doi:10.1038/nn717

Mateos JM, Luthi A, Savic N, Stierli B, Streit P, Gahwiler BH, McKinney RA (2007) Synaptic modifications at the CA3-CA1 synapse after chronic AMPA receptor blockade in rat hippocampal slices. J Physiol 581(Pt 1):129–138. doi:10.1113/jphysiol.2006.120550

Matsuzaki M, Ellis-Davies GC, Nemoto T, Miyashita Y, Iino M, Kasai H (2001) Dendritic spine geometry is critical for AMPA receptor expression in hippocampal CA1 pyramidal neurons. Nat Neurosci 4(11):1086–1092. doi:10.1038/nn736

Matsuzaki M, Honkura N, Ellis-Davies GC, Kasai H (2004) Structural basis of long-term potentiation in single dendritic spines. Nature 429(6993):761–766. doi:10.1038/nature02617

Matus A, Ackermann M, Pehling G, Byers HR, Fujiwara K (1982) High actin concentrations in brain dendritic spines and postsynaptic densities. Proc Natl Acad Sci U S A 79(23):7590–7594

McKinney RA, Capogna M, Durr R, Gahwiler BH, Thompson SM (1999) Miniature synaptic events maintain dendritic spines via AMPA receptor activation. Nat Neurosci 2(1):44–49. doi:10.1038/4548

Migaud M, Charlesworth P, Dempster M, Webster LC, Watabe AM, Makhinson M, He Y, Ramsay MF, Morris RG, Morrison JH, O'Dell TJ, Grant SG (1998) Enhanced long-term potentiation and impaired learning in mice with mutant postsynaptic density-95 protein. Nature 396(6710):433–439. doi:10.1038/24790

Mizui T, Takahashi H, Sekino Y, Shirao T (2005) Overexpression of drebrin A in immature neurons induces the accumulation of F-actin and PSD-95 into dendritic filopodia, and the formation of large abnormal protrusions. Mol Cell Neurosci 30(4):630–638

Niethammer M, Kim E, Sheng M (1996) Interaction between the C terminus of NMDA receptor subunits and multiple members of the PSD-95 family of membrane-associated guanylate kinases. J Neurosci 16(7):2157–2163

Nimchinsky EA, Sabatini BL, Svoboda K (2002) Structure and function of dendritic spines. Annu Rev Physiol 64:313–353. doi:10.1146/annurev.physiol.64.081501.160008

Okabe S, Miwa A, Okado H (2001) Spine formation and correlated assembly of presynaptic and postsynaptic molecules. J Neurosci 21(16):6105–6114

Okamoto K, Nagai T, Miyawaki A, Hayashi Y (2004) Rapid and persistent modulation of actin dynamics regulates postsynaptic reorganization underlying bidirectional plasticity. Nat Neurosci 7(10):1104–1112. doi:10.1038/nn1311

Papa M, Segal M (1996) Morphological plasticity in dendritic spines of cultured hippocampal neurons. Neuroscience 71(4):1005–1011
Papa M, Bundman MC, Greenberger V, Segal M (1995) Morphological analysis of dendritic spine development in primary cultures of hippocampal neurons. J Neurosci 15(1 Pt 1):1–11
Perez-Otano I, Ehlers MD (2005) Homeostatic plasticity and NMDA receptor trafficking. Trends Neurosci 28(5):229–238. doi:10.1016/j.tins.2005.03.004
Pollard TD, Borisy GG (2003) Cellular motility driven by assembly and disassembly of actin filaments. Cell 112(4):453–465
Portera-Cailliau C, Pan DT, Yuste R (2003) Activity-regulated dynamic behavior of early dendritic protrusions: evidence for different types of dendritic filopodia. J Neurosci 23(18):7129–7142
Rao A, Craig AM (1997) Activity regulates the synaptic localization of the NMDA receptor in hippocampal neurons. Neuron 19(4):801–812
Rochefort NL, Konnerth A (2012) Dendritic spines: from structure to in vivo function. EMBO Rep 13(8):699–708. doi:10.1038/embor.2012.102
Sala C, Piech V, Wilson NR, Passafaro M, Liu G, Sheng M (2001) Regulation of dendritic spine morphology and synaptic function by Shank and Homer. Neuron 31(1):115–130
Saneyoshi T, Fortin DA, Soderling TR (2010) Regulation of spine and synapse formation by activity-dependent intracellular signaling pathways. Curr Opin Neurobiol 20(1):108–115. doi:10.1016/j.conb.2009.09.013
Sekino Y, Kojima N, Shirao T (2007) Role of actin cytoskeleton in dendritic spine morphogenesis. Neurochem Int 51(2–4):92–104. doi:10.1016/j.neuint.2007.04.029
Shim KS, Lubec G (2002) Drebrin, a dendritic spine protein, is manifold decreased in brains of patients with Alzheimer's disease and Down syndrome. Neurosci Lett 324(3):209–212
Shirao T, González-Billault C (2013) Actin filaments and microtubules in dendritic spines. J Neurochem 126(2):155–164
Star EN, Kwiatkowski DJ, Murthy VN (2002) Rapid turnover of actin in dendritic spines and its regulation by activity. Nat Neurosci 5(3):239–246. doi:10.1038/nn811
Svitkina TM (2013) Ultrastructure of protrusive actin filament arrays. Curr Opin Cell Biol 25(5):574–581. doi:10.1016/j.ceb.2013.04.003
Takahashi H, Sekino Y, Tanaka S, Mizui T, Kishi S, Shirao T (2003) Drebrin-dependent actin clustering in dendritic filopodia governs synaptic targeting of postsynaptic density-95 and dendritic spine morphogenesis. J Neurosci 23(16):6586–6595
Takahashi H, Mizui T, Shirao T (2006) Down-regulation of drebrin A expression suppresses synaptic targeting of NMDA receptors in developing hippocampal neurones. J Neurochem 97(Suppl 1):110–115. doi:10.1111/j.1471-4159.2005.03536.x
Takahashi H, Yamazaki H, Hanamura K, Sekino Y, Shirao T (2009) Activity of the AMPA receptor regulates drebrin stabilization in dendritic spine morphogenesis. J Cell Sci 122(Pt 8):1211–1219. doi:10.1242/jcs.043729
Takumi Y, Ramirez-Leon V, Laake P, Rinvik E, Ottersen OP (1999) Different modes of expression of AMPA and NMDA receptors in hippocampal synapses. Nat Neurosci 2(7):618–624. doi:10.1038/10172
Tomita S, Fukata M, Nicoll RA, Bredt DS (2004) Dynamic interaction of stargazin-like TARPs with cycling AMPA receptors at synapses. Science 303(5663):1508–1511. doi:10.1126/science.1090262
Tomita S, Adesnik H, Sekiguchi M, Zhang W, Wada K, Howe JR, Nicoll RA, Bredt DS (2005) Stargazin modulates AMPA receptor gating and trafficking by distinct domains. Nature 435(7045):1052–1058. doi:10.1038/nature03624
Turrigiano GG (2008) The self-tuning neuron: synaptic scaling of excitatory synapses. Cell 135(3):422–435. doi:10.1016/j.cell.2008.10.008
Wang Y, Durkin JP (1995) alpha-Amino-3-hydroxy-5-methyl-4-isoxazolepropionic acid, but not N-methyl-D-aspartate, activates mitogen-activated protein kinase through G-protein beta gamma subunits in rat cortical neurons. J Biol Chem 270(39):22783–22787

Wang Y, Small DL, Stanimirovic DB, Morley P, Durkin JP (1997) AMPA receptor-mediated regulation of a Gi-protein in cortical neurons. Nature 389(6650):502–504. doi:10.1038/39062
Yoshihara Y, De Roo M, Muller D (2009) Dendritic spine formation and stabilization. Curr Opin Neurobiol 19(2):146–153. doi:10.1016/j.conb.2009.05.013
Zhou Q, Homma KJ, Poo MM (2004) Shrinkage of dendritic spines associated with long-term depression of hippocampal synapses. Neuron 44(5):749–757. doi:10.1016/j.neuron.2004.11.011
Ziv NE, Smith SJ (1996) Evidence for a role of dendritic filopodia in synaptogenesis and spine formation. Neuron 17(1):91–102

Chapter 11
Role of Drebrin in Synaptic Plasticity

Yuko Sekino, Noriko Koganezawa, Toshiyuki Mizui, and Tomoaki Shirao

Abstract Synaptic plasticity underlies higher brain function such as learning and memory, and the actin cytoskeleton in dendritic spines composing excitatory postsynaptic sites plays a pivotal role in synaptic plasticity. In this chapter, we review the role of drebrin in the regulation of the actin cytoskeleton during synaptic plasticity, under long-term potentiation (LTP) and long-term depression (LTD). Dendritic spines have two F-actin pools, drebrin-decorated stable F-actin (DF-actin) and drebrin-free dynamic F-actin (FF-actin). Resting dendritic spines change their shape, but are fairly constant over time at steady state because of the presence of DF-actin. Accumulation of DF-actin is inversely regulated by the intracellular Ca^{2+} concentration. However, LTP and LTD stimulation induce Ca^{2+} influx through *N*-methyl-D-aspartate (NMDA) receptors into the potentiated spines, resulting in drebrin exodus via myosin II ATPase activation. The potentiated spines change to excited state because of the decrease in DF-actin and thus change their shape robustly. In LTP, the Ca^{2+} increase via NMDA receptors soon returns to the basal level, and α-amino-3-hydroxy-5-methyl-4-isoxazolepropionic acid receptor (AMPAR) expression at the postsynaptic membrane is increased. The Ca^{2+} recovery and AMPAR increase coordinately induce the re-accumulation of DF-actin and change the dendritic spines from the excited state to steady state during LTP maintenance. During LTD, the prolonged intracellular Ca^{2+} increase inhibits the re-accumulation of DF-actin, resulting in facilitation of AMPAR endocytosis. Because of the positive feedback loop of the AMPAR decrease and drebrin re-accumulation inhibition, the dendritic spines are instable during LTD maintenance. Taken together, we propose the presence of resilient spines at steady state and plastic spines at excited state and discuss the physiological and pathological relevance of the two-state model to synaptic plasticity.

Y. Sekino (✉)
Department of Neurobiology and Behavior, Gunma University Graduate School of Medicine, Maebashi 371-8511, Japan

Laboratory of Chemical Pharmacology, Graduate School of Pharmaceutical Sciences, The University of Tokyo, Tokyo 113-0033, Japan
e-mail: yukos@mol.f.u-tokyo.ac.jp

N. Koganezawa • T. Mizui • T. Shirao
Department of Neurobiology and Behavior, Gunma University Graduate School of Medicine, Maebashi 371-8511, Japan

T. Shirao, Y. Sekino (eds.), *Drebrin*, Advances in Experimental Medicine and Biology 1006, DOI 10.1007/978-4-431-56550-5_11

Keywords Dendritic spine • Drebrin exodus • Long-term potentiation • Long-term depression • Actin • Myosin • Ca^{2+}

11.1 Introduction

The dendritic spine shape is believed to closely correlate with higher brain functions such as learning and memory (Dailey and Smith 1996; Desmond and Levy 1983). Golgi staining of neurons shows a variety of spine shapes such as thin headless, stubby, and mushroom shapes. Because these spine shapes are believed to link to higher brain functions, the molecular base of morphological regulation was extensively studied. However, Ramon y Cajal proposed over a hundred years ago that spines dynamically move and change according to activity (Yuste 2015). In fact, learning does not link to the shapes per se, but to the shape changes. While the smaller spine is generally preferable for learning, the larger spine is not (Matsuzaki et al. 2004).

Actin is highly concentrated in dendritic spines (Matus et al. 1982). Although the average spine size within a population is constant while there is no external stimulus (Yasumatsu et al. 2008), dendritic spine shapes are not static but dynamic (Fischer et al. 1998; Maletic-Savatic et al. 1999), and the presence of dynamic and stable F-actin pools in dendritic spines was postulated to regulate dynamic spine morphology (Halpain 2000). However, how actin regulates spine morphology has not been fully elucidated.

Drebrin is a key actin-binding protein that forms the stable F-actin pool in dendritic spines. In 1999, we reported that drebrin immunostaining in neuronal somata and dendritic shafts was much weaker than in dendritic spines when cortical neurons and astrocytes were cocultured for 3 weeks, indicating that drebrin is characteristically concentrated in dendritic spines (Hayashi and Shirao 1999). However, we subsequently noticed that when we changed the culture medium from a conditioned medium to a fresh non-conditioned medium, an intense staining of cell somata and dendritic shafts with faint staining of dendritic spines was observed. This raised the hypothesis that drebrin localization changes when neurons are hyperactive in a fresh culture medium. In 2006, we found the activity-dependent translocation of drebrin between dendritic spines and the parent dendrite (Sekino et al. 2006).

In this chapter, we first review the two F-actin pools, dynamic and stable pools, in resting dendritic spines and introduce the Ca^{2+}-dependent regulation of the drebrin-decorated F-actin stable pool. Then, we review the molecular mechanism of drebrin exodus induced in potentiated spines and the role of drebrin exodus and re-accumulation in long-term potentiation (LTP). Furthermore, we discuss the role of spine stability and instability in LTP and long-term depression (LTD), particularly focusing on the neutral dynamic morphology of resting spines and N-methyl-D-aspartate receptor (NMDAR)-dependent morphological changes of potentiated spines. Finally, we discuss the pathological relevance of drebrin loss to Alzheimer's disease. The role of drebrin in homeostatic synaptic plasticity is not reviewed here, but in Chaps. 8 and 10.

11.2 Actin Cytoskeleton in Resting Dendritic Spines

11.2.1 Two F-actin Pools in Resting Dendritic Spines

Actin is the main cytoskeletal element in dendritic spines that governs the spine morphology (Matus et al. 1982). In addition, dendritic spine motility is completely inhibited by actin-depolymerizing reagents such as latrunculin and cytochalasin (Fischer et al. 1998), suggesting the presence of dynamic F-actin regulating spine motility. However, spine F-actin exhibits marked resistance to cytochalasin D (Allison et al. 1998), suggesting that dendritic spine F-actin is extremely stable and plays a purely structural role. In 2000, Halpain suggested that the dynamic and stable configurations of F-actin regulate the spine shape (Halpain 2000).

Drebrin binding to F-actin results in elongation of the F-actin helical crossover (40 nm) (Sharma et al. 2012). Because drebrin has a cooperative actin-binding activity (for details, see Chap. 5), each F-actin is not covered at all or totally covered by drebrin. The resultant drebrin-decorated F-actin (DF-actin) forms a unique F-actin pool different from other normal helical crossover (36 nm) F-actin pools in dendritic spines. In addition to the structural changes, DF-actin shows a lower treadmilling and a lower depolymerization rate (Mizui et al. 2009; Mikati et al. 2013), and inhibition of actin-activated myosin II ATPase activity (Hayashi et al. 1996). Furthermore, drebrin can cross-link F-actins in a phosphorylation-dependent manner (Worth et al. 2013). Moreover, drebrin can form oligomers (see Chap. 3), which may further increase the stability of the DF-actin complex. Collectively, DF-actin likely exhibits a stable configuration in dendritic spines.

The presence of dynamic and stable F-actin has been detected in dendritic spines using a photoactivatable form of green fluorescent protein (PAGFP) fused to actin monomers. When PAGFP-actin molecules in F-actin are activated in dendritic spines, the fluorescence from the activated molecules decays in two phases (Honkura et al. 2008). This shows the presence of a dynamic F-actin pool with a fast treadmilling rate and a stable pool with a low treadmilling rate, because activated PAGFP-actin turns over by continuous "treadmilling." While the dynamic F-actin is observed in the spine tip, the stable one is largely restricted to the base core of the spine head (Honkura et al. 2008).

We have shown by super resolution microscopy and electron microscopy that drebrin is located at the center of mature dendritic spines (Koganezawa et al. 2017) (Aoki et al. 2005). As mentioned above, DF-actin is more stable than drebrin-free F-actin (FF-actin). Therefore, it is considered that stable F-actin at the core of dendritic spines consists of DF-actin, and the dynamic F-actin at the periphery consists of FF-actin.

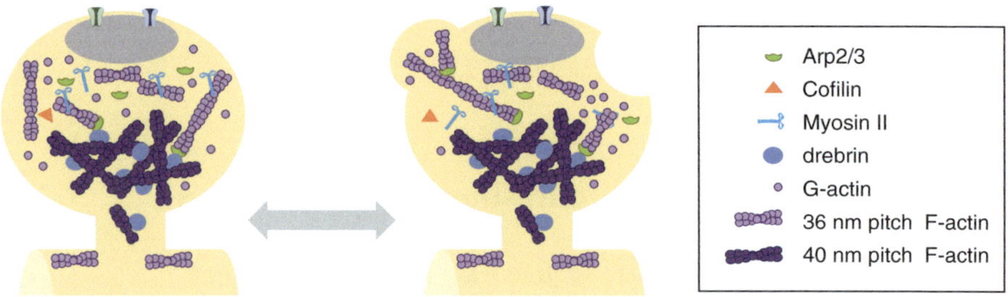

Fig. 11.1 Neutral dynamic morphology model of resting dendritic spines. Drebrin-decorated F-actin (DF-actin, 40 nm helical pitch) and drebrin-free F-actin (FF-actin, 36 nm) are localized at the core and peripheral regions of dendritic spines, respectively. Actin polymerization of FF-actin spontaneously occurs by random filament nucleation and is terminated by filament severing and capping. Despite the spontaneous spine motility, the average spine size of resting spines within the population is constant

11.2.2 Resting Dendritic Spine Motility

At steady state, how is spine motility generated without synaptic input? When cells are cultured on a dish, motile cells form lamellipodia incessantly without any extrinsic stimulation. These lamellipodia consist of a relatively stable actin gel in the bulk and a highly dynamic one at the peripheral edge. Because dendritic spines consist of stable F-actin in the core region and dynamic F-actin at the periphery of the spine head, the underlying mechanism of spontaneous spine motility may be similar to spontaneous lamellipodium formation of motile cells.

Zimmermann and Falcke have proposed a lamellipodium formation model without any extrinsic stimulation (Zimmermann and Falcke 2014). According to this model, individual lamellipodia are formed because of random filament nucleation events amplified by autocatalytic branching and are terminated by filament bundling, severing, and capping. We have recently proposed that spine motility (periodic lamellipodium formation from the spine head) without synaptic stimulation may be similarly determined by random autocatalytic branching of FF-actin located at the periphery of the spine head (Koganezawa et al. 2017). In contrast, the stable actin structure composed of DF-actin provides a stiff substrate in the core region of the spine to push back against, to extend the plasma membrane (Fig. 11.1). Additionally, DF-actin can be stabilized by the interaction of microtubules through the interaction of drebrin with the microtubule plus-end-binding protein, EB3 (Geraldo et al. 2008).

11.2.3 Regulatory Mechanism Underlying DF-Actin Accumulation in Resting Dendritic Spines

Drebrin A expressed in cell soma spontaneously accumulates within dendritic spines after it is transported together with cortical F-actin to dendritic spines in the parent dendrite (Hayashi and Shirao 1999). This suggests a constitutive

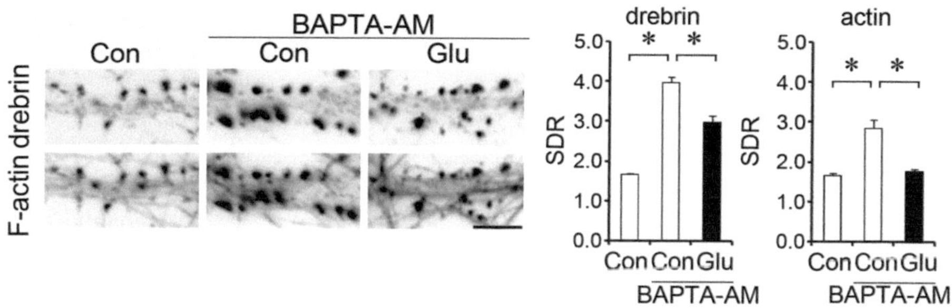

Fig. 11.2 The intracellular Ca^{2+} concentration inversely regulates drebrin-decorated F-actin in resting dendritic spines. Neurons (21 DIV) were preincubated in normal medium containing. 50 μM 1,2-bis(2-aminophenoxy) ethane-*N,N,N,N*-tetraacetic acid acetoxymethyl (BAPTA-AM), a membrane-permeable Ca^{2+} chelator, for 30 min, and stimulated with 100 μM glutamate for an additional 10 min. Images were obtained after double-labeling drebrin and F-actin. Bar graphs represent the spine/dendrite ratios (SDRs) of drebrin and actin. The F-actin images indicate that the spines kept their structure during the experiment although their shape changed. Scale bars, 5 μm. BAPTA-AM pretreatment significantly increased both the drebrin and actin SDR ($n = 30$ cells; $p < 0.01$, Scheffe's test). In the presence of BAPTA-AM, glutamate stimulation significantly decreased both drebrin and actin SDR ($n = 30$ cells; $p < 0.01$, Scheffe's test). Error bars represent S.E.M

accumulation mechanism of DF-actin in dendritic spines. When the intracellular Ca^{2+} in cultured hippocampal neurons was chelated by 1,2-bis(2-aminophenoxy) ethane-*N,N,N,N*-tetraacetic acid acetoxymethyl (BAPTA-AM), drebrin accumulation was more than double that of the control (Fig. 11.2). Therefore, the constitutive DF-actin accumulation mechanism seems inversely regulated by the intracellular Ca^{2+} concentration.

Chelation of extracellular Ca^{2+} with ethylene glycol tetraacetic acid also increased drebrin accumulation. Similarly, blocking NMDAR by the antagonist, D-(−)-2-amino-5-phosphonopentanoic acid (APV), or L-type voltage-dependent calcium channels by nifedipine also significantly increased drebrin accumulation. However, inhibition of Ca^{2+} release from intracellular Ca^{2+} stores with thapsigargin did not affect drebrin accumulation. These findings suggest that the NMDAR and voltage-dependent calcium channels, but not intracellular Ca^{2+} stores, are involved in drebrin accumulation in resting dendritic spines (Mizui et al. 2014).

Ca^{2+} increase in the brain facilitates drebrin degradation (Arai et al. 1991). It has also been reported that overactivation of NMDAR decreased the total amount of drebrin in neurons via calpain-mediated proteolysis (Chimura et al. 2015). This raises the possibility that low Ca^{2+} concentration inhibits drebrin degradation, resulting in DF-actin accumulation. However, this possibility is not very likely because blocking NMDAR increased drebrin accumulation without changing the total amount of drebrin in the neuron.

We have previously reported that α-amino-3-hydroxy-5-methyl-4-isoxazolepropionic acid receptor (AMPAR) activity regulates drebrin A accumulation during development (Takahashi et al. 2009). Fluorescence recovery after photobleaching analysis showed that the Ca^{2+}-impermeable AMPAR activity increased the stable fraction of drebrin A in dendritic spines, consequently promoting

DF-actin accumulation in dendritic spines. AMPAR may regulate drebrin accumulation by activating intracellular signaling pathways other than Ca^{2+} signaling (for details, see Chap. 10).

Intriguingly, extrinsically expressed drebrin A in immature neurons accumulated in filopodia without presynaptic terminals, forming extra-large filopodia (Mizui et al. 2005) or bifurcated abnormal filopodia (Biou et al. 2008). This suggests that AMPAR activity is necessary for accumulating DF-actin in the postsynaptic site, but not in dendritic protrusions, either dendritic filopodia or spines. Drebrin shows bundling activity of DF-actins when cyclin-dependent kinase 5 (Cdk5) phosphorylates serine 142 of drebrin (Worth et al. 2013). Additionally, the small GTPase Ras likely interacts with the C-terminal domain of drebrin, inducing changes in the actin cytoskeleton. Activation of drebrin bundling activity in dendritic spines may catch transported DF-actin along dendrites and sequester them in dendritic spines.

Taken together, DF-actin accumulation in resting dendritic spines is likely regulated by a multi-signaling pathway, such as the AMPAR-mediated intracellular signaling pathway, and by the intracellular Ca^{2+} concentration determined by NMDARs or L-type voltage-dependent calcium channels.

11.3 Activity-Dependent Changes in Drebrin Localization

11.3.1 Drebrin Exodus from Potentiated Synapse

When we stimulated neurons with glutamate, drebrin changed its localization from dendritic spines to the parent dendrite and cell soma (Fig. 11.3a). Interestingly, when neurons were stimulated with a low concentration of glutamate, a robust drebrin loss from dendritic spines within a partial population was observed. This weak glutamate stimulation did not induce a partial loss of drebrin from dendritic spines within a whole population. Additionally, drebrin translocation (exodus) was completely blocked by an NMDAR antagonist, indicating that the exodus specifically depends on NMDAR activity (Sekino et al. 2006). While we have shown that about 20% of spines contain no drebrin in vivo (Aoki et al. 2005), these spines may have been activated through NMDAR shortly before the analysis, and thus drebrin's reentry was not observed.

In parallel with the drebrin exodus, the F-actin content in dendritic spines decreases (Ouyang et al. 2005; Mizui et al. 2014). Does NMDAR activation specifically induce DF-actin translocation or does it generally induce both DF- and FF-actin? To answer this question, we used the neuronal culturing technique, called the Banker method (Kaech and Banker 2006; Goslin and Banker 1989), because it is difficult to discriminate the neuronal F-actin from the glial one in a cortical mixed culture.

In the Banker method, embryonic hippocampal neurons prepared from an embryonic day (ED) 18 rat brain are plated on a coverslip coated with poly-L-lysine. After attachment, the neurons are cultured on a glial sheet in a dish for 21 days in vitro

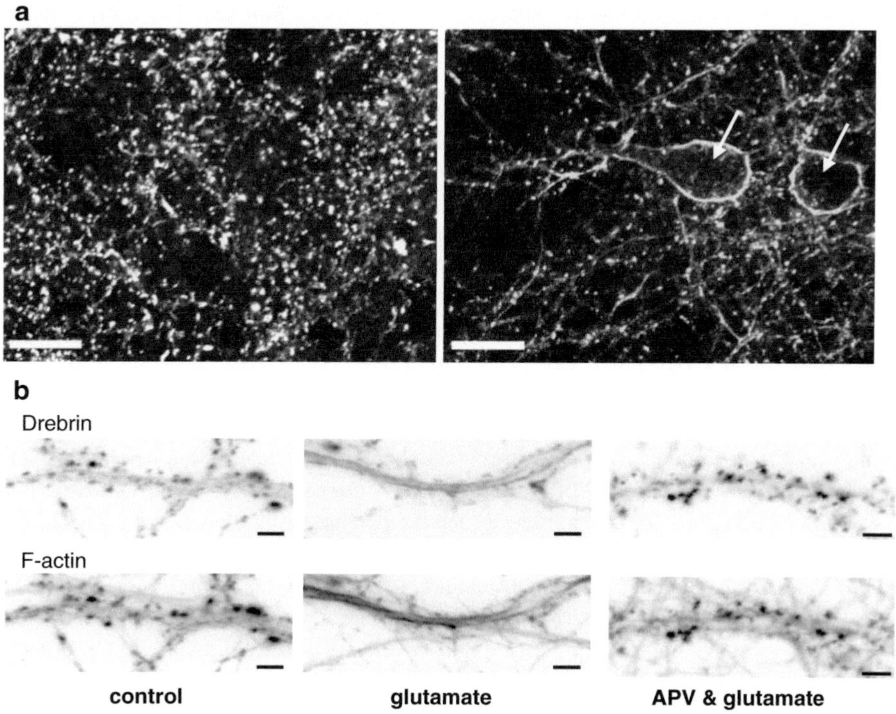

Fig. 11.3 (**a**) Confocal images of drebrin distribution in 21-DIV cultured cortical neurons stimulated with 100 μM glutamate (*right image*) and control neurons (*left image*). Note that the puncta immunostained with an anti-drebrin antibody are scattered as a dot-like pattern in the control neurons (*left image*), but not in the stimulated neurons (*right image*). Arrows indicate sites of neuronal somata. (**b**) Images were obtained from 21-DIV neurons double-labeled for drebrin and F-actin. Neurons were stimulated with glutamate in the absence and presence of APV. Scale bars in (**a**) and (**b**) are 10 μm and 3 μm, respectively

(DIV). By analyzing the pure neuronal culture on the coverslip, we could determine the distribution of drebrin and F-actin in neurons without interference of the glial F-actin (Takahashi et al. 2003).

To quantify the accumulation of DF-actin and FF-actin within a single dendritic spine, we double-stained the neurons with an anti-drebrin antibody (mAb M2F6) and rhodamine-labeled phalloidin and determined the spine/dendrite ratio (SDR) of drebrin and F-actin. The accumulation levels of DF-actin can be deduced by the drebrin SDR, and the accumulation level of total F-actin (DF-actin + FF-actin) can be deduced by the actin SDR.

In mature hippocampal neurons, both drebrin SDR and actin SDR were about 1.7 (Mizui et al. 2014). After glutamate stimulation, the drebrin and actin SDRs significantly decreased to about 0.5, namely, the intensity of dendritic spines was lower than that of dendritic shafts. Conversely, this stimulation did not affect the density of spines or presynaptic terminals. When NMDARs on neurons were inhibited by the antagonist, APV, DF-actin exodus was completely blocked, but total F-actin exodus was only partially blocked (Fig. 11.3b). This suggests that DF-actin exodus is specifically regulated by NMDARs, whereas FF-actin exodus is likely regulated by another signaling pathway (Mizui et al. 2014), although we cannot exclude the possibility that FF-actin is also partly regulated by NMDARs.

We then examined whether the activation of synaptic activity is involved in drebrin exodus. When γ-aminobutyric acid $(GABA)_A$ receptors were blocked by bicuculline, drebrin exodus was induced in parallel to an increase in the spontaneous firing rate of the neuronal network (Sokal et al. 2000). When the presynaptic membrane potential was depolarized by 90 mM KCl, the exodus was also induced in parallel to glutamate release from presynaptic terminals. Thus, synaptically evoked membrane depolarization seems to activate NMDAR, resulting in drebrin exodus.

NMDAR activation is the initial step to induce LTP which is a characteristic synapse function underlying the cellular mechanism of learning and memory. LTP can be induced in cultured neuronal synapses by chemical LTP (cLTP) stimulation (200 μM glycine with 20 μM bicuculline and 1 μM strychnine) (Lu et al. 2001). We used this cLTP stimulation and succeeded to induce drebrin exodus. Interestingly, cLTP-induced drebrin exodus was transient. Drebrin exodus from dendritic spines was prominent at 5 min after the stimulation; however, after 30 min drebrin returned to a level higher than the prestimulated level. In addition, using exogenously expressed GFP-drebrin A as a tracer, we showed that the drebrin SDR began to decrease immediately after the neurons were exposed to a cLTP solution and kept declining throughout the stimulation period. When the stimulation ceased, drebrin SDR began to rise and returned to the prestimulated level within 11 min (Fig. 11.4, open circles) (Mizui et al. 2014).

Conversely, when drebrin exodus was induced by 100 μM glutamate stimulation for 10 min, it needed 24 h to re-accumulate in the dendritic spines (Sekino et al. 2006). Additionally, time-lapse imaging analysis showed that the drebrin SDR began to decrease immediately after chemical LTD stimulation (100 μM glutamate with 10 μM glycine) similarly to LTP stimulation. But even after the stimulation ceased, the drebrin SDR remained low and did not return to the prestimulated level within 11 min (Fig. 11.4, closed circles). These results suggest that LTD stimulation also induces drebrin exodus, but unlike LTP stimulation, the exodus is not transient.

11.3.2 Transient Ca^{2+} Influx Through NMDA Receptors Specifically Induces Drebrin Exodus

Drebrin exodus is not induced when the extracellular Ca^{2+} is chelated (Mizui et al. 2014), but it is induced when the intracellular Ca^{2+} is chelated (Fig. 11.2). In contrast, as mentioned previously, the chelation of either extracellular or intracellular Ca^{2+} increases drebrin content at resting dendritic spines. Additionally, blocking

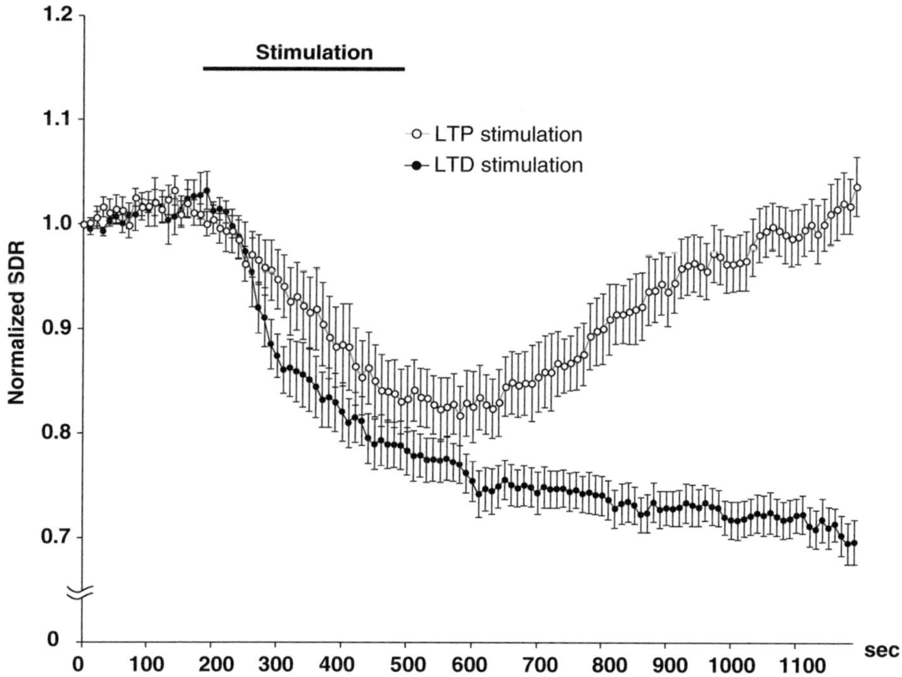

Fig. 11.4 Chemical LTP and LTD stimulations induce transient and prolonged drebrin exodus. We transfected 7-DIV neurons with a GFP-drebrin A expression vector and performed time-lapse imaging at 21 DIV. Open and closed circles represent data obtained by LTP (200 μM glycine with 20 μM bicuculline and 1 μM strychnine) and LTD stimulation (100 μM glutamate with 10 μM glycine) at 30-s intervals. Error bars represent S.E.M. ($n = 7$ neurons)

NMDARs inhibits drebrin exodus, but blocking either L-type voltage-dependent calcium channels or Ca^{2+} release from intracellular Ca^{2+} stores does not inhibit drebrin exodus (Mizui et al. 2014). Taken together, we concluded that the transient Ca^{2+} influx through NMDA receptors specifically induces the drebrin exodus, but the intracellular Ca^{2+} concentration of the dendritic spine at steady state does not affect drebrin exodus following the excitation.

11.3.3 Involvement of Myosin II ATPase in Drebrin Exodus

Myosin II ATPase activity is necessary for LTP formation (Rex et al. 2010). Because inhibition of myosin II ATPase by blebbistatin blocks drebrin exodus, myosin II ATPase activity likely plays a role in LTP formation by inducing drebrin exodus (Mizui et al. 2014).

How does myosin II ATPase activation induce drebrin exodus? Myosin II ATPase activation destabilizes the F-actin network in vitro (Haviv et al. 2008) and in vivo

(Wilson et al. 2010). Murrell and Gardel have proposed an F-actin buckling model (Murrell and Gardel 2012). In disordered nonsarcomeric actomyosin networks, myosin II ATPase activation induces both tension and compression. Consequently, compressive stresses are relieved through F-actin severing. In drebrin exodus, Ca^{2+} influx through NMDARs activates myosin II ATPase, leading to breakup of the stable DF-actin complex via severing DF-actin in the core of dendritic spines. Consequently, tattered short DF-actins spread out evenly into the dendritic shaft.

How does transient Ca^{2+} influx through NMDARs activate myosin II ATPase? As well as tropomyosin, drebrin inhibits myosin II from binding to F-actin in vitro (Hayashi et al. 1996). Therefore, actin-dependent myosin II ATPase is inhibited in the core region of resting dendritic spines where DF-actin is concentrated. In the skeletal muscle, increased intracellular Ca^{2+} disinhibits tropomyosin-dependent myosin II ATPase inhibition. If a similar mechanism is applicable to dendritic spines, transient Ca^{2+} influx through NMDARs disinhibits drebrin-dependent inhibition of myosin II ATPase activity in dendritic spines (Mizui et al. 2014).

11.3.4 Drebrin Re-Accumulates During LTP Maintenance, but not During LTD

As mentioned above, LTP stimulation induces drebrin exodus followed by re-accumulation of DF-actin at dendritic spines (Mizui et al. 2014). Electrical LTP stimulation for 1 s induced progressive Ca^{2+} increases via NMDARs in dendritic spines, which returned to the basal level within 10 s (Petrozzino et al. 1995). Therefore, the Ca^{2+} increase via NMDARs during LTP initiation likely recovers to the basal level soon after it triggers the drebrin exodus. AMPAR levels at the postsynaptic membrane increase during LTP (Lu et al. 2001). If drebrin re-accumulation uses a DF-actin accumulation mechanism similar to that operating in resting dendritic spines, then Ca^{2+} recovery in conjunction with the AMPAR increase may coordinately induce the re-accumulation of DF-actin during LTP maintenance. This is consistent with previous reports that showed an increase in drebrin during LTP in vivo (Fukazawa et al. 2003) and a correlation between drebrin content and the spine head size (Kobayashi et al. 2007).

11.4 Role of Synaptic Stability and Instability in Synaptic Plasticity

11.4.1 Neutral Dynamic Morphology of Resting Dendritic Spines

Spine synapses have a tight structure–function relationship (Matsuzaki et al. 2001; Asrican et al. 2007). For example, larger spines have wider postsynaptic density (PSD) (Harris and Stevens 1989) and more AMPARs (Nusser et al. 1998; Takumi et al. 1999). Additionally, the number of synaptic vesicles in the presynaptic terminal correlates with the PSD area (Harris and Stevens 1989; Knott et al. 2006). Thus, the spine size closely correlates with the synaptic strength.

Although resting spines show structural changes without synaptic input, the average size of dendritic spines is constant within the population (Yasumatsu et al. 2008). According to this neutral dynamic morphology model, the size of dendritic spines is regulated by the F-actin content in dendritic spines. There are two rules we must consider to discuss this model. First, the final F-actin content is determined by the total amount of actin molecules (Korn et al. 1987). Second, the actual polymerization change in the F-actin content depends on the frequency of filament nucleation events.

In resting spines, the presence of the stable F-actin complex blocks the translocation of proteins into the spines (Ouyang et al. 2005). Therefore, the total amount of actin molecules is constant; consequently the final F-actin content remains constant, although the FF-actin content may be increased by random Arp2/3-dependent nucleation or decreased by severing and capping. These are the neutral dynamic morphological changes. In contrast, DF-actin forms a stable actin cytoskeleton in the core region of the spine that does not contribute to the above changes in F-actin content that directly link to spine motility. Therefore, at the risk of oversimplification, the DF-actin content in the dendritic spines inversely regulates the spine motility. Actually, overaccumulation of extrinsically expressed GFP-drebrin A in dendritic filopodia and spines stops their motility (unpublished data).

11.4.2 NMDA Receptor-Mediated Change of Dendritic Spines From Steady to Excited State

The concepts of spine stability and instability are fundamental to understanding the input-specific changes in synaptic strength, namely, synaptic plasticity. Spine stability is maintained by the aforementioned neutral dynamic morphology of resting dendritic spines (Fig. 11.5a). In contrast, spine instability underlying LTP and LTD requires synaptic activation of NMDARs, which leads to Ca^{2+} influx through the NMDAR channel and a rise in Ca^{2+} within the dendritic spines (Malenka and Bear 2004).

Drebrin exodus is specifically triggered by Ca^{2+} influx through synaptic NMDARs during the LTP and LTD induction steps (Fig. 11.5b). Because of the exodus, DF-actin in potentiated dendritic spines decreases, and consequently G-actin may immediately refill the spine by diffusion from the dendritic shaft (Star et al. 2002). This leads to an increase in the total actin content available for polymerization and an increase in FF-actin available for Arp2/3-dependent branching, namely, the two rules of increased F-actin described previously are satisfied (Fig. 11.5c) (Koganezawa et al. 2017). Although LTP-dependent Arp2/3 activation has not been reported, neural Wiskott–Aldrich syndrome protein (N-WASP) and the Arp2/3 complex regulate actin during the development of dendritic spines (Wegner et al. 2008). Thus, at the LTP expression step, Arp2/3-mediated actin nucleation may facilitate F-actin polymerization in dendritic spines and enlarge the spine size. Additionally, the loss of stable F-actin may induce translocation of plasticity-dependent proteins, such as actin, Ca^{2+}/

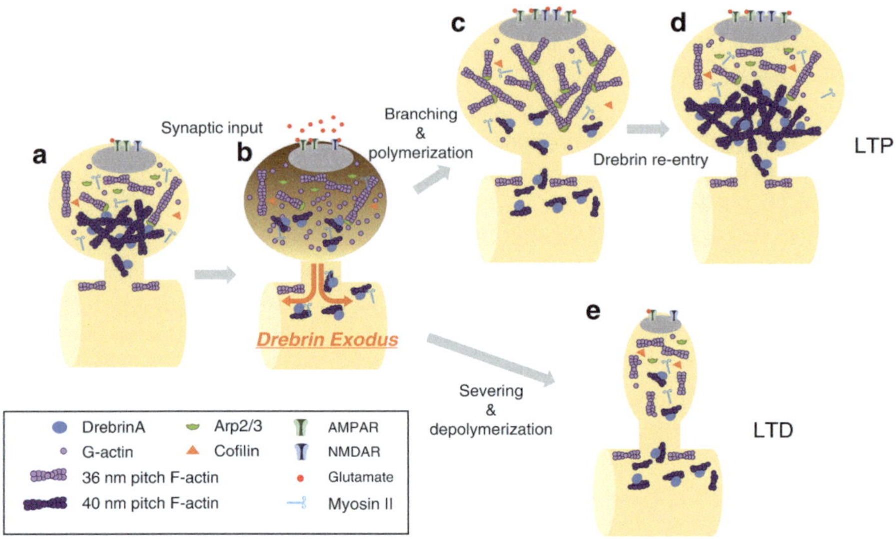

Fig. 11.5 Schematic representation of the actin cytoskeleton during synaptic plasticity. (**a**) Dendritic spines at steady state contain dynamic F-actin (normal helical pitch F-actin) and stable F-actin (long helical pitch F-actin). (**b**) Once Ca^{2+} enters through NMDA receptors (*dark brown*), it activates myosin II ATPase and induces drebrin exodus. (**c**) During LTP, G-actin refills the spine and actin nucleation is promoted. Consequently, polymerization of dynamic F-actin is facilitated and the spine is enlarged. (**d**) The Ca^{2+} concentration returns to a normal level (*light brown*), and the AMPA receptor accumulates; the reconstruction of drebrin A-decorated stable F-actin is induced, contributing to the maintenance of enlarged spine morphology. (**e**) During LTD, the prolonged intracellular Ca^{2+} increase inhibits drebrin re-accumulation. Then, low drebrin facilitates AMPAR endocytosis, resulting in decreased AMPAR density at the postsynaptic membrane. Because AMPAR activity is necessary for the re-accumulation of drebrin, AMPAR decrease and drebrin re-accumulation inhibition form a positive feedback loop. Additionally, cofilin, which triggers depolymerization of F-actin, is activated. Consequently, these changes lead to spine shrinkage

calmodulin-dependent protein kinase II, and AMPAR. Moreover, drebrin exodus may facilitate AMPAR trafficking to postsynaptic membranes by disinhibiting myosin V, which is required for AMPAR trafficking during LTP (Correia et al. 2008).

NMDAR-mediated facilitation of F-actin polymerization was observed between 3 min and 18 min after the synaptic activation and then ended (Okamoto et al. 2004). In contrast, NMDAR-mediated spine enlargement continued over 30 min. Furthermore, electrophysiological studies have shown that NMDAR-dependent LTP lasts hours, days, or even weeks and the so-called late phases or longer-lasting components of LTP require new protein synthesis and gene transcription (Abraham and Williams 2003). How can enlarged spines keep their size even after the facilitation of F-actin polymerization ends? LTP stimulation induces drebrin re-accumulation at the LTP maintenance step. As mentioned previously, drebrin accumulation in dendritic spines is inversely regulated by the intracellular Ca^{2+} concentration and positively by AMPAR activity. Therefore, both the Ca^{2+} concentration recovery to steady state and the increase in AMPAR insertion into postsynaptic membranes likely induce the drebrin re-accumulation in dendritic spines

(Fig. 11.5d). Hence, the dendritic spines changes from the excited state to steady state, making it possible for the dendritic spines to keep their enlarged size with a secure increase in synaptic strength until newly synthesized proteins are delivered to these spines for the "late phases" of LTP.

Drebrin is classified into a drebrin E isoform which is widely observed in various tissues and a neuron-specific drebrin A isoform. Although both isoforms can make long helical crossover F-actin, drebrin A, but not E, is crucial for LTP formation because drebrin A-knockout mice, in which drebrin E continues to be expressed throughout life instead of drebrin A (Kojima et al. 2010), show impaired hippocampal LTP and fear learning in an age-dependent manner (Kojima et al. 2016).

11.4.3 Spine Instability During LTD Formation

LTD stimulation also induces a similar drebrin exodus to LTP stimulation. However, LTD stimulation downregulates Arp2/3-mediated actin nucleation (Rocca et al. 2013). Thus, because of the absence of Arp2/3-dependent F-actin branching, F-actin polymerization is not facilitated even when the total actin content available for polymerization is increased by drebrin exodus. Instead, LTD stimulation induces AMPAR internalization (Mulkey et al. 1994; Malenka and Bear 2004) and spine shrinkage (Wang et al. 2007). Drebrin exodus may increase the AMPAR internalization because endocytosis such as human immunodeficiency virus (HIV)-1 entry is suppressed by drebrin (Gordon-Alonso et al. 2013). Because of the decrease in AMPAR activity at the LTD expression stage, drebrin re-accumulation is not observed in dendritic spines (Fig. 11.5e). The absence of a stable DF-actin core makes it difficult for dynamic FF-actin to push back against and extend the plasma membrane. Additionally, cofilin, which triggers depolymerization of F-actin, is activated (Zhou et al. 2004; Cao et al. 2017). Consequently, these changes lead to spine shrinkage. This idea is consistent with our previous study demonstrating that the drebrin content correlates with the dendritic spine size (Kobayashi et al. 2007). Furthermore, according to this model, the dendritic spines during LTD maintenance are not in steady state, but in the excited state.

In fact, the preferential sites for morphological changes are small spines (Matsuzaki et al. 2004) that express a small number of AMPARs (Matsuzaki et al. 2001). Conversely, large spines express AMPARs abundantly and are stable for months in vivo (Trachtenberg et al. 2002).

11.5 Pathological Relevance of Drebrin Loss to Alzheimer's Disease

Alzheimer's disease (AD) is a neurodegenerative disorder accompanied by severe progressive memory and cognitive impairment. Drebrin decrease is a hallmark of the AD brain (Harigaya et. al 1996; Kojima and Shirao 2007; Ma et al. 2015), as well as amyloid plaques (senile plaques) and neurofibrillary tangles (Selkoe and Hardy 2016). Drebrin is markedly decreased in dendritic spines before significant presynaptic changes appear.

As drebrin loss confers a loss of drebrin-decorated stable F-actin, the balance between the stability and instability of spines may be disrupted in AD. According to the drebrin exodus model, dendritic spines without drebrin accumulation should be constitutively instable because of the lack of stable F-actin. Therefore, although LTP stimulation can enlarge the dendritic spine in the AD brain, it is likely to soon be restored to the previous size, resulting in failure of LTP formation. This AD model has been reviewed in detail by Shirao et al. (2017).

11.6 Conclusion

In this chapter, we discussed the role of the actin cytoskeleton in dendritic spines, particularly focusing on synaptic plasticity, and concluded that synaptic plasticity begins with drebrin exodus, by which dendritic spines change from steady state to excited state. Here, we proposed a "two-state model of synaptic plasticity" (Fig. 11.6).

The resting spine at steady state has a high drebrin content. According to the aforementioned neutral dynamic morphology model, dendritic spines at steady state

Fig. 11.6 Two-state model of synaptic plasticity. (**a**) LTP formation. Dendritic spines without tetanic stimulation are at steady state with high drebrin. These spines can spontaneously change their size by random actin nucleation even at steady state, but they are resilient and seldom change into the excited state. Once LTP stimulation induces Ca^{2+} influx, drebrin is decreased because of the drebrin exodus. Consequently, dendritic spines change into the excited state and become plastic. Shortly after LTP stimulation, the Ca^{2+} concentration returns to normal, and F-actin polymerization is facilitated by the high actin nucleation activity, resulting in the enlargement of spine size and the increase in AMPAR density. Normal Ca^{2+} and increased AMPAR coordinately increase drebrin, and then the spines return to steady state. (**b**) LTD formation. Once LTD stimulation induces Ca^{2+} influx, drebrin is decreased because of the drebrin exodus. After LTD stimulation, the Ca^{2+} concentration remains elevated, and the AMPAR density decreases because of the facilitated endocytosis of AMPAR, inhibiting the re-accumulation of drebrin. Even after the Ca^{2+} concentration returns to normal, the low drebrin content confers high instability to dendritic spines. The diameter of the circles shows the dendritic spine head size. The AMPAR density on the postsynaptic membrane is shown by a *red line*. The Ca^{2+} and drebrin content in the spine head are shown by *black and blue lines*, respectively

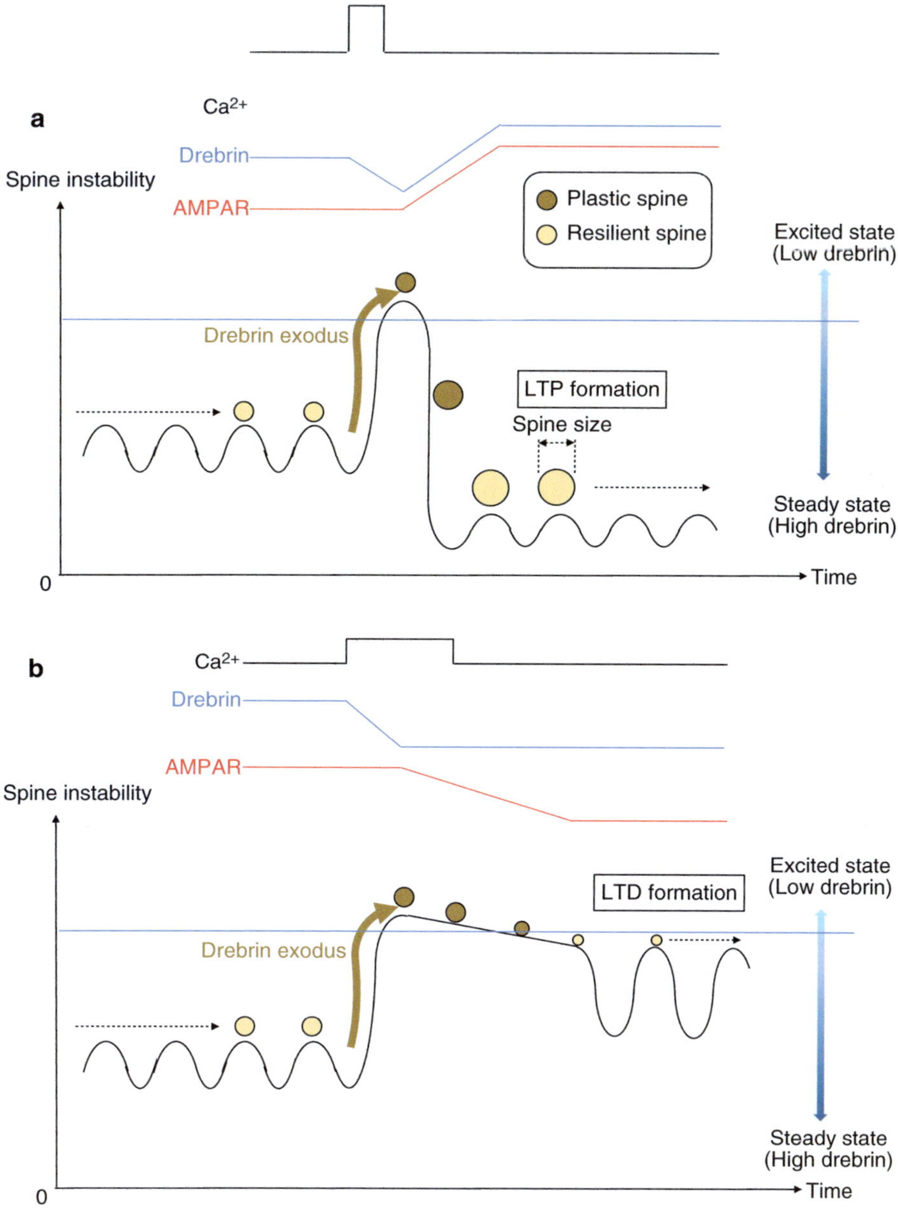
a
Ca2+
Drebrin
AMPAR
Spine instability
Plastic spine
Resilient spine
Excited state
(Low drebrin)
Drebrin exodus
LTP formation
Spine size
Steady state
(High drebrin)
0
Time
b
Ca2+
Drebrin
AMPAR
Spine instability
LTD formation
Excited state
(Low drebrin)
Drebrin exodus
Steady state
(High drebrin)
0
Time

are resilient, namely, the average size of dendritic spines does not change at steady state because of the presence of stable F-actin, although they spontaneously increase their size by random nucleation and decrease it by severing and capping.

In the initial process of either LTP or LTD, the Ca^{2+} influx through NMDARs triggers myosin-driven drebrin exodus, resulting in drebrin decrease. Thus, potentiated dendritic spines change from steady state to excited state, where dynamic F-actin predominates and F-actin begins to polymerize by activation of actin nucleation.

Because of the increase in dynamic F-actin, potentiated spines become plastic and exhibit LTP or LTD as a result of two different processes. During LTP, the increased Ca^{2+} soon returns to the basal level, and AMPAR density and the spine size increase. The increase in AMPAR density in conjunction with Ca^{2+} recovery coordinately accumulates drebrin in dendritic spines, leading the spine to steady state. Thus, the resilient spines keep their enlarged size with a secure increase in synaptic strength during LTP maintenance.

In contrast during LTD, the intracellular Ca^{2+} increase is prolonged, which inhibits drebrin re-accumulation. Then, low drebrin facilitates AMPAR endocytosis, resulting in decreased AMPAR density at the postsynaptic membrane. Because AMPAR activity is necessary for the re-accumulation of drebrin, AMPAR decrease and drebrin re-accumulation inhibition form a positive feedback loop, keeping high plasticity of the dendritic spines. This is consistent with the observation that small spines are preferable for synaptic plasticity.

According to this model, the resilient spines in Alzheimer's disease likely show high spine instability because of low drebrin. This may cause defects in LTP maintenance in Alzheimer's disease. Thus, it is important to examine the pathological relevance of spine instability to various types of dementia.

References

Abraham WC, Williams JM (2003) Properties and mechanisms of LTP maintenance. Neuroscientist 9:463–474

Allison DW, Gelfand VI, Spector I, Craig AM (1998) Role of actin in anchoring postsynaptic receptors in cultured hippocampal neurons: differential attachment of NMDA versus AMPA receptors. J Neurosci 18:2423–2436

Aoki C, Sekino Y, Hanamura K, Fujisawa S, Mahadomrongkul V, Ren Y, Shirao T (2005) Drebrin A is a postsynaptic protein that localizes in vivo to the submembranous surface of dendritic sites forming excitatory synapses. J Comp Neurol 483:383–402

Arai H, Sato K, Uto A, Yasumoto Y (1991) Effect of Transient cerebral ischemia in mongolian gerbils on synaptic vesicle protein (SVP-38) and developmentally regulated brain protein (drebrin). Neurosci Res Commun 9:143–150

Asrican B, Lisman J, Otmakhov N (2007) Synaptic strength of individual spines correlates with bound Ca^{2+}-calmodulin-dependent kinase II. J Neurosci 27:14007–14011

Biou V, Brinkhaus H, Malenka RC, Matus A (2008) Interactions between drebrin and Ras regulate dendritic spine plasticity. Eur J Neurosci 27:2847–2859

Cao F, Zhou Z, Pan X, Leung C, Xie W, Collingridge G, Jia Z (2017) Developmental regulation of hippocampal long-term depression by cofilin-mediated actin reorganization. Neuropharmacology 112:66–75

Chimura T, Launey T, Yoshida N (2015) Calpain-mediated degradation of drebrin by excitotoxicity in vitro and in vivo. PLoS One 10:e0125119

Correia SS, Bassani S, Brown TC, Lise MF, Backos DS, El-Husseini A, Passafaro M, Esteban JA (2008) Motor protein-dependent transport of AMPA receptors into spines during long-term potentiation. Nat Neurosci 11:457–466

Dailey ME, Smith SJ (1996) The dynamics of dendritic structure in developing hippocampal slices. J Neurosci 16:2983–2994

Desmond NL, Levy WB (1983) Synaptic correlates of associative potentiation/depression: an ultrastructural study in the hippocampus. Brain Res 265:21–30

Fischer M, Kaech S, Knutti D, Matus A (1998) Rapid actin-based plasticity in dendritic spines. Neuron 20:847–854

Fukazawa Y, Saitoh Y, Ozawa F, Ohta Y, Mizuno K, Inokuchi K (2003) Hippocampal LTP is accompanied by enhanced F-actin content within the dendritic spine that is essential for late LTP maintenance in vivo. Neuron 38:447–460

Geraldo S, Khanzada UK, Parsons M, Chilton JK, Gordon-Weeks PR (2008) Targeting of the F-actin-binding protein drebrin by the microtubule plus-tip protein EB3 is required for neuritogenesis. Nat Cell Biol 10:1181–1189

Gordon-Alonso M, Rocha-Perugini V, Alvarez S, Ursa A, Izquierdo-Useros N, Martinez-Picado J, Munoz-Fernandez MA, Sanchez-Madrid F (2013) Actin-binding protein drebrin regulates HIV-1-triggered actin polymerization and viral infection. J Biol Chem 288:28382–28397

Goslin K, Banker G (1989) Experimental observations on the development of polarity by hippocampal neurons in culture. J Cell Biol 108(4):1507–1516

Halpain S (2000) Actin and the agile spine: how and why do dendritic spines dance? Trends Neurosci 23:141–146

Harigaya Y, Shoji M, Shirao T, Hirai S (1996) Disappearance of actin-binding protein, drebrin, from hippocampal synapses in Alzheimer's disease. J Neurosci Res 43(1):87–92

Harris KM, Stevens JK (1989) Dendritic spines of CA 1 pyramidal cells in the rat hippocampus: serial electron microscopy with reference to their biophysical characteristics. J Neurosci 9:2982–2997

Haviv L, Gillo D, Backouche F, Bernheim-Groswasser A (2008) A cytoskeletal demolition worker: myosin II acts as an actin depolymerization agent. J Mol Biol 375:325–330

Hayashi K, Shirao T (1999) Change in the shape of dendritic spines caused by overexpression of drebrin in cultured cortical neurons. J Neurosci 19:3918–3925

Hayashi K, Ishikawa R, Ye LH, He XL, Takata K, Kohama K, Shirao T (1996) Modulatory role of drebrin on the cytoskeleton within dendritic spines in the rat cerebral cortex. J Neurosci 16:7161–7170

Honkura N, Matsuzaki M, Noguchi J, Ellis-Davies GC, Kasai H (2008) The subspine organization of actin fibers regulates the structure and plasticity of dendritic spines. Neuron 57:719–729

Kaech S, Banker G (2006) Culturing hippocampal neurons. Nat Protoc 1:2406–2415

Knott GW, Holtmaat A, Wilbrecht L, Welker E, Svoboda K (2006) Spine growth precedes synapse formation in the adult neocortex in vivo. Nat Neurosci 9(9):1117–1124. Epub 2006 Aug 6

Kobayashi C, Aoki C, Kojima N, Yamazaki H, Shirao T (2007) Drebrin a content correlates with spine head size in the adult mouse cerebral cortex. J Comp Neurol 503:618–626

Koganezawa N, Hanamura K, Sekino Y, Shirao T (2017) The role of drebrin in dendritic spines. Mol Cell Neurosci. doi:10.1016/j.mcn.2017.01.004

Kojima N, Hanamura K, Yamazaki H, Ikeda T, Itohara S, Shirao T (2010) Genetic disruption of the alternative splicing of drebrin gene impairs context-dependent fear learning in adulthood. Neuroscience 165:138–150

Kojima N, Shirao T (2007) Synaptic dysfunction and disruption of postsynaptic drebrin-actin complex: a study of neurological disorders accompanied by cognitive deficits. Neurosci Res 58(1):1–5. Epub 2007 Feb 11

Korn ED, Carlier MF, Pantaloni D (1987) Actin polymerization and ATP hydrolysis. Science 238(4827):638–644

Kojima N, Yasuda H, Hanamura K, Ishizuka Y, Sekino Y, Shirao T (2016) Drebrin A regulates hippocampal LTP and hippocampus-dependent fear learning in adult mice. Neuroscience 324:218–226

Lu W, Man H, Ju W, Trimble WS, MacDonald JF, Wang YT (2001) Activation of synaptic NMDA receptors induces membrane insertion of new AMPA receptors and LTP in cultured hippocampal neurons. Neuron 29:243–254

Ma Q, Ruan YY, Xu H, Shi XM, Wang ZX, Hu YL (2015) Safflower yellow reduces lipid peroxidation, neuropathology, tau phosphorylation and ameliorates amyloid β-induced impairment of learning and memory in rats. Biomed Pharmacother 76:153–164. doi: 10.1016/j.biopha.2015.10.004. Epub 2015 Nov 19

Malenka RC, Bear MF (2004) LTP and LTD: an embarrassment of riches. Neuron 44:5–21

Maletic-Savatic M, Malinow R, Svoboda K (1999) Rapid dendritic morphogenesis in CA1 hippocampal dendrites induced by synaptic activity. Science 283:1923–1927

Matsuzaki M, Honkura N, Ellis-Davies GC, Kasai H (2004) Structural basis of long-term potentiation in single dendritic spines. Nature 429:761–766

Matsuzaki M, Ellis-Davies GC, Nemoto T, Miyashita Y, Iino M, Kasai H (2001) Dendritic spine geometry is critical for AMPA receptor expression in hippocampal CA1 pyramidal neurons. Nat Neurosci 4:1086–1092

Matus A, Ackermann M, Pehling G, Byers HR, Fujiwara K (1982) High actin concentrations in brain dendritic spines and postsynaptic densities. Proc Natl Acad Sci U S A 79:7590–7594

Mikati MA, Grintsevich EE, Reisler E (2013) Drebrin-induced stabilization of actin filaments. J Biol Chem 288:19926–19938

Mizui T, Takahashi H, Sekino Y, Shirao T (2005) Overexpression of drebrin A in immature neurons induces the accumulation of F-actin and PSD-95 into dendritic filopodia, and the formation of large abnormal protrusions. Mol Cell Neurosci 30:149–157

Mizui T, Kojima N, Yamazaki H, Katayama M, Hanamura K, Shirao T (2009) Drebrin E is involved in the regulation of axonal growth through actin-myosin interactions. J Neurochem 109:611–622

Mizui T, Sekino Y, Yamazaki H, Ishizuka Y, Takahashi H, Kojima N, Kojima M, Shirao T (2014) Myosin II ATPase activity mediates the long-term potentiation-induced exodus of stable F-actin bound by drebrin A from dendritic spines. PLoS One 9:e85367

Mulkey RM, Endo S, Shenolikar S, Malenka RC (1994) Involvement of a calcineurin/inhibitor-1 phosphatase cascade in hippocampal long-term depression. Nature 369:486–488

Murrell MP, Gardel ML (2012) F-actin buckling coordinates contractility and severing in a biomimetic actomyosin cortex. Proc Natl Acad Sci U S A 109:20820–20825

Nusser Z, Lujan R, Laube G, Roberts JD, Molnar E, Somogyi P (1998) Cell type and pathway dependence of synaptic AMPA receptor number and variability in the hippocampus. Neuron 21(3):545–559

Okamoto K, Nagai T, Miyawaki A, Hayashi Y (2004) Rapid and persistent modulation of actin dynamics regulates postsynaptic reorganization underlying bidirectional plasticity. Nat Neurosci 7:1104–1112

Ouyang Y, Wong M, Capani F, Rensing N, Lee CS, Liu Q, Neusch C, Martone ME, Wu JY, Yamada K, Ellisman MH, Choi DW (2005) Transient decrease in F-actin may be necessary for translocation of proteins into dendritic spines. Eur J Neurosci 22:2995–3005

Petrozzino JJ, Pozzo Miller LD, Connor JA (1995) Micromolar Ca2+ transients in dendritic spines of hippocampal pyramidal neurons in brain slice. Neuron 14:1223–1231

Rex CS, Gavin CF, Rubio MD, Kramar EA, Chen LY, Jia Y, Huganir RL, Muzyczka N, Gall CM, Miller CA, Lynch G, Rumbaugh G (2010) Myosin IIb regulates actin dynamics during synaptic plasticity and memory formation. Neuron 67:603–617

Rocca DL, Amici M, Antoniou A, Blanco Suarez E, Halemani N, Murk K, McGarvey J, Jaafari N, Mellor JR, Collingridge GL, Hanley JG (2013) The small GTPase Arf1 modulates Arp2/3-mediated actin polymerization via PICK1 to regulate synaptic plasticity. Neuron 79:293–307

Selkoe DJ, Hardy J (2016) The amyloid hypothesis of Alzheimer's disease at 25 years. EMBO Mol Med 8(6):595–608. doi: 10.15252/emmm.201606210

Sekino Y, Tanaka S, Hanamura K, Yamazaki H, Sasagawa Y, Xue Y, Hayashi K, Shirao T (2006) Activation of N-methyl-D-aspartate receptor induces a shift of drebrin distribution: disappearance from dendritic spines and appearance in dendritic shafts. Mol Cell Neurosci 31:493–504

Sharma S, Grintsevich EE, Hsueh C, Reisler E, Gimzewski JK (2012) Molecular cooperativity of drebrin1-300 binding and structural remodeling of F-actin. Biophys J 103:275–283
Shirao T, Hanamura K, Koganezawa N, Ishizuka Y, Yamazaki H, Sekino Y (2017) The role of drebrin in neurons. J Neurochem. doi:10.1111/jnc.13988
Sokal DM, Mason R, Parker TL (2000) Multi-neuronal recordings reveal a differential effect of thapsigargin on bicuculline- or gabazine-induced epileptiform excitability in rat hippocampal neuronal networks. Neuropharmacology 39:2408–2417
Star EN, Kwiatkowski DJ, Murthy VN (2002) Rapid turnover of actin in dendritic spines and its regulation by activity. Nat Neurosci 5(3):239–246
Takahashi H, Yamazaki H, Hanamura K, Sekino Y, Shirao T (2009) Activity of the AMPA receptor regulates drebrin stabilization in dendritic spine morphogenesis. J Cell Sci 122:1211–1219
Takahashi H, Sekino Y, Tanaka S, Mizui T, Kishi S, Shirao T (2003) Drebrin-dependent actin clustering in dendritic filopodia governs synaptic targeting of postsynaptic density-95 and dendritic spine mor phogenesis. J Neurosci 23:6586–6595
Takumi Y, Ramírez-León V, Laake P, Rinvik E, Ottersen OP (1999) Different modes of expression of AMPA and NMDA receptors in hippocampal synapses. Nat Neurosci 2(7):618–624
Trachtenberg JT, Chen BE, Knott GW, Feng G, Sanes JR, Welker E, Svoboda K (2002) Long-term in vivo imaging of experience-dependent synaptic plasticity in adult cortex. Nature 420:788–794
Wang XB, Yang Y, Zhou Q (2007) Independent expression of synaptic and morphological plasticity associated with long-term depression. J Neurosci 27:12419–12429
Wegner AM, Nebhan CA, Hu L, Majumdar D, Meier KM, Weaver AM, Webb DJ (2008) N-wasp and the arp2/3 complex are critical regulators of actin in the development of dendritic spines and synapses. J Biol Chem 283:15912–15920
Wilson CA, Tsuchida MA, Allen GM, Barnhart EL, Applegate KT, Yam PT, Ji L, Keren K, Danuser G, Theriot JA (2010) Myosin II contributes to cell-scale actin network treadmilling through network disassembly. Nature 465:373–377
Worth DC, Daly CN, Geraldo S, Oozeer F, Gordon-Weeks PR (2013) Drebrin contains a cryptic F-actin-bundling activity regulated by Cdk5 phosphorylation. J Cell Biol 202:793–806
Yasumatsu N, Matsuzaki M, Miyazaki T, Noguchi J, Kasai H (2008) Principles of long-term dynamics of dendritic spines. J Neurosci 28:13592–13608
Yuste R (2015) The discovery of dendritic spines by Cajal. Front Neuroanat 9:18
Zhou Q, Homma KJ, Poo MM (2004) Shrinkage of dendritic spines associated with long-term depression of hippocampal synapses. Neuron 44:749–757
Zimmermann J, Falcke M (2014) Formation of transient lamellipodia. PLoS One 9:e87638

Chapter 12
Drebrin in Alzheimer's Disease

Yuta Ishizuka and Kenji Hanamura

Abstract Alzheimer's disease (AD) is a neurodegenerative disorder accompanied by severe progressive memory and cognitive impairment. The brain of AD patients has an abundance of two abnormal structures, amyloid plaques (senile plaques) and neurofibrillary tangles. In addition, drebrin loss is another hallmark of AD brains, which is a common feature in the brain of both AD patients and AD mouse models. Strong evidence from human genetics and transgenic mouse models has indicated that amyloid β (Aβ) is part of the etiology and pathogenesis of AD. Recently, it has become clear that synaptic dysfunction, including reduced synaptic transmission and loss of dendritic spines, occurs prior to the formation of amyloid plaques and neuronal cell loss. Furthermore, immunohistochemistry using postmortem human brains and AD mouse models has shown that drebrin loss in postsynaptic sites occurs earlier than the presynaptic change in AD brains. In addition, dysregulation of glutamate receptor trafficking and the p21-activated kinase/LIM kinase pathway has been observed in AD brains. It is now believed that soluble Aβ oligomers, namely, Aβ-derived diffusible ligands (ADDLs), but not insoluble Aβ aggregation mediates Aβ toxicity. ADDLs bind to the postsynaptic site and induce the aberrant morphology and density of dendritic spines. Consistent with the AD mouse models, the surface expression of glutamate receptors decreases after ADDL exposure. Importantly, the ADDL-induced drebrin loss in dendritic spines occurs prior to aberrations in dendritic spine morphology and density. These observations indicate that drebrin loss in dendritic spines occurs at the prodromal stage of AD, before the density and morphology of dendritic spines change. Quantitation of drebrin may be a possible tool for diagnosing the prodromal stage of AD, before dementia development in AD.

Keywords Drebrin • Cofilin • α-Amino-3-hydroxy-5-methyl-4-isoxaxole propionic acid receptor • *N*-Methyl-D-aspartate receptor • Alzheimer's disease • Aβ-derived diffusible ligands • Synaptic dysfunction

---

Y. Ishizuka (✉) • K. Hanamura
Department of Neurobiology and Behavior, Gunma University Graduate School of Medicine, Maebashi 371-8511, Japan
e-mail: yishizuka@gunma-u.ac.jp; yishizuka1110@gmail.com

T. Shirao, Y. Sekino (eds.), *Drebrin*, Advances in Experimental Medicine and Biology 1006, DOI 10.1007/978-4-431-56550-5_12

12.1 Introduction

Alzheimer's disease (AD) is a neurodegenerative disorder affecting numerous senior citizens worldwide (Prince et al. 2016). Its main clinical symptoms include mental changes such as progressive memory and cognitive impairment, personality changes, and some behavioral disorders, which may substantially lower the quality of life of elderly individuals (Sosa-Ortiz et al. 2012). The pathology of AD is caused by excessive production of amyloid β (Aβ). Amyloid plaques in the extracellular space, which are generated with Aβ, were originally thought to contribute to neuronal cell death, resulting in memory and cognitive defects in AD (Hardy and Higgins 1992). However, the loss of synapses correlates more closely with the severity of dementia in AD (DeKosky and Scheff 1990; Terry et al. 1991).

Drebrin is an actin-binding protein and is classified into two major isoforms in mammals: drebrin E (ubiquitous isoform) and drebrin A (neuron-specific isoform). Drebrin A accumulates in the mature dendritic spines of excitatory synapses (Sekino et al. 2007). In the past decades, drebrin has been drawing intense research interest, because of the robust decrease in drebrin level in several neurodegenerative and neuropsychiatric disorders that show cognitive defects. In this chapter, we review the drebrin loss underlying the pathology of AD both in vitro and in vivo and propose using drebrin as a marker for estimating synaptic dysfunction in AD.

12.2 Pathological Changes in Drebrin Level in Alzheimer's Disease and Mild Cognitive Impairment

The cognitive impairment observed in AD was considered to correlate with the degree of brain atrophy including neuronal cell loss. However, the loss of synapses, which correlates more robustly with the degree of cognitive impairment than the number of amyloid plaques, tangles, and neuronal loss does (Terry et al. 1991), was observed in AD brains. In addition, several AD mouse models showed abnormal long-term potentiation (LTP) and learning defects well before amyloid plaque formation (Chapman et al. 1999; Hsia et al. 1999; Jacobsen et al. 2006). These observations indicate that synaptic failure is likely to occur prior to neuronal cell loss in the pathogenesis of AD. The synapse consists of presynaptic terminals containing neurotransmitters and postsynaptic sites containing neurotransmitter receptors, postsynaptic density, and actin cytoskeleton. Alterations in the synaptic protein stoichiometry in each synapse may induce synaptic dysfunction and synapse loss, because several studies have shown that the expression levels of various pre- and postsynaptic proteins change in various neurological disorders. Drebrin is an actin-binding protein that accumulates in dendritic spines, which are typically postsynaptic receptive regions of excitatory synapses (Sekino et al. 2007). Several studies have revealed that the protein level of drebrin is decreased in the brain of AD patients and the decreased drebrin level correlates with cognitive dysfunction.

12.2.1 Drebrin in Alzheimer's Disease Patients

Immunohistochemical analysis revealed that drebrin expression is decreased in the hippocampus of AD patients (Harigaya et al. 1996). Western blot analysis also showed an approximately 70–80% decrease in the levels of drebrin protein. Furthermore, the decrease in drebrin was not restricted to the hippocampus, as it was also observed throughout the cerebral cortex (Hatanpää et al. 1999). In contrast to the marked postsynaptic changes in drebrin level in AD, the content and distribution of synaptophysin, a presynaptic protein, were barely changed (Harigaya et al. 1996). These asymmetric changes in pre- and postsynaptic proteins may result in synaptic dysfunction and cognitive impairment in AD. However, Counts et al. have shown that synaptophysin levels were decreased by ~35% even in severe AD brains (Counts et al. 2006). Therefore, drebrin quantification is more sensitive and consistent than the classical markers for synaptopathy evaluation in AD. Drebrin is also decreased in the brain of Down syndrome (Shim and Lubec 2002) and bipolar disorder (Kim et al. 2010) patients, both of which are associated with cognitive defects (Cotrena et al. 2016; Head et al. 2012). It is known that because of the extra copy of chromosome 21 harboring the amyloid precursor protein (*APP*) gene, many Down syndrome patients develop AD-like neuropathology by the age of 40 years (Head et al. 2012). Overproduction of Aβ may be involved in the decrease in drebrin in Down syndrome.

12.2.2 Drebrin in Mild Cognitive Impairment Patients

Mild cognitive impairment (MCI) is a condition in which an individual has mild but measurable changes in thinking abilities that are noticeable to the affected person, family members, and friends. However, MCI patients have the ability to carry out everyday activities (Petersen and Morris 2005). MCI is thought to be a putative prodromal stage of AD, because people with MCI, especially those with memory problems, are more likely to develop AD than people without MCI (Fischer et al. 2007; Busse et al. 2006). In MCI, drebrin is significantly decreased in the superior temporal cortex, although it remains unchanged in most cortical regions (Counts et al. 2006). However, in AD, the drebrin protein level further declines, as it has been observed that in the superior temporal cortex, the decrease in drebrin protein level from no cognitive impairment to AD was greater (an approximate decrease of 57%) than that from no cognitive impairment to MCI (an approximate decrease of 35%). The cognitive function of subjects was also measured by Mini-Mental State Examination (MMSE), which is widely used to measure general cognitive function (Folstein et al. 1975). This study, combining pre- and postmortem analyses, demonstrated that the drebrin level is drastically decreased in AD patients who show an MMSE score lower than 25. Intriguingly, in MCI patients, drebrin is decreased although the MMSE score is almost normal. These findings suggest that drebrin levels begin to decline before the development of the severe cognitive decline in AD. Drebrin may be a useful marker for estimating cognitive functions. Interestingly, a significant decrease in the drebrin level also occurs during normal aging (Hatanpää et al. 1999). The cognitive function declines with aging; thus, it is reasonable to speculate that drebrin loss underlies synaptic dysfunction accompanied by cognitive defects in AD.

12.3 Drebrin in Mouse Models of Alzheimer's Disease

As previously stated, the AD brain is characterized by massive neuronal cell and synapse loss in specific regions of the brain (Selkoe 2002), as well as amyloid plaques and neurofibrillary lesions. The major protein component of amyloid plaques is the Aβ polypeptide, which is derived from APP. Many genetically modified mice that mimic the clinical and pathological phenotype of AD have been generated (Guo et al. 2012), because a detailed analysis of the molecular mechanisms underlying synaptic dysfunction in AD is difficult to perform using human postmortem brains. Using these mouse models revealed several molecular mechanisms underlying the pathology of AD. Drebrin loss may cause synaptic dysfunction in AD, because the decrease in drebrin is a common feature among several mouse models exhibiting memory and cognitive impairment.

12.3.1 Disruption of the Balance Between Drebrin and Cofilin Induces Actin Disassembly in Dendritic Spines, Resulting in Cognitive Defects in Tg2576 Mice

The Tg2576 transgenic mouse is one of the most well-characterized and widely used AD mouse models (Hsiao et al. 1996). It overexpresses a mutant form of human APP (isoform 695) containing the double mutation, K670N and M671L (Lys670 → Asn, Met671 → Leu, Swedish mutation), which is observed in familial AD (FAD). Tg2576 mice develop normally, but the loss of dendritic spine is observed after 4 months of age in the CA1 region and dentate gyrus of the hippocampus (Lanz et al. 2003; Jacobsen et al. 2006). However, neuronal cell loss and neurofibrillary tangles are not found in these mice at 16 months of age (Irizarry et al. 1997). Tg2576 mice develop amyloid plaques by 12 months of age and exhibit impaired learning and memory at 10 months of age (Hsiao et al. 1996). Additionally, the basal synaptic transmission is normal at both 2–8 months of age (no amyloid plaques) and 15–17 months of age (many amyloid plaques). However, the hippocampal LTP becomes severely impaired at the latter age (Chapman et al. 1999). A separate study has demonstrated that contextual memory deficits are observed in Tg2576 mice from 4–5 months of age (Jacobsen et al. 2006).

Drebrin is decreased in the cortex of Tg2576 mice without a significant decrease in the synaptophysin level at 22 months of age (Calon et al. 2004). This asymmetric synaptic marker loss is consistent with human AD brains. In contrast to drebrin, the cofilin level is increased around amyloid plaques in Tg2576 mice at 22 months of age (Zhao et al. 2006). Cofilin is an actin filament (F-actin)-severing protein. The cofilin activity is regulated by phosphorylation of Ser3 through the p21-activated kinase (PAK)/LIM-kinase (LIMK) pathway (Arber et al. 1998; Yang et al. 1998b). Additionally, PAK inhibition causes drebrin loss and memory deficits in normal mice at 11 months of age (Zhao et al. 2006). Thus, actin depolymerization may be upregulated in AD brains, because the PAK/LIMK activity decreases both in the brain of aged Tg2576 mice and AD patients (Zhao et al. 2006; Bamburg and Wiggan 2002). The details of the molecular mechanism of the drebrin-cofilin interaction in Aβ-derived diffusible ligands (ADDL)-exposed neurons are mentioned in Sect. 12.4 of this chapter.

Fractin (fragment of actin) also increases in the aged Tg2576 mouse brain (Calon et al. 2004) and in AD brains (Yang et al. 1998a; Rossiter et al. 2000). Thus, it is probable that an increased fractin level indicates the disruption of actin assembly induced by cofilin hyperactivation in AD. The reciprocal relationship between drebrin and cofilin may be a result of the competitive binding to F-actin. This hypothesis is supported by a study that revealed that drebrin inhibits severing of F-actin induced by cofilin (Grintsevich and Reisler 2014). Drebrin-decorated F-actin is resistant to the depolymerization activity of cofilin. Therefore, imbalanced actin dynamics regulated by drebrin and cofilin may cause the morphological change of dendritic spines and synaptic dysfunction in AD.

12.3.2 Drebrin Loss Is Involved in Cognitive Deficit in DHA-Depleted Tg2576 Mice

In addition to genetically induced vulnerability to AD, a number of environmental risk factors are thought to be involved in the pathogenesis and progression of AD. One risk factor candidate is docosahexaenoic acid (DHA). DHA is an essential omega-3 polyunsaturated fatty acid (PUFA), which comprises approximately 15% of the total fatty acids in the brain and is enriched in synapses (Salem et al. 2001). Epidemiological studies have shown that DHA levels are significantly lower in AD (Conquer et al. 2000) and correlate with the severity of dementia (Tully et al. 2003). Lipid peroxidation, which is enhanced in the AD brain, accelerates the degradation of PUFAs, including DHA (Montine and Morrow 2005). These findings indicate that the DHA level is a significant risk factor for the development of AD. DHA-depleted Tg2576 mice show decreased postsynaptic drebrin and postsynaptic density-95 (PSD-95) and increased fractin without altering the synaptophysin level (Calon et al. 2004). Additionally, dietary depletion of *n*-3 PUFA evokes impaired spatial memory in Tg2576 mice. Note that drebrin and PSD-95 levels and memory impairment are restored by supplementing the mice's diet with DHA. Overexpression of PSD-95 promotes synaptic maturation and enhances synaptic strength (El-Husseini et al. 2000; Schnell et al. 2002), whereas its knockdown causes synaptic dysfunctions, implying a role in synaptic plasticity and dendritic spine formation (Beique et al. 2006; Ehrlich et al. 2007; Migaud et al. 1998). Importantly, decreased PSD-95 level in DHA-depleted Tg2576 may be secondary to drebrin loss, because drebrin regulates the synaptic targeting of PSD-95 (Takahashi et al. 2003). In cultured neurons, antisense suppression of drebrin prevents the synaptic clustering of PSD-95. In addition, drebrin A-specific knockout mice, in which isoform conversion from drebrin E to drebrin A is disrupted (Kojima et al. 2010), exhibit abnormal dendritic spine morphology, impaired LTP, and hippocampus-dependent memory impairment (Kojima et al. 2016). Taking these findings into account, drebrin loss is a defect that can exert a fatal influence on synaptic dysfunction and cognitive defects in AD. Furthermore, we would like to emphasize that the maintenance of drebrin content in the brain may be a crucial factor for cognition.

12.3.3 Impaired AMPAR Activity Reduces the Drebrin Level in Dendritic Spines in APP and Presenilin-1-Double-Knockin Mice

Another useful AD mouse model is the $APP^{NLh/NLh}/PS1^{P264L/P264L}$-double-knockin (DKI) mouse, which harbors both mutant *APP* and mutant *presenilin* (*PS*)*-1* (Flood et al. 2002). $APP^{NLh/NLh}/PS1^{P264L/P264L}$-DKI mice were generated by crossing $APP^{NLh/NLh}$-KI mice with $PS1^{P264L/P264L}$-KI mice. The $APP^{NLh/NLh}$ KI mouse is a FAD-KI line, as it has human Aβ in the endogenous mouse *APP* gene (Reaume et al. 1996). Compared with transitional transgenic models such as Tg2576, these KI mice have several advantages. The expression level of APP remains at the physiological level, because the KI allele is controlled by the native *APP* promoter. In addition, $APP^{NLh/NLh}$-KI mice bear the Swedish mutation (K679/M671L) with the human sequence, resulting in amyloidogenic β-secretase cleavage of this mutant APP, which is accurate and enhanced, but does not show Aβ deposition. $PS1^{P264L/P264L}$-KI mice were generated by insertion of the P264L mutation into the mouse *PS1* gene, because this mutation (Pro264 → Leu) causes an onset of FAD in humans (Mann et al. 2001). However, neuropathology is not observed in $PS1^{P264L/P264L}$-KI mice, and primary cortical neurons derived from these mice exhibit basal levels of neurodegeneration comparable to wild-type mice (Siman et al. 2000). In contrast to these single KI mice, $APP^{NLh/NLh}/PS1^{P264L/P264L}$-DKI mice exhibit Aβ deposition beginning at 6 months of age and reaching robust levels by 15 months of age and show the age-related impairment of memory flexibility (Chang et al. 2006). Electrophysiological analysis revealed that $APP^{NLh/NLh}/PS1^{P264L/P264L}$-DKI mice show decreased α-amino-3-hydroxy-5-methyl-4-isoxaxole propionic acid receptor (AMPAR) activity in the CA1 region of the hippocampus at 9–12 months of age with low amyloid plaques and age-related impaired long-lasting synaptic plasticity, such as LTP and long-term depression (LTD) (Chang et al. 2006). These findings are supported by anatomical data obtained from quantitative immunoelectron microscopy analysis showing a decrease in the number of synaptic AMPARs in CA1 pyramidal cells. Thus, these results suggest that the regulation of AMPAR trafficking on the postsynaptic membrane is impaired in $APP^{NLh/NLh}/PS1^{P264L/P264L}$-DKI mice. Deficit in endocytotic and lysosomal trafficking of the AMPAR pathway is known to significantly contribute to AD pathogenesis (Pimplikar et al. 2010). Consistent with this notion, several in vitro studies using organotypic hippocampal slice cultures and dissociated neuronal cultures have demonstrated that APP overexpression and exposure to Aβ induce endocytosis of surface and synaptic AMPAR, leading to synaptic depression and LTP inhibition (Hsieh et al. 2006; Minano-Molina et al. 2011).

A decrease in the drebrin level is also observed in this model. Quantitative immunoelectron microscopy revealed that these mice have fewer drebrin-positive spines than wild-type mice by 6 months of age (Mahadomrongkul et al. 2005; Aoki et al. 2007). It is possible that this drebrin loss is due to the impairment of AMPAR activity, because drebrin accumulation and stability within dendritic spines depend on AMPAR activity (Takahashi et al. 2009). In addition, chronic AMPAR blockade induces abnormal morphology of dendritic spines. Importantly, several studies have demonstrated that Aβ promotes AMPAR endocytosis and hence synaptic depression (Hsieh et al. 2006; Zhao et al. 2010; Liu et al. 2010; Minano-Molina et al. 2011). As mentioned above, drebrin loss induces impairment of PSD-95 clustering in dendritic spines. Taken together, it is beyond doubt that the impairment of AMPAR activity induces drebrin loss, resulting in morphological changes of dendritic spines in AD.

12.3.4 Drebrin Is Involved in the Dysregulation of NMDAR Distribution Induced by Presenilin Loss

PS1 and *PS2* have been identified as pathogenic loci involved in the majority of early onset, autosomal dominant AD. Many mutations are located in genes encoding PS1 and PS2 in FAD patients (Sherrington et al. 1995; Rogaev et al. 1995). PS1 mutations have been demonstrated to increase the production of the pathogenic Aβ (Moehlmann et al. 2002). PS conditional double knockout (cDKO) mice lack both PS1 and PS2. PS-cDKO mice were generated by crossing floxed PS1, αCaMKII-Cre transgenic mice (Yu et al. 2001), and $PS2^{-/-}$ mice (Steiner et al. 1999). PS-cDKO mice have *PS1* conditional deletion in excitatory neurons of the postnatal forebrain from about 1 month of age along with *PS2* germline deletion (Saura et al. 2004). At 2 months of age, PS-cDKO mice exhibit normal brain morphology, including a normal neuron number, volume, and spine density. In addition, open-field and rotarod tests revealed no significant alterations in behavior, motor coordination, and exploratory anxiety. However, at 2 months of age, mice exhibit mild impairments in hippocampal learning and memory, contextual fear learning, LTP, and *N*-methyl-D-aspartate receptor (NMDAR) response (Saura et al. 2004). By 6 months of age, PS-cDKO mice exhibit severer symptoms. These mice fail to learn the water maze and contextual fear conditioning tasks and exhibit deficits in the open-field and rotarod tests. By 9 months of age, the loss of dendritic spines is observed in CA1 pyramidal neurons.

At 2 months of age, drebrin A expression decreases both near the PSD and far from the PSD in the CA1 hippocampus of PS-cDKO mice (Lee and Aoki 2012). In contrast to drebrin, the GluN2A subunit of NMDARs is abnormally increased at pre- and postsynaptic sites (Aoki et al. 2009b). Importantly, these alterations are observed at 2 months of age, even though other severe symptoms such as spine loss and neurodegeneration are not observed until 6 months of age. Drebrin loss from dendritic spines may lead to abnormal trafficking of NMDAR to postsynapses at the earlier stage of AD, because drebrin is essential for the homeostatic synaptic scaling of NMDARs (Takahashi et al. 2006; Aoki et al. 2009a). Taken together, the loss of both PS1 and PS2 may result in impaired regulation of NMDARs through a mechanism involving drebrin A. Additionally, it is possible that these early events eventually lead to severer phenomena such as decreased spine density, neuronal cell loss, and cognitive defects in AD.

12.4 Molecular Mechanisms of ADDL-Induced Drebrin Loss in Dendritic Spines in Alzheimer's Disease

As mentioned above, synapse loss shows the strongest correlation with AD-associated cognitive and memory impairments. Accordingly, advances in the past decades have highlighted the molecular mechanisms of synapse loss in AD. Conventionally, it has been thought that synapse loss depends on the toxicity of insoluble amyloid fibrils, because amyloid plaques are found in AD postmortem brains and in many mouse models of AD exhibiting AD-like cognitive deficits. Although many transgenic AD mouse models generally develop amyloid plaques, some AD mouse models show synaptic dysfunction, such as abnormal LTP and cognitive deficits, well in advance of plaque formation. Recently, it has been revealed that soluble Aβ oligomers, known as ADDLs, are potent central nervous system neurotoxins that accumulate in the AD brain (Gong et al. 2003; Lambert et al. 1998). Based on these observations, it is considered that Aβ toxicity is mediated not only by the insoluble amyloid fibrils but also by ADDLs, and synaptic failure is likely to be one of the earliest events that occur in the pathogenesis of AD prior to amyloid plaque formation and neuronal cell loss. Moreover, several studies performed on human subjects have revealed deficits in LTP-like cortical plasticity in mild-to-moderate AD patients (Inghilleri et al. 2006; Battaglia et al. 2007; Koch et al. 2012). Application of ADDLs induces a decrease in the drebrin level accompanied by aberrant dendritic spine morphology and a reduction in spine density (Lacor et al. 2007; Ishizuka et al. 2014). Additionally, ADDLs facilitate the internalization of AMPAR and NMDAR (Minano-Molina et al. 2011; Snyder et al. 2005), dendritic spine shrinkage (Lacor et al. 2007), and eventual synaptic loss (Shankar et al. 2007). As mentioned in the previous section, drebrin loss may cause dysregulation of NMDAR distribution, resulting in synaptic dysfunction and hence cognitive defects. This section focuses on the molecular mechanisms of ADDL-induced aberrations in the dendritic spine architecture such as drebrin loss, cofilin hyperactivation, and impaired glutamate receptor trafficking.

12.4.1 ADDLs Induce NMDAR-Mediated Drebrin Loss in Dendritic Spines

The effects of ADDLs on dendritic spines may be mediated, at least in part, by selective binding of ADDLs to certain receptors, such as α7 nicotinic acetylcholine receptor (α7 nAChR) (Wang et al. 2000), leukocyte immunoglobulin-like receptor B2 (Kim et al. 2013), and cellular prion protein (PrP^c) (Lauren et al. 2009). ADDLs bind to PrP^c, inducing Fyn kinase activation and subsequent phosphorylation of the GluN2B subunit of NMDARs at Tyr1472 (Freir et al. 2011; Um et al. 2012) (Fig. 12.1). This phosphorylation induces an initial increase in NMDAR activity at the postsynaptic site, which induces an increase in Ca^{2+} influx (De Felice et al. 2007). Transient activation of NMDARs induced by the ADDLs-PrP^c interaction is thought to be a trigger of AD pathology, because inhibition of NMDAR or PrP^c prevents the toxicity of ADDLs (De Felice et al. 2007; Decker et al. 2010; Lauren

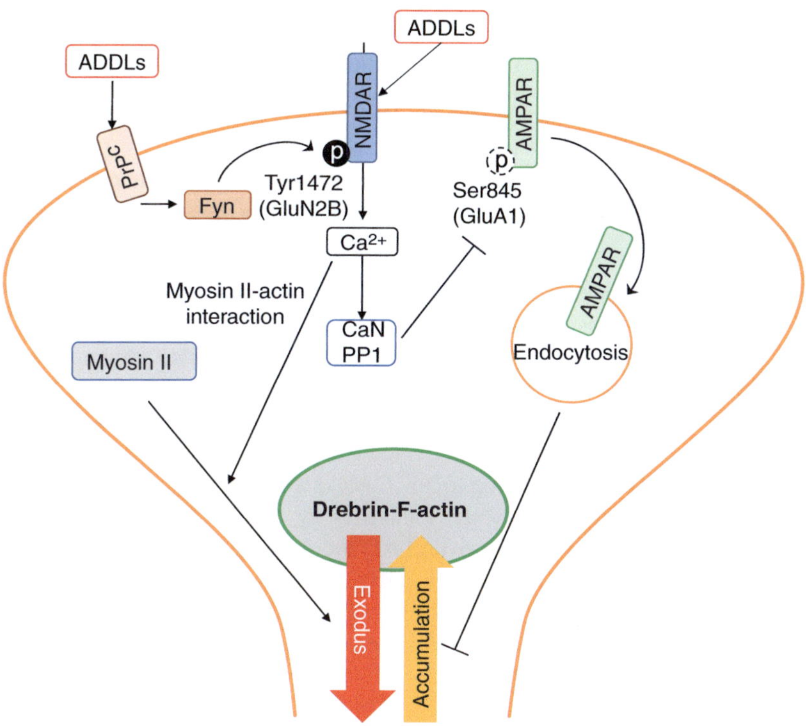

Fig. 12.1 PrP_c-mediated excessive Ca^{2+} influx induces drebrin exodus and AMPAR internalization. ADDL-PrP_c interaction initially activates GluN2B-containing NMDAR by phosphorylation of GluN2B at Tyr1472. Ca^{2+} influx through NMDAR induces an interaction between myosin II and F-actin, leading to drebrin exodus by myosin II ATPase activation. This Ca^{2+} influx also induces calcineurin (CaN)/PP1 activation, leading to dephosphorylation of the GluA1 subunit of AMPARs at Ser845. Dephosphorylated AMPARs are internalized by endocytosis. The internalization of AMPAR suppresses the localization of drebrin in dendritic spines

et al. 2009; Shankar et al. 2007). ADDLs reduce the amount of drebrin as well as the dendritic spine density in cultured neurons (Ishizuka et al. 2014; Lacor et al. 2007; Zhao et al. 2006). As previously described, decreased drebrin is consistent with the observation in AD patients and AD mouse models. Collectively, there is little doubt that ADDLs induce drebrin loss and hence synaptic failure in the progression of AD. We have previously shown that NMDAR activation induces the disappearance of drebrin from dendritic spines and its appearance in dendritic shafts (Sekino et al. 2006). We have also revealed that activation of myosin II ATPase by Ca^{2+} influx through NMDARs induces the translocation of drebrin A from dendritic spines to the shaft, so-called drebrin exodus (Mizui et al. 2014) (Fig. 12.1). Based on the above observations, it is clear that ADDL-induced drebrin loss is mediated by excessive Ca^{2+} influx through abnormal activation of NMDARs, because drebrin loss induced by ADDLs can be prevented by memantine, an NMDAR antagonist (Lacor et al. 2007). It should be noted that low exposure to ADDLs induces drebrin loss in dendritic spines without affecting the morphology and density of dendritic spines, or the protein level of drebrin (Ishizuka et al. 2014). Therefore, we are certain that drebrin loss in dendritic spines occurs prior to the changes in the dendritic spine morphology during the progression of AD.

12.4.2 Possible Involvement of Drebrin in ADDL-Induced Impairment of NMDAR Trafficking to the Postsynaptic Site

Following ADDL-induced initial activation of NMDARs, the surface expression of NMDARs decreases (Lacor et al. 2007; Snyder et al. 2005). ADDLs also bind to α7 nAChR, leading to Ca^{2+} influx and calcineurin activation (also known as protein phosphatase 2B), which in turn leads to protein phosphatase 1 (PP1) activation (Fig. 12.2). Activated PP1 dephosphorylates and activates striatal-enriched protein tyrosine phosphatase 61 ($STEP_{61}$). Activated $STEP_{61}$ promotes the dephosphorylation of GluN2B at Tyr1472, leading to NMDAR internalization from the postsynaptic membranes (Snyder et al. 2005; Zhang et al. 2011). Consistent with these findings, $STEP_{61}$ is upregulated in several AD mouse models and in the brain of AD patients (Snyder et al. 2005; Kurup et al. 2010). ADDL-induced drebrin dysfunction may impair the reinsertion of NMDARs into the postsynaptic membrane, resulting in impaired synaptic transmission, because drebrin is essential for the synaptic targeting of NMDARs (Aoki et al. 2009a).

Under physiological conditions, drebrin loss in dendritic spines is a temporary phenomenon. Drebrin reenters into dendritic spines after its exodus, which is induced by NMDAR activation. This bidirectional translocation of drebrin is thought to be essential for synaptic plasticity (Mizui et al. 2014). LTP is associated with drebrin influx into dendritic spines in vivo (Fukazawa et al. 2003; Bosch et al. 2014). However, ADDLs eventually reduce the expression level of drebrin in cultured neurons (Lacor et al. 2007; Zhao et al. 2006; Ishizuka et al. 2014). This discrepancy raised the following question: what are the molecular mechanisms involved in the ADDL-induced irreversible drebrin loss observed in AD?

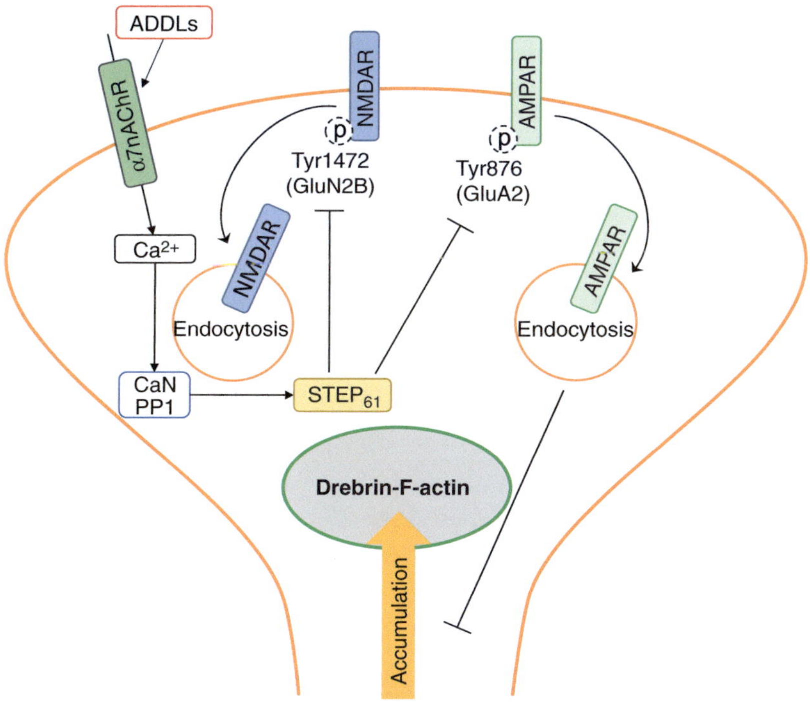

Fig. 12.2 α7 nAChR-mediated AMPAR internalization inhibits the accumulation of drebrin in dendritic spines, resulting in impairment of NMDAR trafficking to postsynaptic sites. ADDL-induced Ca^{2+} influx through α7 nAChR activates CaN/PP1, leading to $STEP_{61}$ activation. $STEP_{61}$ dephosphorylates both AMPARs (GluA2 subunit at Tyr876) and NMDARs (GluN2B at Try1472). The dephosphorylation of these receptors induces their internalization by endocytosis. The internalization of AMPAR suppresses the localization of drebrin in dendritic spines. This drebrin loss in dendritic spines induces the impairment of NMDAR reinsertion into postsynaptic membranes

12.4.3 AMPAR Internalization Induces Drebrin Loss in Dendritic Spines

In cultured neurons and acute brain slices, ADDLs induce excessive Ca^{2+} influx through NMDARs, which subsequently activates the calcineurin/PP1 pathway (Shankar et al. 2007; De Felice et al. 2007). Calcineurin/PP1 activation mediates the ADDL-induced internalization of AMPARs by dephosphorylating the GluA1 subunit of AMPARs at Ser845 (Hsieh et al. 2006; Minano-Molina et al. 2011) (Fig. 12.1). The GluA1 subunit at Ser845 is known to be phosphorylated by protein kinase A (PKA) and is crucial for maintaining the stability of AMPARs at synaptic sites (He et al. 2009; Oh et al. 2006; Man et al. 2007). Thus, ADDLs induce AMPAR internalization and impair the synaptic scaling of AMPARs by activating calcineurin/PP1. $STEP_{61}$ dephosphorylates the GluA2 subunit at Tyr876, leading to internalization of AMPARs (Moult et al. 2006; Zhang et al. 2008) (Fig. 12.2). $STEP_{61}$ activity is required for ADDL-induced internalization of AMPARs. Genetic and pharmacological inhibition of $STEP_{61}$ restores the number of AMPARs on the post-synaptic membrane, enhances LTP, and improves cognitive function in AD mice (Zhang et al. 2011, 2010; Xu et al. 2014). This $STEP_{61}$-mediated AMPAR internalization is consistent with several AD mouse models. As described above, $APP^{NLh/NLh}/PS1^{P264L/P264L}$-DKI mice show decreased AMPAR activity accompanied by a decrease in synaptic AMPAR number in the CA1 region of the hippocampus (Chang et al. 2006). Additionally, $APP_{swe, Ind}$ transgenic mice, which harbor both Swedish (K670N/M670L) and Indiana (V717F, valine717 → phenylalanine) mutations in human APP (Chishti et al. 2001), display lower levels of Ser845 phosphorylation, a decrease in surface AMPARs expression and impaired learning and memory (Minano-Molina et al. 2011). Thus, we presume that drebrin loss is partly due to the impairment of AMPAR activity induced by ADDLs, because drebrin accumulation and stability within dendritic spines require AMPAR activity (Takahashi et al. 2009).

12.4.4 ADDL-Induced Cofilin Activation Causes Actin Disassembly in Dendritic Spines

The ADDL-induced initial increase in NMDAR activity induces calcineurin-mediated dephosphorylation and activation of cofilin (Wang et al. 2005; Wu et al. 2010) (Fig. 12.3). This activation of cofilin may also induce actin disassembly in dendritic spines, resulting in synaptic dysfunction. Consistent with this model, both calcineurin and cofilin are required for the ADDL-induced loss of dendritic spines (Shankar et al. 2007). As discussed above, the cofilin level is elevated in the brain of AD patients and Tg2576 mice. ADDL exposure induces a rapid and persistent reduction in the PAK activity and an abnormal distribution of phosphorylated PAK in cultured hippocampal neurons (Zhao et al. 2006). These effects of ADDLs are protected by pretreatment with the oligomer-specific A11 antibody, which does not recognize contaminating fibrils or monomers (Kayed et al. 2003). Exogenously overexpressed PAK prevents ADDL-induced drebrin loss in cultured neurons, although ADDL exposure induces drebrin loss in control neurons. On the contrary, dominant-negative PAK does not prevent ADDL toxicity. These in vitro experiments' results support the hypotheses that ADDLs cause PAK signaling defects and drebrin loss. Therefore, disruption of the balance between drebrin and cofilin may induce actin disassembly in dendritic spines, resulting in synaptic dysfunction in AD.

12.4.5 ADDLs Induce Proteolysis of Drebrin

Disturbance of intracellular Ca^{2+} promotes the activation of calcium-dependent papain-like enzyme (calpain) (Murachi et al. 1980). A recent study has reported that NMDAR-mediated calpain activation elicits the degradation of drebrin and F-actin

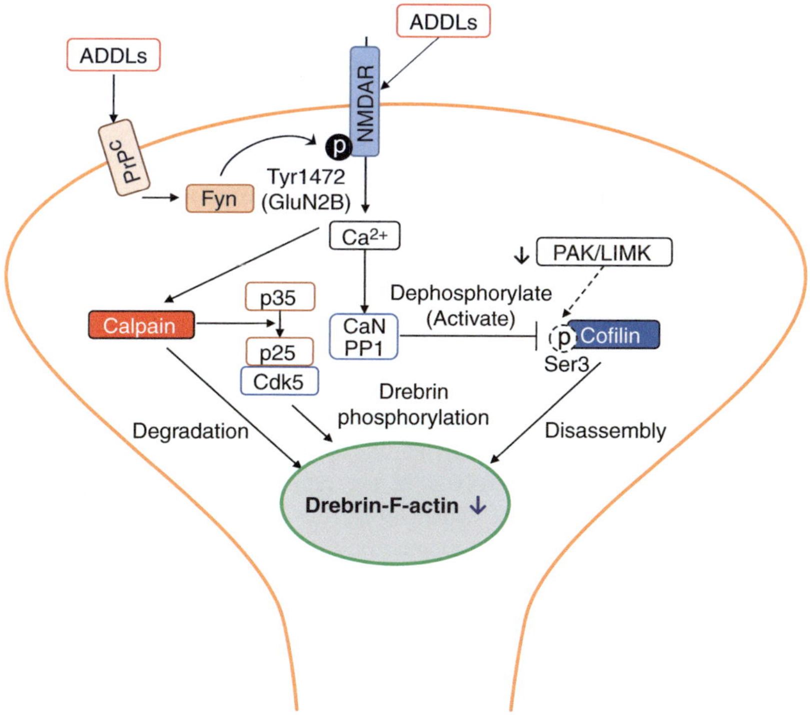

Fig. 12.3 ADDL-induced calpain and cofilin activation induce drebrin degradation. Ca^{2+} influx through NMDAR activates both the calpain and CaN/PP1 pathways. Activated calpain decomposes drebrin and F-actin and induces the conversion of p35 to p25 for the activation of Cdk5. Activated Cdk5 phosphorylates drebrin at Ser142. Cofilin is also activated by CaN/PP1-induced dephosphorylation at Ser3. The activation of calpain and cofilin induces actin disassembly in dendritic spines

in cultured neurons (Chimura et al. 2015) (Fig. 12.3). Thus, ADDL exposure may induce calpain-mediated drebrin degradation in AD, because ADDLs induce both NMDAR hyperactivation and a decrease in drebrin. In agreement with this finding, calpain activity is elevated in AD (Saito et al. 1993). Similarly, calpastatin, an endogenous calpain inhibitor, is decreased in AD (Rao et al. 2008). Therefore, this calpain-induced degradation of drebrin and F-actin may also contribute to the loss of dendritic spines in AD. Additionally, calpain activation induces the proteolytic conversion of p35, which is a regulatory subunit of cyclin-dependent kinase 5 (Cdk5), to p25. p25 is a neuron-specific Cdk5 activator and a constitutively active form that enhances Cdk5 activity in AD (Patrick et al. 1999; Lew et al. 1994). Intriguingly, Cdk5 phosphorylates drebrin at Ser142, possibly resulting in a conformational change of drebrin (Worth et al. 2013). If the phosphorylated form of drebrin is more vulnerable to proteolysis than the non-phosphorylated form, the CDK5-mediated phosphorylation of this site may be involved in the ADDL-induced drebrin degradation.

12.4.6 HDAC Mediates ADDL-Induced Drebrin Loss in Dendritic Spines

Recently, we have revealed that histone deacetylase (HDAC) mediates drebrin loss in dendritic spines induced by ADDL exposure (Ishizuka et al. 2014). HDACs are a class of enzymes that remove acetyl groups from lysine amino acids on histones, allowing the histones to wrap the DNA more tightly and downregulate gene expression (Gräff and Tsai 2013). It is clear that increased HDAC activity correlates with the pathology and the etiology of AD, because HDAC activity is elevated in the AD mouse models (Gräff et al. 2012) and some HDAC inhibitors can improve memory and cognitive function, as well as reduced dendritic spine density in Tg2576 (Ricobaraza et al. 2009, 2012) and APP/PS1 (Kilgore et al. 2010; Francis et al. 2009) mouse models of AD. Intriguingly, drebrin loss induced by ADDL exposure is ameliorated by the HDAC inhibitor, suberoylanilide hydroxamic acid (SAHA). Thus, epigenetic modification may be involved in the regulation of drebrin localization. Although the precise molecular mechanism of this process is not known, yet, the inhibition of HDAC may induce the expression of a protein that inhibits Ca^{2+} signal transduction, such as a Ca^{2+} sensor that triggers the disinhibition of myosin II ATPase involved in the drebrin exodus. Alternatively, HDAC inhibition may induce the expression of a protein that promotes drebrin accumulation in dendritic spines. On the contrary, HDAC activation may suppress the expression of a protein involved in drebrin accumulation in dendritic spines in AD brains. These findings suggest that HDAC is involved in ADDL-induced synaptic defects and that the regulation of histone acetylation/deacetylation is crucial for synaptic functions under cellular stress conditions, such as ADDL exposure. Taken together, drebrin disappearance from dendritic spines followed by ADDL-induced excessive NMDAR activation underlies the synaptic dysfunction in AD, and HDAC may be a potential therapeutic target for AD patients.

12.5 Conclusions and Outlook

In this chapter, we discussed the aberrant behaviors of drebrin in AD by focusing on three main topics: (1) human AD brain, (2) mouse models of AD, and (3) the molecular mechanism of ADDL-induced drebrin loss. Drebrin loss is a common feature in the brain of AD patients and AD mouse models. The study performed by combining ante- and postmortem analyses of human subjects revealed that the drebrin level closely correlates with the severity of cognitive impairment. Additionally, the AD mouse models showed a decreased drebrin level and impaired glutamate receptor distribution. It is also clear that ADDLs induce an excess influx of Ca^{2+}, leading to glutamate receptor internalization and drebrin loss. It should be noted that acute exposure to a low dose of ADDLs induces drebrin loss in dendritic spines without reducing both the density of dendritic spines and the expression level of drebrin. This observation strongly suggests that drebrin loss in dendritic spines occurs at the prodromal stage of AD, in which the patients begin to display slight cognitive impairments prior to the development of MCI, because the amount of drebrin is decreased in MCI patients. HDAC also contributes to drebrin loss in AD, because HDAC inhibitor attenuates ADDL-induced drebrin loss. These findings provide insight into the mechanisms of ADDL-toxicity and suggest that HDAC inhibitors may hold therapeutic potential against AD.

In conclusion, by focusing on drebrin, our understanding of the molecular mechanisms underlying synaptic dysfunction accompanied by cognitive defects in AD has been greatly improved. Increasing evidence has uncovered the molecular mechanisms of synaptic dysfunction in AD, such as drebrin loss in dendritic spines, impaired glutamate receptor trafficking, and decreased PAK/LIMK signal. These signals act cooperatively to mediate the toxic effects of ADDLs; there is no doubt that the aberration in drebrin behavior in dendritic spines is one of the direct causes of dendritic spine abnormality. Therefore, further study of drebrin and its associated signaling mechanisms would be of value to develop more effective therapeutic strategies for AD treatment.

Acknowledgment We would like to thank Dr. Masaki Ikeda for his comments and useful discussion of the manuscript.

References

Aoki C, Mahadomrongkul V, Fujisawa S, Habersat R, Shirao T (2007) Chemical and morphological alterations of spines within the hippocampus and entorhinal cortex precede the onset of Alzheimer's disease pathology in double knock-in mice. J Comp Neurol 505:352–362

Aoki C, Kojima N, Sabaliauskas N et al (2009a) Drebrin a knockout eliminates the rapid form of homeostatic synaptic plasticity at excitatory synapses of intact adult cerebral cortex. J Comp Neurol 517:105–121

Aoki C, Lee J, Nedelescu H, Ahmed T, Ho A, Shen J (2009b) Increased levels of NMDA receptor NR2A subunits at pre- and postsynaptic sites of the hippocampal CA1: an early response to conditional double knockout of presenilin 1 and 2. J Comp Neurol 517:512–523

Arber S, Barbayannis FA, Hanser H, Schneider C, Stanyon CA, Bernard O, Caroni P (1998) Regulation of actin dynamics through phosphorylation of cofilin by LIM-kinase. Nature 393:805–809

Bamburg JR, Wiggan OP (2002) ADF/cofilin and actin dynamics in disease. Trends Cell Biol 12:598–605

Battaglia F, Wang HY, Ghilardi MF, Gashi E, Quartarone A, Friedman E, Nixon RA (2007) Cortical plasticity in Alzheimer's disease in humans and rodents. Biol Psychiatry 62:1405–1412

Beique JC, Lin DT, Kang MG, Aizawa H, Takamiya K, Huganir RL (2006) Synapse-specific regulation of AMPA receptor function by PSD-95. Proc Natl Acad Sci U S A 103:19535–19540

Bosch M, Castro J, Saneyoshi T, Matsuno H, Sur M, Hayashi Y (2014) Structural and molecular remodeling of dendritic spine substructures during long-term potentiation. Neuron 82:444–459

Busse A, Hensel A, Guhne U, Angermeyer MC, Riedel-Heller SG (2006) Mild cognitive impairment: long-term course of four clinical subtypes. Neurology 67:2176–2185

Calon F, Lim GP, Yang F et al (2004) Docosahexaenoic acid protects from dendritic pathology in an Alzheimer's disease mouse model. Neuron 43:633–645

Chang EH, Savage MJ, Flood DG, Thomas JM, Levy RB, Mahadomrongkul V, Shirao T, Aoki C, Huerta PT (2006) AMPA receptor downscaling at the onset of Alzheimer's disease pathology in double knockin mice. Proc Natl Acad Sci U S A 103:3410–3415

Chapman PF, White GL, Jones MW et al (1999) Impaired synaptic plasticity and learning in aged amyloid precursor protein transgenic mice. Nat Neurosci 2:271–276

Chimura T, Launey T, Yoshida N (2015) Calpain-mediated degradation of Drebrin by Excitotoxicity in vitro and in vivo. PLoS One 10:e0125119

Chishti MA, Yang DS, Janus C et al (2001) Early-onset amyloid deposition and cognitive deficits in transgenic mice expressing a double mutant form of amyloid precursor protein 695. J Biol Chem 276:21562–21570

Conquer JA, Tierney MC, Zecevic J, Bettger WJ, Fisher RH (2000) Fatty acid analysis of blood plasma of patients with Alzheimer's disease, other types of dementia, and cognitive impairment. Lipids 35:1305–1312

Cotrena C, Branco LD, Shansis FM, Fonseca RP (2016) Executive function impairments in depression and bipolar disorder: association with functional impairment and quality of life. J Affect Disord 190:744–753

Counts SE, Nadeem M, Lad SP, Wuu J, Mufson EJ (2006) Differential expression of synaptic proteins in the frontal and temporal cortex of elderly subjects with mild cognitive impairment. J Neuropathol Exp Neurol 65:592–601

De Felice FG, Velasco PT, Lambert MP, Viola K, Fernandez SJ, Ferreira ST, Klein WL (2007) Abeta oligomers induce neuronal oxidative stress through an N-methyl-D-aspartate receptor-dependent mechanism that is blocked by the Alzheimer drug memantine. J Biol Chem 282:11590–11601

Decker H, Jurgensen S, Adrover MF et al (2010) N-methyl-D-aspartate receptors are required for synaptic targeting of Alzheimer's toxic amyloid-beta peptide oligomers. J Neurochem 115:1520–1529

DeKosky ST, Scheff SW (1990) Synapse loss in frontal cortex biopsies in Alzheimer's disease: correlation with cognitive severity. Ann Neurol 27:457–464

Ehrlich I, Klein M, Rumpel S, Malinow R (2007) PSD-95 is required for activity-driven synapse stabilization. Proc Natl Acad Sci U S A 104:4176–4181

El-Husseini AE, Schnell E, Chetkovich DM, Nicoll RA, Bredt DS (2000) PSD-95 involvement in maturation of excitatory synapses. Science 290:1364–1368

Fischer P, Jungwirth S, Zehetmayer S, Weissgram S, Hoenigschnabl S, Gelpi E, Krampla W, Tragl KH (2007) Conversion from subtypes of mild cognitive impairment to Alzheimer dementia. Neurology 68:288–291

Flood DG, Reaume AG, Dorfman KS et al (2002) FAD mutant PS-1 gene-targeted mice: increased a beta 42 and a beta deposition without APP overproduction. Neurobiol Aging 23:335–348

Folstein MF, Folstein SE, McHugh PR (1975) Mini-mental state. A practical method for grading the cognitive state of patients for the clinician. *J Psychiatr Res* 12:189–198

Francis YI, Fà M, Ashraf H, Zhang H, Staniszewski A, Latchman DS, Arancio O (2009) Dysregulation of histone acetylation in the APP/PS1 mouse model of Alzheimer's disease. J Alzheimers Dis 18:131–139

Freir DB, Nicoll AJ, Klyubin I et al (2011) Interaction between prion protein and toxic amyloid beta assemblies can be therapeutically targeted at multiple sites. Nat Commun 2:336

Fukazawa Y, Saitoh Y, Ozawa F, Ohta Y, Mizuno K, Inokuchi K (2003) Hippocampal LTP is accompanied by enhanced F-actin content within the dendritic spine that is essential for late LTP maintenance in vivo. Neuron 38:447–460

Gong Y, Chang L, Viola KL, Lacor PN, Lambert MP, Finch CE, Krafft GA, Klein WL (2003) Alzheimer's disease-affected brain: presence of oligomeric a beta ligands (ADDLs) suggests a molecular basis for reversible memory loss. Proc Natl Acad Sci U S A 100:10417–10422

Gräff J, Tsai LH (2013) Histone acetylation: molecular mnemonics on the chromatin. Nat Rev Neurosci 14:97–111

Gräff J, Rei D, Guan JS et al (2012) An epigenetic blockade of cognitive functions in the neurodegenerating brain. Nature 483:222–226
Grintsevich EE, Reisler E (2014) Drebrin inhibits cofilin-induced severing of F-actin. Cytoskeleton 71:472–483
Guo Q, Wang Z, Li H, Wiese M, Zheng H (2012) APP physiological and pathophysiological functions: insights from animal models. Cell Res 22:78–89
Hardy JA, Higgins GA (1992) Alzheimer's disease: the amyloid cascade hypothesis. Science 256:184–185
Harigaya Y, Shoji M, Shirao T, Hirai S (1996) Disappearance of actin-binding protein, drebrin, from hippocampal synapses in Alzheimer's disease. J Neurosci Res 43:87–92
Hatanpää K, Isaacs KR, Shirao T, Brady DR, Rapoport SI (1999) Loss of proteins regulating synaptic plasticity in normal aging of the human brain and in Alzheimer disease. J Neuropathol Exp Neurol 58:637–643
He K, Song L, Cummings LW, Goldman J, Huganir RL, Lee HK (2009) Stabilization of Ca^{2+}–permeable AMPA receptors at perisynaptic sites by GluR1-S845 phosphorylation. Proc Natl Acad Sci U S A 106:20033–20038
Head E, Silverman W, Patterson D, Lott IT (2012) Aging and down syndrome. Curr Gerontol Geriatr Res 2012:412536
Hsia AY, Masliah E, McConlogue L et al (1999) Plaque-independent disruption of neural circuits in Alzheimer's disease mouse models. Proc Natl Acad Sci U S A 96:3228–3233
Hsiao K, Chapman P, Nilsen S, Eckman C, Harigaya Y, Younkin S, Yang F, Cole G (1996) Correlative memory deficits, Abeta elevation, and amyloid plaques in transgenic mice. Science 274:99–102
Hsieh H, Boehm J, Sato C, Iwatsubo T, Tomita T, Sisodia S, Malinow R (2006) AMPAR removal underlies Abeta-induced synaptic depression and dendritic spine loss. Neuron 52:831–843
Inghilleri M, Conte A, Frasca V, Scaldaferri N, Gilio F, Santini M, Fabbrini G, Prencipe M, Berardelli A (2006) Altered response to rTMS in patients with Alzheimer's disease. Clin Neurophysiol 117:103–109
Irizarry MC, McNamara M, Fedorchak K, Hsiao K, Hyman BT (1997) APPSw transgenic mice develop age-related a beta deposits and neuropil abnormalities, but no neuronal loss in CA1. J Neuropathol Exp Neurol 56:965–973
Ishizuka Y, Shimizu H, Takagi E, Kato M, Yamagata H, Mikuni M, Shirao T (2014) Histone deacetylase mediates the decrease in drebrin cluster density induced by amyloid beta oligomers. Neurochem Int 76:114–121
Jacobsen JS, Wu CC, Redwine JM et al (2006) Early-onset behavioral and synaptic deficits in a mouse model of Alzheimer's disease. Proc Natl Acad Sci U S A 103:5161–5166
Kayed R, Head E, Thompson JL, McIntire TM, Milton SC, Cotman CW, Glabe CG (2003) Common structure of soluble amyloid oligomers implies common mechanism of pathogenesis. Science 300:486–489
Kilgore M, Miller CA, Fass DM, Hennig KM, Haggarty SJ, Sweatt JD, Rumbaugh G (2010) Inhibitors of class 1 histone deacetylases reverse contextual memory deficits in a mouse model of Alzheimer's disease. Neuropsychopharmacology 35:870–880
Kim HW, Rapoport SI, Rao JS (2010) Altered expression of apoptotic factors and synaptic markers in postmortem brain from bipolar disorder patients. Neurobiol Dis 37:596–603
Kim T, Vidal GS, Djurisic M, William CM, Birnbaum ME, Garcia KC, Hyman BT, Shatz CJ (2013) Human LilrB2 is a beta-amyloid receptor and its murine homolog PirB regulates synaptic plasticity in an Alzheimer's model. Science 341:1399–1404
Koch G, Di Lorenzo F, Bonni S, Ponzo V, Caltagirone C, Martorana A (2012) Impaired LTP- but not LTD-like cortical plasticity in Alzheimer's disease patients. J Alzheimers Dis 31:593–599
Kojima N, Hanamura K, Yamazaki H, Ikeda T, Itohara S, Shirao T (2010) Genetic disruption of the alternative splicing of drebrin gene impairs context-dependent fear learning in adulthood. Neuroscience 165:138–150

Kojima N, Yasuda H, Hanamura K, Ishizuka Y, Sekino Y, Shirao T (2016) Drebrin a regulates hippocampal LTP and hippocampus-dependent fear learning in adult mice. Neuroscience 324:218–226

Kurup P, Zhang Y, Xu J, Venkitaramani DV, Haroutunian V, Greengard P, Nairn AC, Lombroso PJ (2010) Abeta-mediated NMDA receptor endocytosis in Alzheimer's disease involves ubiquitination of the tyrosine phosphatase STEP61. J Neurosci 30:5948–5957

Lacor PN, Buniel MC, Furlow PW, Clemente AS, Velasco PT, Wood M, Viola KL, Klein WL (2007) Abeta oligomer-induced aberrations in synapse composition, shape, and density provide a molecular basis for loss of connectivity in Alzheimer's disease. J Neurosci 27:796–807

Lambert MP, Barlow AK, Chromy BA et al (1998) Diffusible, nonfibrillar ligands derived from Abeta1-42 are potent central nervous system neurotoxins. Proc Natl Acad Sci U S A 95:6448–6453

Lanz TA, Carter DB, Merchant KM (2003) Dendritic spine loss in the hippocampus of young PDAPP and Tg2576 mice and its prevention by the ApoE2 genotype. Neurobiol Dis 13:246–253

Lauren J, Gimbel DA, Nygaard HB, Gilbert JW, Strittmatter SM (2009) Cellular prion protein mediates impairment of synaptic plasticity by amyloid-beta oligomers. Nature 457:1128–1132

Lee D, Aoki C (2012) Presenilin conditional double knockout mice exhibit decreases in drebrin a at hippocampal CA1 synapses. Synapse 66:870–879

Lew J, Huang QQ, Qi Z, Winkfein RJ, Aebersold R, Hunt T, Wang JH (1994) A brain-specific activator of cyclin-dependent kinase 5. Nature 371:423–426

Liu SJ, Gasperini R, Foa L, Small DH (2010) Amyloid-beta decreases cell-surface AMPA receptors by increasing intracellular calcium and phosphorylation of GluR2. J Alzheimers Dis 21:655–666

Mahadomrongkul V, Huerta PT, Shirao T, Aoki C (2005) Stability of the distribution of spines containing drebrin a in the sensory cortex layer I of mice expressing mutated APP and PS1 genes. Brain Res 1064:66–74

Man HY, Sekine-Aizawa Y, Huganir RL (2007) Regulation of {alpha}-amino-3-hydroxy-5-methyl-4-isoxazolepropionic acid receptor trafficking through PKA phosphorylation of the Glu receptor 1 subunit. Proc Natl Acad Sci U S A 104:3579–3584

Mann DM, Pickering-Brown SM, Takeuchi A, Iwatsubo T, Members of the Familial Alzheimer's Disease Pathology Study Group (2001) Amyloid angiopathy and variability in amyloid beta deposition is determined by mutation position in presenilin-1-linked Alzheimer's disease. Am J Pathol 158:2165–2175

Migaud M, Charlesworth P, Dempster M et al (1998) Enhanced long-term potentiation and impaired learning in mice with mutant postsynaptic density-95 protein. Nature 396:433–439

Minano-Molina AJ, Espana J, Martin E, Barneda-Zahonero B, Fado R, Sole M, Trullas R, Saura CA, Rodriguez-Alvarez J (2011) Soluble oligomers of amyloid-beta peptide disrupt membrane trafficking of alpha-amino-3-hydroxy-5-methylisoxazole-4-propionic acid receptor contributing to early synapse dysfunction. J Biol Chem 286:27311–27321

Mizui T, Sekino Y, Yamazaki H, Ishizuka Y, Takahashi H, Kojima N, Kojima M, Shirao T (2014) Myosin II ATPase activity mediates the long-term potentiation-induced exodus of stable F-actin bound by drebrin a from dendritic spines. PLoS One 9:e85367

Moehlmann T, Winkler E, Xia X et al (2002) Presenilin-1 mutations of leucine 166 equally affect the generation of the notch and APP intracellular domains independent of their effect on Abeta 42 production. Proc Natl Acad Sci U S A 99:8025–8030

Montine TJ, Morrow JD (2005) Fatty acid oxidation in the pathogenesis of Alzheimer's disease. Am J Pathol 166:1283–1289

Moult PR, Gladding CM, Sanderson TM, Fitzjohn SM, Bashir ZI, Molnar E, Collingridge GL (2006) Tyrosine phosphatases regulate AMPA receptor trafficking during metabotropic glutamate receptor-mediated long-term depression. J Neurosci 26:2544–2554

Murachi T, Tanaka K, Hatanaka M, Murakami T (1980) Intracellular Ca^{2+}–dependent protease (calpain) and its high-molecular-weight endogenous inhibitor (calpastatin). Adv Enzym Regul 19:407–424

Oh MC, Derkach VA, Guire ES, Soderling TR (2006) Extrasynaptic membrane trafficking regulated by GluR1 serine 845 phosphorylation primes AMPA receptors for long-term potentiation. J Biol Chem 281:752–758

Patrick GN, Zukerberg L, Nikolic M, de la Monte S, Dikkes P, Tsai LH (1999) Conversion of p35 to p25 deregulates Cdk5 activity and promotes neurodegeneration. Nature 402:615–622

Petersen RC, Morris JC (2005) Mild cognitive impairment as a clinical entity and treatment target. Arch Neurol 62:1160–1163. discussion 1167

Pimplikar SW, Nixon RA, Robakis NK, Shen J, Tsai LH (2010) Amyloid-independent mechanisms in Alzheimer's disease pathogenesis. J Neurosci 30:14946–14954

Prince M, Comas-Herrera A, Knapp M, Guerchet M, Karagiannidou M (2016) Improving healthcare for people living with dementia: coverage, quality and costs now and in the future. In: World Alzheimer Report 2016

Rao MV, Mohan PS, Peterhoff CM et al (2008) Marked calpastatin (CAST) depletion in Alzheimer's disease accelerates cytoskeleton disruption and neurodegeneration: neuroprotection by CAST overexpression. J Neurosci 28:12241–12254

Reaume AG, Howland DS, Trusko SP, Savage MJ, Lang DM, Greenberg BD, Siman R, Scott RW (1996) Enhanced amyloidogenic processing of the beta-amyloid precursor protein in gene-targeted mice bearing the Swedish familial Alzheimer's disease mutations and a humanized Abeta sequence. J Biol Chem 271:23380–23388

Ricobaraza A, Cuadrado-Tejedor M, Perez-Mediavilla A, Frechilla D, Del Rio J, Garcia-Osta A (2009) Phenylbutyrate ameliorates cognitive deficit and reduces tau pathology in an Alzheimer's disease mouse model. Neuropsychopharmacology 34:1721–1732

Ricobaraza A, Cuadrado-Tejedor M, Marco S, Perez-Otano I, Garcia-Osta A (2012) Phenylbutyrate rescues dendritic spine loss associated with memory deficits in a mouse model of Alzheimer disease. Hippocampus 22:1040–1050

Rogaev EI, Sherrington R, Rogaeva EA et al (1995) Familial Alzheimer's disease in kindreds with missense mutations in a gene on chromosome 1 related to the Alzheimer's disease type 3 gene. Nature 376:775–778

Rossiter JP, Anderson LL, Yang F, Cole GM (2000) Caspase-cleaved actin (fractin) immunolabelling of Hirano bodies. Neuropathol Appl Neurobiol 26:342–346

Saito K, Elce JS, Hamos JE, Nixon RA (1993) Widespread activation of calcium-activated neutral proteinase (calpain) in the brain in Alzheimer disease: a potential molecular basis for neuronal degeneration. Proc Natl Acad Sci U S A 90:2628–2632

Salem N Jr, Litman B, Kim HY, Gawrisch K (2001) Mechanisms of action of docosahexaenoic acid in the nervous system. Lipids 36:945–959

Saura CA, Choi SY, Beglopoulos V et al (2004) Loss of presenilin function causes impairments of memory and synaptic plasticity followed by age-dependent neurodegeneration. Neuron 42:23–36

Schnell E, Sizemore M, Karimzadegan S, Chen L, Bredt DS, Nicoll RA (2002) Direct interactions between PSD-95 and stargazin control synaptic AMPA receptor number. Proc Natl Acad Sci U S A 99:13902–13907

Sekino Y, Tanaka S, Hanamura K, Yamazaki H, Sasagawa Y, Xue Y, Hayashi K, Shirao T (2006) Activation of N-methyl-D-aspartate receptor induces a shift of drebrin distribution: disappearance from dendritic spines and appearance in dendritic shafts. Mol Cell Neurosci 31:493–504

Sekino Y, Kojima N, Shirao T (2007) Role of actin cytoskeleton in dendritic spine morphogenesis. Neurochem Int 51:92–104

Selkoe DJ (2002) Alzheimer's disease is a synaptic failure. Science 298:789–791

Shankar GM, Bloodgood BL, Townsend M, Walsh DM, Selkoe DJ, Sabatini BL (2007) Natural oligomers of the Alzheimer amyloid-beta protein induce reversible synapse loss by modulating an NMDA-type glutamate receptor-dependent signaling pathway. J Neurosci 27:2866–2875

Sherrington R, Rogaev EI, Liang Y et al (1995) Cloning of a gene bearing missense mutations in early-onset familial Alzheimer's disease. Nature 375:754–760

Shim KS, Lubec G (2002) Drebrin, a dendritic spine protein, is manifold decreased in brains of patients with Alzheimer's disease and down syndrome. Neurosci Lett 324:209–212

Siman R, Reaume AG, Savage MJ, Trusko S, Lin YG, Scott RW, Flood DG (2000) Presenilin-1 P264L knock-in mutation: differential effects on abeta production, amyloid deposition, and neuronal vulnerability. J Neurosci 20:8717–8726

Snyder EM, Nong Y, Almeida CG et al (2005) Regulation of NMDA receptor trafficking by amyloid-beta. Nat Neurosci 8:1051–1058

Sosa-Ortiz AL, Acosta-Castillo I, Prince MJ (2012) Epidemiology of dementias and Alzheimer's disease. Arch Med Res 43:600–608

Steiner H, Duff K, Capell A et al (1999) A loss of function mutation of presenilin-2 interferes with amyloid beta-peptide production and notch signaling. J Biol Chem 274:28669–28673

Takahashi H, Sekino Y, Tanaka S, Mizui T, Kishi S, Shirao T (2003) Drebrin-dependent actin clustering in dendritic filopodia governs synaptic targeting of postsynaptic density-95 and dendritic spine morphogenesis. J Neurosci 23:6586–6595

Takahashi H, Mizui T, Shirao T (2006) Down-regulation of drebrin a expression suppresses synaptic targeting of NMDA receptors in developing hippocampal neurones. J Neurochem 97(Suppl 1):110–115

Takahashi H, Yamazaki H, Hanamura K, Sekino Y, Shirao T (2009) Activity of the AMPA receptor regulates drebrin stabilization in dendritic spine morphogenesis. J Cell Sci 122:1211–1219

Terry RD, Masliah E, Salmon DP, Butters N, DeTeresa R, Hill R, Hansen LA, Katzman R (1991) Physical basis of cognitive alterations in Alzheimer's disease: synapse loss is the major correlate of cognitive impairment. Ann Neurol 30:572–580

Tully AM, Roche HM, Doyle R, Fallon C, Bruce I, Lawlor B, Coakley D, Gibney MJ (2003) Low serum cholesteryl ester-docosahexaenoic acid levels in Alzheimer's disease: a case-control study. Br J Nutr 89:483–489

Um JW, Nygaard HB, Heiss JK, Kostylev MA, Stagi M, Vortmeyer A, Wisniewski T, Gunther EC, Strittmatter SM (2012) Alzheimer amyloid-beta oligomer bound to postsynaptic prion protein activates Fyn to impair neurons. Nat Neurosci 15:1227–1235

Wang HY, Lee DH, Davis CB, Shank RP (2000) Amyloid peptide Abeta(1–42) binds selectively and with picomolar affinity to alpha7 nicotinic acetylcholine receptors. J Neurochem 75:1155–1161

Wang Y, Shibasaki F, Mizuno K (2005) Calcium signal-induced cofilin dephosphorylation is mediated by slingshot via calcineurin. J Biol Chem 280:12683–12689

Worth DC, Daly CN, Geraldo S, Oozeer F, Gordon-Weeks PR (2013) Drebrin contains a cryptic F-actin-bundling activity regulated by Cdk5 phosphorylation. J Cell Biol 202:793–806

Wu HY, Hudry E, Hashimoto T et al (2010) Amyloid beta induces the morphological neurodegenerative triad of spine loss, dendritic simplification, and neuritic dystrophies through calcineurin activation. J Neurosci 30:2636–2649

Xu J, Chatterjee M, Baguley TD et al (2014) Inhibitor of the tyrosine phosphatase STEP reverses cognitive deficits in a mouse model of Alzheimer's disease. PLoS Biol 12:e1001923

Yang F, Sun X, Beech W, Teter B, Wu S, Sigel J, Vinters HV, Frautschy SA, Cole GM (1998a) Antibody to caspase-cleaved actin detects apoptosis in differentiated neuroblastoma and plaque-associated neurons and microglia in Alzheimer's disease. Am J Pathol 152:379–389

Yang N, Higuchi O, Ohashi K, Nagata K, Wada A, Kangawa K, Nishida E, Mizuno K (1998b) Cofilin phosphorylation by LIM-kinase 1 and its role in Rac-mediated actin reorganization. Nature 393:809–812

Yu H, Saura CA, Choi SY et al (2001) APP processing and synaptic plasticity in presenilin-1 conditional knockout mice. Neuron 31:713–726

Zhang Y, Venkitaramani DV, Gladding CM, Zhang Y, Kurup P, Molnar E, Collingridge GL, Lombroso PJ (2008) The tyrosine phosphatase STEP mediates AMPA receptor endocytosis after metabotropic glutamate receptor stimulation. J Neurosci 28:10561–10566

Zhang Y, Kurup P, Xu J et al (2010) Genetic reduction of striatal-enriched tyrosine phosphatase (STEP) reverses cognitive and cellular deficits in an Alzheimer's disease mouse model. Proc Natl Acad Sci U S A 107:19014–19019

Zhang Y, Kurup P, Xu J, Anderson GM, Greengard P, Nairn AC, Lombroso PJ (2011) Reduced levels of the tyrosine phosphatase STEP block beta amyloid-mediated GluA1/GluA2 receptor internalization. J Neurochem 119:664–672

Zhao L, Ma QL, Calon F et al (2006) Role of p21-activated kinase pathway defects in the cognitive deficits of Alzheimer disease. Nat Neurosci 9:234–242

Zhao WQ, Santini F, Breese R et al (2010) Inhibition of calcineurin-mediated endocytosis and alpha-amino-3-hydroxy-5-methyl-4-isoxazolepropionic acid (AMPA) receptors prevents amyloid beta oligomer-induced synaptic disruption. J Biol Chem 285:7619–7632

Chapter 13
Drebrins and Connexins: A Biomedical Perspective

Irina V. Majoul, Justus S. Ernesti, Eugenia V. Butkevich, and Rainer Duden

Abstract In this chapter we summarize knowledge on the role of drebrin in cell–cell communications. Specifically, we follow drebrin-connexin-43 interactions and drebrin behavior at the cell–cell interface described earlier. Drebrin is a part of the actin cytoskeleton which is a target of numerous bacteria and viruses invading mammalian cells. Drebrin phosphorylation, self-inhibition and transition between filaments, particles, and podosomes underlie cellular mechanisms involved in diseases and cognitive disorders. Cytoskeletal rearrangements influence the state of gap junction contacts which regulate cell signaling and metabolic flow of information across cells in tissues. Taking into account that connexin-43 (Cx43) (together with Cx30) is heavily expressed in astrocytes and that drebrin supports cell–cell contacts, the understanding of details of how brain cells live and die reveals molecular pathology involved in neurodegeneration, Alzheimer's disease (AD), other cognitive disorders, and aging.

Bidirectional connexin channels are permeable to Ca^{2+} ions, IP3, ATP, and cAMP. Connexin hemichannels are important for paracrine regulation and can release and exchange energy with other cells using ATP to transfer information and to support damaged cells. Connexin channels, hemichannels, and adhesion plaques are regulated by assembly and disassembly of the actin cytoskeleton. Drebrin degradation can alter gap junction communication, and drebrin level is decreased in the brain of AD patients. The diversity of drebrin functions in neurons, astrocytes, and non-neuronal cells still remains to be revealed. We believe that the knowledge on drebrin summarized here will contribute to key questions, "covering the gap" between cell–cell communications and the submembrane cytoskeleton.

Keywords Drebrin • Actin • Myosin • Connexins • CDRs • Calcium • cAMP • Small GTPases • Astrocytes • Neurodegeneration

---

I.V. Majoul (✉) • J.S. Ernesti • R. Duden
Institute of Biology, Centre for Structural and Cell Biology in Medicine, University of Lübeck, Ratzeburger Allee 160, 23538 Lübeck, Germany
e-mail: irina.majoul@bio.uni-luebeck.de

E.V. Butkevich
Drittes Physikalisches Institut, Georg-August-Universität, Friedrich-Hund-Platz 1 37077 Göttingen, Germany

T. Shirao, Y. Sekino (eds.), *Drebrin*, Advances in Experimental Medicine and Biology 1006, DOI 10.1007/978-4-431-56550-5_13

13.1 Introduction

Brain cells use two main ways to communicate with each other: one way is through the synaptic release of neuromediators (transmitters and modulators) and the other is through direct cell–cell interactions that are mediated by gap junction (GJ) channels (reviewed by Guthrie and Gilula 1989; Goodenough et al. 1996; Bennett and Zukin 2004). Among the 37 trillion cells existing in our body (Bianconi et al. 2013), there are approximately 100 billion neurons and ~1 trillion astroglial cells (Pakkenberg and Gundersen 1988 Herculano-Houzel 2009), (see also Fig. 13.1). Astrocytes form GJ contacts with up to ten partners in the three-dimensional space of the brain, thus providing thousands of highly permeable gap junction contacts that are constantly being built de novo, reorganized, and degraded in accord with physiological signals, moreover, astrocytes control the amount of synaptic contacts (Ullian et al. 2001).

In contrast to synaptic vesicles, which require time to fuse in order to release their signal mediators, gap junction channels and connexin hemichannels operate by the non-vesicular direct release of signal molecules and are thus much faster than synaptic vesicle-mediated signal transmission (Moore et al. 2014). Gap junction channels mediate transmission of signals between the cytoplasms of connected cells. This way, paracrine and regulatory signals propagate information encoded in the gradients of ATP, cAMP, IP3, and Ca^{2+} through gap junctions that may facilitate or terminate signals directed to the distant parts of the brain (Dell'Acqua et al. 2006; Berridge et al. 2012). Thus, a complex relationship between neurons and the surrounding astroglial networks supports brain physiology in health and disease (Alvarez-Maubecin et al. 2000; Orellana et al. 2012). Connexin hemichannels, on the other hand, can release ATP, Ca^{2+}, and other messenger molecules into the extracellular space, thus affording yet another mode of paracrine cell–cell signaling in the brain.

13.2 Connexin-43 Identified as a *Bona Fide* Interactor of the Actin-Binding Protein Drebrin

Which molecules can structurally modify cell–cell interfaces, regulate formation of spines and gap junction contacts, tune their function in accord with activity-dependent tasks, and operate in the extremely complex architectural network (as one shown in Fig. 1) composed of contacting lipid-based cell membranes?

In search of molecules used by brain cells to regulate highly permeable cell–cell contacts, we analyzed interacting partners of connexin-43 (Cx43), one of the more ubiquitous connexins expressed in astrocytes and in many other human tissues including heart muscle (Duffy et al. 2002). Toward this end, we cloned the cytosolic C-terminus of Cx43 into a GST-tagging plasmid, expressed and purified the chimeric protein, and used it as a bait to find binding partners. We fractionated homogenates of mouse brain and tried a series of different concentrations of ATP in our in vitro binding reactions. This approach recovered a number of proteins that were resolved/identified by MALDI-TOF analyses. In different fractions we obtained, in addition to the already known interaction partners of Cx43 molecule ZO-1 (Toyofuku et al. 1998, 2001 Giepmans and Moolenaar 1998) and tubulin (Giepmans et al. 2001), a previously undescribed binding partner- which proved to be drebrin. Indeed, in primary astrocytic cell cultures, probed with antibodies by -immunofluorescence, drebrin was colocalized with Cx43 at cell–cell interfaces, and downregulation of drebrin using specific RNAi resulted in dramatic connexin internalization (Butkevich et al. 2004). We observed a close molecular proximity of these two proteins at cell–cell interfaces in transfected living cells by donor-acceptor FRET signals after co-expression of drebrin-CFP and Cx43-YFP-tagged proteins (Butkevich et al. 2004). The role of drebrin interactions with Cx43 channels was reviewed by Stout et al. (2004).

Drebrin was first described by Tomoaki Shirao (Shirao and Obata 1985) and originally identified in neuronal cells. Soon they recognized the enormous membrane modifying the morphogenic potential of this protein, when it was shown to be able to form neurite-like outgrowths upon overexpression in cells (Shirao et al. 1992). A decade later it was established that three isoforms of drebrin are present in mammalian cells: neuron-specific drebrin A, ubiquitous drebrin E (Shirao and Obata, 1986; Shirao et al. 1988), and finally the so-called s-drebrin A, which represents a form of a C-terminally truncated drebrin A (Jin et al. 2002). Drebrin A was later shown to be involved in the formation of dendritic spines (Takahashi et al. 2003; Mizui et al. 2005) and in synaptic plasticity (Sekino et al. 2006; Takahashi et al. 2006; Aoki et al. 2009; Mizui et al. 2014). During development all, not only neuronal drebrin A, contribute to a triangle of protein-protein interacting partners—actin, drebrin E, and myosin II and V that have been shown to be necessary to drive axon growth (Mizui et al. 2009).

Drebrin was of interest to researchers from both cytoskeletal and gap junction fields. The group of Werner Franke working on desmosomes and other cytoskeletal proteins described drebrin E as an important protein necessary to anchor actin filaments (Peitsch et al. 1999), similar to drebrin A (Ikeda et al. 1995, 1996). Moreover, an excess of drebrin in cells was shown to be deposited in the form of drebrin particles (Asada et al. 1994; Peitsch et al. 2001). At the same time, the group of Daniel Goodenough working on connexins reported that drebrin E was specifically bound to F-actin in the stomach and in kidney cells (Keon et al. 2000). After we discovered that drebrin is a Cx43-interacting protein (Butkevich et al. 2004), it was then shown that apart from the Cx43 permeability function, unopposed membranes with Cx43 hemichannels are necessary to mediate cell–cell adhesion function important for correct neuronal precursor migration and layer formation in early development (Schaar and McConnell 2005; Elias et al. 2007; Marin et al. 2010) (see Fig. 13.1).

The ability of drebrin to support cell–cell adhesion and cell–cell interfaces may be vitally important for plasticity-dependent tasks. The localization of drebrin at cell–cell interfaces or outgrowths often seen in living cells suggests that this molecule may function to integrate intracellular and extracellular signals to accordingly align cell–cell contacts (cartoon Fig. 13.2) and reorganize actin networks in response to cell activity (see Figs. 13.3 and 13.4).

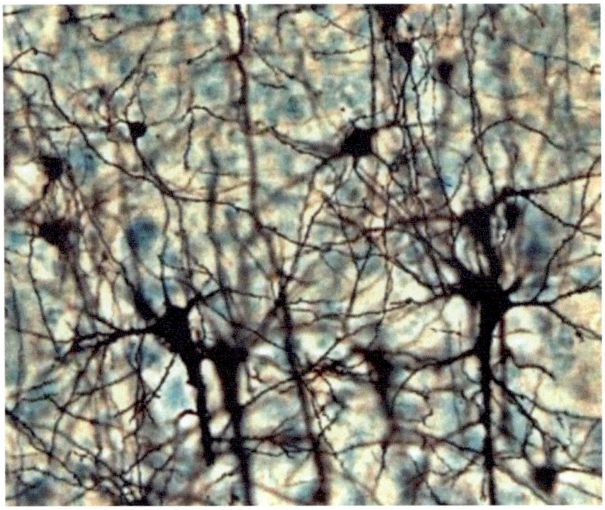

Fig. 13.1 Cortex slice impregnated with silver Golgi reagent reveals neurons (seen in *black*), while astroglial nuclei (seen in *blue*) were stained with Nissl cresyl violet (*left*), and intra-filopodial distribution of expressed drebrin-A (*right*)

The dual nature of connexin-drebrin interactions resembles the opposing force of yin-yang black-white complementary. Drebrin's task together with Cx43 is not only to align opposite membranes to form highly permeable GJ channel domains (i.e. GJ plaques) but also to constantly reposition these domains in 3-D space dictated by cellular activity.

13.3 Drebrin and Connexin at Cell–Cell Interfaces

Astrocytes in the brain express both drebrin E and Cx43. Submembrane drebrin is often seen accumulated in cell–cell contact zones and is colocalized there with connexin channels in GJ plaques (Figs. 13.2 and 13.3). During formation of cell–cell interfaces, the delivery of connexin hemichannels from the Golgi is regulated by complex trafficking steps (Majoul et al. 2009) and involves a microtubular transport machinery (for review see Akhmanova et al. 2009). Interestingly, on the way to the PM, connexins transiently accumulate in the Golgi; drebrin that is seen on the bottom of the cell with actin fibers in the upper 3-D Z-sections appeared again around the Golgi (Fig. 13.4), where its function is less known. Both molecules are often colocalized at cell–cell interfaces repeating similar complex geometry that can be visualized after fixation. We used the mouse monoclonal anti-drebrin antibody M2F6 (created by T. Shirao's group and recognizing both the drebrin A and E isoforms) and our anti-Cx43 polyclonal epitope-specific C-terminal antibody to reveal their colocalization (Fig. 13.3). Live cell analyses confirmed that within less than 6 h after

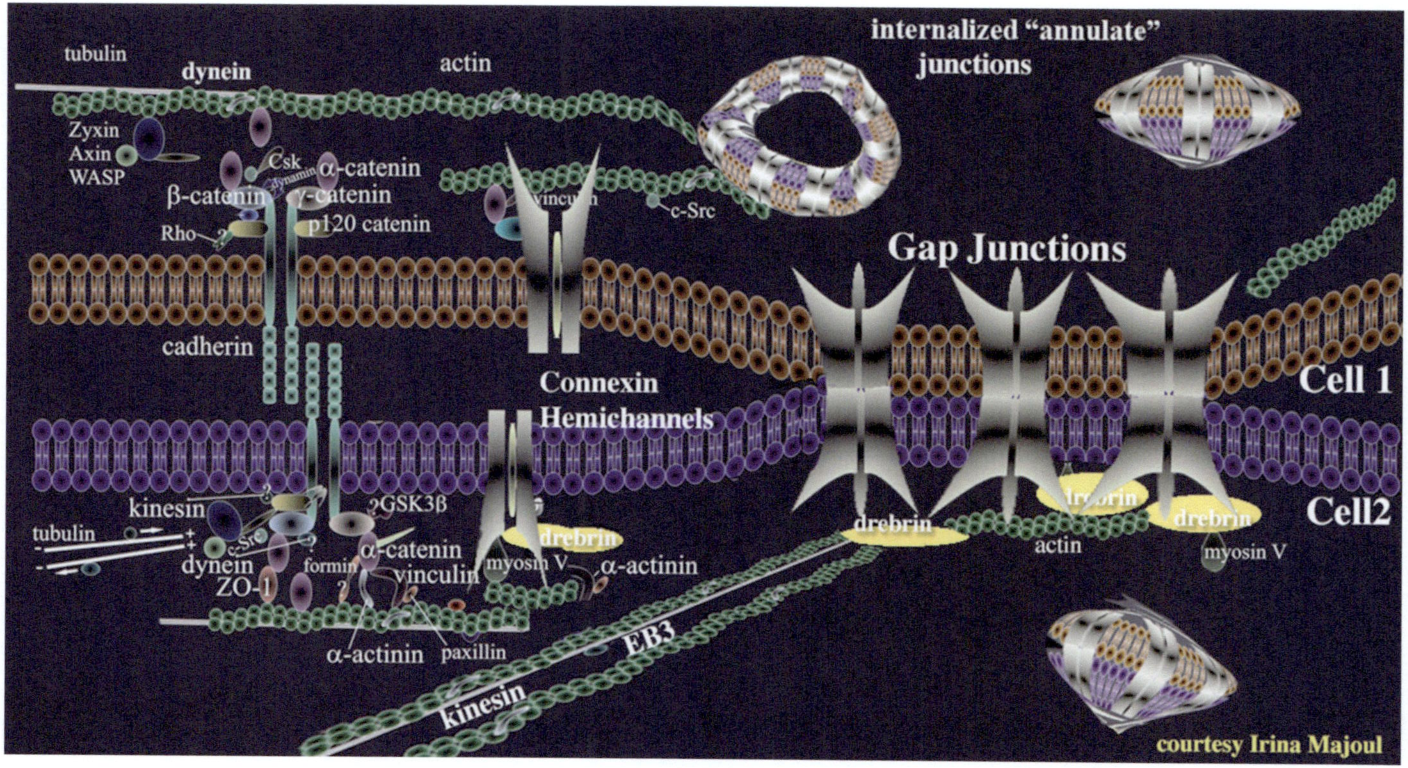

Fig. 13.2 Schematic drawing of a cell–cell interface. Initial proximity of cell–cell contacts is probably mediated by cadherins operating in a calcium-dependent manner. Next, drebrin may assist the connexin hemichannels from two opposing plasma membranes to establish highly-permeable GJ contacts. The accumulation of Cx43 hemichannels will require formation of multi-subunit complexes composed of molecules interacting with the cytosolic part of connexin tails. The drebrin that participated in the lateral alignment of connexin hemichannels is not present after Cx43 is internalized in the form of annular junctions (see Fig.13.4)

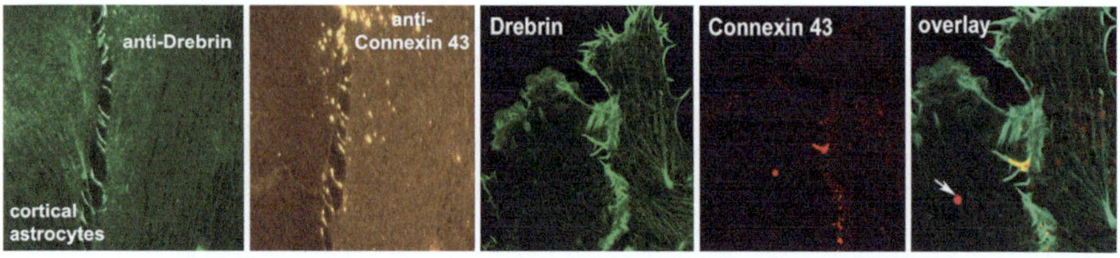

Fig. 13.3 (*Left*) Native distribution of drebrin and Cx43 seen in a primary culture of rat cortical astrocytes co-stained with anti-drebrin (monoclonal) and anti-Cx43 (polyclonal) antibodies. Both molecules are present at specific points of cell–cell interface. (*Right*) Living Vero cells express GFP-tagged drebrin E that colocalized with Cx43-RFP at cell–cell interfaces. No drebrin fluorescence is seen around internalized annular junctions of Cx43 (overlay, *white arrow*)

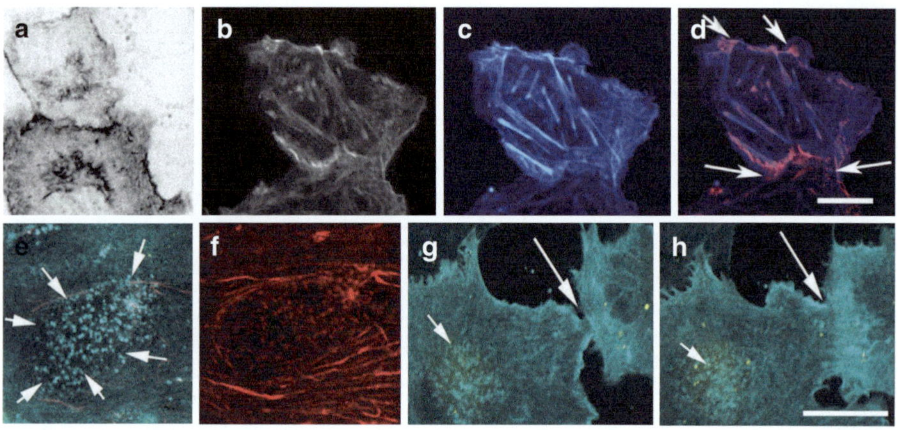

Fig. 13.4 (**a**) Drebrin E reactivity is often present in the Golgi region. (**b, c**) Distribution of drebrin E-CFP (seen in **b**) at the cell–cell interface in cells co-transfected with actin-red (seen in **c**). (**d**) Overlay: actin (*blue*), drebrin (*red*). In contrast to actin, drebrin is preferentially present at the cell-cell interface and at the leading edge. (**e**) Highly resolved image of "Drebrin particles" accumulated in the cytoplasm in the perinuclear region indicated by *arrows*. (**f**) actin-red present in some of "drebrin particles" but most of it fluorescence seen in the surrounding fibers. (**g**) Drebrin particles are less prominent in cells during formation cell–cell contacts, big arrows. (**g–h**) Time-dependent images of drebrin show the development of the cell–cell contact, images are taken with intervals of 20 min (connexin is not shown). Scale bar, 20 μm

co-expression drebrin can be colocalized with Cx43. Their dynamic proximity remained even during dramatic cytoskeletal rearrangements of GJ plaques at cell–cell interfaces (Fig. 13.3, right panel). In contrast to cell–cell interfaces where drebrin is colocalized with Cx43, internalized Cx43 in the form of annular junctions is usually depleted from drebrin immunoreactivity (Fig. 13.2, overlay). Inside the cell, drebrin remains organized along the actin stress fibers (seen in Fig. 13.4).

Live cell imaging revealed how drebrin participates in the reorganization of new GJ Cx43 interfaces, and drebrin-directed RNAi oligo treatment resulted in the disorganization and partial removal of GJ contacts (Butkevich et al. 2004). Drebrin is

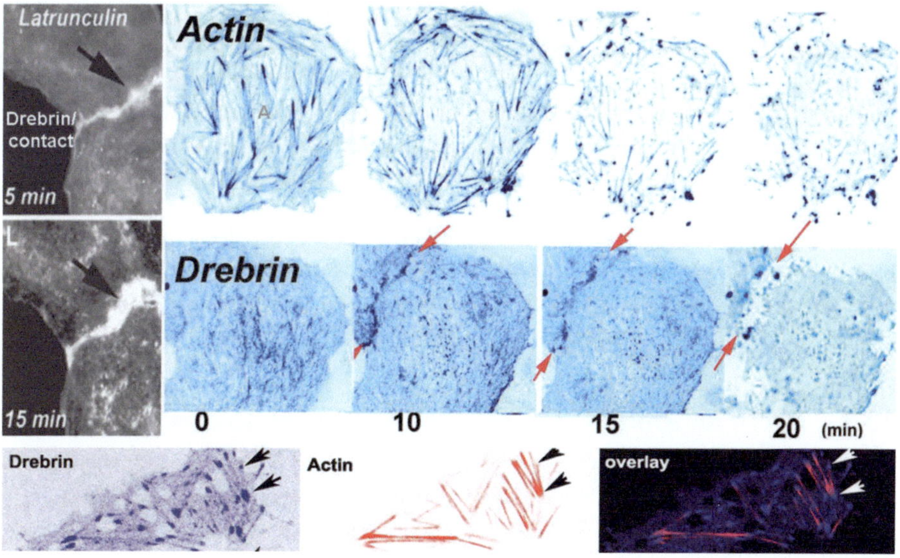

Fig. 13.5 (*Upper left panel*) Accumulation of drebrin-GFP at cell–cell interface seen after the disassembly of actin by latrunculin A (original data stack used in Butkevich et al. 2004). (*Upper right panel*) Actin and drebrin were co-transfected into the same cell to follow the disassembly of actin by latrunculin A. Upon the disassembly of actin, drebrin is accumulated at cell–cell interfaces (*red arrows*) and apparently prevents the rupture of cell–cell contacts (seen in the last image around 20 min). (*Lower panel*) Cells co-expressing tagged drebrin-CFP and actin-red were treated with cholera toxin (wt) to increase cAMP/PKA-dependent phosphorylation that accumulated drebrin in dense clusters. Actin filaments were disassembled by a short application of latrunculin A that was then reversibly washed away. After washout and addition of fresh media, actin filaments (*red*) are growing to reconnect the remaining clusters of drebrin (*blue*). Drebrin nodes remained densely packed throughout the experiment (overlay)

participating in other protein-protein interactions, thus fulfilling multiple other cellular functions (Fig. 13.5).

Two different drebrin isoforms, E and A, in the brain have different distributions and functions (Shirao and Obata 1986; Shirao et al. 1989). Both the ubiquitously expressed drebrin E and the neuron-specific drebrin A have the same N-terminal ADF (actin-depolymerizing factor) domain that is followed by a coiled-coil domain, and both domains are believed to be involved in the interaction with actin (Hayashi et al. 1999; Grintsevich et al. 2010) (see Fig. 1.3 in Chap. 1). Drebrin interaction with actin filaments is likely to provide a dynamic cellular plasticity different from actin-tropomyosin interactions. In different parts of the cell, actin seems to have a variety of interacting partners, and drebrin may act by disrupting actin interaction with other proteins such as tropomyosin, cofilin, and/or α-actinin (Ishikawa et al. 1994; Grinstevich and Reisler 2014). It remains unclear how at different cellular locations tropomyosin and troponin from one side and drebrin from the other form different and physiologically distinct complexes that are controlled in a domain-specific and phosphorylation-dependent manner by other interacting partners. Cytoskeletal molecules in drebrin proximity include actin, tropomyosin, α-actinin, actin motors, myosin II, myosin V, myosin VI, and postsynaptic scaffolds (Shirao and Sekino 2001). The N-terminal domain of drebrin provides interaction with submembrane partners, e.g., Cx43 and other proteins of the cellular machinery (Ambrosi et al. 2016). The morphogenic property of drebrin can allow this molecule to participate in fast vital cellular decisions that provide plasticity to the submembrane cytoskeleton and regulatory systems important for the early development in neuronal precursor migration, layer formation, neurite outgrowth, cell–cell contact formation, and control of growth cones. At later stages of cellular expansion, drebrin seems to be involved in synaptic connectivity, and many other activities of the developing brain (Song et al. 2008; Mizui et al. 2009; Sonego et al. 2015).

13.4 Drebrin, Small GTPases, and Bacterial Toxins

It is not surprising that the bacterial toxins that co-evolved with mammalian cells developed direct approaches to deplete cellular energy and to disrupt cell–cell communications. Thus, internalization and intracellular transport of cholera toxin (a member of the AB5 toxin family) into mammalian cells leads to conversion of cellular ATP energy into cAMP the concentration of which can be increased up to 20-fold (Majoul et al. 1998). Both Cx43 and drebrin can be phosphorylated by protein kinase A during toxin transport that will modify cell–cell contacts.

Rho GTPases are well-known regulators of the submembrane cytoskeleton (Hall 1998), but interestingly they may also provide a supplementary link between the cytoskeleton and Ca^{2+} signaling (Uehata et al. 1997). A better understanding of Rho GTPase action on the submembrane actin cytoskeleton was obtained from the study of bacterial toxins that invade cells and act on Rho GTPases (Aktories and Just 1995). These data revealed multiple layers of complexity in the action of bacterial toxins that influence both the regulation of the actin cytoskeleton and cellular junctions.

Both live cell analyses of drebrin behavior, and drebrin depletion analyses, suggested that drebrin at cell–cell contacts is rather dependent on the GTPase activity of Cdc42 and from RhoA at noncontacting membranes (our unpublished observations). These data are in agreement with the known functions for Cdc42 and Rho-family GTPases as critical regulators of cell polarity, actin cytoskeleton, and cell–cell communication (Ridley 2000; Etienne-Manneville and Hall 2002; Derangeon et al. 2008).

Drebrin expression in living cells is strongly influenced by signaling mechanisms and the GTPase activities of Rho, Rac, and Cdc42 that are believed to act primarily on actin. Without properly assembled actin, cells cannot rearrange cell–cell contacts or migrate and remain attached to the substrate. Specific dependence of drebrin on Cdc42 and RhoA may explain correlation between the cellular distribution of drebrin at cell–cell interfaces and at noncontacting membranes, where drebrin protrusion-repulsion activity is also dependent on a RhoA gradient. Other drebrin interacting partners at cell–cell interfaces may be critical for GJ-mediated cell–cell communication.

RhoA activity also affects Cx43 permeability without a significant change in the geometry of the cell–cell interface or the distribution of Cx43 phosphorylation bands on blots. Nevertheless, a high level of RhoA expression can disrupt submembrane Cx43 interactions with drebrin (our unpublished observations). Because cell–cell communications rely on the integrity of the drebrin-actin cytoskeleton, the remodeling of cortical actin by small GTPases and/or bacterial toxins can directly influence the state of GJ communications. Dye transfer and gating of Cx43 gap junctions are usually disrupted by bacterial toxin internalization, and both actin and RhoA GTPases are targets of many toxins.

Disruption of cell–cell communications was shown by Suh et al. (2012) when they grew cells on glass coated with laminin-111. It turned out that decreased permeability of gap junctions between cells growing on laminin-111 was due to an increased activity of the small GTPase RhoA. The application of C3 (a RhoA inhibitor) reversed the effect of laminin-111, increased Cx43/drebrin/ZO-1 integrity and Cx43 phosphorylation, and recovered the control levels of dye transfer. These data reveal a perspective picture where the stability of a drebrin-Cx43/ZO-1 complex favors the existence of GJ coupled cell populations with defined levels of dye transfer. The reduction of Cx43/ZO-1/drebrin complex resulted either in an increase of cell proliferation (in some cells) or in apoptosis with a specific pattern of Cx43 phosphorylation in other clusters (Suh et al. 2012).

Small GTPases change spine morphologies similar to those like the phenotypes that result from the application of GEFs (GTP-for-GDP exchange factors) or specific bacterial toxins. We began to understand how bacterial toxins act on small GTPases specifically toward the Rho family, with the discovery that bacterial toxins are able to deamidate Gln 63 of RhoA (Schmidt et al. 1997). A modification of the cytoskeleton by bacterial toxins seems to include the modification of gap junctions and dendritic spines that are particularly vulnerable during development (Matsuzaki et al. 2004).

13.5 Drebrin Self-Inhibition

We and other groups observed that many cell types natively regulate an excess of drebrin by forming drebrin particles, sometimes seen in the cytoplasm or by forming ring-like podosomes (Asada et al. 1994; Peitsch et al. 1999). Formation of drebrin complexes may reflect a native ability of cells to control drebrin levels to control its action on actin which would otherwise induce the formation of outgrowths and other uncontrolled changes of cell shape. Analyzing the action of drebrin in both cells and solutions, we hypothesize that conformation-dependent self-inhibition of drebrin or the formation of cytoplasmic multimeric drebrin-partner complexes is a way for cells to deal with the high morphogenic potential of this molecule (see Fig. 13.6).

The hypothetical mechanism of the conformational self-inhibition of drebrin still remains controversial, and complexity in the crystallization of this molecule has prevented further detailed structural analyses. Molecular machinery of reversible phosphorylation (or other modifications) of drebrin, its inactivation of actin-modifying ADF, middle 233-366 and coil-coil domains, and the role of the C-terminal remain to be analyzed in detail (Fig. 13.7).

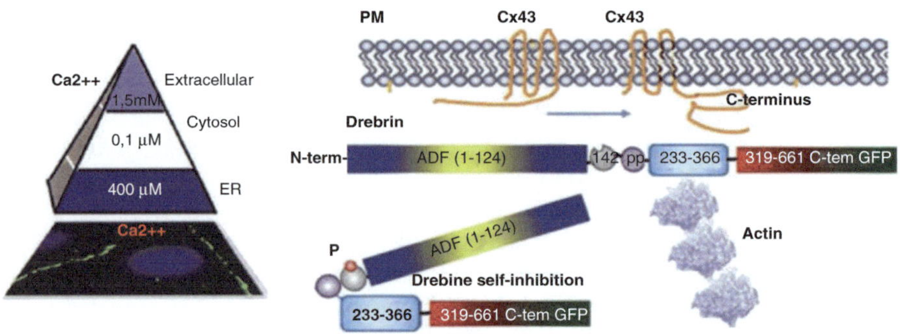

Fig. 13.6 Healthy cellular membranes maintain high gradients of Ca^{2+}concentration between the PM and cytoplasmic space (15,000:1) and between the endoplasmic reticulum (ER) and the cytosol (4,000:1). Cx43 channels and hemichannels (permeable for Ca^{2+}) keeping these gradients can regulate its in-coming concentration by channel closure. Drebrin domains, its self-inhibition and interaction with Cx43 and actin are depicted (*right*)

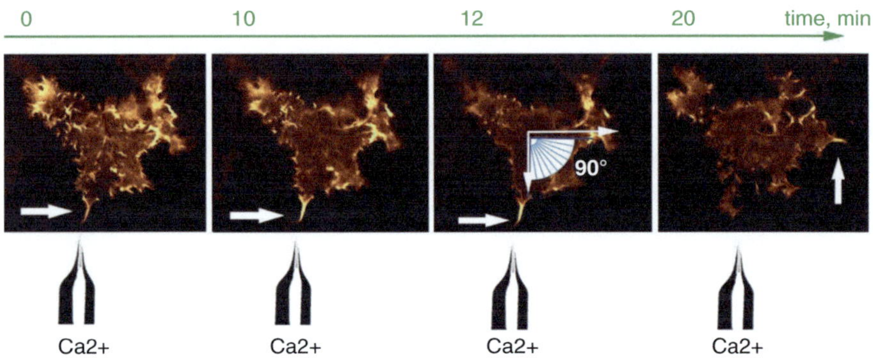

Fig. 13.7 Drebrin E-YFP is expressed in cells growing on glass coverslips. Time-dependent dynamic movement of drebrin was imaged in live cells. Some cells form rather immobile outgrowths with connexin hemichannels acting as an attachment tail. In this experiment Ca^{2+} concentration of the media was set to 2 mM, while the microelectrode positioned close to the immobile tail has a [Ca^{2+}] concentration 3 times higher. Within the next 20 min, the "immobile" tail of drebrin (indicated by *arrow*) was retracted and re-appeared away from the direction of Ca^{2+} flux in the 90° position corner (as indicated by the last *arrow*)

13.6 Similarity Between Two Molecular Complexes: Cx43-Drebrin and the Glycine Receptor-Gephyrin

Cx43-drebrin interactions have some degree of functional similarity with another well-known protein complex formed by the protein gephyrin and the glycine receptor. Gephyrin is expressed in both excitatory neurons (which release glutamate) and in inhibitory neurons which release GABA or glycine or both neuromediators. However, gephyrin localizes specifically to inhibitory synapses (that should contain stronger binding partners) on the postsynaptic side and is similar to drebrin in that it forms a complex with a transmembrane channel molecule, the glycine receptor. Gephyrin is critical for glycine receptor clustering at inhibitory synapses and was also found in a complex with another inhibitory receptor—$GABA_A$, nevertheless, $GABA_A$ is functionally less dependent on gephyrin than the glycine receptor (Kneussel et al. 2001).

Next, similar to the dependence of drebrin on small GTPases described above, gephyrin in the PSD interacts with collybistin, which is a GEF (guanine nucleotide exchange factor) for the small GTPase Cdc42 (Kins et al. 2000).

13.7 Drebrin, Cx43, and Second Messengers

Gap junctions represent cellular platforms for the effective exchange of second messengers such as Ca^{2+}, IP3, cAMP, and cGMP between cells. Under normal conditions, these permeable "hot spots" of cellular connectivity are necessary to keep control over cell–cell connectivity and organ physiology. Gap junctions regulate cell–cell communications and can support the metabolic state of impaired cells in organs by releasing ATP. Live cell FRET analyses revealed that direct interaction between drebrin and the Cx43-C-terminus is ATP (i.e. phosphorylation) dependent. It can be that only healthy cells provide regulatory impact on channel stability and permeability because drebrin is involved in the rearrangements of Cx43 gap junction channels at different stages of contact formation (Butkevich et al. 2004; Majoul et al. 2009). Drebrin-dependent rearrangement of Cx43 channels on the plasma membrane is regulated by membrane phosphatidylinositol 4,5-bisphosphate (PIP2) and IP3 signaling to the ER. Residing in the inner leaflet of the plasma membrane, PIP2 is involved in signaling and is also a substrate for phospholipase C that generates diacylglycerol (DAG) and inositol triphosphate (IP3) (Rebecchi and Pentyala 2000; Berridge 2010). When surrounding GJ plaques, PIP2 may also dock clusters of drebrin, helping this molecule to interact with the C-termini of connexins in a signal- and activity-dependent manner. As we observed earlier, Cx43 permeability and dye transfer are dependent on the external ATP level, which upon physiological stimuli can modulate GJ permeability by affecting the open-closed state of GJ channels. Drebrin can filter cAMP, IP3, and Ca^{2+} information between coupled cells by supporting contacts, redirecting membrane GJ domain, or just removing them.

The IP3-dependent Ca^{2+} signaling was recently studied in astrocytes after a constitutive genetic receptor deletion in IP3R2 KO mice, where no significant changes were observed (Jego et al. 2014). Nevertheless, the effects of abnormal Ca^{2+} homeostasis, compensatory mechanisms, and inositol 1,4,5-trisphosphate receptor 1 (IP3 R1) which represents the main caspase-3-sensitive channel receptor for Ca^{2+} have not been evaluated. Indeed, in living cells we observed that the behavior of gap junctions at cell–cell interfaces is extremely sensitive to the level of Ca^{2+} and ATP in the surrounding media (our unpublished observations) and in in vitro reconstruction experiments of the binding of the Cx43-C-terminus to drebrin required at least 1.5 mM ATP in the binding solution. Cell–cell permeability is also dependent on the ATP concentration and when we lowered extracellular ATP to 1 mM binding decayed. This energy-dependent decline of cell–cell communications and the instability of drebrin may also include other submembrane factors such as the cortical actin cytoskeleton, myosin motors, and microtubular transport activity. Noticeably, in young cells with undepleted energy level, drebrin modifications regulate numerous processes of migration and layer formation (Sonego et al. 2015).

13.8 Drebrin and Ca^{2+}

Ca^{2+} is involved in the regulation of numerous processes in astrocytes, neuronal and nonneuronal cells (Berridge et al. 2000; Berridge 2010), and the release of Ca^{2+} from the ER is important for many signaling events (Berridge 2012). Small dendritic spines operate with Ca^{2+} signaling (Sabatini et al. 2001). Very low background levels of Ca^{2+} in the cytoplasm of healthy cells (70–100 nano-molar range) allow signals to propagate via Ca^{2+} entry through the plasma membrane or via the release of Ca^{2+} from the endoplasmic reticulum, an important intracellular calcium store. Most of the Ca^{2+} ions propagating signals in non-excitable cells come from the ER, which is true as well for excitable muscle and secretory cells. In contrast, neurons, rely mainly on Ca^{2+} entry domains on the PM, and neuronal ER stores calcium for their local modular needs. This interesting fact may explain neuronal dependence on well-developed propagation of Ca^{2+} waves in astrocytes. Moreover, during complex cellular events under both physiological and pathophysiological conditions, Ca^{2+} is necessary to regulate mitochondrial activity and the production of cellular ATP. Cellular loss of Ca^{2+} is associated with the induction of the unfolded protein response (UPR), which results in cell death because the energy supply for the ER requires both Ca^{2+} homeostasis and mitochondrial ATP production (Berridge et al. 2000; Walter and Ron 2011).

Drebrin is sensitive to the Ca^{2+} concentration, and the release of Ca^{2+} from the ER is directly coupled to the plasma membrane channels known as store-operated Ca^{2+} entry (SOCE). SOCEs are composed of stromal interacting molecule 1 (STIM1) and the Ca^{2+} channel pore-forming unit ORAI1 positioned in the plasma membrane (reviewed by Hogan and Rao 2007). The ER Ca^{2+} sensor STIM1 binds directly to ORAI1 proteins in the plasma membrane, which comprise the pore-forming subunit of the Ca^{2+} release-activated Ca^{2+} channel. Ca^{2+} influx through the channels activates the key functional cellular responses, metabolism, cell movement, and gene expression. Gap junctions can spatially restrict Ca^{2+} signals by forming PM microdomains and can sense intracellular Ca^{2+} being activated by the emptying of intracellular Ca^{2+} stores. The same proteins can act to restore Ca^{2+} in the ER after global and local Ca^{2+} dependent cellular events. During the resting state, cells can accumulate up to 95% of the intracellular Ca^{2+} in the ER lumen. Surprisingly, it was discovered that 3,5-bis(trifluoromethyl)pyrazole (BTP) compound regulates store-operated calcium channels via the actin-binding protein drebrin (Mercer et al. 2010).

Connexin channels and hemichannels control gradients of signaling molecules, and specifically of Ca^{2+} (Benninger et al. 2008). Finally, astrocytes can use connexins to buffer Ca^{2+} ions as was shown recently by Mark Yeager group (Zhang et al. 2016) who solved the crystal structure of the first extracellular loop of Cx26 bound to Ca^{2+}. This discovery demonstrated that extracellular loops of connexins can bind Ca^{2+} ions. Ca^{2+} oscillations in primary cultures of astrocytes and neurons can be monitored in living cells with the dye Fluo-4. We recorded Ca^{2+} oscillations with a Nikon Ti imaging system using an Andor Ixon EM-CCD camera and time-lapse imaging of cells excited with a 488 nm laser and reading the Ca^{2+} fluorescence with dichroic emission filters 505 nm beamsplitter and 530/20 nm barrier filter.

How can drebrin sense extracellular and intracellular (ER) Ca^{2+} and how may Ca^{2+} influence the behavior of drebrin? To understand whether drebrin is responding to Ca^{2+} changes, we tested live cell behavior of the drebrin A isoform expressed as a fluorescently tagged molecule. This experiment reveals how drebrins may respond to an elevation of Ca^{2+} level (Fig. 13.7). This way, drebrin rearrangements may help GJ channels and connexin hemichannels to respond to gradients of extracellular Ca^{2+} and to provide bidirectional control of cell–cell communications. Drebrin may also help astrocytes to outline connexin protrusions in proximity to neurons keeping control over the cellular territories. The recently resolved crystal structure of the extracellular loop of connexins revealed that the bound Ca^{2+}ion acted to close the channel electrostatically (Zhang et al. 2016). Connexins expressed in astrocytes may buffer extracellular calcium released by neurons via a similar mechanism.

Although gap junctions provide direct pathways for the exchange of Ca^{2+} between coupled cells, it would be important to know mechanisms of how extracellular information is transferred to the intracellular molecules, which regulate migration, cell–cell communication, and connectivity.

13.9 Degradation of Drebrin by Calpain

The direct relationship between drebrin and Ca^{2+} remained unexplored until recently when it was shown that drebrin can be proteolytically cleaved by calpain, a calcium-dependent, non-lysosomal cysteine protease although it was reported that drebrin degradation is calcium-dependent (Arai et al. 1991). It happens during excitotoxic neuronal death caused by activation of NMDA-type glutamate receptors (NMDARs), but interestingly, pharmacological inhibition of F-actin only facilitated the degradation of drebrin (Chimura et al. 2015). The authors concluded that degradation of drebrin during neurodegeneration is triggered by similar mechanisms and may modulate NMDAR-mediated signal transductions at different stages.

13.10 Cx43 and Drebrin in Astrocytes

In astrocytes Cx43 and Cx30 are the two main GJ channel-forming proteins. Cx43 is stabilized by ZO-1 (Giepmans and Moolenaar 1998; Giepmans et al. 2001) and modified by drebrin (Butkevich et al. 2004). The mobility of astrocytic outgrowth and the rearrangement of GJ plaques in astrocytes are mediated by drebrin (as seen in Fig. 13.3), because not ZO-1 but drebrin can disassemble actin or bundle F-actin thus providing a tool to direct submembrane activity. Drebrin is one of the most strong morphogenic molecules known (Shirao et al. 1992, 1994; Majoul et al. 2007). In in vitro reconstruction experiments, we showed that drebrin-like protein can bind actin and fold it into ring structures (Butkevich et al. 2015).

In the brain, GJ contacts as well as synapses are constantly modified by specific electrical and metabolic patterns of neuronal activity (Rao et al. 2005; Robel and Sontheimer 2016). Drebrins are crucial for the support of both cell–cell communications and neuronal activity and drebrin-dependent morphological and functional changes contribute to both electrical and chemical signal propagation and to the astrocytic connectivity. Neuronal drebrin A-dependent changes are shown to be involved in learning, memory formation, and other cognitive functions that require spine assembly (Kojima et al. 2016). The dynamic redistribution of drebrin toward the plasma membranes is responsible for dendritic spine maturation (Aoki et al. 2005; Takahashi et al. 2003, 2006).

Drebrin-formed outgrowths help astrocytes to come in a close proximity to active synapses; similar to podocytes in the kidney, astrocytes in the brain form highly branched outgrowths that can sense distant gradients of passing ions (as, e.g., Ca^{2+}), electrolytes, and fluid. Structure-function relationships seen with electron microscopy reveal the distribution of neurons and astrocytes in brain slices and suggest that connexin hemichannels may support neurons by absorbing small metabolites in close proximity to neuronal terminals, dendritic spines, and clefts (Fig.13.8). Drebrin property to move connexins in a close proximity to the terminals (as shown here for Cx43) is likely to master Cx30 adhesive insertion of astrocytic processes into the synaptic cleft for a better control of the glutamate level (Pannasch et al. 2014). During synaptic stimulation, ions and toxic metabolite molecules are released by neurons in the surrounding space (Swanson et al. 2004). Close proximity of astrocytes can facilitate the absorbance of Na^+, K^+, and other molecules; moreover, astrocytes can release ATP in a close proximity to neurons (Farahani et al. 2005). Remarkably, the astrocytic Ca^{2+} is elevated only when the acetyl cholinergic synapses are activated, not when the glutamatergic one; moreover, only mGluR activation can induce Ca^{2+} signaling in astrocytes (Perea et al. 2014).

13.11 Cx43 and Brain Pathology

During the transition into pathological states, astrocytes undergo molecular and morphological changes. Although cases of reactive astrogliosis are clinically ubiquitous, the cellular and molecular mechanisms of astrocyte behavior remain poorly understood. Remarkably, human pathologies of CNS (central nervous system) involve neurons only rarely (neuroblastoma) and even this is mediated via astrocytes, with the most common type of reactive astrogliosis known as glioblastomas. Interestingly, neuroblastoma cells in culture constitutively express drebrin, and drebrin reactivity was attenuated with antisense to drebrin (Toda et al. 1999).

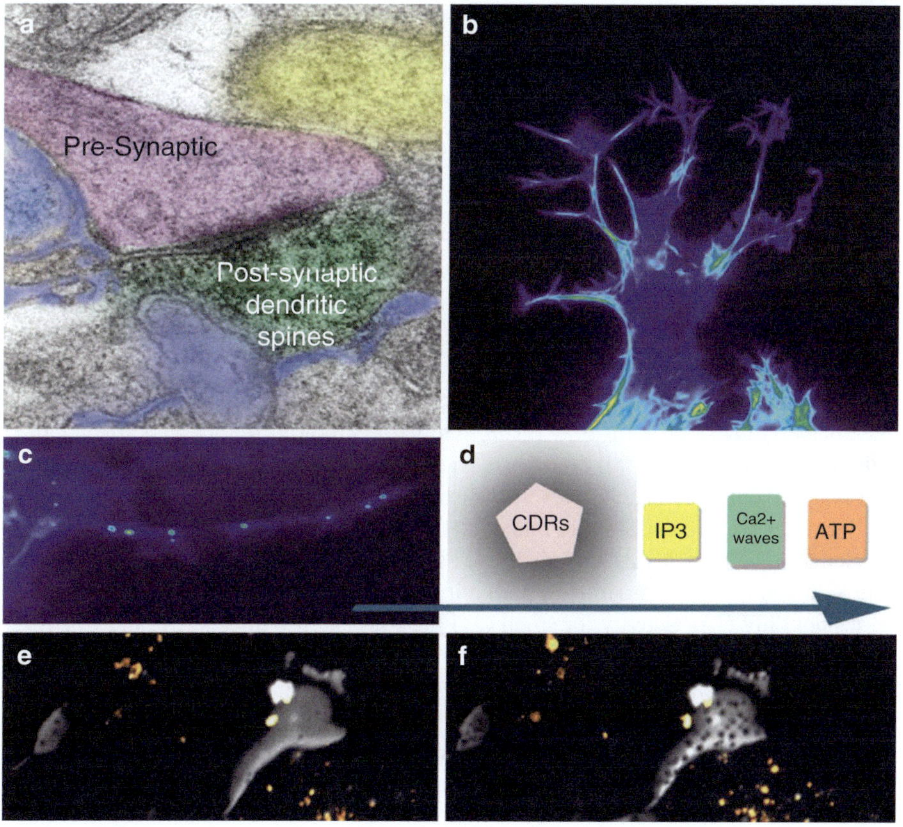

Fig. 13.8 (**a**) Close proximity of glial cells (in *blue*) to the synaptic terminals can be seen by electron microscopy (EM image courtesy of Dr. Vadim Rogachevsky, RAN, Synapse.ru). (**b**) Primary cortical astrocytes were co-transfected with drebrin-CFP and Cx43. (**c**) Cx43 is seen in dotted structures propagating along the drebrin-formed outgrowth. (**e, f**) Cx43 gap junction plaque before and during CDR responses (**d**). CDR-induced signaling from the GJ domain may include changes in the second messenger (IP3) release from the PM, Ca^{2+} homeostasis, and ATP level by which astrocytes can sense and respond to the changes in the molecular gradients in neuronal proximity

Reactive astrogliosis and scar formation should complete the neural repair by protecting neurons and regulating the inflammatory CNS responses. Instead, in many cases glial scar formation rather inhibits axonal regeneration. In an attempt to decrease reactive gliosis, conditional Cx43 KO animals have been studied. Reactive astrogliosis should regulate the uptake of glutamate that is toxic for neurons by producing glutathione and reducing oxidative stress and by releasing ATP and adenosine (Chen et al. 2001), and Cx43 expressed in astrocytes is functioning in form of channels and hemichannels upon hypoxic preconditioning (Bosch et al. 2014). In contrast to expected results, astrocytes from Cx43 KO mice show a decreased neuroprotective effect of hypoxic preconditioning. Instead, the KO of another channel molecule, the water-transporting aquaporin AQP4, in astrocytes reduced cytotoxic swelling and improved the outcome after stroke (Zador et al. 2009). A possible role of drebrin was not addressed in these studies.

13.12 Drebrin and Neurologic Disorders

Cellular analyses revealed that small GTPases and their regulators are involved in disruption of cell–cell contacts and in dendritic spine pathology (Lan et al. 2005). Mutations in guanine nucleotide exchange factors (GEFs), whose crucial functions are to exchange GTP for GDP on these GTPases, are known to cause disease states. Taking into account the described above action of bacterial toxins on Rho GTPases and the dependence of drebrin on RhoA, described above, suggests that the disbalance between these proteins may result in human pathologies. Indeed, kalirin, a protein directly interacting with drebrin, is an exchange factor for Rho (i.e. a Rho-GEF) that is modifying dendritic spines in control and disease states (Penzes et al. 2003).

Even more severe diseases are related to the GTPase-activating proteins (GAPs) that accelerate GTP hydrolysis on small GTPases that regulate the actin cytoskeleton. The X-linked mental retardation protein oligophrenin-1 is a Rho-GAP protein and possesses not only GAP activity for small GTPase Rho but also affects Rac and Cdc42 as well. Oligophrenin-1 is found in both neuronal and astroglial cells, and it colocalizes with actin at the leading edges (Fauchereau et al. 2003). Interestingly, even more than actin, such leading edges are found to carry drebrin (as, e.g., seen in Figs. 13.4b, d and 13.5). GTPase activity is important for the spine formation and if blocked would increase unbalanced activity of Rho GTPases resulting in disruption of spines. However, the brain pathology and cellular function of oligophrenin-1 still remain to be elucidated in details.

Decreased levels of drebrin have been shown in the brains of Alzheimer patients (AD) (Harigaya et al. 1996; Shim and Lubec 2002). Analyzed at postmortem pathology levels, drebrin decay was restricted to specific brain regions. The decay of drebrin alters spine morphology and changes numerous functions of axons and dendrites. Drebrin decay disrupts connexin-mediated astrocytic connectivity thus decreasing metabolic support of synaptic connectivity. Drebrin was shown to be involved in both astrocytic action and neuronal experience-dependent plasticity of synaptic transmission, the process that represents the cellular basis of learning. Neurodegeneration and aging reveal morphologic tissue disruption, decreased spine density, loss of pyramidal neurons, and metabolic changes that are believed to be dependent on astroglial connectivity supported by drebrin.

13.13 Drebrin-Mediated Plasticity of Gap Junction Communications in Brain

High-resolution electron microscopy images revealed that astrocytes are often surrounding neuronal terminals. The functional significance of astrocytic gap junctions in a close proximity to synaptic terminals is of great interest (Jeanson et al. 2016;

Genoud et al. 2015). 3-D reconstructions of the hippocampal neurons revealed that astrocytes are covering newly formed spines rather than the larger established ones (Medvedev et al. 2014). It was also shown that astrocytes direct microglial cells to phagocytose weak synapses in the retino-geniculate system, thereby participating in the elimination of "wrong" and "weak" signaling (Stevens et al. 2007). Gap junctions not only connect cell interiors but can also unite cells in syncytium-like clusters that can provide additional help in establishment of common properties such as collective cell polarization and ion-dependent signal propagation. Indeed connexin-mediated astrocytic Ca^{2+} wave propagations spread much more efficiently than that produced by the firing of individual neurons. Astrocytic connectivity is based mainly on connexins Cx43 and Cx30 which provide fast bidirectional communications important for transmitting neurons Ca^{2+}, IP3, cAMP, cGMP, and metabolites. How drebrin activity influences Cx43 channels in different tissues remains unclear, but the downregulation of drebrin in Alzheimer's disease may contribute not only to deregulation of spines but also to a reduction in astrocytic connectivity of both channels and hemichannels. Massive opening of astrocytic hemichannels induced in hippocampal slices of AD mice correlates with the loss of ATP and resulted in cell death (Orellana et al. 2013; Bosch and Kielian 2014). Modulation of connexin hemichannels seems to impair memory in Alzheimer's mice (reviewed by Koulakoff et al. 2012). Except Cx43, drebrin activity may influence other connexins as, e.g., Cx36 that is upregulated in early development and acts before neuronal receptors are expressed (Arumugam et al. 2005).

Live cell reconstruction experiments allow us to hypothesize that highly dynamic drebrin-formed outgrowths of astrocytes may bring Cx43 GJ channels and hemichannels into close proximity with dendrites and synaptic terminals. We hypothesize that this proximity has functional implications that are based on signaling and on metabolic cooperation between the two main types of brain cells. Moreover, we discovered millisecond-fast rearrangements of connexin channel clusters—so-called CDRs—formed inside the GJ plaques in response to stress agents (Majoul et al. 2013). Similar fast mechanisms can be used by the astrocytic gap junctions to sense "hot spots" caused by neuronal terminals. This new aspect allows to suggest that close proximity of connexin channels can dramatically influence neuronal behavior (Alvarez-Maubecin et al. 2000; Malchiodi-Albedi et al. 2012). Based on these discoveries and other hypothetical considerations, we suggest a role of drebrin-connexin interactions important for brain function (Fig. 13.8).

Different types of neurons release specific neurotransmitters such as glutamate, GABA, ACh, and 5-HT to activate common Ca^{2+} and ATP-dependent responses in glial cells. Astrocytic connexin channels and hemichannels can induce and propagate IP3 and Ca^{2+} gradients across the 3-D space of astroglial circuits thereby providing a supplement layer of brain connectivity that is, as yet, less well understood (Bani-Yaghoub et al. 1999; Belliveau et al. 1997, 2006). In this way, astroglial connexins can contribute to dynamic neuro-glial interactions. Although still poorly understood, it seems likely that the drebrin-Cx43 interactions will assume an ever-larger role in our understanding of the conceptually exciting, structure - function relationships between astrocytes and neurons on brain function.

13.14 Perspectives and Future Directions

Drebrin's role in supporting cell–cell contacts in astrocytes is as important as its role affecting neuronal spines or post-synaptic density. The cytoskeletal rearrangements leading to cell–cell contacts are necessary to regulate cell signaling and the metabolic flow of information. Cx43 together with Cx30 are the two main isoforms expressed in astrocytes, and drebrin seems to modify the plasticity of cell–cell contacts to either, increase or limit the flow of information through the astrocytic GJs. The drebrin-dependent mobility of connexin hemichannels in the PM is vital to support neuronal activity by releasing the key cellular energy molecule ATP, where and when it is most needed (Belliveau et al. 2006). Although many questions covering neuronal receptors, gap junctions, membrane receptors, ion channels, effector molecules, and the submembrane cytoskeleton still remain open (Vargas et al. 2008), we believe that future studies of drebrin's role in brain connectivity will uncover new molecular targets for drugs designed to limit the loss of brain function associated with aging, neurodegeneration and Alzheimer's disease, and other cognitive disorders. New discoveries should bring light to these molecules and reveal drugs for treatment of cognitive disorders.

Acknowledgments We are grateful to Michael Berridge for inspiration and novel insights into Ca^{2+}- dependent regulation mechanisms, and to Daniel K. Hartline and James Pawley for critical reading of our MS. We wish to apologize that we could not cite many important original articles due to space limitations.

References

Akhmanova A, Stehbens SJ, Yap AS (2009) Touch, grasp, deliver and control: functional cross-talk between microtubules and cell adhesions. Traffic 10:268–274. doi:10.1111/j.1600-0854.2008.00869.x

Aktories K, Just I (1995) Monoglucosylation of low-molecular mass GTP-binding rho proteins by clostridial cytotoxins. Trends Cell Biol 5:441–443

Alvarez-Maubecin V, Garcia-Hernandez F, Williams JT, Van Bockstaele EJ (2000) Functional coupling between neurons and glia. J Neurosci 20:4091–4098

Ambrosi C, Ren C, Spagnol G, Cavin G, Cone A, Grintsevich EE, Sosinsky GE, Sorgen PL (2016) Connexin43 forms supramolecular complexes through non-overlapping binding sites fro drebrin, tubulin, and ZO-1. PLoS One 11(6):e0157073. doi:10.1371/journal.pone.0157073

Aoki C, Sekino Y, Hanamura K, Fujisawa S, Mahadomrongkul V, Ren Y, Shirao T (2005) Drebrin a is a postsynaptic protein that localizes in vivo to the submembranous surface of dendritic sites forming excitatory synapses. J Comp Neurol 483(4):383–402

Aoki C, Kojima N, Sabaliauskas N, Shah L, Ahmed TH, Oakford J, Ahmed T, Yamazaki H, Hanamura K, Shirao T (2009) Drebrin a knockout eliminates the rapid form of homeostatic synaptic plasticity at excitatory synapses of intact adult cerebral cortex. J Comp Neurol 517:105–121

Arai H, Sato K, Uto A, Yasumoto Y (1991) Effect of transient cerebral ischemia in mongolian gerbils on synaptic vesicle protein (SVP-38) and developmentally regulated brain protein (drebrin). Neurosci Res Commun 9:143–150

Arumugam H, Liu X, Colombo PJ, Corriveau RA, Belousov AB (2005) NMDA receptors regulate developmental gap junction uncoupling via CREB signaling. Nat Neurosci 8:1720–1726
Asada H, Uyemura K, Shirao T (1994) Actin-binding protein, drebrin, accumulates in submembranous regions in parallel with neuronal differentiation. J Neurosci Res 38(2):149–159
Bani-Yaghoub M et al (1999) Gap junction blockage interferes with neuronal and astroglial differentiation of mouse P19 embryonal carcinoma cells. Dev Genet 24:69–81
Belliveau DJ et al (1997) Differential expression of gap junctions in neurons and astrocytes derived from P19 embryonal carcinoma cells. Dev Genet 21:187–200
Belliveau DJ et al (2006) Enhanced neurite outgrowth in PC12 cells mediated by connexin hemichannels and ATP. J Biol Chem 281:20920–20931
Bennett MVL, Zukin RS (2004) Electrical coupling and neuronal synchronization in the mammalian brain. Neuron 41(4):495–511
Benninger RK, Zhang M, Head WS, Satin LS, Piston DW (2008) Gap junction coupling and calcium waves in the pancreatic islet. Biophys J 95:5048–5061
Berridge MJ (2010) Calcium signalling remodelling and disease. Biochem Soc Trans 40:297–309
Berridge MJ (2012) Neural calcium signalling. Neuron 21:13–26
Berridge MJ, Lipp P, Bootman MD (2000) The versatility and universality of calcium signalling. Nat Rev Mol Cell Biol 1(1):11–21
Bianconi E, Piovesan A, Facchin F, Beraudi A, Casadei R, Frabetti F, Vitale L, Pelleri MC, Tassani S, Piva F, Perez-Amodio S, Strippoli P, Canaider S (2013) An estimation of the number of cells in the human body. Ann Hum Biol 40(6):463–471. doi:10.3109/03014460.2013.807878
Bosch M, Kielian T (2014) Hemichannels in neurodegenerative diseases: is there a link to pathology? Front Cell Neurosci 8:242. doi:10.3389/fncel.2014.00242
Butkevich E, Hülsmann S, Wenzel D, Shirao T, Duden R, Majoul I (2004) Drebrin is a novel connexin-43 binding partner that links gap junctions to the submembrane cytoskeleton. Curr Biol 14:650–658
Butkevich E, Bodensiek K, Fakhri N, von Roden K, Schaap IA, Majoul I, Schmidt CF, Klopfenstein DR (2015) Drebrin-like protein DBN-1 is a sarcomere component that stabilizes actin filaments during muscle contraction. Nat Commun 6:7523. doi:10.1038/ncomms8523
Chen Y et al (2001) Astrocytes protect neurons from nitric oxide toxicity by a glutathione-dependent mechanism. J Neurochem 77:1601–1610
Chimura T, Launey T, Yoshida N (2015) Calpain-mediated degradation of Drebrin by Excitotoxicity In vitro and In vivo. PLoS One 10(4):e0125119. doi:10.1371/journal.pone.0125119
Dell'Acqua ML, Smith KE, Gorski JA, Horne EA, Gibson ES, Gomez LL (2006) Regulation of neuronal PKA signaling through AKAP targeting dynamics. Eur J Cell Biol 85(7):627–633
Derangeon M, Bourmeyster N, Plaisance I, Pinet-Charvet C, Chen Q, Duthe F, Popoff MR, Sarrouilhe D, Hervé JC (2008) RhoA GTPase and F-actin dynamically regulate the permeability of Cx43-made channels in rat cardiac myocytes. J Biol Chem 283:30754–30765
Duffy HS, Delmar M, Spray DC (2002) Formation of the gap junction nexus: binding partners for connexins. J Physiol Paris 96:243–249. doi:10.1016/S0928-4257(02)00012-8
Elias LA, Wang DD, Kriegstein AR (2007) Gap junction adhesion is necessary for radial migration in the neocortex. Nature 448(7156):901–907. Epub 2007/08/24. nature06063
Etienne-Manneville S, Hall A (2002) Rho GTPases in cell biology. Nature 420:629–635
Farahani R, Pina-Benabou MH, Kyrozis A, Siddiq A, Barradas PC, Chiu FC, Cavalcante LA, Lai JCK, Stanton PK, Rozental R (2005) Alterations in metabolism and gap junction expression may determine the role of astrocytes as "good samaritans" or executioners. Glia 50:351–361. doi:10.1002/glia.20213
Fauchereau F et al (2003) The RhoGAP activity of OPHN1, a new F-actin-binding protein is negatively controlled by its amino-terminal domain. Mol Cell Neurosci 23:574–586
Genoud C, Houades V, Kraftsik R, Welker E, Giaume C (2015) Proximity of excitatory synapses and astroglial gap junctions in layer IV of the mouse barrel cortex. Neuroscience 291:241–249. doi:10.1016/j.neuroscience.2015.01.051
Giepmans BNG, Moolenaar WH (1998) The gap junction protein connexin43 interacts with the second PDZ domain of the zona occludens-1 protein. Curr Biol 8(16):931–934

Giepmans BNG, Verlaan I, Hengeveld T, Janssen H, Calafat C, Falk MM, Moolenaar WH (2001) Gap junction protein connexin-43 interacts directly with microtubules. Curr Biol 11:1364–1368
Goodenough DA, Goliger JA, Paul DL (1996) Connexins, connexons, and intercellular communication. Annu Rev Biochem 65:475–502
Grinstevich EE, Reisler E (2014) Drebrin inhibits cofilin-induced severing of F-actin. Cytoskeleton 71:472–483
Grintsevich EE, Galkin VE, Orlova A, Ytterberg AJ, Mikati MM, Kudryashov DS, Loo JA, Egelman EH, Reisler E (2010) Mapping of drebrin binding site on F-actin. J Mol Biol 398:542–554
Guthrie SC, Gilula NB (1989) Gap junctional communication and development. Trends Neurosci 12:12–16
Hall A (1998) Rho GTPases and the actin cytoskeleton. Science 279:509–514
Harigaya Y, Shoji M, Shirao T, Hirai S (1996) Disappearance of actin-binding protein, drebrin, from hippocampal synapses in Alzheimer's disease. J Neurosci Res 43(1):87–92
Hayashi K, Ishikawa R, Kawai-Hirai R, Takagi T, Taketomi A, Shirao T (1999) Domain analysis of the actin-binding and actin-remodeling activities of drebrin. Exp Cell Res 253:673–680
Herculano-Houzel S (2009) The human brain in numbers: a linearly scaled-up primate brain. Front Hum Neurosci 3:31. doi:org/10.3389/neuro.09.031.2009
Hogan PG, Rao A (2007) Dissecting ICRAC, a store-operated calcium current. Trends Biochem Sci 32:235–245
Ikeda K, Shirao T, Toda M, Asada H, Toya S, Uyemura K (1995) Effect of a neuron-specific actin-binding protein, drebrin a, on cell-substratum adhesion. Neurosci Lett 194(3):197–200
Ikeda K, Kaub PA, Asada H, Uyemura K, Toya S, Shirao T (1996) Stabilization of adhesion plaques by the expression of drebrin a in fibroblasts. Brain Res Dev Brain Res 91(2):227–236
Ishikawa R, Hayashi K, Shirao T, Xue Y, Takagi T, Sasaki Y, Kohama K (1994) Drebrin, a development-associated brain protein from rat embryo, causes the dissociation of tropomyosin from actin filaments. J Biol Chem 269:29928–29933
Jeanson T, Pondaven A, Ezan P, Mouthon F, Charvériat M, Giaume C (2016) Antidepressants impact Connexin 43 channel functions in astrocytes. Front Cell Neurosci 9:495. doi:10.3389/fncel.2015.00495. eCollection
Jego P, Pacheco-Torres J, Araque A, Canals S (2014) Functional MRI in mice lacking IP3-dependent calcium signaling in astrocytes. J Cereb Blood Flow Metab 34(10):1599–1603
Jin MS, Tanaka Y, Sekino Y, Ren H, Yamazaki R, Kawai-Hirai N, Kojima ST (2002) A novel, brain-specific mouse drebrin: cDNA cloning, chromosomal mapping, genomic structure, expression, and functional characterization. Genomics 79:686–692
Keon BH, Jedrzejewski PT, Paul DL, Goodenough DA (2000) Isoform specific expression of the neuronal F-actin binding protein, drebrin, in specialized cells of stomach and kidney epithelia. J Cell Sci 113(2):325–326
Kins S, Betz H, Kirsch J (2000) Collybistin, a newly identified brain-specific GEF, induces submembrane clustering of gephyrin. Nat Neurosci 3(1):22–29
Kneussel M, Brandstatter JH, Gasnier B, Feng G, Sanes JR, Betz H (2001) Gephyrin-independent clustering of postsynaptic GABAA receptor subtypes. Mol Cell Neurosci 17:973–982
Kojima N, Yasuda H, Hanamura K, Ishizuka Y, Sekino Y, Shirao T (2016) Drebrin a regulates hippocampal LTP and hippocampus-dependent fear learning in adult mice. Neuroscience 324:218–226
Lan Z, Kurata WE, Martyn KD, Jin C, Lau AF (2005) Novel rab GAP-like protein, CIP85, interacts with connexin43 and induces its degradation. Biochemistry 44:2385–2396
Majoul I, Sohn K, Wieland FT, Pepperkok R, Pizza M, Hillemann J, Söling HD (1998) KDEL receptor (Erd2p)-mediated retrograde transport of the cholera toxin a subunit from the Golgi involves COPI, p23, and the COOH terminus of Erd2p. J Cell Biol 143:601–612
Majoul I, Shirao T, Sekino Y, Duden R (2007) Many faces of drebrin: from building dendritic spines and stabilizing gap junctions to shaping neurite-like cell processes. Histochem Cell Biol 127:355–361

Majoul IV, Onichtchouk D, Butkevich E, Wenzel D, Chailakhyan LM, Duden R (2009) Limiting transport steps and novel interactions of Connexin-43 along the secretory pathway. Histochem Cell Biol 132(3):263–280

Majoul IV, Gao L, Betzig E, Onichtchouk D, Butkevich E, Kozlov Y, Bukauskas F, Bennett MLV, Lippincott-Schwartz J, Duden R (2013) Fast structural responses of gap junction membrane domains to AB5 toxins. PNAS 110:E4125–E4133. doi:10.1073/pnas.1315850110

Malchiodi-Albedi F, Paradisi S, Di Nottia M, Simone D, Travaglione S, Falzano L, Fiorentini C (2012) CNF1 improves Astrocytic ability to support neuronal growth and differentiation in vitro. PLoS One 7(4):e34115. doi:org/10.1371/journal.pone.0034115

Marin O, Valiente M, Ge X, Tsai LH (2010) Guiding neuronal cell migrations. Cold Spring Harb Perspect Biol 2:a001834

Matsuzaki M, Honkura N, Ellis-Davies GC, Kasai H (2004) Structural basis of long-term potentiation in single dendritic spines. Nature 429:761–766

Medvedev N, Popov V, Henneberger C, Kraev I, Rusakov DA, Stewart MG (2014) Glia selectively approach synapses on thin dendritic spines. Philos Trans R Soc Lond Ser B Biol Sci 369(1654):20140047. doi:10.1098/rstb.2014.0047

Mercer JC, Qi Q, Mottram LF, Law M, Bruce D, Iyer A, Shirao T, August A (2010) Chemico-genetic identification of Drebrin as a regulator of calcium responses. Intl J Biochem & Cell Biol 42(2):337–345. doi:doi. org/10.1016/j.biocel.2009.11.019

Mizui T, Takahashi H, Sekino Y, Shirao T (2005) Overexpression of drebrin a in immature neurons induces the accumulation of F-actin and PSD-95 into dendritic filopodia, and the formation of large abnormal protrusions. Mol Cell Neurosci 30:149–157

Mizui T, Kojima N, Yamazaki H, Katayama M, Hanamura K, Shirao T (2009) Drebrin E is involved in the regulation of axonal growth through actin-myosin interactions. J Neurochem 109:611–622

Mizui T, Sekino Y, Yamazaki1 YIH, Takahashi H, Kojima N, Kojima M, Shirao T (2014) Myosin II ATPase activity mediates the long-term potentiation-induced exodus of stable F-actin bound by drebrin a from dendritic spines. PLoS One 9(1):e85367

Moore AR, Zhou WL, Sirois CL, Belinsky GS, Zecevic N, Antic SD (2014) Connexin hemichannels contribute to spontaneous electrical activity in the human fetal cortex. PNAS 111(37):E3919–E3928

Orellana JA, von Bernhardi R, Giaume C, Sáez JC (2012) Glial hemichannels and their involvement in aging and neurodegenerative diseases. Rev Neurosci 23:163–177. doi:10.1515/revneuro-2011-0065.

Orellana JA, Martinez AD, Retamal MA (2013) Gap junction channels and hemichannels in the CNS: regulation by signaling molecules. Neuropharmacology 75:567

Pakkenberg B, Gundersen HJG (1988) Total number of neurons and glial cells in human brain nuclei estimated by the disector and the fractionator. J Microsc 150:1–20

Peitsch WK, Grund C, Kuhn C (1999) Drebrin is a widespread actin-associating protein enriched at junctional plaques, defining a specific microfilament anchorage system in polar epithelial cells. Eur J Cell Biol 78:767–778

Peitsch WK, Hofmann I, Prätzel S et al (2001) Drebrin particles: components in the ensemble of proteins regulating actin dynamics of lamellipodia and filopodia. Eur J Cell Biol 80:567–579

Penzes P, Beeser A, Chernoff J, Schiller MR, Eipper BA, Mains RE, Huganir RL (2003) Rapid induction of dendritic spine morphogenesis by trans-synaptic ephrinB-EphB receptor activation of the rho-GEF kalirin. Neuron 37:263–274

Perea G, Sur M, Araque A (2014) Neuron-glia networks: integral gear of brain function. Front Cell Neurosci 8:378

Pannasch U et al (2014) Connexin 30 sets synaptic strength by controlling astroglial synapse invasion. Nat Neurosci 17:549–558

Rao KV et al (2005) Astrocytes protect neurons from ammonia toxicity. Neurochem Res 30:1311–1318

Rebecchi RMJ, Pentyala SN (2000) Structure, function, and control of phosphoinositide-specific phospholipase C. Physiol Rev 80:1291–1335

Ridley A (2000) Rho. In: Hall A (ed) GTPases, vol 24. Oxford University Press, Oxford, pp 89–136

Robel S, Sontheimer H (2016) Glia as drivers of abnormal neuronal activity. Nat Neurosci 19:28–33. doi:10.1038/nn.4184

Sabatini BL, Maravall M, Svoboda K (2001) Ca($^{2+}$) signaling in dendritic spines. Curr Opin Neurobiol 11:349–356

Schmidt G, Sehr P, Wilm M, Selzer J, Mann M, Aktories K (1997) Gln63 of rho is deamidated by *Escherichia coli* cytotoxic necrotizing factor 1. Nature 387:725–729

Schaar BT, McConnell SK (2005) Cytoskeletal coordination during neuronal migration. Proc Natl Acad Sci U S A 102(38):13652–13657

Sekino Y, Tanaka S, Hanamura K, Yamazaki H, Sasagawa Y, Xue Y, Hayashi K, Shirao T (2006) Activation of N-methyl-d-aspartate receptor induces a shift of drebrin distribution: disappearance from dendritic spines and appearance in dendritic shafts. Mol Cell Neurosci 31:493–504

Shim KS, Lubec G (2002) Drebrin, a dendritic spine protein, is manifold decreased in brains of patients with Alzheimer's disease and down syndrome. Neurosci Lett 324:209–212

Shirao T, Obata K (1985) Two acidic proteins associated with brain development in chick embryo. J Neurochem 44:1210–1216

Shirao T, Obata K (1986) Immunochemical homology of 3 developmentally regulated brain proteins and their developmental change in neuronal distribution. Brain Res 394:233–244

Shirao T, Sekino Y (2001) Clustering and anchoring mechanisms of molecular constituents of postsynaptic scaffolds in dendritic spines. Neurosci Res 40:1–7

Shirao T, Kojima N, Kato Y, Obata K (1988) Molecular cloning of a cDNA for the developmentally regulated brain protein, drebrin. Brain Res 464:71–74

Shirao T, Kojima N, Nabeta Y, Obata K (1989) Two forms of drebrins, developmentally regulated brain proteins in rat. Proc Japan Acad 65:169–172

Shirao T, Kojima N, Obata K (1992) Cloning of drebrin a and induction of neurite-like processes in drebrin-transfected cells. Neuroreport 3:109–112

Shirao T, Hayashi K, Ishikawa R, Isa K, Asada H, Ikeda K, Uyemura K (1994) Formation of thick, curving bundles of actin by drebrin a expressed in fibroblasts. Exp Cell Res 215:145–153

Sonego M, Oberoi M, Stoddart J, Gajendra S, Hendricusdottir R, Oozeer F, Lalli G (2015) Drebrin regulates neuroblast migration in the postnatal mammalian brain. PLoS ONE 10(5):e0126478. doi:doi. org/10.1371/journal.pone.0126478

Song M, Kojima N, Hanamura K, Sekino Y, Inoue HK, Mikuni M et al (2008) Expression of drebrin E in migrating neuroblasts in adult rat brain: coincidence between drebrin E disappearance from cell body and cessation of migration. Neuroscience 152(3):670–682. doi:10.1016/j.neuroscience.2007.10.068

Stevens B, Allen N, Vazquez LE, Howell GR, Christopherson KS, Nouri N, et al (2007) The Classical Complement Cascade Mediates CNS Synapse Elimination. Cell 131 (6):1164–1178

Stout C, Goodenough D, Paul D (2004) Connexins: functions without junctions. Curr Opin Cell Biol 16:507–512

Suh HN, Kim MO, Han HJ (2012) Laminin-111 stimulates proliferation of mouse embryonic stem cells through a reduction of gap Junctional intercellular communication via RhoA-mediated Cx43 phosphorylation and dissociation of Cx43/ZO-1/Drebrin complex. Stem Cells Dev 21(11):2058–2070

Swanson RA et al (2004) Astrocyte influences on ischemic neuronal death. Curr Mol Med 4:193–205

Takahashi H, Sekino S, Tanaka S, Mizui T, Kishi S, Shirao T (2003) Drebrin-dependent actin clustering in dendritic filopodia governs synaptic targeting of postsynaptic density-95 and dendritic spine morphogenesis. J Neurosci 23:6586–6595

Takahashi H, Mizui T, Shirao T (2006) Down-regulation of drebrin a expression suppresses synaptic targeting of NMDA receptors in developing hippocampal neurons. J Neurochem 97:110–115

Toda M, Shirao T, Uyemura K (1999) Suppression of an actin-binding protein, drebrin, by antisense transfection attenuates neurite outgrowth in neuroblastoma B104 cells. Dev Brain Res 114(2):193–200

Toyofuku T, Yabuki M, Otsu M, Kuzuya K, Hori M, Tada M (1998) Direct association of the gap junction protein connexin-43 with ZO+-1 in cardiac myocytes. J Biol Chem 273(21):12725–12731

Toyofuku T, Akamatsu Y, Zhang H, Kuzuya T, Tada M, Hori M (2001) C-Src regulates the interaction between connexin-43 and ZO-1 in cardiac myocytes. J Biol Chem 276(3):1780–1788

Uehata M, Ishizaki T, Satoh H, Ono T, Kawahara T, Morishita T, Tamakawa H, Yamagami K, Inui J, Maekawa M, Narumiya S (1997) Calcium sensitization of smooth muscle mediated by a rho-associated protein kinase in hypertension. Nature 389:990–994

Ullian EM, Sapperstein SK, Christopherson, Barres BA (2001) Control of Synapse Number by Glia. Science 291(5504):657–661

Vargas MR et al (2008) Nrf2 activation in astrocytes protects against neurodegeneration in mouse models of familial amyotrophic lateral sclerosis. J Neurosci 28:13574–13581

Walter P, Ron D (2011) The unfolded protein response: from stress pathway to homeostatic regulation. Science 334(6059):1081–1086

Zador Z et al (2009) Role of aquaporin-4 in cerebral edema and stroke. Handb Exp Pharmacol 190:159–170

Zhang Q, Harris AL, Abagyan R, Yeager M (2016) An electrostatic mechanism for Ca^{2+}-mediated regulation of gap junction channels. Nat Commun 7:8770. doi:10.1038/ncomms9770

Chapter 14
Homer, Spikar, and Other Drebrin-Binding Proteins in the Brain

Hiroyuki Yamazaki and Tomoaki Shirao

Abstract Drebrin is a major F-actin-binding protein in the brain. In the past two decades, many drebrin-binding proteins in addition to F-actin have been identified in several research fields including neuroscience, oncology, and immunology. Among the drebrin-binding proteins, there are various kinds of proteins including scaffold proteins, nuclear proteins, phosphatases, microtubule-binding proteins, G-actin-binding proteins, gap junction proteins, chemokine receptors, and cell-adhesion-related proteins. The interaction between drebrin and its binding partners seems to play important roles in higher brain functions, because drebrin is involved in the pathogenesis of some neurological diseases with cognitive defects. In this chapter, we will first review the interaction of Homer and spikar with drebrin, particularly focusing on spine morphogenesis and synaptic function. Homer contributes to spine morphogenesis by cooperating with shank and activated Cdc42 small GTPase, suggesting a novel signaling pathway comprising Homer, drebrin, shank, and Cdc42 for spine morphogenesis. Drebrin sequesters spikar in the cytoplasm and stabilizes it in dendritic spines, leading to spine formation. Finally, we will introduce some other drebrin-binding proteins including end-binding protein 3 (EB3), profilin, progranulin, and phosphatase and tensin homologue (PTEN). These proteins are involved in Alzheimer's disease and cancer. Therefore, further studies on drebrin and its binding proteins will be of great importance to elucidate the pathologies of various diseases and may contribute to their medical treatment and diagnostics development.

Keywords Homer • Spikar/ZMYND8 • EB3 • Profilin • Progranulin • PTEN • Ras • Dendritic spine • Drebrin-binding protein

---

H. Yamazaki (✉) • T. Shirao
Department of Neurobiology and Behavior, Gunma University Graduate School of Medicine, Maebashi 371-8511, Japan
e-mail: spikar@gunma-u.ac.jp

T. Shirao, Y. Sekino (eds.), *Drebrin*, Advances in Experimental Medicine and Biology 1006, DOI 10.1007/978-4-431-56550-5_14

14.1 Introduction

Drebrin is a typical F-actin-binding protein in the dendritic spines in the brain of vertebrates. Additionally, flies and nematodes express a drebrin-like protein, although they contain a characteristic SH3 domain in the C-terminal region (Butkevich et al. 2015). They may be considered partly homologues to drebrin and mouse actin-binding protein 1 (mAbp1/SH3P7) (Lappalainen et al. 1998; Yamazaki et al. 2001). Therefore, the interaction between drebrin and its binding partners seems critical for higher brain functions such as learning and memory, which are observed in vertebrates. We elucidated various drebrin characters as an F-actin-binding protein between the 1980s and the 1990s (Shirao and Obata 1985; Ishikawa et al. 1994); however, there have only been a few reports about drebrin-binding proteins, even after the discovery that drebrin is a component of the actin cytoskeleton. In 1996, we found myosin and gelsolin in the drebrin-bound actin complex using co-immunoprecipitation analysis with anti-drebrin antibody (mAb M2F6) (Hayashi et al. 1996). Soon after, Mammoto et al. have found that the G-actin-binding protein, profilin, binds to drebrin (Mammoto et al. 1998). In the twenty-first century, the identification reports of novel drebrin-binding proteins including Homer and spikar rapidly increased (Fig. 14.1). Homer is involved in various

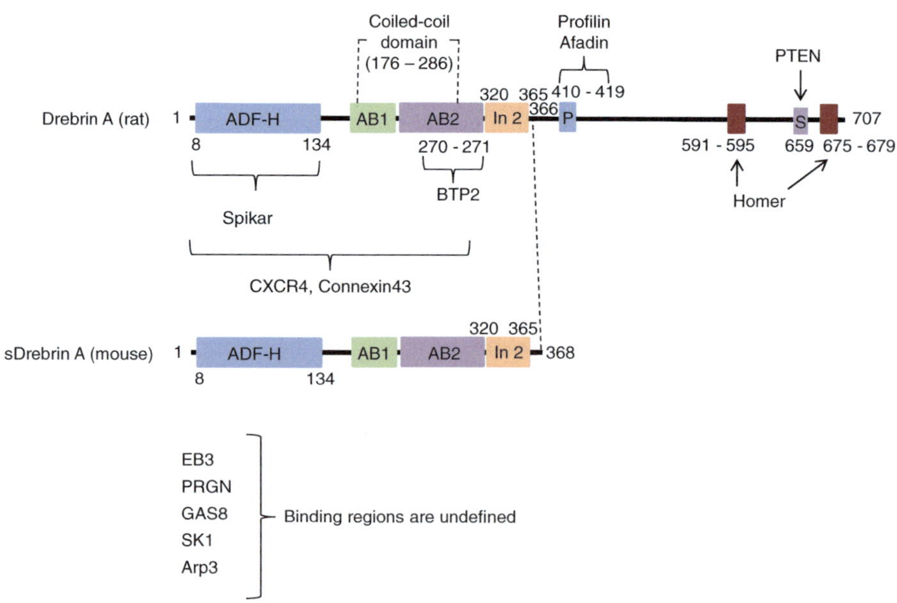

Fig. 14.1 Schematic representation of drebrin and the binding sites of drebrin-binding proteins. Spikar binds to the ADF-H domain in the N-terminal region. Homer binds to the conserved sequence, PPXXF, in the C-terminal region. Profilin and afadin bind to the proline-rich region. CXCR4 binds to the N-terminal region (1–271). PTEN regulates the phosphorylation of Ser 659 of rat drebrin A (Ser 647 in the human homologue) by activity-dependent phosphatase activity. The binding sites on drebrin of EB3, connexin 43, progranulin, GAS8, SK1, and Arp3 are unknown. *ADF-H* ADF homology domain, *AB*1 Actin-binding region 1, *AB*2 Actin-binding region 2, *In*2 Ins 2 (drebrin A-specific sequence), *P* Proline-rich region

brain functions including synaptic functions, neuronal development, neurological diseases, and behavior (for review, see Szumlinski et al. 2006; Foa and Gasperini 2009; Pouliquin and Dulhunty 2009; Luo et al. 2012). Spikar is a transcription cofactor that is also involved in dendritic spine formation. We identified spikar as a drebrin-binding protein (Yamazaki et al. 2014), which was registered in the DDBJ/EMBL/GenBank databases in 2001. In this chapter, we first focus on Homer and spikar and introduce their function in synapse formation. In the latter half, we introduce other drebrin-binding proteins reported in the past two decades and discuss the relationship and implications of their interactions regarding brain functions and disorders.

14.2 Homer

Homer protein family members are encoded by three genes (*Homer1*, *2* and *3*), and many splicing variants have been identified (Shiraishi-Yamaguchi and Furuichi 2007). The subtypes are also called Vesl, PSD-Zip45, and Cupidin. Each Homer gene expresses long- and short-type Homers; the long types are Homer1b–d, Homer2a and b, and Homer3a and b, and the short types are Homer1a (Vesl-1s), Homer2c and d, and Homer3c and d. Short-type Homers lack self-assembly region, which contains coiled-coil domain and leucine zipper motif. Homer1a is a short-type protein that was originally identified as an immediate early gene induced by seizures in rat hippocampus (Kato et al. 1997; Brakeman et al. 1997). Homer1a protein was also isolated as a metabotropic glutamate receptor (mGluR)-binding protein during the same period (Brakeman et al. 1997). Since then, several binding partners of Homer have been identified. It has been elucidated that the Ena/VASP homology 1 (EVH1) domain of Homer recognizes a proline-rich consensus sequence (PPXXF or FPPPP) in several proteins including mGluR, Shank, inositol trisphosphate receptor (IP3R), transient receptor potential canonical (TRPC) channels, dynamin3, and drebrin (Xiao et al. 1998; Tu et al. 1999, 1998; Yuan et al. 2003; Gray et al. 2003; Shiraishi et al. 2003). Homer forms a mesh-like matrix structure with shank, which causes spine enlargement and spine recruitment of several proteins including IP3R, postsynaptic density (PSD)-95, F-actin, guanylate kinase-associated protein (GKAP), and glutamate receptors (Sala et al. 2001; Hayashi et al. 2009).

The drebrin E and drebrin A isoforms, both of which have two PPXXF motifs in their C-terminal region, have been shown to bind to Homer2b (Shiraishi-Yamaguchi et al. 2009). Double mutation of the PPXXF motifs causes loss of the Homer2b-binding activity of drebrin, although it is not clear which motif is crucial for the drebrin-Homer2b binding (Shiraishi-Yamaguchi et al. 2009). The long subtype Homer proteins form an intermolecular disulfide bond between cysteine residues of C-terminal coiled-coil domain by oxidation, resulting in dimerization (Nepliouev et al. 2011). Disruption of the disulfide bond with a reducing reagent such as dithiothreitol (DTT) attenuates the binding of Homer and drebrin (Nepliouev et al. 2011). Another assay using Homer1a showed that either of the drebrin mutants, F543A or F621A, could disrupt the interaction between drebrin and Homer1a (Nepliouev et al. 2011), indicating that both of the PPXXF motifs of drebrin bind to Homer.

Thus, it has been suggested that drebrin can bind to the long subtype Homer dimers more efficiently than to the short subtype Homer. Because the EVH1 domains of Homer are highly conserved between Homer family proteins, drebrin is supposed to be able to bind to Homer3 as well. Conversely, the brain-specific short-type drebrin isoform, s-drebrin A, contains actin-binding regions but no PPXXF motif (Jin et al. 2002), suggesting that s-drebrin A regulates the actin cytoskeleton in the spine without being implicated in a Homer-protein complex (Fig. 14.1).

What is the physiological role of the interactions between drebrin and Homer? Although Homer can bind to several PSD proteins, it localizes not only at the PSD but also in the cytoplasmic area of dendritic spines (Shiraishi et al. 2004), where it seems to interact with IP3R on the endoplasmic reticulum (ER). Drebrin is concentrated in the cytoplasmic area of dendritic spines rather than in the PSD area (Kobayashi et al. 2007). Thus, drebrin may interact with Homer in the center of the spines. Drebrin attenuates the interaction between actin and myosin V (Ishikawa et al. 2007), and myosin Va, concentrated at the tip of the ER tubule, plays a pivotal role in the ER targeting. Therefore, it has been suggested that drebrin is involved in the localization of the Homer-IP3R-ER complex at the center of dendritic spines (Fig. 14.2).

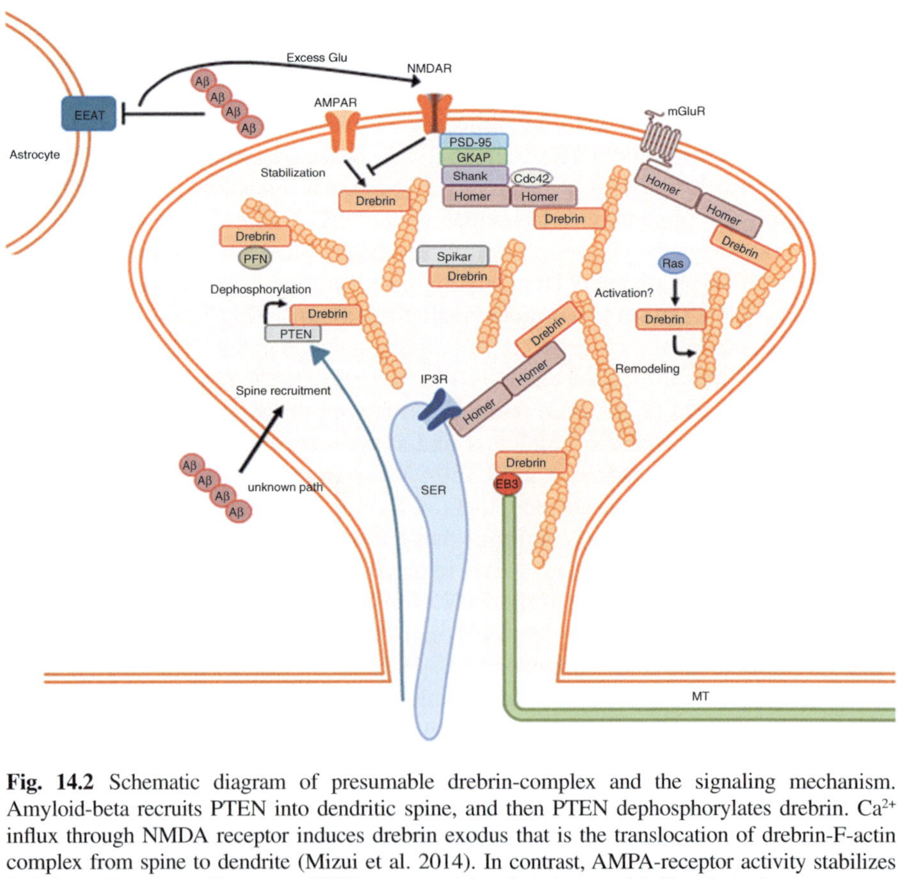

Fig. 14.2 Schematic diagram of presumable drebrin-complex and the signaling mechanism. Amyloid-beta recruits PTEN into dendritic spine, and then PTEN dephosphorylates drebrin. Ca^{2+} influx through NMDA receptor induces drebrin exodus that is the translocation of drebrin-F-actin complex from spine to dendrite (Mizui et al. 2014). In contrast, AMPA-receptor activity stabilizes drebrin within spines. Biou et al. (2008) suggested that drebrin remodels F-actin in the downstream of Ras signaling. Drebrin-EB3 complex plays a role as an adaptor between F-actin and microtubules

Drebrin is known to be involved in spine morphogenesis and formation (Takahashi et al. 2003; Ivanov et al. 2009). Homer2 interacts with activated Cdc42, and a Homer2 mutant lacking the Cdc42-binding region impaired synapse formation and function (Shiraishi-Yamaguchi et al. 2009). Cdc42 and its regulators such as guanine nucleotide exchange factors (GEFs) play important roles in spine morphogenesis and formation (Moon and Zheng 2003; Chen et al. 2012). The Cdc42-binding region of Homer is separate from the EVH1 domain, which may enable drebrin association with the Homer-Cdc42 complex. These data raise the possibility of a novel signaling pathway for spine morphogenesis involving drebrin and activated small G proteins.

14.3 Spikar/ZMYND8

Spikar (also called KIAA1125, CTCL, tumor antigen Se14-3, ZMYND8, Prkcbp1, or RACK7) was first isolated as a drebrin-binding protein by a yeast two-hybrid screening of a rat brain library (Yamazaki et al. 2014). Previously, this gene was reported as a protein kinase C-binding protein (Fossey et al. 2000). Many splice variants of spikar are found in various database sites including NCBI. We have classified spikar into three major isoforms: spikar A, spikar B, and spikar delta-C (Fig. 14.3). Spikar A is similar to spikar B but lacks 20 residues in the N-terminal and 28 residues in the C-terminal region. Spikar delta-C lacks the C-terminal region of spikar A/B and contains a specific sequence in its C-terminal. There are no reports about the function of spikar delta-C so far.

Spikar contains a PHD domain, a bromodomain, a PWWP domain, a MYND domain, and a nuclear-receptor recognition motif (LSYLL) (Fig. 14.3). The PHD

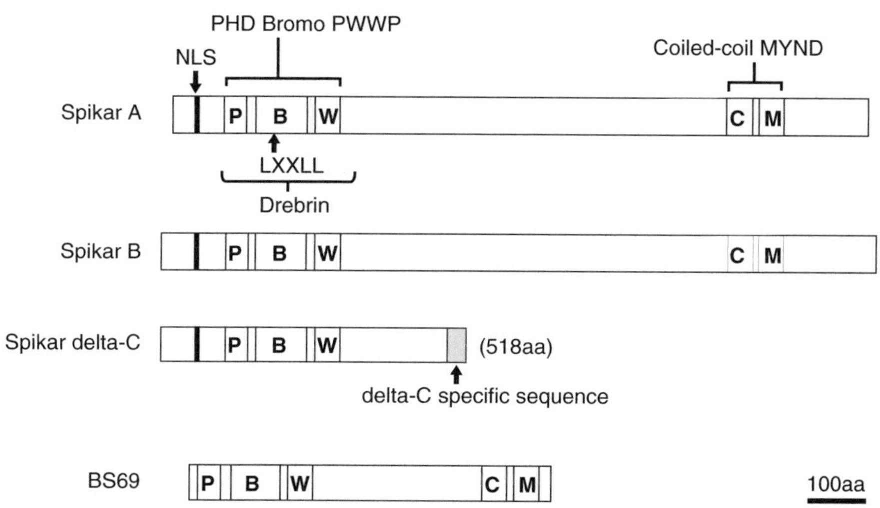

Fig. 14.3 Schematic diagram of spikar isoforms and BS69. *P* PHD domain, *Bro* Bromodomain, *W* PWWP domain, *C* Coiled-coil, *M* MYND domain

domain consists of two zinc fingers, and many PHD-containing proteins associate with chromatin and regulate its activities (Bienz 2006; Baker et al. 2008). The bromodomain associates with acetylated lysine residues found in the C-terminal region of histones (Sanchez and Zhou 2009). Bromodomain-containing proteins are associated with obesity, inflammation, cancer, and neurological disorders (Padmanabhan et al. 2016; Sanchez and Zhou 2009; Belkina and Denis 2012) and have become new therapeutic targets for these diseases. The nuclear-receptor recognition motif of spikar in the bromodomain binds to nuclear receptors including thyroid hormones and steroid receptors (Savkur and Burris 2004). The PWWP domain is named after a conserved Pro-Trp-Trp-Pro motif; however, the first three amino acids are not strictly conserved (Stec et al. 2000; Qin and Min 2014). The PWWP motif of spikar is Pro-Phe-Trp-Pro. The PWWP domain has a DNA- and histone-binding ability (Qiu et al. 2002; Wu et al. 2011). Like the PHD domain, the MYND domain contains a zinc finger motif, which can fold two zinc atoms (Lutterbach et al. 1998). The most studied MYND domain-containing proteins are the SMYD family proteins, which consist of SMYD1–5 and function as a lysine methyltransferase (Spellmon et al. 2015). Taken together, spikar is likely a nuclear protein.

BS69 is likely to be a spikar family protein, because it contains a PHD domain, a bromodomain, a PWWP domain, and a MYND domain in the same order within the molecule. BS69 is an adenovirus E1A-binding protein that interacts with the nuclear-receptor corepressor, N-CoR, and functions as a transcription corepressor (Masselink and Bernards 2000). However, we found that spikar functions as a coactivator of thyroid hormone receptor, glucocorticoid receptor, and estrogen receptor (Yamazaki et al. 2014). Interestingly, it has recently been reported that spikar acts as a transcription corepressor of histone H3 lysine 4 (H3K4) demethylase (*JARID1D*) (Li et al. 2016). Moreover, spikar is associated with the nucleosome remodeling and deacetylase (NuRD) complex in the nucleus, where it plays a role as a DNA damage response (DDR) factor, recruiting the NuRD complex to damaged chromatin, to repress transcription and promote double strand break repair by homologous recombination (Kloet et al. 2015; Gong et al. 2015; Adhikary et al. 2016).

The drebrin-binding region of spikar was assessed by a yeast two-hybrid assay and a GST pull-down assay, which showed that the N-terminal region of spikar interacts with the ADF-H domain of drebrin (Yamazaki et al. 2014). The ADF-H domain is found in all drebrin isoforms. In addition, the drebrin-binding region of spikar is found in spikar A, spikar B, and spikar delta-C. In the drebrin-binding region of spikar, there are nucleosome-associating domains, suggesting that drebrin may interfere with the association between spikar and the nucleosome.

Spikar is ubiquitously expressed in various tissues and concentrated in the cellular nuclei, although the expression level is higher in the spleen and thymus (Yamazaki et al. 2014). Spikar contains a nuclear localization signal (NLS, KKKKK) in the N-terminal region and is localized in the nucleus, but a spikar mutant (mNLS-spikar) with a mutated NLS (KTTKK) does not localize in the nucleus. In the rat brain, spikar is highly expressed in the cerebellum, although it is expressed throughout the brain (Yamazaki et al. 2014). In cultured astrocytes, however, spikar was hardly detected by western blotting. However, immunocytochemical studies and observation of GFP-spikar localization indicated that spikar is concentrated in the nucleus in HEK293 cells and cultured neurons (Yamazaki et al. 2014). Interestingly, in neurons, spikar is found in the cytoplasm in addition to the nucleus. Particularly, spikar is localized in dendritic spines, where drebrin is concentrated (Yamazaki et al. 2014). Furthermore, spikar tends to localize in drebrin-rich spines (Yamazaki et al. 2014). These findings suggest that spikar is localized in dendritic spines in a drebrin-dependent manner. In contrast, the localization of drebrin in dendritic spines was unaffected in spikar-knockdown neurons (Yamazaki et al. 2014). In addition to the postsynaptic localization, our subcellular fractionation data suggest that spikar is localized in presynapses associated with synaptic vesicles (Fig. 14.4).

Knockdown of spikar using shRNA resulted in decreased spine density without alteration of spine morphology (Yamazaki et al. 2014). We conducted rescue experiments using wild-type spikar and mNLS-spikar, which showed that not only wild-type spikar but also mNLS-spikar were able to recover the spine density (Yamazaki et al. 2014). Importantly, the fact that cytoplasmic spikar could rescue the spine density indicates that spikar plays multifunctional roles in both the nucleus and cytoplasm. In HEK293 cells, GFP-spikar was localized in the nucleus and rarely observed in the cytoplasm. In contrast, in HEK293 cells overexpressing drebrin, GFP-spikar tended to localize in the cytoplasm in addition to the nucleus (Yamazaki et al. 2014). Moreover, in drebrin-knockdown neurons, the spine localization of spikar was weak (Yamazaki et al. 2014). These observations indicate that the cytoplasmic localization of spikar, in particular the spine localization in neurons, depends on drebrin. Live-cell imaging of spikar-knockdown cultured neurons showed that spikar plays a role in spine maintenance and de novo spine sprouting (Yamazaki et al. 2014). Conversely, overexpression of mNLS-spikar increased the spine and filopodium number. Furthermore, drebrin knockdown abolished the effect of the mNLS-spikar overexpression. These data suggest that the localization and the function of spikar depend on drebrin (Yamazaki et al. 2014). Because drebrin accumulates in filopodia from an early stage of neuronal development (Takahashi et al. 2003), we propose that drebrin sequesters and stabilizes spikar in dendritic spines and contributes to spine formation and maintenance (Fig. 14.4). This idea is consistent with the fact that the spine localization of drebrin does not depend on spikar.

Several studies have shown a relationship between spikar and cancer. For example, spikar expression is upregulated in squamous cell cancer and cervical intraepithelial neoplasia (Bierkens et al. 2013). In colorectal cancer, a high mutation frequency was found in the spikar gene (Park et al. 2002). Additionally, spikar plays a role in promoting tumor angiogenesis (Kuroyanagi et al. 2014). Because drebrin is highly expressed in some cancer cells and is involved in tumor formation and invasion (Vaskova et al. 2011; Terakawa et al. 2013; Lin et al. 2014; Mizutani et al. 2014; Xu et al. 2015), the interaction between drebrin and spikar may be critical for cancer growth. Investigation of the role of the interaction between drebrin and spikar in cancer cells may provide new insight into the mechanism of cancer migration and growth.

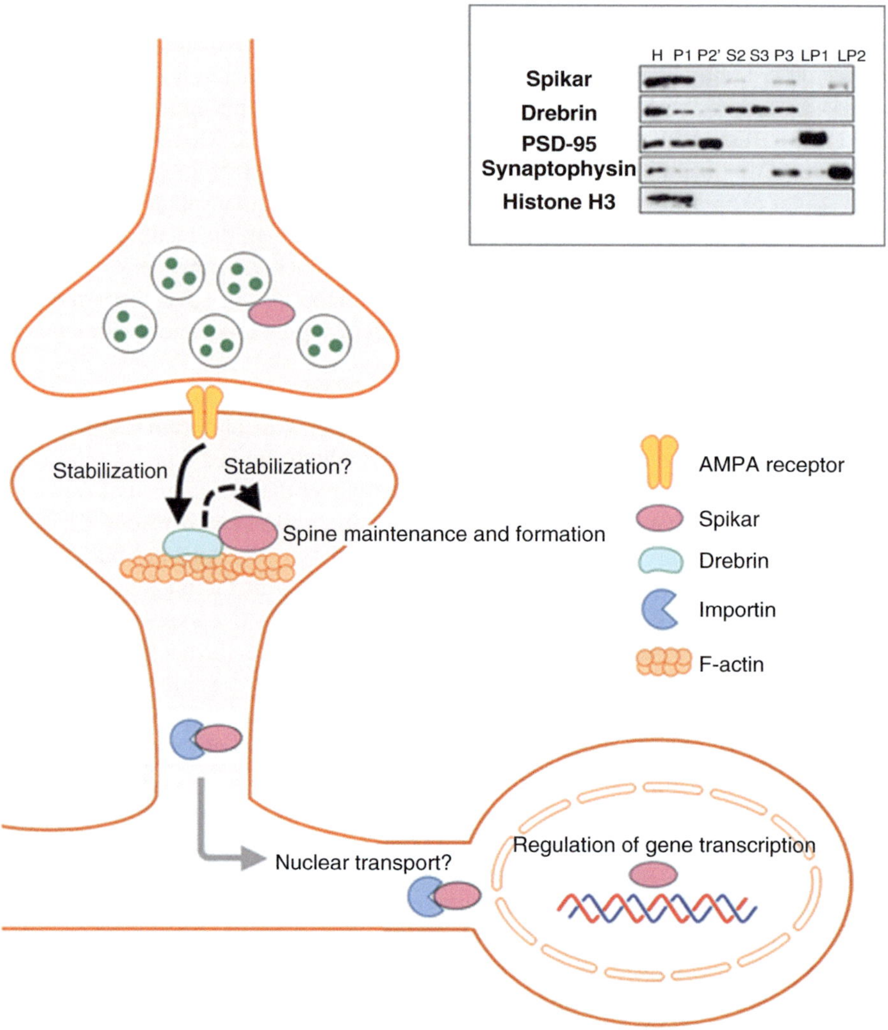

Fig. 14.4 Schematic diagram illustrating the anchoring of spikar in dendritic spines and its function in the nucleus. AMPA-receptor activity stabilizes drebrin in dendritic spines. Drebrin sequesters spikar in dendritic spines, and then spikar accumulates in drebrin-rich spines. Spikar function depends on drebrin. mNLS-spikar concentrates in dendritic spines, suggesting that spikar transports into the nucleus from dendritic spines with the importin complex. In the nucleus, spikar is involved in the regulation of gene transcription. Inset immunoblot image shows the subcellular distribution of spikar. The protein extract (20 μg protein) from each fraction was analyzed by western blotting with antibodies against spikar, drebrin, PSD-95, synaptophysin, and histone H3. The fractions were as follows: *H* homogenate, *P*1 cell nuclei and debris, *P*2 synaptosomal fraction, *S*2 non-synaptosomal fraction, *S*3 cytosolic fraction, *P*3 microsomal fraction, *LP*1 synaptosomal plasma-membrane fraction, *LP*2 synaptic vesicle fraction

14.4 EB3

EB3 was isolated as a molecule that binds to the C-terminal region of APCL, which is a brain-specific homologue of the tumor suppressor gene APC, by yeast two-hybrid screening. The name is based on its significant homology to the microtubule-binding protein, EB1 (Su et al. 1995). EB1 and EB3 are expressed ubiquitously in many tissues, although EB3 is particularly highly expressed in the brain and skeletal muscle (Nakagawa et al. 2000). EB3 and EB1 bind to the plus-tip of microtubules and regulate its dynamics (Straube and Merdes 2007). Drebrin was found in a protein complex that was pulled down with GST-EB3 from the growth cone cytosol but not with GST-EB1 (Geraldo et al. 2008). This indicates that drebrin specifically binds to EB3 but not to EB1. The binding site for drebrin is in the middle region of EB3, between the microtubule-biding region and the coiled-coil region. This middle region is absent from the EB1 sequence (Geraldo et al. 2008). In growth cones, drebrin interacts with EB3 at the tips of microtubules in the proximal region of the filopodia. Drebrin knockdown or overexpression of dominant negative EB3 causes an inhibition of neurite elongation. These findings indicate that the drebrin-EB3 interaction is critical for growth cone mobility by regulating the association between actin-filaments and microtubules.

Recently, we and others have shown that CDK5 phosphorylates drebrin at Ser 142 (pan drebrin) and Ser 342 (drebrin A) (Tanabe et al. 2014; Worth et al. 2013). Gordon-Weeks and colleagues proposed a model in which the phosphorylation at Ser 142 causes a conformational change of drebrin, consequently exposing the coiled-coil domain and EB3 binding region to the outside of the molecule (see Fig. 4.3 in Chap. 4) (Worth et al. 2013). It has been suggested that phosphorylated drebrin can more easily interact with EB3 than nonphosphorylated drebrin. Additionally, the phosphorylation-induced conformational change of drebrin enables bundling of two discrete actin-filaments. However, at present, this model needs to be verified by more studies.

While it has previously been believed that microtubules are not present in dendritic spines, imaging technology using fluorescent proteins and the RNAi technique have revealed that dynamic microtubules can transiently protrude into dendritic spines and regulate their development (Gu et al. 2008). The protrusion of microtubules into spines depends on Ca^{2+} influx through synaptic NMDA receptors (Merriam et al. 2013). Conversely, excessive Ca^{2+} influx induced by 50 μM glutamate removes EB3 from the growing microtubule plus-ends in dendrites (Kapitein et al. 2011). In addition, we have reported that the Ca^{2+} influx through NMDA receptors induced drebrin exodus from dendritic spines (Mizui et al. 2014). Therefore, drebrin may function as an adaptor that links the EB3 complex of microtubule plus-ends to the actin cytoskeleton in dendritic spines and enables a transient protrusion of microtubules into dendritic spines (Fig. 14.2). It is consistent with a recent report that drebrin knockdown decreased spine invasion of microtubules (Merriam et al. 2013).

The interaction of EB3 and drebrin has also been suggested in non-neuronal cells. Drebrin E is localized in the apical domain of columnar epithelial cells (Bazellières et al. 2012). Drebrin knockdown influences the apicobasal elongation during cell development, a phenotype that is observed in EB3-knockdown cells (Bazellières et al. 2012). The drebrin-EB3 complex regulates the cooperation between the actomyosin apical network and the apico-lateral microtubule network. Disruption of this interaction results in destabilization of the actin-based terminal web.

14.5 Profilin

Profilin is a classical actin-binding protein, which was isolated as an inhibitor molecule of actin polymerization in 1977 (Carlsson et al. 1977). Profilin sequesters G-actin, which results in inhibition of actin polymerization (Carlsson et al. 1977; Mockrin and Korn 1980). However, profilin also has an actin polymerization function (Tilney et al. 1983; Pring et al. 1992; Pantaloni and Carlier 1993). The interaction between the profilin I isoform and drebrin was first identified by affinity column chromatography with GST-profilin from rat brain cytosol (Mammoto et al. 1998) and was confirmed by a membrane overlay assay. Because profilin I can be purified with poly-proline agarose beads (Tanaka and Shibata 1985), it is likely to bind to the proline-rich region of drebrin (Fig. 14.1). Profilin II is specifically expressed in the brain (Honore et al. 1993), although both profilin I and profilin II are found in the brain. The interaction between profilin II and drebrin has not been reported, yet; however, drebrin is expected to bind to profilin II, because profilin II binds to several ligands (poly-proline, actin, phosphoinositides) that profilin I binds to (Lambrechts et al. 2000). Profilin II accumulates in dendritic spines after induction of long-term potentiation (LTP) in vitro and in vivo (Ackermann and Matus 2003). In addition, it has been indicated that emotional excitation such as fear conditioning drives profilin into dendritic spines in rat amygdala (Lamprecht et al. 2006). Similarly, drebrin immunoreactivity increased in the middle molecular layer of the hippocampus concomitantly with enhancement in F-actin content after in vivo LTP induction (Fukazawa et al. 2003). These observations suggest that the interaction between drebrin and profilin is enhanced after LTP induction.

14.6 Progranulin

Progranulin, also known as granulin-epithelin precursor, proepithelin, acrogranin, and GP88/PC-cell-derived growth factor, is a ubiquitously expressed, secreted protein throughout the body, which was originally identified as a growth factor-like molecule (Shoyab et al. 1990; Anakwe and Gerton 1990; Plowman et al. 1992;

Bhandari et al. 1993; Zanocco-Marani et al. 1999; Zhu et al. 2002; Bateman and Bennett 2009). Progranulin is a multifunctional protein that is involved in several types of physiological phenomena including inflammation, cell proliferation, wound healing, development, and tumorigenesis (Konopka et al. 2014; Bateman and Bennett 2009). The interaction between progranulin and drebrin has recently been discovered in the course of cancer research. Drebrin was found in the eluted proteins of a pull-down assay with recombinant progranulin and protein extracts of 5637 bladder cancer cells (Xu et al. 2015). Drebrin modulated the progranulin-induced actin cytoskeleton remodeling and motility of cancer cells. Drebrin-depletion in UMUC-3 bladder cancer cells inhibited tumor formation in implanted mice in vivo (Xu et al. 2015).

The best-known important characteristic of progranulin in neurology is its implication in frontotemporal dementia (FTD) and Alzheimer's disease (AD). Mutations in the *progranulin* gene are found in familial FTD patients (Baker et al. 2006), and the mutations are risk factors for AD (Carecchio et al. 2009; Perry et al. 2013). Progranulin knockdown in cultured neurons results in a decrease in dendritic spine density and an increase in filopodium-like protrusion (Tapia et al. 2011). Interestingly, drebrin is decreased in the brain of patients with Alzheimer's disease (Harigaya et al. 1996) and Down syndrome (Shim and Lubec 2002). In addition, the loss of drebrin function by an antisense oligonucleotide treatment caused a spine density decrease (Takahashi et al. 2006). Therefore, drebrin may be involved in a progranulin-related signal transduction pathway for spine formation and development.

14.7 PTEN

Phosphatase and tensin homologue (PTEN), which is a tumor suppressor, functions as a phosphatase for phosphatidylinositol (3,4,5)-trisphosphate (PIP3) and a protein phosphatase (Song et al. 2012). Drebrin was found in PTEN-immunoprecipitates from rat brain and liver by mass spectrometry analysis (Kreis et al. 2013). PTEN dephosphorylates drebrin at Ser 647 (Ser 659 in rat drebrin A) in an activity-dependent manner (Kreis et al. 2013). The Ser residue is conserved among chicken, human, rabbit, mouse, rat, and xenopus (Kreis et al. 2013). It has recently been reported that PTEN is essential for Alzheimer's disease pathology (Knafo et al. 2016). As mentioned above, drebrin is decreased in Alzheimer's disease. Furthermore, amyloid-β oligomer induced PTEN recruitment into dendritic spines (Knafo et al. 2016) and drebrin decreased in dendritic spines (Lacor et al. 2007; Ishizuka et al. 2014). Moreover, PTEN is known to be essential for NMDA-receptor-dependent long-term depression (Arendt et al. 2014). It has been reported that over-activation of NMDA receptor induces calpain-mediated proteolysis of drebrin (Chimura et al. 2015). Taken together, the dephosphorylation of drebrin at Ser 659 by PTEN may be involved in the pathology of Alzheimer's disease, possibly through drebrin degradation.

14.8 Ras

Ras is a small G protein family, which consists of three major members: H-Ras, N-Ras, and K-Ras (Ellis et al. 1981; Shimizu et al. 1983). They were first identified as oncogenes, and its signal transduction pathways including extracellular and intracellular signals have been studied in detail (Malumbres and Barbacid 2003). In addition to tumor formation driven by Ras mutation, Ras proteins are engaged on several vital phenomena such as cell proliferation and differentiation (Feramisco et al. 1984; Hagag et al. 1986). In synapses, Ras is involved in neurotransmitter release and synaptic plasticity such as LTP and LTD (Brambilla et al. 1997; Jovanovic et al. 2000; Komiyama et al. 2002; Kim et al. 2003; Hou and Klann 2004; Schenk et al. 2005; Qin et al. 2005; Li et al. 2006; Rumbaugh et al. 2006; Banko et al. 2006). The activity of Ras is regulated by exchanging GTP binding (active form) or GDP binding (inactive form), and this conversion is mediated by their upstream regulators GAP/GEF (Shih et al. 1980; Calés et al. 1988; Adari et al. 1988; West et al. 1990; Crechet et al. 1990; Huang et al. 1990). Among a number of downstream signaling pathway of Ras, MAPK and PI3K cascades are assumed to be associated with synaptic function such as memory formation (Atkins et al. 1998; Blum et al. 1999; Chen et al. 2005; Peineau et al. 2007; Kim et al. 2011; Choi et al. 2014).

It is not clear whether drebrin forms a protein complex with Ras. However, Biou et al. (2008) have reported that Ras modulates drebrin function in dendritic spines. The overexpression of constitutively active Ras resulted in a similar phenotype as the drebrin overexpression in cultured neurons, and the effect on dendritic spine morphology was dependent on drebrin (Biou et al. 2008). Additionally, dominant negative Ras abolished the phenotype of drebrin overexpression (Biou et al. 2008). The downstream effectors of Ras play significant roles in spine growth and morphogenesis. For instance, PI3K regulates the activity of Rho family small G proteins, which are closely associated with dendritic spine morphogenesis (Sjolander et al. 1991; Rodriguez-Viciana et al. 1994; Nakayama et al. 2000; Penzes et al. 2001; Tashiro and Yuste 2008). Moreover, constitutively active Ras binds to Rac GEF Tiam1, and the interaction is essential for Rac1 activation (Yamauchi et al. 2005). Taken together, these findings suggest that Ras activation is involved in drebrin-mediated dendritic spine morphogenesis.

14.9 Other Drebrin-Binding Proteins

In addition to the drebrin-binding proteins described above, there are various other studies about drebrin-related proteins, which form protein-complexes with drebrin (Fig. 14.1). Rufy3, GAS8, SK1, Arp3, and myosin IIB were found in immunoprecipitated complexes with drebrin (Wei et al. 2014; Zhao et al. 2009; Yagoub et al. 2014; Li et al. 2011; Cheng et al. 2000); however, so far it is unknown whether drebrin directly binds to these proteins. Nonetheless, it is known that drebrin binds to connexin 43 (Butkevich et al. 2004), which is a well-known component of gap junctions in glial cells (Rash et al. 2000), and stabilizes gap junctions to the sub-membrane actin cytoskeleton (Butkevich et al. 2004). Afadin is a cell-adhesion-related protein, which binds to the poly-proline region of drebrin with its PR1-2 region (Rehm et al. 2013). CXCR4, a chemokine receptor, binds to the N-terminal region (1–271) of drebrin (Perez-Martinez et al. 2010). In T lymphocytes, drebrin recruits CXCR4 to the immune synapses upon super-antigen stimulation (Perez-Martinez et al. 2010). The relationships between drebrin and connexin 43, afadin, and CXCR4 are reviewed in detail in other chapters of this book.

14.10 Conclusion

In this chapter, we reviewed drebrin-binding proteins with regard to neuronal functions. Identification of novel drebrin-binding proteins has increased across various disciplines in recent years, although the implications of their association with drebrin have not been fully determined. In some cases, such as connexin 43, drebrin acts like a stabilizer for binding partners on the actin cytoskeleton, and its knockdown leads to alteration of their subcellular localization. Based on these observations, drebrin may also function as a scaffold protein tethering its binding partners to the actin cytoskeleton in the cytoplasmic area of dendritic spines. We suppose that this interaction is regulated by synaptic activity. Because drebrin is involved in the mechanism of neurological diseases, further studies on the drebrin interactome will be of great importance to determining the pathologies of various diseases and may contribute to their medical treatment and diagnostics development.

References

Ackermann M, Matus A (2003) Activity-induced targeting of profilin and stabilization of dendritic spine morphology. Nat Neurosci *6*:1194–1200

Adari H, Lowy DR, Willumsen BM, Der CJ, McCormick F (1988) Guanosine triphosphatase activating protein (GAP) interacts with the p21 ras effector binding domain. Science 240:518–521

Adhikary S, Sanyal S, Basu M, Sengupta I, Sen S, Srivastava DK, Roy S, Das C (2016) Selective recognition of H3.1K36 dimethylation/H4K16 acetylation facilitates the regulation of all-trans-retinoic acid (ATRA)-responsive genes by putative chromatin reader ZMYND8. J Biol Chem *291*:2664–2681

Anakwe OO, Gerton GL (1990) Acrosome biogenesis begins during meiosis: evidence from the synthesis and distribution of an acrosomal glycoprotein, acrogranin, during guinea pig spermatogenesis. Biol Reprod 42:317–328

Arendt KL, Benoist M, Lario A, Draffin JE, Munoz M, Esteban JA (2014) PTEN counteracts PIP3 upregulation in spines during NMDA-receptor-dependent long-term depression. J Cell Sci 127:5253–5260

Atkins CM, Selcher JC, Petraitis JJ, Trzaskos JM, Sweatt JD (1998) The MAPK cascade is required for mammalian associative learning. Nat Neurosci 1:602–609

Baker M et al (2006) Mutations in progranulin cause tau-negative frontotemporal dementia linked to chromosome 17. Nature 442:916–919

Baker LA, Allis CD, Wang GG (2008) PHD fingers in human diseases: disorders arising from misinterpreting epigenetic marks. Mutat Res 647:3–12

Banko JL, Hou L, Poulin F, Sonenberg N, Klann E (2006) Regulation of eukaryotic initiation factor 4E by converging signaling pathways during metabotropic glutamate receptor-dependent long-term depression. J Neurosci 26:2167–2173

Bateman A, Bennett HP (2009) The granulin gene family: from cancer to dementia. Bioessays 31:1245–1254

Bazellières E, Massey-Harroche D, Barthelemy-Requin M, Richard F, Arsanto JP, Le Bivic A (2012) Apico-basal elongation requires a drebrin-E-EB3 complex in columnar human epithelial cells. J Cell Sci 125:919–931

Belkina AC, Denis GV (2012) BET domain co-regulators in obesity, inflammation and cancer. Nat Rev Cancer 12:465–477

Bhandari V, Giaid A, Bateman A (1993) The complementary deoxyribonucleic acid sequence, tissue distribution, and cellular localization of the rat granulin precursor. Endocrinology 133:2682–2689

Bienz M (2006) The PHD finger, a nuclear protein-interaction domain. Trends Biochem Sci 31:35–40

Bierkens M, Krijgsman O, Wilting SM, Bosch L, Jaspers A, Meijer GA, Meijer CJ, Snijders PJ, Ylstra B, Steenbergen RD (2013) Focal aberrations indicate EYA2 and hsa-miR-375 as oncogene and tumor suppressor in cervical carcinogenesis. Genes Chromosomes Cancer 52:56–68

Biou V, Brinkhaus H, Malenka RC, Matus A (2008) Interactions between drebrin and Ras regulate dendritic spine plasticity. Eur J Neurosci 27(11):2847–2859

Blum S, Moore AN, Adams F, Dash PK (1999) A mitogen-activated protein kinase cascade in the CA1/CA2 subfield of the dorsal hippocampus is essential for long-term spatial memory. J Neurosci 19:3535–3544

Brakeman PR, Lanahan AA, O'Brien R, Roche K, Barnes CA, Huganir RL, Worley PF (1997) Homer: a protein that selectively binds metabotropic glutamate receptors. Nature 386:284–288

Brambilla R, Gnesutta N, Minichiello L, White G, Roylance AJ, Herron CE, Ramsey M, Wolfer DP, Cestari V, Rossi-Arnaud C, Grant SG, Chapman PF, Lipp HP, Sturani E, Klein R (1997) A role for the Ras signalling pathway in synaptic transmission and long-term memory. Nature 390:281–286

Butkevich E, Hulsmann S, Wenzel D, Shirao T, Duden R, Majoul I (2004) Drebrin is a novel connexin-43 binding partner that links gap junctions to the submembrane cytoskeleton. Curr Biol 14:650–658

Butkevich E, Bodensiek K, Fakhri N, von Roden K, Schaap IA, Majoul I, Schmidt CF, Klopfenstein DR (2015) Drebrin-like protein DBN-1 is a sarcomere component that stabilizes actin filaments during muscle contraction. Nat Commun 6:7523

Calés C, Hancock JF, Marshall CJ, Hall A (1988) The cytoplasmic protein GAP is implicated as the target for regulation by the ras gene product. Nature 332:548–551

Carecchio M, Fenoglio C, De Riz M, Guidi I, Comi C, Cortini F, Venturelli E, Restelli I, Cantoni C, Bresolin N, Monaco F, Scarpini E, Galimberti D (2009) Progranulin plasma levels as potential biomarker for the identification of GRN deletion carriers. A case with atypical onset as clinical amnestic mild cognitive impairment converted to Alzheimer's disease. J Neurol Sci 287:291–293

Carlsson L, Nystrom LE, Sundkvist I, Markey F, Lindberg U (1977) Actin polymerizability is influenced by profilin, a low molecular weight protein in non-muscle cells. J Mol Biol 115:465–483

Chen X, Garelick MG, Wang H, Lil V, Athos J, Storm DR (2005) PI3 kinase signaling is required for retrieval and extinction of contextual memory. Nat Neurosci 8:925–931

Chen C, Wirth A, Ponimaskin E (2012) Cdc42: an important regulator of neuronal morphology. Int J Biochem Cell Biol 44:447–451

Cheng XT, Hayashi K, Shirao T (2000) Non-muscle myosin IIB-like immunoreactivity is present at the drebrin-binding cytoskeleton in neurons. Neurosci Res 36:167–173

Chimura T, Launey T, Yoshida N (2015) Calpain-mediated degradation of drebrin by excitotoxicity in vitro and in vivo. PLoS One 10:e0125119

Choi JH, Park P, Baek GC, Sim SE, Kang SJ, Lee Y, Ahn SH, Lim CS, Lee YS, Collingridge GL, Kaang BK (2014) Effects of PI3Kgamma overexpression in the hippocampus on synaptic plasticity and spatial learning. Mol Brain 7:78

Crechet JB, Poullet P, Mistou MY, Parmeggiani A, Camonis J, Boy-Marcotte E, Damak F, Jacquet M (1990) Enhancement of the GDP-GTP exchange of RAS proteins by the carboxyl-terminal domain of SCD25. Science 248:866–868

Ellis RW, Defeo D, Shih TY, Gonda MA, Young HA, Tsuchida N, Lowy DR, Scolnick EM (1981) The p21 src genes of Harvey and Kirsten sarcoma viruses originate from divergent members of a family of normal vertebrate genes. Nature 292:506–511

Feramisco JR, Gross M, Kamata T, Rosenberg M, Sweet RW (1984) Microinjection of the oncogene form of the human H-ras (T-24) protein results in rapid proliferation of quiescent cells. Cell 38:109–117

Foa L, Gasperini R (2009) Developmental roles for Homer: more than just a pretty scaffold. J Neurochem 108:1–10

Fossey SC, Kuroda S, Price JA, Pendleton JK, Freedman BI, Bowden DW (2000) Identification and characterization of PRKCBP1, a candidate RACK-like protein. Mamm Genome 11:919–925

Fukazawa Y, Saitoh Y, Ozawa F, Ohta Y, Mizuno K, Inokuchi K (2003) Hippocampal LTP is accompanied by enhanced F-actin content within the dendritic spine that is essential for late LTP maintenance in vivo. Neuron 38:447–460

Geraldo S, Khanzada UK, Parsons M, Chilton JK, Gordon-Weeks PR (2008) Targeting of the F-actin-binding protein drebrin by the microtubule plus-tip protein EB3 is required for neuritogenesis. Nat Cell Biol 10:1181–1189

Gong F, Chiu LY, Cox B, Aymard F, Clouaire T, Leung JW, Cammarata M, Perez M, Agarwal P, Brodbelt JS, Legube G, Miller KM (2015) Screen identifies bromodomain protein ZMYND8 in chromatin recognition of transcription-associated DNA damage that promotes homologous recombination. Genes Dev 29:197–211

Gray NW, Fourgeaud L, Huang B, Chen J, Cao H, Oswald BJ, Hemar A, McNiven MA (2003) Dynamin 3 is a component of the postsynapse, where it interacts with mGluR5 and Homer. Curr Biol 13:510–515

Gu J, Firestein BL, Zheng JQ (2008) Microtubules in dendritic spine development. J Neurosci 28:12120–12124

Hagag N, Halegoua S, Viola M (1986) Inhibition of growth factor-induced differentiation of PC12 cells by microinjection of antibody to ras p21. Nature 319:680–682

Harigaya Y, Shoji M, Shirao T, Hirai S (1996) Disappearance of actin-binding protein, drebrin, from hippocampal synapses in Alzheimer's disease. J Neurosci Res 43:87–92

Hayashi K, Ishikawa R, Ye LH, He XL, Takata K, Kohama K, Shirao T (1996) Modulatory role of drebrin on the cytoskeleton within dendritic spines in the rat cerebral cortex. J Neurosci 16:7161–7170

Hayashi MK, Tang C, Verpelli C, Narayanan R, Stearns MH, Xu RM, Li H, Sala C, Hayashi Y (2009) The postsynaptic density proteins Homer and Shank form a polymeric network structure. Cell 137:159–171

Honore B, Madsen P, Andersen AH, Leffers H (1993) Cloning and expression of a novel human profilin variant, profilin II. FEBS Lett 330:151–155

Hou L, Klann E (2004) Activation of the phosphoinositide 3-kinase-Akt-mammalian target of rapamycin signaling pathway is required for metabotropic glutamate receptor-dependent long-term depression. J Neurosci 24:6352–6361

Huang YK, Kung HF, Kamata T (1990) Purification of a factor capable of stimulating the guanine nucleotide exchange reaction of ras proteins and its effect on ras-related small molecular mass G proteins. Proc Natl Acad Sci U S A 87:8008–8012

Ishikawa R, Hayashi K, Shirao T, Xue Y, Takagi T, Sasaki Y, Kohama K (1994) Drebrin, a development-associated brain protein from rat embryo, causes the dissociation of tropomyosin from actin filaments. J Biol Chem 269:29928–29933

Ishikawa R, Katoh K, Takahashi A, Xie C, Oseki K, Watanabe M, Igarashi M, Nakamura A, Kohama K (2007) Drebrin attenuates the interaction between actin and myosin-V. Biochem Biophys Res Commun 359:398–401

Ishizuka Y, Shimizu H, Takagi E, Kato M, Yamagata H, Mikuni M, Shirao T (2014) Histone deacetylase mediates the decrease in drebrin cluster density induced by amyloid beta oligomers. Neurochem Int 76:114–121

Ivanov A, Esclapez M, Pellegrino C, Shirao T, Ferhat L (2009) Drebrin A regulates dendritic spine plasticity and synaptic function in mature cultured hippocampal neurons. J Cell Sci 122:524–534

Jin M, Tanaka S, Sekino Y, Ren Y, Yamazaki H, Kawai-Hirai R, Kojima N, Shirao T (2002) A novel, brain-specific mouse drebrin: cDNA cloning, chromosomal mapping, genomic structure, expression, and functional characterization. Genomics 79:686–692

Jovanovic JN, Czernik AJ, Fienberg AA, Greengard P, Sihra TS (2000) Synapsins as mediators of BDNF-enhanced neurotransmitter release. Nat Neurosci 3:323–329

Kapitein LC, Yau KW, Gouveia SM, van der Zwan WA, Wulf PS, Keijzer N, Demmers J, Jaworski J, Akhmanova A, Hoogenraad CC (2011) NMDA receptor activation suppresses microtubule growth and spine entry. J Neurosci 31(22):8194–8209

Kato A, Ozawa F, Saitoh Y, Hirai K, Inokuchi K (1997) Vesl, a gene encoding VASP/Ena family related protein, is upregulated during seizure, long-term potentiation and synaptogenesis. FEBS Lett 412:183–189

Kim JH, Lee HK, Takamiya K, Huganir RL (2003) The role of synaptic GTPase-activating protein in neuronal development and synaptic plasticity. J Neurosci 23:1119–1124

Kim JI et al (2011) PI3Kgamma is required for NMDA receptor-dependent long-term depression and behavioral flexibility. Nat Neurosci 14:1447–1454

Kloet SL, Baymaz HI, Makowski M, Groenewold V, Jansen PW, Berendsen M, Niazi H, Kops GJ, Vermeulen M (2015) Towards elucidating the stability, dynamics and architecture of the nucleosome remodeling and deacetylase complex by using quantitative interaction proteomics. FEBS J 282:1774–1785

Knafo S et al (2016) PTEN recruitment controls synaptic and cognitive function in Alzheimer's models. Nat Neurosci 19:443–453

Kobayashi C, Aoki C, Kojima N, Yamazaki H, Shirao T (2007) Drebrin a content correlates with spine head size in the adult mouse cerebral cortex. J Comp Neurol 503:618–626

Komiyama NH, Watabe AM, Carlisle HJ, Porter K, Charlesworth P, Monti J, Strathdee DJ, O'Carroll CM, Martin SJ, Morris RG, O'Dell TJ, Grant SG (2002) SynGAP regulates ERK/MAPK signaling, synaptic plasticity, and learning in the complex with postsynaptic density 95 and NMDA receptor. J Neurosci 22:9721–9732

Konopka J, Richbourgh B, Liu C (2014) The role of PGRN in musculoskeletal development and disease. Front Biosci (Landmark Ed) 19:662–671

Kreis P, Hendricusdottir R, Kay L, Papageorgiou IE, van Diepen M, Mack T, Ryves J, Harwood A, Leslie NR, Kann O, Parsons M, Eickholt BJ (2013) Phosphorylation of the actin binding protein Drebrin at S647 is regulated by neuronal activity and PTEN. PLoS One 8:e71957

Kuroyanagi J, Shimada Y, Zhang B, Ariyoshi M, Umemoto N, Nishimura Y, Tanaka T (2014) Zinc finger MYND-type containing 8 promotes tumour angiogenesis via induction of vascular endothelial growth factor-A expression. FEBS Lett 588:3409–3416

Lacor PN, Buniel MC, Furlow PW, Clemente AS, Velasco PT, Wood M, Viola KL, Klein WL (2007) Abeta oligomer-induced aberrations in synapse composition, shape, and density provide a molecular basis for loss of connectivity in Alzheimer's disease. J Neurosci 27:796–807

Lambrechts A, Braun A, Jonckheere V, Aszodi A, Lanier LM, Robbens J, Van Colen I, Vandekerckhove J, Fassler R, Ampe C (2000) Profilin II is alternatively spliced, resulting in profilin isoforms that are differentially expressed and have distinct biochemical properties. Mol Cell Biol 20:8209–8219

Lamprecht R, Farb CR, Rodrigues SM, LeDoux JE (2006) Fear conditioning drives profilin into amygdala dendritic spines. Nat Neurosci 9:481–483

Lappalainen P, Kessels MM, Cope MJ, Drubin DG (1998) The ADF homology (ADF-H) domain: a highly exploited actin-binding module. Mol Biol Cell 9:1951–1959

Li S, Tian X, Hartley DM, Feig LA (2006) Distinct roles for Ras-guanine nucleotide-releasing factor 1 (Ras-GRF1) and Ras-GRF2 in the induction of long-term potentiation and long-term depression. J Neurosci 26:1721–1729

Li MW, Xiao X, Mruk DD, Lam YL, Lee WM, Lui WY, Bonanomi M, Silvestrini B, Cheng CY (2011) Actin-binding protein drebrin E is involved in junction dynamics during spermatogenesis. Spermatogenesis 1:123–136

Li N, Li Y, Lv J, Zheng X, Wen H, Shen H, Zhu G, Chen TY, Dhar SS, Kan PY, Wang Z, Shiekhattar R, Shi X, Lan F, Chen K, Li W, Li H, Lee MG (2016) ZMYND8 reads the dual histone mark H3K4me1-H3K14ac to antagonize the expression of metastasis-linked genes. Mol Cell 63:470–484

Lin Q, Tan HT, Lim TK, Khoo A, Lim KH, Chung MC (2014) iTRAQ analysis of colorectal cancer cell lines suggests Drebrin (DBN1) is overexpressed during liver metastasis. Proteomics 14:1434–1443

Luo P, Li X, Fei Z, Poon W (2012) Scaffold protein Homer 1: implications for neurological diseases. Neurochem Int 61:731–738

Lutterbach B, Sun D, Schuetz J, Hiebert SW (1998) The MYND motif is required for repression of basal transcription from the multidrug resistance 1 promoter by the t(8;21) fusion protein. Mol Cell Biol 18:3604–3611

Malumbres M, Barbacid M (2003) RAS oncogenes: the first 30 years. Nat Rev Cancer 3:459–465

Mammoto A, Sasaki T, Asakura T, Hotta I, Imamura H, Takahashi K, Matsuura Y, Shirao T, Takai Y (1998) Interactions of drebrin and gephyrin with profilin. Biochem Biophys Res Commun 243:86–89

Masselink H, Bernards R (2000) The adenovirus E1A binding protein BS69 is a corepressor of transcription through recruitment of N-CoR. Oncogene 19:1538–1546

Merriam EB, Millette M, Lumbard DC, Saengsawang W, Fothergill T, Hu X, Ferhat L, Dent EW (2013) Synaptic regulation of microtubule dynamics in dendritic spines by calcium, F-actin, and drebrin. J Neurosci 33:16471–16482

Mizui T, Sekino Y, Yamazaki H, Ishizuka Y, Takahashi H, Kojima N, Kojima M, Shirao T, Sato M (2014) Myosin II ATPase activity mediates the long-term potentiation-induced exodus of stable F-actin bound by drebrin A from dendritic spines. PLoS One 9(1), e85367

Mizutani Y, Iwamoto I, Kanoh H, Seishima M, Nagata K (2014) Expression of drebrin, an actin binding protein, in basal cell carcinoma, trichoblastoma and trichoepithelioma. Histol Histopathol 29:757–766

Mockrin SC, Korn ED (1980) Acanthamoeba profilin interacts with G-actin to increase the rate of exchange of actin-bound adenosine 5′-triphosphate. Biochemistry 19:5359–5362

Moon SY, Zheng Y (2003) Rho GTPase-activating proteins in cell regulation. Trends Cell Biol 13:13–22

Nakagawa H, Koyama K, Murata Y, Morito M, Akiyama T, Nakamura Y (2000) EB3, a novel member of the EB1 family preferentially expressed in the central nervous system, binds to a CNS-specific APC homologue. Oncogene 19:210–216

Nakayama AY, Harms MB, Luo L (2000) Small GTPases Rac and Rho in the maintenance of dendritic spines and branches in hippocampal pyramidal neurons. J Neurosci 20:5329–5338

Nepliouev I, Zhang ZS, Stiber JA (2011) Effect of oxidative stress on homer scaffolding proteins. PLoS One 6:e26128

Padmanabhan B, Mathur S, Manjula R, Tripathi S (2016) Bromodomain and extra-terminal (BET) family proteins: new therapeutic targets in major diseases. J Biosci 41:295–311

Pantaloni D, Carlier MF (1993) How profilin promotes actin filament assembly in the presence of thymosin beta 4. Cell 75:1007–1014

Park J, Betel D, Gryfe R, Michalickova K, Di Nicola N, Gallinger S, Hogue CW, Redston M (2002) Mutation profiling of mismatch repair-deficient colorectal cancers using an in silico genome scan to identify coding microsatellites. Cancer Res 62:1284–1288

Peineau S, Taghibiglou C, Bradley C, Wong TP, Liu L, Lu J, Lo E, Wu D, Saule E, Bouschet T, Matthews P, Isaac JT, Bortolotto ZA, Wang YT, Collingridge GL (2007) LTP inhibits LTD in the hippocampus via regulation of GSK3beta. Neuron 53:703–717

Penzes P, Johnson RC, Sattler R, Zhang X, Huganir RL, Kambampati V, Mains RE, Eipper BA (2001) The neuronal Rho-GEF Kalirin-7 interacts with PDZ domain-containing proteins and regulates dendritic morphogenesis. Neuron 29:229–242

Perez-Martinez M, Gordon-Alonso M, Cabrero JR, Barrero-Villar M, Rey M, Mittelbrunn M, Lamana A, Morlino G, Calabia C, Yamazaki H, Shirao T, Vazquez J, Gonzalez-Amaro R, Veiga E, Sanchez-Madrid F (2010) F-actin-binding protein drebrin regulates CXCR4 recruitment to the immune synapse. J Cell Sci 123:1160–1170

Perry DC, Lehmann M, Yokoyama JS, Karydas A, Lee JJ, Coppola G, Grinberg LT, Geschwind D, Seeley WW, Miller BL, Rosen H, Rabinovici G (2013) Progranulin mutations as risk factors for Alzheimer disease. JAMA Neurol 70:774–778

Plowman GD, Green JM, Neubauer MG, Buckley SD, McDonald VL, Todaro GJ, Shoyab M (1992) The epithelin precursor encodes two proteins with opposing activities on epithelial cell growth. J Biol Chem 267:13073–13078

Pouliquin P, Dulhunty AF (2009) Homer and the ryanodine receptor. Eur Biophys J 39:91–102

Pring M, Weber A, Bubb MR (1992) Profilin-actin complexes directly elongate actin filaments at the barbed end. Biochemistry 31:1827–1836

Qin Y, Zhu Y, Baumgart JP, Stornetta RL, Seidenman K, Mack V, van Aelst L, Julius Zhu J (2005) State-dependent Ras signaling and AMPA receptor trafficking. Genes Dev 19(17):2000–2015

Qin S, Min J (2014) Structure and function of the nucleosome-binding PWWP domain. Trends Biochem Sci 39:536–547

Qiu C, Sawada K, Zhang X, Cheng X (2002) The PWWP domain of mammalian DNA methyltransferase Dnmt3b defines a new family of DNA-binding folds. Nat Struct Biol 9:217–224

Rash JE, Staines WA, Yasumura T, Patel D, Furman CS, Stelmack GL, Nagy JI (2000) Immunogold evidence that neuronal gap junctions in adult rat brain and spinal cord contain connexin-36 but not connexin-32 or connexin-43. Proc Natl Acad Sci U S A 97:7573–7578

Rehm K, Panzer L, van Vliet V, Genot E, Linder S (2013) Drebrin preserves endothelial integrity by stabilizing nectin at adherens junctions. J Cell Sci 126:3756–3769

Rodriguez-Viciana P, Warne PH, Dhand R, Vanhaesebroeck B, Gout I, Fry MJ, Waterfield MD, Downward J (1994) Phosphatidylinositol-3-OH kinase as a direct target of Ras. Nature 370:527–532

Rumbaugh G, Adams JP, Kim JH, Huganir RL (2006) SynGAP regulates synaptic strength and mitogen-activated protein kinases in cultured neurons. Proc Natl Acad Sci U S A 103:4344–4351

Sala C, Piech V, Wilson NR, Passafaro M, Liu G, Sheng M (2001) Regulation of dendritic spine morphology and synaptic function by Shank and Homer. Neuron 31:115–130

Sanchez R, Zhou MM (2009) The role of human bromodomains in chromatin biology and gene transcription. Curr Opin Drug Discov Devel 12:659–665

Savkur RS, Burris TP (2004) The coactivator LXXLL nuclear receptor recognition motif. J Pept Res 63:207–212

Schenk U, Menna E, Kim T, Passafaro M, Chang S, De Camilli P, Matteoli M (2005) A novel pathway for presynaptic mitogen-activated kinase activation via AMPA receptors. J Neurosci 25:1654–1663

Shih TY, Papageorge AG, Stokes PE, Weeks MO, Scolnick EM (1980) Guanine nucleotide-binding and autophosphorylating activities associated with the p21src protein of Harvey murine sarcoma virus. Nature 287:686–691

Shim KS, Lubec G (2002) Drebrin, a dendritic spine protein, is manifold decreased in brains of patients with Alzheimer's disease and Down syndrome. Neurosci Lett 324:209–212

Shimizu K, Goldfarb M, Perucho M, Wigler M (1983) Isolation and preliminary characterization of the transforming gene of a human neuroblastoma cell line. Proc Natl Acad Sci U S A 80:383–387

Shiraishi Y, Mizutani A, Mikoshiba K, Furuichi T (2003) Coincidence in dendritic clustering and synaptic targeting of homer proteins and NMDA receptor complex proteins NR2B and PSD95 during development of cultured hippocampal neurons. Mol Cell Neurosci 22:188–201

Shiraishi Y, Mizutani A, Yuasa S, Mikoshiba K, Furuichi T (2004) Differential expression of Homer family proteins in the developing mouse brain. J Comp Neurol 473:582–599

Shiraishi-Yamaguchi Y, Furuichi T (2007) The Homer family proteins. Genome Biol 8:206
Shiraishi-Yamaguchi Y, Sato Y, Sakai R, Mizutani A, Knopfel T, Mori N, Mikoshiba K, Furuichi T (2009) Interaction of Cupidin/Homer2 with two actin cytoskeletal regulators, Cdc42 small GTPase and Drebrin, in dendritic spines. BMC Neurosci 10:25
Shirao T, Obata K (1985) Two acidic proteins associated with brain development in chick embryo. J Neurochem 44:1210–1216
Shoyab M, McDonald VL, Byles C, Todaro GJ, Plowman GD (1990) Epithelins 1 and 2: isolation and characterization of two cysteine-rich growth-modulating proteins. Proc Natl Acad Sci U S A 87:7912–7916
Sjolander A, Yamamoto K, Huber BE, Lapetina EG (1991) Association of p21ras with phosphatidylinositol 3-kinase. Proc Natl Acad Sci U S A 88:7908–7912
Song MS, Salmena L, Pandolfi PP (2012) The functions and regulation of the PTEN tumour suppressor. Nat Rev Mol Cell Biol 13:283–296
Spellmon N, Holcomb J, Trescott L, Sirinupong N, Yang Z (2015) Structure and function of SET and MYND domain-containing proteins. Int J Mol Sci 16:1406–1428
Stec I, Nagl SB, van Ommen GJ, den Dunnen JT (2000) The PWWP domain: a potential protein-protein interaction domain in nuclear proteins influencing differentiation? FEBS Lett 473:1–5
Straube A, Merdes A (2007) EB3 regulates microtubule dynamics at the cell cortex and is required for myoblast elongation and fusion. Curr Biol 17:1318–1325
Su LK, Burrell M, Hill DE, Gyuris J, Brent R, Wiltshire R, Trent J, Vogelstein B, Kinzler KW (1995) APC binds to the novel protein EB1. Cancer Res 55:2972–2977
Szumlinski KK, Kalivas PW, Worley PF (2006) Homer proteins: implications for neuropsychiatric disorders. Curr Opin Neurobiol 16:251–257
Takahashi H, Sekino Y, Tanaka S, Mizui T, Kishi S, Shirao T (2003) Drebrin-dependent actin clustering in dendritic filopodia governs synaptic targeting of postsynaptic density-95 and dendritic spine morphogenesis. J Neurosci 23:6586–6595
Takahashi H, Mizui T, Shirao T (2006) Down-regulation of drebrin A expression suppresses synaptic targeting of NMDA receptors in developing hippocampal neurones. J Neurochem 97(Suppl 1):110–115
Tanabe K, Yamazaki H, Inaguma Y, Asada A, Kimura T, Takahashi J, Taoka M, Ohshima T, Furuichi T, Isobe T, Nagata K, Shirao T, Hisanaga S (2014) Phosphorylation of drebrin by cyclin-dependent kinase 5 and its role in neuronal migration. PLoS One 9:e92291
Tanaka M, Shibata H (1985) Poly(L-proline)-binding proteins from chick embryos are a profilin and a profilactin. Eur J Biochem 151:291–297
Tapia L, Milnerwood A, Guo A, Mills F, Yoshida E, Vasuta C, Mackenzie IR, Raymond L, Cynader M, Jia W, Bamji SX (2011) Progranulin deficiency decreases gross neural connectivity but enhances transmission at individual synapses. J Neurosci 31:11126–11132
Tashiro A, Yuste R (2008) Role of Rho GTPases in the morphogenesis and motility of dendritic spines. Methods Enzymol 439:285–302
Terakawa Y, Agnihotri S, Golbourn B, Nadi M, Sabha N, Smith CA, Croul SE, Rutka JT (2013) The role of drebrin in glioma migration and invasion. Exp Cell Res 319:517–528
Tilney LG, Bonder EM, Coluccio LM, Mooseker MS (1983) Actin from Thyone sperm assembles on only one end of an actin filament: a behavior regulated by profilin. J Cell Biol 97:112–124
Tu JC, Xiao B, Yuan JP, Lanahan AA, Leoffert K, Li M, Linden DJ, Worley PF (1998) Homer binds a novel proline-rich motif and links group 1 metabotropic glutamate receptors with IP3 receptors. Neuron 21:717–726
Tu JC, Xiao B, Naisbitt S, Yuan JP, Petralia RS, Brakeman P, Doan A, Aakalu VK, Lanahan AA, Sheng M, Worley PF (1999) Coupling of mGluR/Homer and PSD-95 complexes by the Shank family of postsynaptic density proteins. Neuron 23:583–592
Vaskova M, Kovac M, Volna P, Angelisova P, Mejstrikova E, Zuna J, Brdicka T, Hrusak O (2011) High expression of cytoskeletal protein drebrin in TEL/AML1pos B-cell precursor acute lymphoblastic leukemia identified by a novel monoclonal antibody. Leuk Res 35:1111–1113
Wei Z, Sun M, Liu X, Zhang J, Jin Y (2014) Rufy3, a protein specifically expressed in neurons, interacts with actin-bundling protein Fascin to control the growth of axons. J Neurochem 130:678–692

West M, Kung HF, Kamata T (1990) A novel membrane factor stimulates guanine nucleotide exchange reaction of ras proteins. FEBS Lett 259:245–248

Worth DC, Daly CN, Geraldo S, Oozeer F, Gordon-Weeks PR (2013) Drebrin contains a cryptic F-actin-bundling activity regulated by Cdk5 phosphorylation. J Cell Biol 202:793–806

Wu H, Zeng H, Lam R, Tempel W, Amaya MF, Xu C, Dombrovski L, Qiu W, Wang Y, Min J (2011) Structural and histone binding ability characterizations of human PWWP domains. PLoS One 6:e18919

Xiao B, Tu JC, Petralia RS, Yuan JP, Doan A, Breder CD, Ruggiero A, Lanahan AA, Wenthold RJ, Worley PF (1998) Homer regulates the association of group 1 metabotropic glutamate receptors with multivalent complexes of homer-related, synaptic proteins. Neuron 21:707–716

Xu SQ, Buraschi S, Morcavallo A, Genua M, Shirao T, Peiper SC, Gomella LG, Birbe R, Belfiore A, Iozzo RV, Morrione A (2015) A novel role for drebrin in regulating progranulin bioactivity in bladder cancer. Oncotarget 6:10825–10839

Yagoub D, Wilkins MR, Lay AJ, Kaczorowski DC, Hatoum D, Bajan S, Hutvagner G, Lai JH, Wu W, Martiniello-Wilks R, Xia P, McGowan EM (2014) Sphingosine kinase 1 isoform-specific interactions in breast cancer. Mol Endocrinol 28:1899–1915

Yamauchi J, Miyamoto Y, Tanoue A, Shooter EM, Chan JR (2005) Ras activation of a Rac1 exchange factor, Tiam1, mediates neurotrophin-3-induced Schwann cell migration. Proc Natl Acad Sci U S A 102(41):14889–14894

Yamazaki H, Takahashi H, Aoki T, Shirao T (2001) Molecular cloning and dendritic localization of rat SH3P7. Eur J Neurosci 14:998–1008

Yamazaki H, Kojima N, Kato K, Hirose E, Iwasaki T, Mizui T, Takahashi H, Hanamura K, Roppongi RT, Koibuchi N, Sekino Y, Mori N, Shirao T (2014) Spikar, a novel drebrin-binding protein, regulates the formation and stabilization of dendritic spines. J Neurochem 128:507–522

Yuan JP, Kiselyov K, Shin DM, Chen J, Shcheynikov N, Kang SH, Dehoff MH, Schwarz MK, Seeburg PH, Muallem S, Worley PF (2003) Homer binds TRPC family channels and is required for gating of TRPC1 by IP3 receptors. Cell 114:777–789

Zanocco-Marani T, Bateman A, Romano G, Valentinis B, He ZH, Baserga R (1999) Biological activities and signaling pathways of the granulin/epithelin precursor. Cancer Res 59:5331–5340

Zhao L, Xu L, Zhou X, Zhu Q, Yang Z, Zhang C, Zhu X, Yu M, Zhang Y, Zhao X, Huang P (2009) Interaction of influenza virus NS1 protein with growth arrest-specific protein 8. Virol J 6:218

Zhu J, Nathan C, Jin W, Sim D, Ashcroft GS, Wahl SM, Lacomis L, Erdjument-Bromage H, Tempst P, Wright CD, Ding A (2002) Conversion of proepithelin to epithelins: roles of SLPI and elastase in host defense and wound repair. Cell 111:867–878

Part IV
Drebrin in Non-Nervous System

Chapter 15
Role of Drebrin at the Immunological Synapse

Vera Rocha-Perugini, Mónica Gordon-Alonso, and Francisco Sánchez-Madrid

Abstract Although drebrin was first described in neurons, it is also expressed in cells of the immune system, such as T lymphocytes and mast cells. Another member of the drebrin family of proteins, mammalian actin-binding protein 1 (mAbp-1) is more widely expressed and plays important roles in the function of macrophages, polymorphonuclear neutrophils, and B lymphocytes. We will briefly discuss on the function of mAbp-1 and drebrin in immune cells with emphasis on T cells. Specifically, drebrin enables the immune responses of CD4$^+$ T lymphocytes. T cells are activated after the recognition of an antigen presented by antigen-presenting cells through cognate cell-cell contacts called immunological synapses (IS). In CD4$^+$ T cells, drebrin associates with the chemokine receptor CXCR4, and both molecules redistribute to the IS displaying similar dynamics. Through its interaction with CXCR4 and the actin cytoskeleton, drebrin regulates T cell activation. CD4$^+$ T cells are one of the main targets for the human immunodeficiency virus (HIV)-1. This virus utilizes the IS structure to be transmitted to uninfected cells, forming cell-cell contacts called virological synapses (VS). Interestingly, drebrin negatively regulates HIV-1 infection of CD4$^+$ T lymphocytes, by regulating actin polymerization at the VS.

Keywords Drebrin • Immune system • T lymphocytes • T cell activation Immunological synapse • CXCR4 • HIV-1 • Virological synapse

---

Vera Rocha-Perugini and Mónica Gordon-Alonso contributed equally to this work.

V. Rocha-Perugini • F. Sánchez-Madrid (✉)
Servicio de Inmunología, Hospital de la Princesa, Instituto de Investigación Sanitaria La Princesa, Madrid, Spain

Vascular Pathophysiology Research Area, Centro Nacional de Investigaciones Cardiovasculares Carlos III (CNIC), Madrid, Spain
e-mail: fsmadrid@salud.madrid.org

M. Gordon-Alonso
Institute de Duve, Universite Catholique de Louvain, Brussels, Belgium

T. Shirao, Y. Sekino (eds.), *Drebrin*, Advances in Experimental Medicine and Biology 1006, DOI 10.1007/978-4-431-56550-5_15

15.1 Introduction

Drebrin was firstly described in neurons (Shirao and Obata 1985; Sekino et al. 2007; Majoul et al. 2007); however, it was later revealed that it is also widely expressed in nonneuronal cells (Keon et al. 2000; Peitsch et al. 1999, 2001). In the immune system, drebrin is expressed in T lymphocytes and mast cells but not in B lymphocytes. Drebrin modulates T cell activation (Perez-Martinez et al. 2010) and histamine release and degranulation in mast cells (Law et al. 2015). However, whether drebrin plays any role in other immune cells remains to be determined. Mammalian actin-binding protein 1 (mAbp-1), also called SH3P7, HPK1-interacting protein of 55 kDa (HIP-55), or drebrin-like protein (DBL) is the other member of the drebrin family. mAbp-1, which is ubiquitously expressed (Kessels et al. 2000; Yamazaki et al. 2001), is also linked to leukocyte functions (Schymeinsky et al. 2011). In this chapter, we will describe in detail the roles of drebrin in the regulation of immune cells. We will also briefly address what is known about mAbp-1 in the immune system.

In mast cells, drebrin modulates actin reorganization, regulating efficient FcεRI-dependent degranulation and calcium influx (Law et al. 2015). Drebrin is also necessary for the actin reorganization associated with the activation of store-operated channels, thus modulating calcium entry in T cells (Mercer et al. 2010). The specific activation of store-operated calcium channels by drebrin will be further discussed in the next chapter.

Macrophages engulf and digest foreign particles and dying cells, a process called phagocytosis. The activation of the central nervous system tissue-resident macrophages (microglia) results in reduced expression of drebrin in neuronal-glial cocultures, suggesting an increased neuronal synaptic degeneration during inflammation (Jebelli et al. 2014). Drebrin expression in macrophages outside the nervous system has not been explored. However, in murine macrophages, the drebrin-like protein mAbp-1 is expressed and interacts with Fyb (also termed SLAP-130 or ADAP), an immune cell-specific protein involved in phagocytosis and macrophage activation (Yuan et al. 2005).

In primary murine neutrophils, mAbp-1 is necessary for integrin αMβ2-dependent bacterial phagocytosis (Schymeinsky et al. 2009). During an acute inflammatory response against bacterial and fungal pathogens, neutrophils are rapidly recruited to the inflammation site in a tightly controlled manner that involves rolling along the activated endothelium, neutrophil activation, firm adhesion, adhesion strengthening, spreading, intraluminal crawling, and extravasation to the inflammation site (Mocsai 2013). β2 integrins play an essential role in these steps of the extravasation process through tight modulation of its affinity (Evans et al. 2009). β2 integrin activation triggers intracellular signaling cascades including the activation of the Syk kinase (Willeke et al. 2003), which is known to phosphorylate mAbp-1 in lymphocytes (Han et al. 2003; Larbolette et al. 1999). Upon β2-mediated adhesion and migration of neutrophil-like differentiated HL-60 cells, mAbp-1 is enriched at the leading edge in a Syk-dependent manner, co-localizing with F-actin and hematopoietic progenitor kinase 1 (HPK1) (Hepper et al. 2012; Jakob et al. 2013). Moreover, mAbp-1 controls neutrophil adhesion, spreading and intraluminal crawling in the in vivo inflamed cremaster muscle model and under physiological flow conditions in vitro, through the regulation of the transmission of mechanical forces from β2 integrins to the actin cytoskeleton (Schymeinsky et al. 2009; Hepper et al. 2012).

B lymphocyte activation is triggered by antigen binding to B-cell receptors (BCR), leading to BCR signaling (Harwood and Batista 2010). After recognition of an antigen, BCR mediates subsequent antigen internalization and processing in specialized lysosomal compartments (Yuseff and Lennon-Dumenil 2015). Antigens are then presented at the B-cell plasma membrane through the complexes formed with the major histocompatibility class II (MHC II), allowing cell-cell interaction with T lymphocytes (Yuseff and Lennon-Dumenil 2015). The drebrin-like protein mAbp-1 enables BCR-mediated antigen internalization, processing, and presentation through the interaction with dynamin 2 and the actin cytoskeleton (Onabajo et al. 2008). In mouse B cells, mAbp-1 is phosphorylated by Src and Syk kinases after BCR stimulation (Larbolette et al. 1999). In those cells, mAbp-1 negatively regulates BCR signaling by recruiting inhibitory signaling proteins SHIP-1 and HPK1 to BCR clusters (Seeley-Fallen et al. 2014). Although drebrin expression has not been reported in B lymphocytes (Machtaler et al. 2014), it is aberrantly expressed in some leukemias and lymphomas, e.g., mantle cell lymphoma (Wang et al. 2010), B-cell precursor acute lymphoblastic leukemia (BCP-ALL) (Vaskova et al. 2011), and in some immortalized cell lines, such as the B-cell precursor leukemia BV113 and promyelocytic leukemia HL-60, but not in acute myeloid leukemia KG1a (Peitsch et al. 2001). ALL is a heterogeneous disease displaying a number of subtypes genetically different that differ in their capacity to respond to chemotherapy (Yeoh et al. 2002). Analyses of BCP-ALL by gene expression profiling reveal a correlation between high drebrin expression and the presence of the genetic lesion TEL/AML1 (chromosomal fusion 12p13-21q22) (Shurtleff et al. 1995), confirmed by RT-qPCR and flow cytometry (Vaskova et al. 2011). Analyses by gene expression profiling also show lower drebrin expression in patients who relapsed or failed to achieve complete clinical remission (Yeoh et al. 2002; Bhojwani et al. 2008).

15.2 Drebrin at T Cell Immunological Synapses

In resting human T cells, drebrin co-localizes with F-actin along the cortical area beneath the plasma membrane (Perez-Martinez et al. 2010). Resting $CD4^+$ and $CD8^+$ T lymphocytes are activated upon close contact with antigen-presenting cells (APCs). These intercellular contacts are characterized by the formation of a highly organized molecular structure called the immunological synapse (IS) (Monks et al. 1998). IS formation requires the rearrangement of transmembrane and intracellular proteins into concentric rings and constitutes an architectural requirement for the subsequent activation of T cells (Dustin et al. 2010). The IS structure displays a central ring, named central supramolecular activation cluster (cSMAC), surrounded by a peripheral ring (pSMAC), and an outer ring called distal supramolecular activation cluster

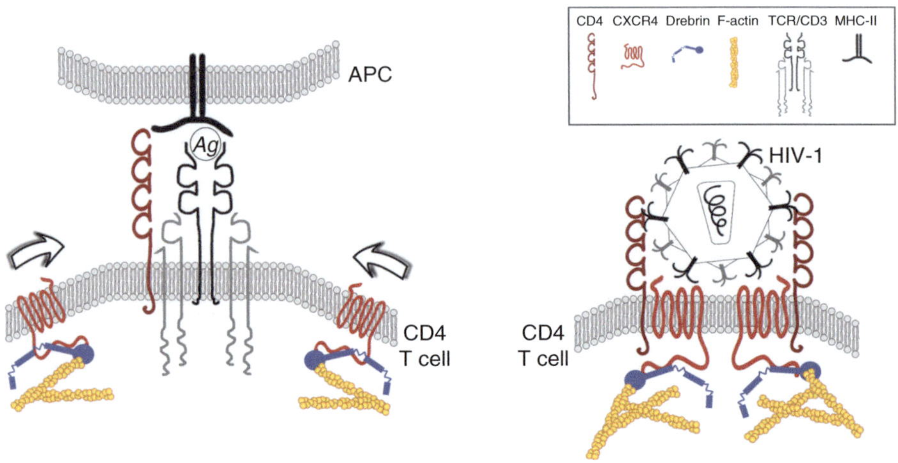

Fig. 15.1 Role of drebrin during CD4⁺ T cell activation and HIV-1 infection. *Left panel*: CD4⁺ T cells recognize an antigenic peptide (Ag) loaded in the MHC II of an antigen-presenting cell (APC) and organize its molecules forming the immunological synapse (IS). Drebrin is recruited to the periphery of the IS together with F-actin and is responsible for the accumulation of CXCR4 in this area. *Right panel*: HIV-1 binds to CD4 and CXCR4 on the surface of CD4⁺ T cells, promoting protein rearrangements that generate the virological synapse (VS), similar to the IS. Drebrin is recruited to the VS and participates in the regulation of local actin polymerization. Proteins and structures depicted are not to scale

(dSMAC). The actin cytoskeleton is involved in the segregation of molecules during IS formation and is needed for the maintenance of the IS (Ritter et al. 2013; Vicente-Manzanares and Sanchez-Madrid 2004). Actin polymerization is triggered by antigen engagement, allowing the T cell to project a lamellipodium that embraces the APC. When the contact is secured, polymerized actin gets cleared from the cSMAC (particularly in CD8⁺ cells) and accumulates at the pSMAC and dSMAC, where it functions mainly as a scaffold for the adhesion molecules. Many actin-related proteins, such as α4β1 integrin (VLA-4), talin, cofilin, and Arp2/3, are enriched in the pSMAC and dSMAC (Eibert et al. 2004; Mittelbrunn et al. 2004; Nolz et al. 2007; Perez-Martinez et al. 2010; Rocha-Perugini et al. 2014a). During IS formation, drebrin is also recruited to these concentric areas and co-localizes with F-actin (Fig. 15.1, left panel). Independently of its actin-binding domain, drebrin associates through its N-terminal region with the chemokine receptor CXCR4 (Perez-Martinez et al. 2010). Chemokine receptors regulate cell migration by sensing the gradients of released chemokines (Rossi and Zlotnik 2000). CXCR4 signaling modulates T cell adhesion proteins in response to chemokine CXCL12 (also called SDF-1a) and might also influence T cell activation as it has been found associated with the T cell receptor (TCR) (Kumar et al. 2006). We have shown that drebrin and CXCR4 are constitutively associated. The association of drebrin and CXCR4 occurs in CD4⁺ T cells. CD4⁺ T cells and this association is enhanced by antigenic stimulation. Thus, during IS formation between superantigen-loaded B cells and CD4⁺ T cells, both drebrin and CXCR4 molecules redistribute to the pSMAC and dSMAC with identical dynamics (Fig. 15.1, left panel).

Drebrin association with CXCR4 is needed for CXCR4 redistribution to the IS, since knocking down drebrin expression by siRNA impairs CXCR4 recruitment to the IS, without affecting CXCR4 expression levels (Perez-Martinez et al. 2010). Drebrin silencing in CD4⁺ T cells impacts the IS formation and, consequently, T cell activation. Human CD4⁺ T cells silenced for drebrin conjugate normally with superantigen-loaded B cells and upregulate early activation markers such as CD69 (Perez-Martinez et al. 2010). However, they show a decrease in F-actin polymerization at the IS and a reduced IL-2 production (mRNA) and secretion (Perez-Martinez et al. 2010). Therefore, drebrin enables the complete activation of the CD4⁺ T cell.

Drebrin-related protein mAbp-1 is also expressed in human and murine T cells and is recruited to the IS during antigen presentation (Le Bras et al. 2004). Nevertheless, mAbp-1 does not affect F-actin polymerization or conjugate formation among antigen-loaded B cells and T cells. In fact, Le Bras and colleagues reported that mAbp-1 negatively regulates T cell activation by promoting TCR internalization and NFAT inhibition (Le Bras et al. 2004). Strikingly, in murine T cells, the role of mAbp-1 seems completely different (Han et al. 2005). Although T cell development is normal in mAbp-1 knock-out mice, T cell activation through the TCR is decreased, showing reduced LAT/PLCγ1 phosphorylation, and impaired HPK1 and JNK kinase activities (Han et al. 2005). Subsequently, T cell proliferation is reduced as well as IL-2 production, and the upregulation of activation markers such as CD69 (Han et al. 2005). Moreover, the authors could not detect any significant decrease in TCR internalization in these murine T cells upon stimulation with anti-CD3 antibodies. In rat T cells, mAbp-1 has been involved in the modulation of RhoA and the actin cytoskeleton (Skelton et al. 2010). Skelton and co-workers show that, in their system, the immunosuppressive activity that reduces T cell infiltration is due to reduced T cell motility and activation, which correlates with the dissociation of mAbp-1 from F-actin.

15.3 Drebrin at Virological Synapses

Transmission of human immunodeficiency virus (HIV)-1 can take place either by direct viral contact with target cells or through different types of cell-cell contacts with uninfected target cells, as filopodial bridges, tunnels, membrane nanotubes, and virological synapses (VS) (Permanyer et al. 2010; Zhong et al. 2013; Sagar 2014). VS are polarized cell-cell structures similar to IS, showing specific recruitment of receptors (CD4 and CXCR4) and adhesion molecules, and local actin polymerization (Haller and Fackler 2008; Piguet and Sattentau 2004). At any rate, HIV-1 entry into target cells depends on the interaction between the viral envelope glycoprotein complex gp120/gp41 (Env) and the cellular receptor CD4 and co-receptors CXCR4 and CCR5 (Blumenthal et al. 2012; Dale et al. 2013; Agosto et al. 2015). In addition to transmembrane receptors, several host cytoskeletal proteins play a role throughout the viral replication cycle (Fackler and Krausslich 2006; Lehmann et al. 2011; Rocha-Perugini et al. 2014b). Among them, drebrin negatively regulates HIV-1 infection of CD4⁺ T lymphocytes (Gordon-Alonso et al. 2013). In agreement with its direct association with the HIV-1 co-receptor CXCR4 (Perez-Martinez et al. 2010), drebrin is specifically enriched at VS (Fig. 15.1, right panel), and it co-localizes partly with the internalized virus at the cytoplasm of target $CD4^+$ T lymphocytes (Gordon-Alonso et al. 2013). Using small interfering RNAs to knock down drebrin expression, Gordon-Alonso and colleagues showed an increase in HIV-1 entry and Env-driven membrane fusion, while viral attachment and reverse transcription remained unaffected (Gordon-Alonso et al. 2013).

In order to enter its target cells, HIV-1 triggers a dynamic reorganization of the actin cytoskeleton (Permanyer et al. 2013; Rocha-Perugini et al. 2014b; Vorster et al. 2011; Yoder et al. 2008). Evidences suggest that actin polymerization is necessary for receptor clustering and stabilization, allowing viral-induced membrane fusion (Gordon-Alonso et al. 2013; Harmon et al. 2010; Kadiu et al. 2007; Komano et al. 2004). After the formation of the viral pore at the plasma membrane, local actin depolymerization is induced, probably because the highly polymerized actin structure presents a physical barrier to the entry of the viral core into the cell (Garcia-Exposito et al. 2013; Vasiliver-Shamis et al. 2009; Yoder et al. 2008). Drebrin silencing in $CD4^+$ T lymphocytes impairs the accumulation of polymerized actin at VS (Gordon-Alonso et al. 2013). Interestingly, profilin recruitment toward VS was enhanced in drebrin knocked down $CD4^+$ T cells (Gordon-Alonso et al. 2013), indicating that drebrin interaction with profilin is important in this process. Profilin enhances actin dynamics and at high concentrations prevents actin polymerization (Bubb et al. 2003). Therefore, profilin increased accumulation at VS might account for the diminished levels of polymerized actin, as well as for the increased viral entry, observed in drebrin-silenced cells.

Using drebrin truncated forms, Gordon-Alonso and colleagues could determine which regions of the protein are involved in the regulation of HIV-1 infection. The overexpression of the drebrin truncated form Dreb(1–366)-GFP, which lacks the Pro-rich motives necessary for its interaction with profilin and other molecules, results in similar effects to the ones observed when silencing drebrin, while the overexpression of the truncated form Dreb(319–707)-GFP, which lacks the actin-binding domain, does not affect HIV-1 infection (Gordon-Alonso et al. 2013). Therefore, drebrin association with CXCR4 is essential for its negative role on HIV-1 infection, since Dreb(319–707)-GFP barely associates with the co-receptor (Perez-Martinez et al. 2010). Moreover, results obtained with Dreb(1–366)-GFP confirm that drebrin negatively regulates HIV-1 infection by limiting profilin accumulation and by stabilizing polymerized actin at HIV-1 entry sites (Gordon-Alonso et al. 2013).

Acknowledgments The authors thank Dr. M. Vicente-Manzanares (Hospital de la Princesa, Universidad Autónoma de Madrid, Spain) for the critical reading of the manuscript. This study was supported by grants: SAF2011-25834 and SAF2014-55579-R from the Spanish Ministry of Economy and Competitiveness, INDISNET-S2011/BMD-2332 from the Comunidad de Madrid, Cardiovascular Network RD12-0042-0056 from the Instituto Salud Carlos III, and ERC-2011-AdG 294340-GENTRIS. M. Gordon-Alonso was supported by Institute de Duve, Universite Catholique de Louvain.

References

Agosto LM, Uchil PD, Mothes W (2015) HIV cell-to-cell transmission: effects on pathogenesis and antiretroviral therapy. Trends Microbiol 23(5):289–295. doi:10.1016/j.tim.2015.02.003

Bhojwani D, Kang H, Menezes RX, Yang W, Sather H, Moskowitz NP, Min DJ, Potter JW, Harvey R, Hunger SP, Seibel N, Raetz EA, Pieters R, Horstmann MA, Relling MV, den Boer ML, Willman CL, Carroll WL (2008) Gene expression signatures predictive of early response and outcome in high-risk childhood acute lymphoblastic leukemia: a children's oncology group study [corrected]. J Clin Oncol 26(27):4376–4384. doi:10.1200/JCO.2007.14.4519

Blumenthal R, Durell S, Viard M (2012) HIV entry and envelope glycoprotein-mediated fusion. J Biol Chem 287(49):40841–40849. doi:10.1074/jbc.R112.406272

Bubb MR, Yarmola EG, Gibson BG, Southwick FS (2003) Depolymerization of actin filaments by profilin. Effects of profilin on capping protein function. J Biol Chem 278(27):24629–24635. doi:10.1074/jbc.M302796200

Dale BM, Alvarez RA, Chen BK (2013) Mechanisms of enhanced HIV spread through T-cell virological synapses. Immunol Rev 251(1):113–124. doi:10.1111/imr.12022

Dustin ML, Chakraborty AK, Shaw AS (2010) Understanding the structure and function of the immunological synapse. Cold Spring Harb Perspect Biol 2(10):a002311. doi:10.1101/cshperspect.a002311

Eibert SM, Lee KH, Pipkorn R, Sester U, Wabnitz GH, Giese T, Meuer SC, Samstag Y (2004) Cofilin peptide homologs interfere with immunological synapse formation and T cell activation. Proc Natl Acad Sci U S A 101(7):1957–1962. doi:10.1073/pnas.0308282100

Evans R, Patzak I, Svensson L, De Filippo K, Jones K, McDowall A, Hogg N (2009) Integrins in immunity. J Cell Sci 122(Pt 2):215–225. doi:10.1242/jcs.019117

Fackler OT, Krausslich HG (2006) Interactions of human retroviruses with the host cell cytoskeleton. Curr Opin Microbiol 9(4):409–415. doi:10.1016/j.mib.2006.06.010

Garcia-Exposito L, Ziglio S, Barroso-Gonzalez J, de Armas-Rillo L, Valera MS, Zipeto D, Machado JD, Valenzuela-Fernandez A (2013) Gelsolin activity controls efficient early HIV-1 infection. Retrovirology 10:39. doi:10.1186/1742-4690-10-39

Gordon-Alonso M, Rocha-Perugini V, Alvarez S, Ursa A, Izquierdo-Useros N, Martinez-Picado J, Munoz-Fernandez MA, Sanchez-Madrid F (2013) Actin-binding protein drebrin regulates HIV-1-triggered actin polymerization and viral infection. J Biol Chem 288:28382–28397. doi:10.1074/jbc.M113.494906

Haller C, Fackler OT (2008) HIV-1 at the immunological and T-lymphocytic virological synapse. Biol Chem 389(10):1253–1260. doi:10.1515/BC.2008.143

Han J, Kori R, Shui JW, Chen YR, Yao Z, Tan TH (2003) The SH3 domain-containing adaptor HIP-55 mediates c-Jun N-terminal kinase activation in T cell receptor signaling. J Biol Chem 278(52):52195–52202. doi:10.1074/jbc.M305026200

Han J, Shui JW, Zhang X, Zheng B, Han S, Tan TH (2005) HIP-55 is important for T-cell proliferation, cytokine production, and immune responses. Mol Cell Biol 25(16):6869–6878. doi:10.1128/MCB.25.16.6869-6878.2005

Harmon B, Campbell N, Ratner L (2010) Role of Abl kinase and the Wave2 signaling complex in HIV-1 entry at a post-hemifusion step. PLoS Pathog 6(6):e1000956. doi:10.1371/journal.ppat.1000956

Harwood NE, Batista FD (2010) Early events in B cell activation. Annu Rev Immunol 28:185–210. doi:10.1146/annurev-immunol-030409-101216

Hepper I, Schymeinsky J, Weckbach LT, Jakob SM, Frommhold D, Sixt M, Laschinger M, Sperandio M, Walzog B (2012) The mammalian actin-binding protein 1 is critical for spreading and intraluminal crawling of neutrophils under flow conditions. J Immunol 188(9):4590–4601. doi:10.4049/jimmunol.1100878

Jakob SM, Pick R, Brechtefeld D, Nussbaum C, Kiefer F, Sperandio M, Walzog B (2013) Hematopoietic progenitor kinase 1 (HPK1) is required for LFA-1-mediated neutrophil recruitment during the acute inflammatory response. Blood 121(20):4184–4194. doi:10.1182/blood-2012-08-451385

Jebelli J, Hooper C, Pocock JM (2014) Microglial p53 activation is detrimental to neuronal synapses during activation-induced inflammation: implications for neurodegeneration. Neurosci Lett 583:92–97. doi:10.1016/j.neulet.2014.08.049

Kadiu I, Ricardo-Dukelow M, Ciborowski P, Gendelman HE (2007) Cytoskeletal protein transformation in HIV-1-infected macrophage giant cells. J Immunol 178(10):6404–6415

Keon BH, Jedrzejewski PT, Paul DL, Goodenough DA (2000) Isoform specific expression of the neuronal F-actin binding protein, drebrin, in specialized cells of stomach and kidney epithelia. J Cell Sci 113(Pt 2):325–336

Kessels MM, Engqvist-Goldstein AE, Drubin DG (2000) Association of mouse actin-binding protein 1 (mAbp1/SH3P7), an Src kinase target, with dynamic regions of the cortical actin cytoskeleton in response to Rac1 activation. Mol Biol Cell 11(1):393–412

Komano J, Miyauchi K, Matsuda Z, Yamamoto N (2004) Inhibiting the Arp2/3 complex limits infection of both intracellular mature vaccinia virus and primate lentiviruses. Mol Biol Cell 15(12):5197–5207. doi:10.1091/mbc.E04-04-0279

Kumar A, Humphreys TD, Kremer KN, Bramati PS, Bradfield L, Edgar CE, Hedin KE (2006) CXCR4 physically associates with the T cell receptor to signal in T cells. Immunity 25(2):213–224. doi:10.1016/j.immuni.2006.06.015

Larbolette O, Wollscheid B, Schweikert J, Nielsen PJ, Wienands J (1999) SH3P7 is a cytoskeleton adapter protein and is coupled to signal transduction from lymphocyte antigen receptors. Mol Cell Biol 19(2):1539–1546

Law M, Lee Y, Morales JL, Ning G, Huang W, Pabon J, Kannan AK, Jeong AR, Wood A, Carter C, Mohinta S, Song J, August A (2015) Cutting edge: drebrin-regulated actin dynamics regulate IgE-dependent mast cell activation and allergic responses. J Immunol 195:426–430. doi:10.4049/jimmunol.1401442

Le Bras S, Foucault I, Foussat A, Brignone C, Acuto O, Deckert M (2004) Recruitment of the actin-binding protein HIP-55 to the immunological synapse regulates T cell receptor signaling and endocytosis. J Biol Chem 279(15):15550–15560. doi:10.1074/jbc.M312659200

Lehmann M, Nikolic DS, Piguet V (2011) How HIV-1 takes advantage of the cytoskeleton during replication and cell-to-cell transmission. Virus 3(9):1757–1776. doi:10.3390/v3091757

Machtaler S, Choi K, Dang-Lawson M, Falk L, Pournia F, Naus CC, Matsuuchi L (2014) The role of the gap junction protein connexin43 in B lymphocyte motility and migration. FEBS Lett 588(8):1249–1258. doi:10.1016/j.febslet.2014.01.027

Majoul I, Shirao T, Sekino Y, Duden R (2007) Many faces of drebrin: from building dendritic spines and stabilizing gap junctions to shaping neurite-like cell processes. Histochem Cell Biol 127(4):355–361. doi:10.1007/s00418-007-0273-y

Mercer JC, Qi Q, Mottram LF, Law M, Bruce D, Iyer A, Morales JL, Yamazaki H, Shirao T, Peterson BR, August A (2010) Chemico-genetic identification of drebrin as a regulator of calcium responses. Int J Biochem Cell Biol 42(2):337–345. doi:10.1016/j.biocel.2009.11.019

Mittelbrunn M, Molina A, Escribese MM, Yanez-Mo M, Escudero E, Ursa A, Tejedor R, Mampaso F, Sanchez-Madrid F (2004) VLA-4 integrin concentrates at the peripheral supramolecular activation complex of the immune synapse and drives T helper 1 responses. Proc Natl Acad Sci U S A 101(30):11058–11063. doi:10.1073/pnas.0307927101

Mocsai A (2013) Diverse novel functions of neutrophils in immunity, inflammation, and beyond. J Exp Med 210(7):1283–1299. doi:10.1084/jem.20122220

Monks CR, Freiberg BA, Kupfer H, Sciaky N, Kupfer A (1998) Three-dimensional segregation of supramolecular activation clusters in T cells. Nature 395(6697):82–86. doi:10.1038/25764

Nolz JC, Medeiros RB, Mitchell JS, Zhu P, Freedman BD, Shimizu Y, Billadeau DD (2007) WAVE2 regulates high-affinity integrin binding by recruiting vinculin and talin to the immunological synapse. Mol Cell Biol 27(17):5986–6000. doi:10.1128/MCB.00136-07

Onabajo OO, Seeley MK, Kale A, Qualmann B, Kessels M, Han J, Tan TH, Song W (2008) Actin-binding protein 1 regulates B cell receptor-mediated antigen processing and presentation in response to B cell receptor activation. J Immunol 180(10):6685–6695

Peitsch WK, Grund C, Kuhn C, Schnolzer M, Spring H, Schmelz M, Franke WW (1999) Drebrin is a widespread actin-associating protein enriched at junctional plaques, defining a specific microfilament anchorage system in polar epithelial cells. Eur J Cell Biol 78(11):767–778. doi:10.1016/S0171-9335(99)80027-2

Peitsch WK, Hofmann I, Pratzel S, Grund C, Kuhn C, Moll I, Langbein L, Franke WW (2001) Drebrin particles: components in the ensemble of proteins regulating actin dynamics of lamellipodia and filopodia. Eur J Cell Biol 80(9):567–579. doi:10.1078/0171-9335-00194

Perez-Martinez M, Gordon-Alonso M, Cabrero JR, Barrero-Villar M, Rey M, Mittelbrunn M, Lamana A, Morlino G, Calabia C, Yamazaki H, Shirao T, Vazquez J, Gonzalez-Amaro R, Veiga E, Sanchez-Madrid F (2010) F-actin-binding protein drebrin regulates CXCR4 recruitment to the immune synapse. J Cell Sci 123(Pt 7):1160–1170. doi:10.1242/jcs.064238

Permanyer M, Ballana E, Este JA (2010) Endocytosis of HIV: anything goes. Trends Microbiol 18(12):543–551. doi:10.1016/j.tim.2010.09.003

Permanyer M, Pauls E, Badia R, Este JA, Ballana E (2013) The cortical actin determines different susceptibility of naive and memory CD4+ T cells to HIV-1 cell-to-cell transmission and infection. PLoS One 8(11):e79221. doi:10.1371/journal.pone.0079221

Piguet V, Sattentau Q (2004) Dangerous liaisons at the virological synapse. J Clin Invest 114(5):605–610. doi:10.1172/JCI22812

Ritter AT, Angus KL, Griffiths GM (2013) The role of the cytoskeleton at the immunological synapse. Immunol Rev 256(1):107–117. doi:10.1111/imr.12117

Rocha-Perugini V, Gonzalez-Granado JM, Tejera E, Lopez-Martin S, Yanez-Mo M, Sanchez-Madrid F (2014a) Tetraspanins CD9 and CD151 at the immune synapse support T-cell integrin signaling. Eur J Immunol 44(7):1967–1975. doi:10.1002/eji.201344235

Rocha-Perugini V, Gordon-Alonso M, Sanchez-Madrid F (2014b) PIP-2: choreographer of actin-adaptor proteins in the HIV-1 dance. Trends Microbiol 22(7):379–388. doi:10.1016/j.tim.2014.03.009

Rossi D, Zlotnik A (2000) The biology of chemokines and their receptors. Annu Rev Immunol 18:217–242. doi:10.1146/annurev.immunol.18.1.217

Sagar M (2014) Origin of the transmitted virus in HIV infection: infected cells versus cell-free virus. J Infect Dis 210(Suppl 3):S667–S673. doi:10.1093/infdis/jiu369

Schymeinsky J, Gerstl R, Mannigel I, Niedung K, Frommhold D, Panthel K, Heesemann J, Sixt M, Quast T, Kolanus W, Mocsai A, Wienands J, Sperandio M, Walzog B (2009) A fundamental role of mAbp1 in neutrophils: impact on beta(2) integrin-mediated phagocytosis and adhesion in vivo. Blood 114(19):4209–4220. doi:10.1182/blood-2009-02-206169

Schymeinsky J, Sperandio M, Walzog B (2011) The mammalian actin-binding protein 1 (mAbp1): a novel molecular player in leukocyte biology. Trends Cell Biol 21(4):247–255. doi:10.1016/j.tcb.2010.12.001

Seeley-Fallen MK, Liu LJ, Shapiro MR, Onabajo OO, Palaniyandi S, Zhu X, Tan TH, Upadhyaya A, Song W (2014) Actin-binding protein 1 links B-cell antigen receptors to negative signaling pathways. Proc Natl Acad Sci U S A 111(27):9881–9886. doi:10.1073/pnas.1321971111

Sekino Y, Kojima N, Shirao T (2007) Role of actin cytoskeleton in dendritic spine morphogenesis. Neurochem Int 51(2–4):92–104

Shirao T, Obata K (1985) Two acidic proteins associated with brain development in chick embryo. J Neurochem 44(4):1210–1216

Shurtleff SA, Buijs A, Behm FG, Rubnitz JE, Raimondi SC, Hancock ML, Chan GC, Pui CH, Grosveld G, Downing JR (1995) TEL/AML1 fusion resulting from a cryptic t(12;21) is the most common genetic lesion in pediatric ALL and defines a subgroup of patients with an excellent prognosis. Leukemia 9(12):1985–1989

Skelton TS, Tejpal N, Gong Y, Kloc M, Ghobrial RM (2010) Downregulation of RhoA and changes in T cell cytoskeleton correlate with the abrogation of allograft rejection. Transpl Immunol 23(4):185–193. doi:10.1016/j.trim.2010.06.009

Vasiliver-Shamis G, Cho MW, Hioe CE, Dustin ML (2009) Human immunodeficiency virus type 1 envelope gp120-induced partial T-cell receptor signaling creates an F-actin-depleted zone in the virological synapse. J Virol 83(21):11341–11355. doi:10.1128/JVI.01440-09
Vaskova M, Kovac M, Volna P, Angelisova P, Mejstrikova E, Zuna J, Brdicka T, Hrusak O (2011) High expression of cytoskeletal protein drebrin in TEL/AML1pos B-cell precursor acute lymphoblastic leukemia identified by a novel monoclonal antibody. Leuk Res 35(8):1111–1113. doi:10.1016/j.leukres.2011.03.028
Vicente-Manzanares M, Sanchez-Madrid F (2004) Role of the cytoskeleton during leukocyte responses. Nat Rev Immunol 4(2):110–122
Vorster PJ, Guo J, Yoder A, Wang W, Zheng Y, Xu X, Yu D, Spear M, Wu Y (2011) LIM kinase 1 modulates cortical actin and CXCR4 cycling and is activated by HIV-1 to initiate viral infection. J Biol Chem 286(14):12554–12564. doi:10.1074/jbc.M110.182238
Wang X, Bjorklund S, Wasik AM, Grandien A, Andersson P, Kimby E, Dahlman-Wright K, Zhao C, Christensson B, Sander B (2010) Gene expression profiling and chromatin immunoprecipitation identify DBN1, SETMAR and HIG2 as direct targets of SOX11 in mantle cell lymphoma. PLoS One 5(11):e14085. doi:10.1371/journal.pone.0014085
Willeke T, Schymeinsky J, Prange P, Zahler S, Walzog B (2003) A role for Syk-kinase in the control of the binding cycle of the beta2 integrins (CD11/CD18) in human polymorphonuclear neutrophils. J Leukoc Biol 74(2):260–269
Yamazaki H, Takahashi H, Aoki T, Shirao T (2001) Molecular cloning and dendritic localization of rat SH3P7. Eur J Neurosci 14(6):998–1008
Yeoh EJ, Ross ME, Shurtleff SA, Williams WK, Patel D, Mahfouz R, Behm FG, Raimondi SC, Relling MV, Patel A, Cheng C, Campana D, Wilkins D, Zhou X, Li J, Liu H, Pui CH, Evans WE, Naeve C, Wong L, Downing JR (2002) Classification, subtype discovery, and prediction of outcome in pediatric acute lymphoblastic leukemia by gene expression profiling. Cancer Cell 1(2):133–143
Yoder A, Yu D, Dong L, Iyer SR, Xu X, Kelly J, Liu J, Wang W, Vorster PJ, Agulto L, Stephany DA, Cooper JN, Marsh JW, Wu Y (2008) HIV envelope-CXCR4 signaling activates cofilin to overcome cortical actin restriction in resting CD4 T cells. Cell 134(5):782–792. doi:10.1016/j.cell.2008.06.036
Yuan M, Mogemark L, Fallman M (2005) Fyn binding protein, Fyb, interacts with mammalian actin binding protein, mAbp1. FEBS Lett 579(11):2339–2347. doi:10.1016/j.febslet.2005.03.031
Yuseff MI, Lennon-Dumenil AM (2015) B cells use conserved polarity cues to regulate their antigen processing and presentation functions. Front Immunol 6:251. doi:10.3389/fimmu.2015.00251
Zhong P, Agosto LM, Munro JB, Mothes W (2013) Cell-to-cell transmission of viruses. Curr Opin Virol 3(1):44–50. doi:10.1016/j.coviro.2012.11.004

Chapter 16
Drebrin Regulation of Calcium Signaling in Immune Cells

Jonathan Pabon, Man Kit Law, and Avery August

Abstract Store-operated Ca^{2+} channels are plasma membrane channels that are activated by depletion of intracellular Ca^{2+} stores, resulting in an increase in intracellular Ca^{2+}; however, little is known about their regulation. Our work has shown that the immunosuppressant compound BTP2, which blocks Ca^{2+} influx into cells, interacts with the actin-reorganizing protein, drebrin. Here we review the role of drebrin in the regulation of calcium signaling, with a focus on immune cells.

Keywords Store-operated Ca^{2+} channel • BTP2 • Filopodia extension • Jurkat T cells • Acute IgE challenge • FcεRI-mediated actin organization

Abbreviations

BTP2	N-(4-(3,5-*bis*(trifluoromethyl)-1H-pyrazol-1-yl)phenyl)-4-methyl-1,2,3-thiadiazole-5-carboxamide
CRAC	Calcium-release-activated channel
Dbn1	Drebrin
FcεRI	High-affinity receptor for Fc portion of IgE
LatB	Latrunculin B
NFATc1	Nuclear factor of activated T-cell cytoplasmic 1
RBL	Rat basophil leukemia
STIM1	Stromal interaction molecule 1
TRPC	Transient receptor potential channel
WASP	Wiskott-Aldrich syndrome protein
WIP	WASP-interacting protein

J. Pabon • M.K. Law • A. August (✉)
Department of Microbiology and Immunology, Cornell University, Ithaca, NY 14853, USA
e-mail: averyaugust@cornell.edu

T. Shirao, Y. Sekino (eds.), *Drebrin*, Advances in Experimental Medicine and Biology 1006, DOI 10.1007/978-4-431-56550-5_16

16.1 Introduction

A number of physiological processes respond to increases in intracellular Ca^{2+} levels, including the regulation of the immune response. Receptors that stimulate phospholipase C generate the second messengers inositol 1,4,5-trisphosphate (IP_3) and diacylglycerol (DAG) as a result of hydrolysis of phosphoinositol 4,5-bisphosphate (PIP_2). IP_3 induces elevation of intracellular Ca^{2+} via the IP_3 receptor (IP_3R), Ca^{2+} release channels in the endoplasmic reticulum (ER) membrane. Furthermore, when Ca^{2+} stores in the ER is depleted, this stimulates a continued increase in intracellular Ca2+ by activating store-operated channels (SOCs) via store-operated Ca^{2+} entry (SOCE) (Putney 1986, 1990; Smyth et al. 2006), which allows Ca^{2+} to enter into the cell. This store-operated Ca^{2+} entry or SOCE plays critical roles in regulating the function of immune cells (Bergmeier et al. 2013). The mechanism by which this occurs continues to be under intense investigation. In this chapter, we review the evidence that this process is regulated in part by the actin cytoskeleton and that the actin-regulating protein drebrin plays an important role in this process.

16.2 Store-Operated Calcium Channels in Immune Cells

The channel Orai has been identified as the SOC in immune cells and its activity is modulated by the protein STIM1 (Soboloff et al. 2006; Peinelt et al. 2006; Zhang et al. 2005; Liou et al. 2005; Roos et al. 2005; Yeromin et al. 2006; Prakriya et al. 2006). STIM1 modifies Orai1 by translocating near to the plasma membrane, providing a link between the ER, where calcium is released, and the plasma membrane, where Orai allows Ca^{2+} into the cell (Zhang et al. 2005; Hauser and Tsien 2007; Ross et al. 2007; Wu et al. 2006; Spassova et al. 2006; Mercer et al. 2006). STIM1 is a single-spanning membrane protein with a Ca^{2+}-binding EF-hand motif that is anchored in the plasma membrane. STIM1 functions as the sensor of ER luminal Ca^{2+} levels migrates within the ER membrane to sites near the plasma membrane, interacting Ca^{2+} influx channels leading to their activation. The Orai1 protein is also multi-spanning protein localized in the plasma membrane and functions as the pore-forming subunit of the highly selective CRAC channel in the PM.

16.2.1 Role of SOCE in Immune Function

One of the best characterized Ca^{2+}-dependent transcriptional pathways leads to the activation of nuclear factor of activated T cells (NFAT), essential for transcription of many cytokine genes (Srikanth and Gwack 2013). The absence of Orai leads to a defect in Ca^{2+} influx through CRAC channels, which severely compromises activation of a wide range of immune cells (Prakriya and Lewis 2015). Indeed, mutations in Orai1 were initially described in human immunodeficient patients that resulted in the ablation of all CRAC channel activity (Feske et al. 2006). Furthermore, the significance of Orai and SOCE in the pathogenesis of immune-related disease is underscored by its role in hypersensitivity disorders of the immune system, including mast cell activation and the generation of allergic reactions (Feske et al. 2015; Ikeya et al. 2014; Ashmole et al. 2012, 2013). In mast cells, FcεRI stimulation induces the liberation of intracellular Ca^{2+} stores and activation of SOCE, which is essential for the degranulation process, as well as the release of chemokines, and cytokines, which contribute to allergic inflammation. Mice lacking Orai exhibit defective mast cell function and allergic responses (Feske et al. 2015).

In mast cells, binding of the Fc portion of IgE to the high-affinity FcεRI triggers an increase in intracellular Ca^{2+}, an essential step for mast cell activation, degranulation, and the generation of a full mast cell response. This FcεRI-triggered increase in Ca^{2+} occurs via the previously described PLCγ pathway, leading to activation of CRAC channels and influx of extracellular Ca^{2+} (Feske 2007; Gwack et al. 2007; Ishikawa et al. 2003). Ca^{2+} increase is also critical for the degranulation of mast cells, releasing histamine and other preformed pharmacological agents from intracellular vesicles (Di Capite and Parekh 2009; Scharenberg and Kinet 1998; Turner and Kinet 1999).

16.2.2 The Actin Cytoskeleton and SOCE

Actin exists in the cell as globular G-actin and filamentous F-actin, and the polymerization of G-actin into F-actin results in the formation of microfilaments in the actin cytoskeleton (Mattila et al. 2016). The actin cytoskeleton has long been suggested to play an important role in the regulation of intracellular Ca^{2+}, although the evidence is conflicting. While actin cytoskeletal changes appear to be dispensable for ER calcium release, preventing actin depolymerization blocks SOC-mediated increase in intracellular Ca^{2+} (Hao and August 2005; Patterson et al. 1999; Smyth et al. 2007; Rosado et al. 2004; Oka et al. 2002). Cytochalasins, actin-depolymerizing agents, have also been shown to attenuate SOCE and Ca^{2+} (Rueckschloss and Isenberg 2001). Furthermore, in T cells, B cells, and mast cells, the actin-depolymerizing agent Latrunculin B is able to increase in intracellular Ca^{2+} (Hao and August 2005; Nolz et al. 2006; Rivas et al. 2004). However, not much is known about the proteins that may regulate this process. Of particular interest is the finding that mast cells from mice deficient in Wiskott-Aldrich syndrome protein (WASP), a key regulatory protein of F-actin assembly, exhibit by diminished Ca^{2+} mobilization, degranulation, and cytokine secretion (Pivniouk et al. 2003). Similarly, cells from mice deficient in other actin-regulating proteins WASP, WIP, and WAVE2 also exhibit defects in Ca^{2+} mobilization (Nolz et al. 2006; Zhang et al. 1999; Kettner et al. 2004). The actin regulators WASP, WIP, and WAVE2 have been shown to regulate SOCE and Ca^{2+} influx into mast cells and/or T cells (Nolz et al. 2006; Pivniouk et al. 2003; Kettner et al. 2004). These results strongly support the view that the actin cytoskeleton plays an important role in regulating intracellular Ca^{2+} mobilization in immune cells.

16.3 Immunosuppressant BTP Is an Inhibitor of Calcium Signaling

The class of compounds called BTPs (3,5-*bis*(trifluoromethyl)pyrazoles), exemplified by BTP2, have been found to inhibit Ca^{2+} entry into cells (Ishikawa et al. 2003; Zitt et al. 2004). Indeed, BTP2 has been shown to block T-cell receptor (TCR)-induced Ca^{2+} entry and Ca^{2+}-dependent cytokine production (Ishikawa et al. 2003; Zitt et al. 2004; Djuric et al. 2000; Trevillyan et al. 2001; Mercer 2005; Mercer et al. 2010). By inhibiting influx of Ca^{2+} into the cells, BTP2 is able to inhibit a wide variety of processes in immune cells, including inhibition of T-cell production of Th1 and Th2 cytokines (Zitt et al. 2004; Djuric et al. 2000), superoxide generation in neutrophils (Steinckwich et al. 2007), mast cell activation, degranulation, and cytokine production following IgE/FcεRI stimulation (Law et al. 2011). In vivo, BTP2 has been shown to be effective in animal models of allergic asthma, reducing Th2 cytokines and leukotrienes, inhibiting eosinophil infiltration into the lungs, and reducing bronchoconstriction and airway hyperresponsiveness (Law et al. 2011; Ohga et al. 2008).

The mechanism for the effect of BTP2 has been unclear. BTP2 has been reported to decrease Ca^{2+} influx into lymphocytes by enhancing the activity of the TRPM4 (Takezawa et al. 2006) and TRPC3 channels (Kiyonaka et al. 2009). We have shown that BTP2 is able to inhibit mast cell activation and degranulation, which is very dependent on increases in intracellular Ca^{2+} (Melicoff et al. 2009), with an IC_{50} of 23 nM (Law et al. 2011). Furthermore, we have shown that BTP2 is a potent inhibitor of mast cell degranulation in vivo at 10 mg BTP2/kg. BTP2 also inhibited IgE/FcεRI-mediated induction of mast cell production of cytokines IL-3, IL-4, IL-6, TNF-a, and GM-CSF (Law et al. 2011). Using structure-activity relationship analysis, we also showed that the trifluoromethyl group at the C3 of BTP2 is required for its activity, since deleting these trifluoromethyl groups entirely or replacing the trifluoromethyl groups of the BTP ring with less bulky methyl groups completely abrogated the activity of BTP2. By contrast, single replacement of the trifluoromethyl group at the 5- or 3-position led to different effects. While the 5-trifluorome thyl-3-methyl-pyrazole derivate had some activity, the 3-trifluoromethyl-5-methyl-pyrazole derivative of BTP2 had similar activity to the parent BTP2 compound (with an IC_{50} of 25 nM), indicating that the C3 trifluoromethyl group of BTP2 is required for its activity (Law et al. 2011).

These effects of BTP2 are similar to that which has been reported for mast cells lacking either Orai1 or STIM1 (Baba et al. 2008; Vig et al. 2008), characterized by severely impaired degranulation and histamine release, decreased leukotriene production, and cytokine production. These mice also exhibit reduced IgE-mediated allergic response in vivo (Di Capite and Parekh 2009; Baba et al. 2008; Vig et al. 2008). These findings support the conclusion that BTP2 inhibits Ca^{2+} influx into these cells and subsequent downstream functions. Interestingly, our findings on the structure-activity relationships of BTP2 derivatives were similar to that found by Kiyonaka et al. who examined derivatives of pyrazole compounds that target the

TRPC3 channel. Kiyonaka et al. found that bulky functional groups at the 3,5-positions of the pyrazole are important for the inhibition of Ca^{2+} mobilization via the TRPC3 channel (Kiyonaka et al. 2009). Based on these findings, we explored the target of BTP2 as a way to understand the mechanism of action of these compounds and the process by which SOCE and Ca^{2+} influx into cells is regulated.

16.3.1 Drebrin Is a Target of BTP

Using a combination of affinity purification and siRNA approaches, we explored binding partners for BTP2, as a means to identify BTP2-binding proteins. Using a BTP2 affinity column, gel purification, and MALDI/TOF mass-spectrometry-based protein identification, we identified drebrin as a binding partner for BTP2. Of interest is the fact that our BTP2 affinity column purification approach also actin, along with drebrin (Mercer et al. 2010). We found that BTP2 interacted directly with drebrin since drebrin expressed in bacterial cells was also able to interact with BTP2, and soluble BTP2 competed with BTP2 affinity matrix for binding to drebrin.

Further analysis of the BTP2/drebrin interaction revealed that lysines 270 and 271 within drebrin are specifically required for its ability to bind to BTP2. These lysines are found with the central actin-binding domain of drebrin, suggesting that BTP2 may inhibit specific functions of drebrin. When drebrin is overexpressed in adherent cells, it induces the formation of what we have referred to as filopodia-like extensions (FLEs), long, branched extensions and curved, thick actin bundles (Shirao et al. 1994). The formation of these FLEs in adherent cells overexpressing drebrin is significantly reduced by BTP2 treatment. Specifically, BTP2 is able to reduce the number of branch points formed per each FLE but not in the average length of these FLEs, and BTP2 treated cells had significantly reduced FLEs, and remaining FLEs were long and linear as opposed to branched. These findings suggest that BTP affects drebrin's ability to induce plasticity in the actin cytoskeleton (Mercer et al. 2010). Indeed, treatment of neuronal cells with BTP2 also affected the ability of drebrin to alter the dynamics of dendritic spines (Sonego et al. 2015). By contrast, BTP2 is unable to block the formation of these FLEs in cells that express a K270M K271M mutant that no longer interacts with BTP2 (Mercer et al. 2010).

16.4 Drebrin Regulates Mast Cell Activation by Regulating FcεRI-Mediated Increase in Intracellular Ca^{2+}

The identification of drebrin as a target of BTP2, an inhibitor of SOCE and Ca^{2+} into cells, suggested that this protein plays a role in regulating this process. This model was supported by the fact that siRNA-mediated knockdown of drebrin in Jurkat T cells resulted in reduced Ca^{2+} influx following stimulation (Mercer et al. 2010). Furthermore, we generated drebrin-deficient mice and analyzed mast cells from these mice to determine whether drebrin indeed plays a role in the ability of these cells to signal by Ca^{2+} (Law et al. 2015). Analysis of mast-celled development revealed that drebrin deficiency leads to reduced density of skin and numbers of peritoneal mast cells, although phenotypic analysis of the skin tissue-resident mast cells did not reveal any gross or ultrastructural differences from WT mice. Interestingly, development of bone-marrow-derived mast cells (BMMCs) in vitro induced by IL-3 and stem cell factor (SCF) was not affected by the absence of drebrin, suggesting that differentiation of mast cells is independent of drebrin expression. By contrast, culture in IL-3 alone, which can also induce the development of BMMCs, resulted in normal differentiation but reduced numbers of BMMCs in the absence of drebrin (Law et al. 2015). These experiments suggest that while not critical for the development of mast cells, drebrin may play a role in their development under resource-limiting conditions such as in the presence of IL-3 alone.

Analysis of the activation of in vitro derived BMMCs revealed that drebrin is important for Ca^{2+} influx following stimulation with IgE/FcεR1, since they exhibit a significant decrease in this response. This was similar to our findings with mast cells treated with BTP2 (Law et al. 2011, 2015). Furthermore, drebrin-deficient mast cells also exhibit reduced production of cytokines in response to IgE/FcεR1, similar to those treated with BTP2. In addition, similar to mice treated with BTP2, drebrin-deficient mice significantly reduced response to acute IgE challenge in a model of passive systemic anaphylaxis, with lower levels of histamine detected in their serum compared to WT mice, as well as reduced temperature loss upon challenge with IgE and allergen (Law et al. 2015).

16.4.1 Drebrin Regulates FcεRI-Mediated Actin Organization in Mast Cells

Drebrin (Dbn1) is frequently found associated with actin in the dendritic spines of neurons and has been shown to interact with and induce changes in the actin cytoskeleton when overexpressed in cells (Shirao et al. 1994; Biou et al. 2008). Since actin has been suggested to play a role in the regulation of Ca^{2+} influx into cells (Nolz et al. 2006; Rivas et al. 2004; Oka et al. 2004; Rosado et al. 2000), we examined the actin cytoskeleton in WT and $Dbn1^{-/-}$ BMMCs. We found that $Dbn1^{-/-}$ BMMCs have higher levels of F-actin than WT cells, which was more distributed inside the cell in contrast with the localization of F-actin in WT mast cells (Law et al. 2015). Interestingly, altering the F-actin superstructure in these mast cells using Latrunculin B was able to partially rescue the ability of the FcεRI to induce degranulation in vitro in response to antigen crosslinking.

Our previous identification of drebrin as a target for the immunosuppressant BTP, a known regulator of intracellular Ca^{2+} influx into cells (Mercer et al. 2010), along with this work supporting a role for drebrin in regulating actin reorganization,

Ca^{2+} influx, and mast cell function, adds support to the idea that the actin cytoskeleton plays an important role in regulating this process. Our work suggests that drebrin may regulate the actin cytoskeleton.

As discussed previously, a role for the actin cytoskeleton in regulating calcium influx into cells has been previously suggested. Our findings with drebrin suggest that as in the case with the actin-regulating protein WIP, drebrin regulates actin changes downstream of the FcεRI. Altogether, our data supports a model where influx of Ca^{2+} is regulated by modulating actin cytoskeleton in part via drebrin, leading to mast cell activation and degranulation.

Acknowledgments This work was supported in part by grants from the National Institutes of Health (AI51626, AI065566, AI073955, and AI08958) to A.A.

References

Ashmole I, Duffy SM, Leyland ML, Morrison VS, Begg M, Bradding P (2012) CRACM/Orai ion channel expression and function in human lung mast cells. J Allergy Clin Immunol 129:1628–1635.e1622

Ashmole I, Duffy SM, Leyland ML, Bradding P (2013) The contribution of Orai(CRACM)1 and Orai(CRACM)2 channels in store-operated Ca2+ entry and mediator release in human lung mast cells. PLoS One 8:e74895

Baba Y, Nishida K, Fujii Y, Hirano T, Hikida M, Kurosaki T (2008) Essential function for the calcium sensor STIM1 in mast cell activation and anaphylactic responses. Nat Immunol 9:81–88

Bergmeier W, Weidinger C, Zee I, Feske S (2013) Emerging roles of store-operated Ca(2)(+) entry through STIM and ORAI proteins in immunity, hemostasis and cancer. Channels (Austin, TX) 7:379–391

Biou V, Brinkhaus H, Malenka R, Matus A (2008) Interactions between drebrin and Ras regulate dendritic spine plasticity. Eur J Neurosci 27:2847–2859

Di Capite J, Parekh AB (2009) CRAC channels and Ca2+ signaling in mast cells. Immunol Rev 231:45–58

Djuric SW, BaMaung NY, Basha A, Liu H, Luly JR, Madar DJ, Sciotti RJ, Tu NP, Wagenaar FL, Wiedeman PE, Zhou X, Ballaron S, Bauch J, Chen YW, Chiou XG, Fey T, Gauvin D, Gubbins E, Hsieh GC, Marsh KC, Mollison KW, Pong M, Shaughnessy TK, Sheets MP, Smith M, Trevillyan JM, Warrior U, Wegner CD, Carter GW (2000) 3,5-Bis(trifluoromethyl)pyrazoles: a novel class of NFAT transcription factor regulator. J Med Chem 43:2975–2981

Feske S (2007) Calcium signalling in lymphocyte activation and disease. Nat Rev Immunol 7:690–702

Feske S, Gwack Y, Prakriya M, Srikanth S, Puppel SH, Tanasa B, Hogan PG, Lewis RS, Daly M, Rao A (2006) A mutation in Orai1 causes immune deficiency by abrogating CRAC channel function. Nature 441:179–185

Feske S, Wulff H, Skolnik EY (2015) Ion channels in innate and adaptive immunity. Annu Rev Immunol 33:291–353

Gwack Y, Feske S, Srikanth S, Hogan PG, Rao A (2007) Signalling to transcription: store-operated Ca2+ entry and NFAT activation in lymphocytes. Cell Calcium 42:145–156

Hao S, August A (2005) Actin depolymerization transduces the strength of B-cell receptor stimulation. Mol Biol Cell 16:2275–2284

Hauser C, Tsien R (2007) A hexahistidine-Zn2+-dye label reveals STIM1 surface exposure. Proc Natl Acad Sci U S A 104:3693–3697

Ikeya M, Yamanoue K, Mochizuki Y, Konishi H, Tadokoro S, Tanaka M, Suzuki R, Hirashima N (2014) Orai-2 is localized on secretory granules and regulates antigen-evoked Ca(2)(+) mobilization and exocytosis in mast cells. Biochem Biophys Res Commun 451:62–67

Ishikawa J, Ohga K, Yoshino T, Takezawa R, Ichikawa A, Kubota H, Yamada T (2003) A pyrazole derivative, YM-58483, potently inhibits store-operated sustained Ca2+ influx and IL-2 production in T lymphocytes. J Immunol 170:4441–4449

Kettner A, Kumar L, Anton IM, Sasahara Y, de la Fuente M, Pivniouk VI, Falet H, Hartwig JH, Geha RS (2004) WIP regulates signaling via the high affinity receptor for immunoglobulin E in mast cells. J Exp Med 199:357–368

Kiyonaka S, Kato K, Nishida M, Mio K, Numaga T, Sawaguchi Y, Yoshida T, Wakamori M, Mori E, Numata T, Ishii M, Takemoto H, Ojida A, Watanabe K, Uemura A, Kurose H, Morii T, Kobayashi T, Sato Y, Sato C, Hamachi I, Mori Y (2009) Selective and direct inhibition of TRPC3 channels underlies biological activities of a pyrazole compound. Proc Natl Acad Sci U S A 106:5400–5405

Law M, Morales J, Mottram L, Iyer A, Peterson B, August A (2011) Structural requirements for the inhibition of calcium mobilization and mast cell activation by the pyrazole derivative BTP2. Int J Biochem Cell Biol 43:1228–1239

Law M, Lee Y, Morales JL, Ning G, Huang W, Pabon J, Kannan AK, Jeong AR, Wood A, Carter C, Mohinta S, Song J, August A (2015) Cutting edge: drebrin-regulated actin dynamics regulate IgE-dependent mast cell activation and allergic responses. J Immunol 195:426–430

Liou J, Kim M, Heo W, Jones J, Myers J, Ferrell JJ, Meyer T (2005) STIM is a Ca2+ sensor essential for Ca2+-store-depletion-triggered Ca2+ influx. Curr Biol 15:1235–1241

Mattila PK, Batista FD, Treanor B (2016) Dynamics of the actin cytoskeleton mediates receptor cross talk: an emerging concept in tuning receptor signaling. J Cell Biol 212:267–280

Melicoff E, Sansores-Garcia L, Gomez A, Moreira DC, Datta P, Thakur P, Petrova Y, Siddiqi T, Murthy JN, Dickey BF, Heidelberger R, Adachi R (2009) Synaptotagmin-2 controls regulated exocytosis but not other secretory responses of mast cells. J Biol Chem 284:19445–19451

Mercer JC (2005) 3,5-Bistrifluoromethyl pyrazole (BTP) compounds and regulation of store-operated calcium channels by the actin binding protein drebrin. Ph.D. Thesis, The Pennsylvania State University, University Park, p 175

Mercer J, Dehaven W, Smyth J, Wedel B, Boyles R, Bird G, Putney JJ (2006) Large store-operated calcium selective currents due to co-expression of Orai1 or Orai2 with the intracellular calcium sensor, Stim1. J Biol Chem. 281:24979–24990

Mercer J, Qi Q, Mottram L, Law M, Bruce D, Iyer A, Morales J, Yamazaki H, Shirao T, Peterson B, August A (2010) Chemico-genetic identification of drebrin as a regulator of calcium responses. Int J Biochem Cell Biol 42:337–345

Nolz JC, Gomez TS, Zhu P, Li S, Medeiros RB, Shimizu Y, Burkhardt JK, Freedman BD, Billadeau DD (2006) The WAVE2 complex regulates actin cytoskeletal reorganization and CRAC-mediated calcium entry during T cell activation. Curr Biol 16:24–34

Ohga K, Takezawa R, Yoshino T, Yamada T, Shimizu Y, Ishikawa J (2008) The suppressive effects of YM-58483/BTP-2, a store-operated Ca^{2+} entry blocker, on inflammatory mediator release in vitro and airway responses in vivo. Pulm Pharmacol Ther 21:360–369

Oka T, Sato K, Hori M, Ozaki H, Karaki H (2002) FcepsilonRI cross-linking-induced actin assembly mediates calcium signalling in RBL-2H3 mast cells. Br J Pharmacol 136:837–846

Oka T, Hori M, Tanaka A, Matsuda H, Karaki H, Ozaki H (2004) IgE alone-induced actin assembly modifies calcium signaling and degranulation in RBL-2H3 mast cells. Am J Physiol Cell Physiol 286:C256–C263

Patterson R, van Rossum D, Gill D (1999) Store-operated Ca^{2+} entry: evidence for a secretion-like coupling model. Cell 98:487–499

Peinelt C, Vig M, Koomoa DL, Beck A, Nadler MJ, Koblan-Huberson M, Lis A, Fleig A, Penner R, Kinet JP (2006) Amplification of CRAC current by STIM1 and CRACM1 (Orai1). Nat Cell Biol 8:771–773

Pivniouk VI, Snapper SB, Kettner A, Alenius H, Laouini D, Falet H, Hartwig J, Alt FW, Geha RS (2003) Impaired signaling via the high-affinity IgE receptor in Wiskott-Aldrich syndrome protein-deficient mast cells. Int Immunol 15:1431–1440

Prakriya M, Lewis RS (2015) Store-operated calcium channels. Physiol Rev 95:1383–1436

Prakriya M, Feske S, Gwack Y, Srikanth S, Rao A, Hogan P (2006) Orai1 is an essential pore subunit of the CRAC channel. Nature 443:230–233

Putney JW Jr (1986) A model for receptor-regulated calcium entry. Cell Calcium 7:1–12

Putney JW Jr (1990) Capacitative calcium entry revisited. Cell Calcium 11:611–624

Rivas F, O'Keefe J, Alegre M, Gajewski T (2004) Actin cytoskeleton regulates calcium dynamics and NFAT nuclear duration. Mol Cell Biol. 24:1628–1639

Roos J, DiGregorio P, Yeromin A, Ohlsen K, Lioudyno M, Zhang S, Safrina O, Kozak J, Wagner S, Cahalan M, Velicelebi G, Stauderman K (2005) STIM1, an essential and conserved component of store-operated Ca^{2+} channel function. J Cell Biol 169:435–445

Rosado JA, Jenner S, Sage SO (2000) A role for the actin cytoskeleton in the initiation and maintenance of store-mediated calcium entry in human platelets. Evidence for conformational coupling. J Biol Chem 275:7527–7533

Rosado JA, Lopez JJ, Harper AG, Harper MT, Redondo PC, Pariente JA, Sage SO, Salido GM (2004) Two pathways for store-mediated calcium entry differentially dependent on the actin cytoskeleton in human platelets. J Biol Chem 279:29231–29235

Ross K, Whitaker M, Reynolds N (2007) Agonist-induced calcium entry correlates with STIM1 translocation. J Cell Physiol 211:569–576

Rueckschloss U, Isenberg G (2001) Cytochalasin D reduces Ca^{2+} currents via cofilin-activated depolymerization of F-actin in guinea-pig cardiomyocytes. J Physiol 537:363–370

Scharenberg AM, Kinet JP (1998) PtdIns-3,4,5-P3: a regulatory nexus between tyrosine kinases and sustained calcium signals. Cell 94:5–8

Shirao T, Hayashi K, Ishikawa R, Isa K, Asada H, Ikeda K, Uyemura K (1994) Formation of thick, curving bundles of actin by drebrin A expressed in fibroblasts. Exp Cell Res 215:145–153

Smyth JT, Dehaven WI, Jones BF, Mercer JC, Trebak M, Vazquez G, Putney JW Jr (2006) Emerging perspectives in store-operated Ca^{2+} entry: roles of Orai, Stim and TRP. Biochim Biophys Acta 1763:1147–1160

Smyth J, De Haven W, Bird G, Putney JJ (2007) Role of the microtubule cytoskeleton in the function of the store-operated Ca^{2+} channel activator STIM1. J Cell Sci 120:3762–3771

Soboloff J, Spassova MA, Tang XD, Hewavitharana T, Xu W, Gill DL (2006) Orai1 and STIM reconstitute store-operated calcium channel function. J Biol Chem 281:20661–20665

Sonego M, Oberoi M, Stoddart J, Gajendra S, Hendricusdottir R, Oozeer F, Worth DC, Hobbs C, Eickholt BJ, Gordon-Weeks PR, Doherty P, Lalli G (2015) Drebrin regulates neuroblast migration in the postnatal mammalian brain. PLoS One 10:e0126478

Spassova M, Soboloff J, He L, Xu W, Dziadek M, Gill D (2006) STIM1 has a plasma membrane role in the activation of store-operated Ca(2+) channels. Proc Natl Acad Sci U S A 103:4040–4045

Srikanth S, Gwack Y (2013) Orai1-NFAT signalling pathway triggered by T cell receptor stimulation. Mol Cells 35:182–194

Steinckwich N, Frippiat J, Stasia M, Erard M, Boxio R, Tankosic C, Doignon I, Nusse O (2007) Potent inhibition of store-operated Ca^{2+} influx and superoxide production in HL60 cells and polymorphonuclear neutrophils by the pyrazole derivative BTP2. J Leukoc Biol 81:1054–1064

Takezawa R, Cheng H, Beck A, Ishikawa J, Launay P, Kubota H, Kinet JP, Fleig A, Yamada T, Penner R (2006) A pyrazole derivative potently inhibits lymphocyte Ca^{2+} influx and cytokine production by facilitating transient receptor potential melastatin 4 channel activity. Mol Pharmacol 69:1413–1420

Trevillyan J, Chiou X, Chen Y, Ballaron S, Sheets M, Smith M, Wiedeman P, Warrior U, Wilkins J, Gubbins E, Gagne G, Fagerland J, Carter G, Luly J, Mollison K, Djuric S (2001) Potent inhibition of NFAT activation and T cell cytokine production by novel low molecular weight pyrazole compounds. J Biol Chem 276:48118–48126

Turner H, Kinet JP (1999) Signalling through the high-affinity IgE receptor Fc epsilonRI. Nature 402:B24–B30

Vig M, DeHaven WI, Bird GS, Billingsley JM, Wang H, Rao PE, Hutchings AB, Jouvin MH, Putney JW, Kinet JP (2008) Defective mast cell effector functions in mice lacking the CRACM1 pore subunit of store-operated calcium release-activated calcium channels. Nat Immunol 9:89–96

Wu M, Buchanan J, Luik R, Lewis R (2006) Ca^{2+} store depletion causes STIM1 to accumulate in ER regions closely associated with the plasma membrane. J Cell Biol 174:803–813

Yeromin A, Zhang S, Jiang W, Yu Y, Safrina O, Cahalan M (2006) Molecular identification of the CRAC channel by altered ion selectivity in a mutant of Orai. Nature 443:226–229

Zhang J, Shehabeldin A, da Cruz LA, Butler J, Somani AK, McGavin M, Kozieradzki I, dos Santos AO, Nagy A, Grinstein S, Penninger JM, Siminovitch KA (1999) Antigen receptor-induced activation and cytoskeletal rearrangement are impaired in Wiskott-Aldrich syndrome protein-deficient lymphocytes. J Exp Med 190:1329–1342

Zhang S, Yu Y, Roos J, Kozak J, Deerinck T, Ellisman M, Stauderman K, Cahalan M (2005) STIM1 is a Ca^{2+} sensor that activates CRAC channels and migrates from the Ca^{2+} store to the plasma membrane. Nature 437:902–905

Zitt C, Strauss B, Schwarz E, Spaeth N, Rast G, Hatzelmann A, Hoth M (2004) Potent inhibition of Ca^{2+} release-activated Ca^{2+} channels and T-lymphocyte activation by the pyrazole derivative BTP2. J Biol Chem 279:12427–12437

Chapter 17
Drebrin and Spermatogenesis

Haiqi Chen, Michelle W.M. Li, and C. Yan Cheng

Abstract Drebrin is a family of actin-binding proteins with two known members called drebrin A and E. Apart from the ability to stabilize F-actin microfilaments via their actin-binding domains near the N-terminus, drebrin also regulates multiple cellular functions due to its unique ability to recruit multiple binding partners to a specific cellular domain, such as the seminiferous epithelium during the epithelial cycle of spermatogenesis. Recent studies have illustrated the role of drebrin E in the testis during spermatogenesis in particular via its ability to recruit branched actin polymerization protein known as actin-related protein 3 (Arp3), illustrating its involvement in modifying the organization of actin microfilaments at the ectoplasmic specialization (ES) which includes the testis-specific anchoring junction at the Sertoli-spermatid (apical ES) interface and at the Sertoli cell-cell (basal ES) interface. These data are carefully evaluated in light of other recent findings herein regarding the role of drebrin in actin filament organization at the ES. We also provide the hypothetical model regarding its involvement in germ cell transport during the epithelial cycle in the seminiferous epithelium to support spermatogenesis.

Keywords Testis • Spermatogenesis • Drebrin • Blood-testis barrier • Arp2/3 complex • Branched actin polymerization • F-actin • Ectoplasmic specialization • Actin microfilament bundles • Seminiferous epithelial cycle

17.1 Introduction

In the mammalian testes, the functional unit that produces an upward of 200 million sperm per day in each man after puberty at ~12–13 years of age versus 40 million sperm per day in a male rat by 45-day postpartum (dpp) is the seminiferous tubule via

H. Chen • M.W.M. Li • C.Y. Cheng (✉)
The Mary M. Wohlford Laboratory for Male Contraceptive Research, Center for Biomedical Research, Population Council, 1230 York Ave., New York, NY 10065, USA
e-mail: Y-Cheng@popcbr.rockefeller.edu; ccheng@mail.rockefeller.edu

T. Shirao, Y. Sekino (eds.), *Drebrin*, Advances in Experimental Medicine and Biology 1006, DOI 10.1007/978-4-431-56550-5_17

spermatogenesis (Johnson et al. 1980; Amann and Howards 1980). Spermatogenesis takes place in the seminiferous epithelium that encircles the inner surface of seminiferous tubules which is composed of only Sertoli cells and germ cells. The Sertoli cell provides nutritional and structural support for germ cell maturation (Xiao et al. 2014; de Kretser and Kerr 1988; Bardin et al. 1988). In short, spermatogenesis is comprised of four distinctive phases of cellular events, which include (1) self-renewal of undifferentiated spermatogonia and proliferation of spermatogonia via mitosis, and their differentiation to type A and type B spermatogonia, (2) transformation of type B spermatogonia to preleptotene spermatocytes and progression of spermatocytes to undergo meiosis, (3) postmeiotic differentiation of round spermatids to elongated spermatids via spermiogenesis, and (4) the release of sperms at spermiation (Schlatt and Ehmcke 2014; O'Donnell et al. 2011; de Kretser and Kerr 1988; Ehmcke and Schlatt 2006). Since germ cells in particular spermatids are highly differentiated and metabolically quiescent cells, they rely almost exclusively on the Sertoli cell for the provision of nutrients, biomolecules, paracrine, and structural supports (Mruk and Cheng 2004b; Cheng and Mruk 2002). On the other hand, Sertoli cells by 15–17 dpp (day postpartum) in rodents and ~12–13 years of age in humans are differentiated and ceased to divide (Orth 1982; Sharpe et al. 2003) so that Sertoli cells support only a relatively small population of germ cells and most germ cells undergo apoptosis (Billig et al. 1995) in order to maintain a Sertoli/germ cell ratio of ~1:30–1:50 in the seminiferous epithelium in rodents (Weber et al. 1983).

In a cross section of the seminiferous tubule, germ cells at different developmental states are associated with Sertoli cells in the seminiferous epithelium, displaying a unique pattern of Sertoli-germ cell association particularly regarding changes of the spermatid head (e.g., acrosome biogenesis) and elongating of the tail which can be divided into 12, 14, and 6 stages in the mouse, rat, and man, respectively, known as the seminiferous epithelial cycle of spermatogenesis (Clermont and Leblond 1955; Amann 2008; Hess and de Franca 2008). For example, the release of fully developed elongated spermatids (i.e., spermatozoa) from the seminiferous epithelium into the tubule lumen at the luminal edge of the apical compartment and blood-testis barrier (BTB) remodeling near the basal compartment both take place concurrently at stage VIII of the epithelium cycle in the rat and mouse testis, whereas meiosis takes place at stages XIV and XII in the rat and mouse testis, respectively (Parvinen 1982; Hess and de Franca 2008; Leblond and Clermont 1952; Clermont 1972).

17.2 Actin Cytoskeleton, Ectoplasmic Specialization (ES), and Spermatogenesis

While the cellular events that take place in the seminiferous epithelium during the epithelial cycle are known for more than six decades (Parvinen 1982; Hess and de Franca 2008; Leblond and Clermont 1952; Clermont 1972), molecular mechanism(s) that regulate these events remain relatively unknown until recent years. Studies have shown that actin-based cytoskeleton is crucial for these events in particular by conferring structural support to Sertoli cells, maintaining cell polarity and junction integrity as well as by participating in endocytic vesicle-mediated protein trafficking

(O'Donnell 2014; Qian et al. 2014; Cheng and Mruk 2015; Wong and Cheng 2009). Studies in the past two decades have shown that the highly coordinated cellular events that take place in the seminiferous epithelium during the epithelial cycle are mediated at the Sertoli cell-cell and Sertoli-germ cell interface, relying on rapid restructuring of the unique cell junctions at these sites, such as gap junctions (Cheng and Mruk 2010, 2002; Hess and de Franca 2008; Hermo et al. 2010; Mruk and Cheng 2004a, b). Interestingly, the most extensive cell junction noted in the testis is the actin-rich cell-cell anchoring junction known as ectoplasmic specialization (ES), which is found at the Sertoli cell-cell or Sertoli-spermatid interface known as the basal and apical ES, respectively (Cheng and Mruk 2010; Russell and Peterson 1985; Wong et al. 2008a; Yan et al. 2007). The ES is typified by the presence of an array of actin microfilament bundles that lie perpendicular to the Sertoli cell plasma membrane, which are sandwiched between the cisternae of endoplasmic reticulum and the apposing Sertoli-spermatid plasma membranes at the apical ES or the apposing Sertoli cell-cell plasma membranes, thereby containing two arrays versus one array of actin microfilament bundles at the basal ES, respectively (Fig. 17.1). In the rat testis, the basal ES is found to coexist with tight junction (TJ) and gap junction (GJ), which together with the intermediate filament-based desmosome constitutes the BTB (Russell 1979, 1977a; Vogl et al. 2008; Mruk and Cheng 2004a; Wong et al. 2008b). Due to the presence of the actin microfilament bundles on both sides of the Sertoli cells at the basal ES, the BTB is one of the tightest blood-tissue barriers (Cheng and Mruk 2012; Pelletier 2011; Franca et al. 2012). The BTB also divides the seminiferous epithelium into the basal and the apical (adluminal) compartments (Fig. 17.1). Thus, postmeiotic spermatid development takes place in a unique microenvironment supported entirely by the Sertoli cells known as the adluminal (apical) compartment (Fig. 17.1), which is sequestered from the systemic circulation (Cheng and Mruk 2012; Pelletier 2011; Franca et al. 2012). On the other hand, apical ES that shares similar ultrastructural features with the basal ES is found restrictively at the Sertoli-spermatid (step 8–19 spermatids in the rat testis) interface during spermiogenesis (Russell 1977b; Yan et al. 2007; Vogl et al. 2008, 2000). Unlike basal ES which coexists with TJ and GJ, when apical ES appears in step 8 spermatids in the rat testis, it replaces GJ and desmosome at the Sertoli-spermatid interface, becoming the only anchoring device until the release of elongated spermatids at spermiation. Furthermore, there is only a single array of actin microfilament bundles restricted to the side of Sertoli cell at the apical ES. While the network of actin microfilament bundles at the ES is crucial to confer unusual adhesive strength to the BTB and developing spermatids, these microfilaments rapidly reorganize during the epithelial cycle to facilitate the transport of preleptotene spermatocytes across the BTB at stage VIII as well as the progressive transport of developing spermatids across the adluminal compartment so that fully developed spermatids (i.e., spermatozoa) can line up near the tubule lumen to prepare for their eventual release at spermiation. Studies have shown that the rapid reorganization of these actin microfilaments such as their conversion from a bundled to an unbundled/branched configuration and vice versa is mediated via the intriguing actions of three classes of proteins: (1) the actin-bundling proteins such as Eps8 (also an actin barbed end capping protein) (Lie et al. 2009), palladin (also an actin-cross-linking protein) (Qian

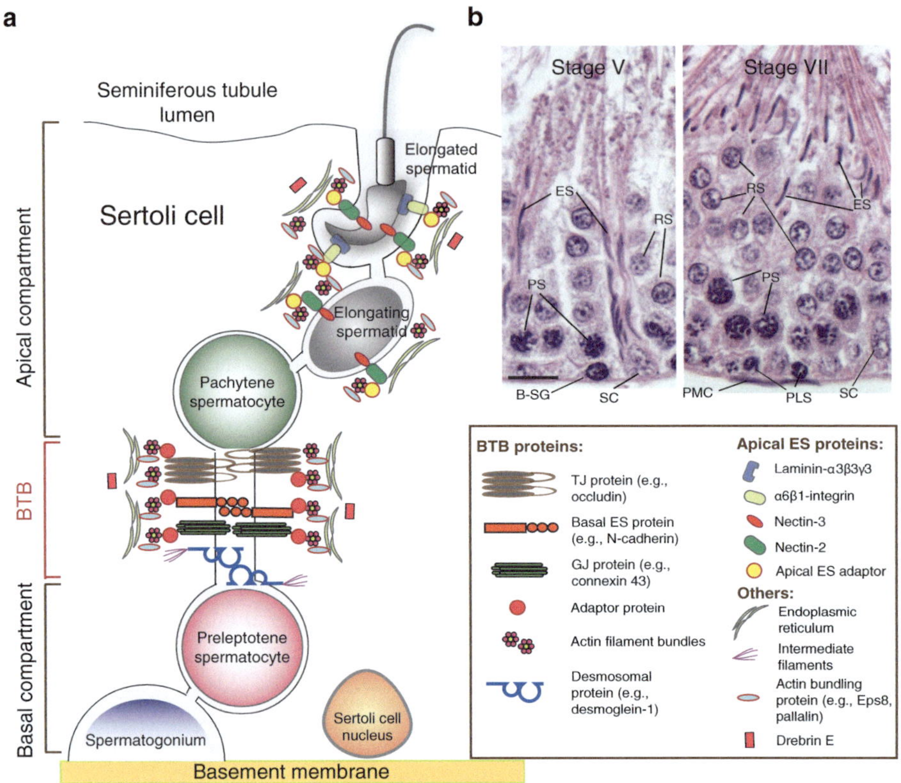

Fig. 17.1 Morphological features and relative location of the actin-rich ES in the seminiferous epithelium of adult rat testes. (**a**) A schematic drawing of the cross section of a seminiferous tubule, illustrating the blood-testis barrier (BTB) that physically divides the seminiferous epithelium into the basal and the adluminal (apical) compartments. Preleptotene spermatocytes derived from type B spermatogonia at stage VII must be transported across the immunological barrier at stage VIII of the epithelial cycle. The BTB integrity is maintained by different cell adhesion complexes of the TJ (e.g., occludin-ZO-1), basal ES (e.g., N-cadherin-β-catenin), and gap junction (e.g., connexin 43/plakofilin-2) which all use actin microfilaments for their attachment, as well as the desmosome (e.g., desmoglein-2/desmocollin-2) which uses vimentin-based intermediate filaments for attachment. Developing spermatids (steps 8–19) in the adluminal compartment remain attached to the Sertoli cell via apical ES by unique adhesion protein complexes such as nectin-afadin and laminin-integrin, of which nectin-3 and laminin-α3β3γ3 are specific to spermatids, whereas necin-2 can be found in both Sertoli cells and spermatids, and α6β1-integrin is Sertoli cell-specific. The typical feature of the ES is the actin microfilament bundles that lie perpendicular to the Sertoli cell plasma membrane, and they are sandwiched between cisternae of endoplasmic reticulum and the apposing plasma membrane of the adjacent Sertoli cells called basal ES *vs.* Sertoli-spermatid called apical ES. These actin microfilament bundles thus confer the ES its unusual adhesive strength to support Sertoli cell adhesion at the BTB and the spermatid adhesion at the ES during spermiogenesis. (**b**) These two micrographs illustrate the typical cell-cell interactions between different germ cells and Sertoli cells at stage V (*left panel*) and stage VII (*right panel*) of the epithelial cycle. For instance in stage V, elongating spermatids (step 17) are found deep inside the base of the adluminal compartment, almost touching the Sertoli cell nucleus (SC), whereas at stage VII, elongated spermatids (step 19) begin to line up near the tubule lumen to prepare for their release into the tubule lumen at stage VIII of the cycle. Scale bar, 60 μm. Abbreviations: *B-SG* type B spermatogonium, *ES* elongating/elongated spermatid, *PLS* preleptotene spermatocyte, *PMC* peritubular myoid cell, *PS* pachytene spermatocyte, *RS* round spermatid, *SC* Sertoli cell

et al. 2013), fascin 1 (Gungor-Ordueri et al. 2014a), plastin 3 (Li et al. 2015c), and ezrin (Gungor-Ordueri et al. 2014b) versus (2) branched actin polymerization (or nucleation) proteins such as the Arp2/3 complex (Lie et al. 2010) and filamin A (Su et al. 2012) and (3) actin nucleation proteins that generate long stretches of microfilaments which can be assembled into actin bundles such as formin 1 (Li et al. 2015b). These proteins coordinate to regulate actin microfilament remodeling at the basal ES and apical ES via their stage-specific and spatiotemporal expression, which, in turn, facilitate the transport of preleptotene spermatocytes across the BTB and the transport of elongating spermatids across the adluminal compartment, respectively. The coordinated efforts of those actin regulatory proteins have recently been reviewed (Li et al. 2015a; Cheng and Mruk 2015; Su et al. 2013). Herein, we focus our discussion on drebrin, a family of proteins known to regulate the action of the Arp2/3 complex which is a barbed end nucleation protein that effectively induces branched actin polymerization, thereby converting actin microfilaments from a linear to a branched configuration. Thus, actin microfilaments can no longer resume a bundled configuration, thereby destabilizing adhesion protein complexes at the Sertoli cell-cell and/or Sertoli-spermatid interface. These changes also facilitate endocytic vesicle-mediated protein trafficking events, such as endocytosis, transcytosis, and recycling, necessary to reassemble "new" basal ES/BTB and apical ES before or following degeneration of the "old" ES.

17.3 Drebrin in the Testis

Drebrin, also known as <u>d</u>evelopmentally <u>re</u>gulated <u>br</u>ain prote<u>in</u>, is an actin-binding protein first identified in chicken brain (Shirao and Obata 1985). Drebrin is a growing protein family which now contains the embryonic type (E) and the adult type (A) in mammals, both of which originate from a single gene (Shirao and Obata 1986; Majoul et al. 2007). As a member of the actin-depolymerizing factor homology (ADF-H) domain-containing family of actin-binding proteins, drebrin binds to actin microfilaments. It alters the helical twist of actin filaments, stabilizes their structures, and strengthens their stiffness (Sharma et al. 2011; Mikati et al. 2013). The coiled-coil (CC) domain and the helical (hel) domain of drebrin, besides the N-terminal ADF-H domain, also contribute to its intrinsic actin-binding activity (Fig. 17.2) (see also Fig. 1.3 in Chap. 1) (Hayashi et al. 1999; Worth et al. 2013; Xu and Stamnes 2006). Multiple binding partners have been identified for drebrin to date (Table 17.1). The drebrin-based protein complex is known to stabilize F-actin, modifies microfilament structure, or moderates the interactions of other actin-binding proteins with actin microfilaments. For instance, drebrin binds to afadin to connect F-actin to nectin-based integral membrane proteins (Rehm et al. 2013). It competes with tropomyosin for F-actin binding and inhibits actin-binding and cross-linking activity of α-actinin (Ishikawa et al. 1994). Furthermore, drebrin is a negative regulator of HIV entry into epithelial cell, as well as HIV-mediated cell fusion by regulating F-actin rearrangement (Gordon-Alonso et al. 2013), and it is also a critical player in glioma cell invasiveness (Terakawa et al. 2013). In addition,

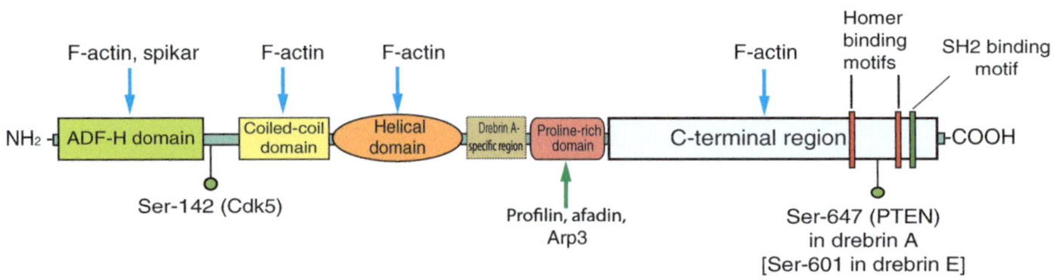

Fig. 17.2 Functional domains in drebrin. Different functional domains along the polypeptide sequence of drebrin, applicable to both drebrin A and drebrin E, are shown herein, except that in drebrin A, there is a drebrin A-specific region located behind the helical domain. Drebrin contains an actin-depolymerizing factor homology (ADF-H) domain, a coiled-coil domain (CC), a helical domain (Hel), as well as a large domain called the blue box which are all capable of interacting with F-actin via their corresponding actin-interaction domain, conferring drebrin its actin-binding protein ability (Worth et al. 2013; Majoul et al. 2007). In addition, spikar is shown to bind to the ADF-H domain (Yamazaki et al. 2014). Drebrin also contains a proline-rich (PP) domain known to bind profilin (Mammoto et al. 1998), afadin (Rehm et al. 2013), and possibly Arp3 (Li et al. 2011). There are also two homer-binding motifs near the C-terminus that interact with cupidin/homer2 (Shiraishi-Yamaguchi et al. 2009) and also an SH2-binding motif. There are two putative phosphorylation sites in drebrin: Ser-142 and Ser-647. Cdk5 is the kinase known to phosphorylate drebrin at Ser-142 (Worth et al. 2013). PTEN is known to dephosphorylate Ser-647 in drebrin A (Kreis et al. 2013), which corresponds to Ser-601 in drebrin E based on sequence homology alignment analysis

Fig. 17.3 A hypothetical model regarding the involvement of drebrin E in the remodeling of the apical ES to facilitate the release of sperms at spermiation. In stage VII tubules, such as the one shown on the *left panel*, the apical ES is supported by the actin microfilament bundles, which are likely maintained through an upregulation of actin-bundling proteins such as Eps8, ezrin, and palladin that confer the actin microfilaments their bundled configuration, which in turn support the functionality of adhesion protein complexes at the site. However, at late stage VII, at the concave (ventral) side of the spermatid head, an upregulation of drebrin E begins to occur (Li et al. 2011). This thus recruits more Arp3 to the site via the high affinity of drebrin E to Arp3 as shown in the testis (Li et al. 2011), so that the intrinsic branched actin polymerization activity of the Arp2/3 complex converts actin microfilaments from a bundled to an unbundled/branched configuration, promoting endocytic vesicle-mediated trafficking events, creating a transient ultrastructure known as apical TBC (tubulobulbar complex) (Vogl et al. 2013), referring to the aggregates of numerous endocytic vesicles readily visible by fluorescence microscopy. The events of endocytosis, transcytosis, and recycling are necessary to recycle apical ES proteins (e.g., integrins, nectins, laminins) from the "old" apical ES to assemble "new" apical ES when step 8 spermatids develop from step 7 arise at stage VIII of the epithelial cycle without requiring de novo synthesis of all the apical ES and pertinent proteins such as peripheral adaptors. This event of actin microfilament reorganization becomes more widespread at stage VIII when virtually all the actin microfilament bundles are replaced by unbundled/branched actin microfilaments to facilitate the release of sperm at spermiation. These changes are also assisted by a downregulation of actin-bundling proteins at the site such as Eps8, ezrin, and palladin, as well as an upregulation of cofilin-1 which cleaves actin microfilaments into shorter fragments to facilitate the breakdown of cell adhesion function conferred by adhesion protein complexes. Collectively, these changes lead to apical ES degeneration to coordinate spermiation although drebrin knockout mice were viable, abnormalities were observed in cortical neurons, mainly associating with dendritic spines (Table 17.2). While drebrin A and E isoforms are both detected in the brain, only drebrin E is found in the rat testis, and it is predominantly expressed by Sertoli cells instead of germ cells (Li et al. 2011). In Sertoli cells cultured in vitro with an established functional TJ-permeability barrier, drebrin E is localized in cell cytoplasm as well as near the cell surface, co-localizing with F-actin (Li et al. 2011). In adult rat testes, drebrin E displays a restrictive, but stage-specific, localization at the ES during the epithelial cycle and co-localizes with F-actin both at the apical ES and the basal ES/BTB. At

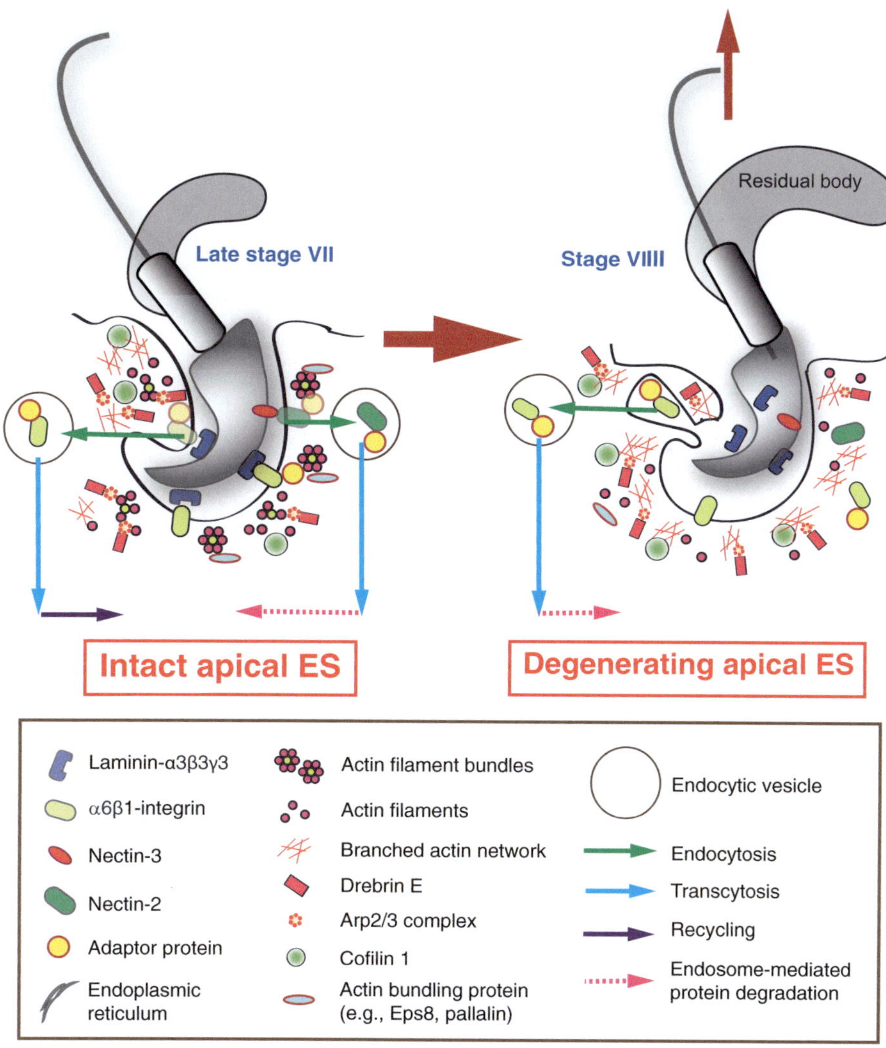

Table 17.1 Binding partners of drebrin

Arp3	Rat Sertoli cells	Recruitment of Arp3 by drebrin E at the Sertoli cell BTB that induces actin branching, converting actin microfilaments in Sertoli cells from a bundled to unbundled/branched configuration that facilitates endocytic vesicle-mediated protein trafficking	n.d. in neither Arp3 nor drebrin E	Li et al. (2011)
Afadin	HUVEC	Afadin serves as a linker between drebrin and nectins in adherens junction	PR domain in drebrin; PR1 and PR2 regions in afadin	Rehm et al. (2013)
Connexin 43	Vero cells; mouse astrocytes	Drebrin acts as a linker between connexin 43 and submembranous actin cytoskeleton	n.d. in drebrin; C-terminal region (amino acids 234–382) in Cx43	Butkevich et al. (2004)
EB3	Rat cortical neurons	Binding of drebrin to EB3 regulates interaction between F-actin and microtubules to support growth cone formation and neuritogenesis	n.d. in drebrin; 3 amino acid inserts in EB3 that are not found in EB1	Geraldo et al. (2008)
Spikar	Rat hippocampal neurons	Extranuclear spikar (a drebrin-binding protein in the brain) accumulation mediated by drebrin that increases dendritic spine density	N-terminal region (amino acids 1–134) in drebrin; N-terminal region (amino acids 88–376) in spikar	Yamazaki et al. (2014)
PTEN	Rat hippocampal or cortical neurons	PTEN negatively regulates S647-phosphorylation of drebrin via dephosphorylation	n.d. in drebrin; C2 domain (amino acids 182–354) in PTEN	Kreis et al. (2013)
Profilin	Rat brain	Drebrin likely regulates actin microfilament via its interaction with profilin	Likely PR domain in drebrin; n.d. in profilin	Mammoto et al. (1998)
Cupidin/ Homer2	Mouse cerebella; rat hippocampal neurons	Deletion of Cdc42-binding region of cupidin disturbs dendritic drebrin distribution in hippocampal neurons	Homer-binding motifs at the C-terminus of drebrin; N-terminal EVH-1 domain in cupidin/ homer2	Shiraishi-Yamaguchi et al. (2009)

[a]This Table is not intended to be exhaustive; the binding partners of drebrin that are selected herein represent those that are found in the testis, or their function is consistent with regulation of spermatogenesis based on our current understanding in the field. However, studies are needed to confirm if many of these proteins are indeed putative binding partners of drebrin in the testis since there are few studies in the literature that explore the role of drebrin in spermatogenesis. *n.d.* not determined, *HUVEC* human umbilical vein endothelial cells, *PR domain* proline-rich domain, *EB3* end-binding protein 3, *PTEN* phosphatase and tensin homolog

Table 17.2 Function of drebrin which is assessed based on studies using genetic models

Drebrin A	Global knockout	Drebrin A-specific KO (drebrin E remained functional in these KO mice) led to a loss of homeostatic synaptic plasticity (HSP) at excitatory synapses of adult cerebral cortex, illustrating drebrin A may be involved in Alzheimer's disease pathogenesis	Cre-Lox P recombination	Aoki et al. (2009)
Drebrins A and E	Global knockout	Total drebrin KO led to a decrease in the number of neurotransmitter receptors, impaired dendritic spine morphogenesis, and reduced memory-related synaptic plasticity in the hippocampus	Cre-Lox P recombination	Jung et al. (2015)
Drebrin A	Global knockout	An impairment of context-dependent fear learning in adult drebrin A KO mice, suggesting a role of drebrin A in cognitive function	Cre-Lox P recombination	Kojima et al. (2010)

KO knockout

the apical ES, drebrin E is first detected at stage V, surrounding the entire head of elongating spermatids; and by stage VII, its localization is shifted mostly to the concave (ventral) side of the spermatid head, co-localizing with two other actin regulatory proteins: actin barbed end capping and bundling protein Eps8 and branched actin nucleation protein Arp3 (Li et al. 2011). However, by stage VIII of the epithelium cycle, drebrin E is downregulated considerably (Li et al. 2011). Drebrin E appears to regulate actin organization in the testis via its association with Apr3 with high affinity, but not Eps8, playing a role in branched actin polymerization. Interestingly, when adult rats were treated with adjudin, a male contraceptive known to induce exfoliation of germ cells in particular elongating/elongated spermatids from the epithelium (Grima et al. 2001; Cheng et al. 2005; Mruk et al. 2006), the localization of drebrin E was grossly affected (Li et al. 2011), similar to Arp3 (Lie et al. 2010) and Eps8 (Lie et al. 2009) following adjudin treatment. This mislocalization and downregulation of drebrin E alongside with Arp3 and Eps8 thus perturb the proper organization of F-actin at the apical ES, thereby causing a disruption of spermatid adhesion that leads to premature spermatid release (Cheng et al. 2011). Drebrin may also be involved in the crosstalk between actin-based and microtubule-based cytoskeleton. For instance, drebrin A was shown to interact directly and specifically with microtubule (MT) end-binding protein 3 (EB3) in growth cone filopodia, and this interaction induces the formation of MT-containing filopodia (Geraldo et al. 2008). Moreover, drebrin A, which is highly concentrated in dendritic spines (Kojima and Shirao 2007; Ferhat 2012), facilitates the entry of MT into dendritic spines following its overexpression in neurons cultured in vitro, resembling the phenotype following an increase in the concentration of stabilized F-actin filaments in neurons (Merriam et al. 2013). Interestingly, a mis-localization of Arp3 and a considerable surge in Arp3/N-WASP interaction were detected following the knockdown of EB1, a plus-end-binding protein of MT, in Sertoli cells with an established TJ-permeability barrier, thereby impeding MT organization as well as proper organization of actin microfilaments of ES at the cell-cell interface to maintain the barrier function (Tang et al. 2015). This finding illustrates that EB1, a close cousin of EB3, is involved in both MT and F-actin regulations. Taken together, these data suggest that drebrin E may be involved in regulating MT dynamics via its interaction with Arp3 and EB1 in Sertoli cells, which will need further verification in future investigation.

17.4 Drebrin E and Ectoplasmic Specialization (ES) Dynamics

Apical ES is a hybrid atypical adherens junction because this cell-cell anchoring junction is constituted by proteins that are found in AJ (e.g., nectins, afadins, N-cadherin), TJ (e.g., JAM-C, CAR), gap junction (e.g., Cx43), and focal adhesion complex (e.g., $\alpha6\beta1$-integrin, laminin-$\alpha3$, laminin-$\beta3$, laminin-$\gamma3$, FAK) in other epithelia, illustrating it is composed of proteins more than just components of adherens junctions (Wong et al. 2008a; Cheng and Mruk 2010). In the rat testis, the stage-specific and spatiotemporal expression of drebrin E at the apical ES resembles that of Arp3 (Li et al. 2011). Arp3 is a component of the Arp2/3 protein complex which when activated by N-WASP upstream triggers barbed end actin nucleation (Weaver et al. 2003), causing branched actin polymerization on preexisting actin filaments. This thus converts existing actin microfilaments from a linear to a branched configuration, destabilizing adhesion protein complexes at the site, facilitating endocytic vesicle-mediated protein trafficking events (Cheng and Mruk 2011). During the epithelial cycle, Arp3 expresses predominantly at the apical ES in stage VII tubules, mostly at the concave side of the head of step 19 spermatids (Lie et al. 2010), converting the site into an ultrastructure known as apical tubulobulbar complex (TBC) (Vogl et al. 2013), which is also the site where endocytic vesicle-mediated trafficking takes place extensively including protein endocytosis, transcytosis, and recycling so that "old" apical ES proteins can be recycled to assemble "new" apical ES when step 8 spermatids arise at stage VIII of the cycle. However, at stage VIII when spermiation takes place and the apical ES undergoes degeneration, Arp3 becomes downregulated considerably and diminished to an almost undetectable level (Lie et al. 2010). Drebrin E, besides expresses similarly to Arp3 spatiotemporally during the epithelial cycle, also has high affinity to bind to Arp3, illustrating that drebrin E may recruit Arp3 to the ES to induce F-actin reorganization and to alter apical ES dynamics to facilitate spermiation. It has been reported that AJ integrity in endothelial cells is stabilized by drebrin since the knockdown of drebrin by RNAi in human umbilical vein endothelial cell (HUVEC), an endothelial cell line, leads to impairment of endothelial cell-cell junctions by causing mis-localization of afadin, thereby impeding the localization of nectin-2 and nectin-3 at the cell-cell interface (Rehm et al. 2013). It is of interest to note that treatment of Sertoli cells with a functional TJ barrier in vitro with TNFα and TGF$\beta3$ that downregulates drebrin E expression also increases structural interaction between drebrin E and Arp3 (Li et al. 2011), illustrating an increase in the intrinsic branched actin polymerization activity. This observation is important since it suggests that cytokines may be playing a role in regulating F-actin organization. Studies have shown that TNFα and TGFβ3 as well as their specific receptors are expressed stage-specifically during the epithelial cycle as reviewed (Li et al. 2009), and they also work in concert with testosterone to regulate endocytic vesicle-mediated protein trafficking events at the basal ES (Yan et al. 2008; Xia et al. 2009). These findings seemingly suggest that drebrin E may also be involved in apical ES endocytic function. This notion is supported by findings of a recent report that drebrin E knockdown by RNAi in Coca2 cells caused mis-localization of apical proteins involved in directing recycling vesicles for lysosome-mediated protein degradation (Vacca et al. 2014), illustrating the involvement of drebrin E in degradation function of lysosomes via its effects on endocytic vesicle-mediated protein trafficking in intestine epithelial cells. In summary, drebrin E serves as a platform to recruit actin regulatory proteins such as Arp3 to the apical ES for actin remodeling. It also participates in the endocytic vesicle-mediated trafficking events to confer plasma membrane plasticity, accommodate changes in spermatid shape, promote germ cell transport, and induce junction restructuring.

17.5 Drebrin, Cofilin, and the Arp2/3 Complex and F-actin Organization

Actin depolymerization factor (ADF, also known as destrin)/cofilin is a family of actin-binding proteins composed of destrin, cofilin-1 (in non-muscle cells, also known as n-cofilin), and cofilin-2 (in muscle cells) (Bowman et al. 2000; Shirao and Gonzalez-Billault 2013). Cofilin has the intrinsic activity to sever and disassemble/depolymerize actin microfilaments, creating new filament fragments with both barbed (+) and pointed (−) ends, so that the Arp2/3 complex can induce branched actin polymerization at the barbed end, also the fast growing end of an actin filament (Ichetovkin et al. 2002), which is essential for rapid turnover of F-actin in mammalian cells for microfilament remodeling (De La Cruz and Gardel 2015; Ohashi 2015; Rust 2015; Bernstein and Bamburg 2010), including Sertoli cells in response to changes in cellular environment in the epithelium during the epithelial cycle. Cofilin also depolymerizes actin monomers from the pointed (−) end of an existing microfilament, which can be recycled to the barbed (+) end for actin nucleation by working in concert with the Arp2/3 complex (Condeelis 2001; DesMarais et al. 2004) to rapidly create a network of branched actin filaments. Studies in synaptic biology have also shown that ADF/cofilin is working in close contact with drebrin for the formation of dendritic spine from dendritic filopodia, involving both actin filaments and microtubules (Shirao and Gonzalez-Billault 2013). This is analogous to Sertoli cells establishing close contacts with developing spermatids during spermiogenesis, such as the establishment of apical ES to confer spermatid transport. In short, emerging evidence has suggested that drebrin is playing a crucial role in mediating changes in the organization of F-actin in specific cellular structures, such as the ES, by working in concert with cofilin and the Arp2/3 complex to modify F-actin organization. For instance, drebrin has been shown to oppose the ability of cofilin to promote actin severing and depolymerization through competitive binding to F-actin and vice versa (Zhao et al. 2006; Grintsevich and Reisler 2014). Additionally, drebrin also contains an ADF homology domain (Lappalainen et al. 1998; Shirao 1995; Kojima et al. 1993; Shirao et al. 1988), making it structurally related to the cofilin. In fact, besides drebrin E (Li et al. 2011), cofilin-1 is a component of the apical ES in the rat testis, restrictively expressed at the concave side of the spermatid head at late stage VII–stage VIII of the epithelial cycle in an ultrastructure called apical tubulobulbar complex (TBC) (Guttman et al. 2004) which is also the site where extensive endocytic vesicle-mediated protein trafficking takes place due to the appearance of multiple giant endocytic vesicles (Vogl et al. 2013, 2014). Thus, it is conceivable that a disruption of cofilin-1 function at the apical ES/apical TBC would impede spermatogenesis. It is of interest to note that knockout of the cofilin-1 gene in mice leads to embryonic lethality (Gurniak et al. 2005). For instance, cofilin-1 mutant embryos had defects in neural crest-derived tissues, lacking actin microfilament bundles in neural crest cells, thereby impairing neural crest cell migration which thus impedes neural development, causing embryonic lethality (Gurniak et al. 2005). As such, the function of cofilin-1 in spermatogenesis is not known. Studies have shown that cofilin can be inactivated via phosphorylation on its Ser-3 residue by LIM kinases (LIMKs, Lin-11/Isl-1/Mec-3 kinases), namely, LIMK1 and LIMK2 (Takahashi et al. 2003), or reactivated via dephosphorylation at Ser-3 residue by phosphatases such as Slingshot (SSH encoded by the *Ssh* gene) which is a family of phosphatases known to dephosphorylate ADF/cofilin, thereby modulating cofilin function (Niwa et al. 2002). In this context, it is of interest to note that LIMK2 is known to be expressed predominantly by germ cells in the mouse (Takahashi et al. 2002), whereas LIMK1 is expressed in the rat testis (Lui et al. 2003). Furthermore, LIMK1 is working in concert with ROCK upstream and β2-integrin at the apical ES to regulate spermatid adhesion in the testis (Lui et al. 2003). However, the role of LIMK1 and LIMK2 in cofilin-1 activation in the testis remains unexplored. Interestingly, LIMK1-deficient mice developed normally and were fertile except displaying abnormalities in hippocampal dendritic spine structure (Meng et al. 2002). On the other hand, the testis of LIMK2-deficient mice when all three LIMK2 isoforms were disrupted was reduced by 20% at 2–4 months of age by weight, and >50% of the tubules were devoid of spermatids (including round, elongating, and elongated spermatids) with fewer spermatocytes and had signs of germ cell apoptosis and necrosis (Takahashi et al. 2002, 2003), illustrating a gross disruption of spermatogenesis. However, these mice remain fertile (Takahashi et al. 2002) since it is noted that rodents remain fertile even with a decline of 90% of spermatogenetic outputs (Robaire 2003). Interestingly, *Ssh-3* KO mice were fertile and healthy (Kousaka et al. 2008). However, it remains to be assessed if the triple KO of the three Slingshot phosphatases SSH1, SSH2, and SSH3 which are found in mammalian tissues (Ohta et al. 2003) would affect spermatogenesis in particular spermiogenesis and ES function. Since specific KO of cofilin-1 in podocytes (also known as glomerular visceral epithelial cells) in the kidney was recently found to cause failure in selective filtration barrier function (Garg et al. 2010), thus, it is noteworthy to examine the phenotype of the Sertoli cell-specific KO of cofilin-1 *vs.* triple KO of *Ssh-1*, *Ssh-2*, and *Ssh-3* in mice.

17.6 Drebrin E, Cofilin-1, the Arp2/3 Complex, and Spermiation

Based on such intimate functional and structural relationship between drebrin, cofilin, and the Arp2/3 complex as briefly discussed herein, we propose a hypothetical model depicted in Fig. 17.3, illustrating their likely involvement in actin remodeling at the apical ES. In brief, at stage VII, the expression of drebrin E at the apical ES, mostly at the concave side of spermatid heads known as apical TBC, is robust and intense, which diminishes somewhat by stages late VII–early VIII and considerably weakened in stage VIII when spermiation begins to take place (Li et al. 2011). On the other hand, cofilin-1 is intensely expressed at the apical ES/apical TBC in late stage VII–stage VIII tubules at the same site (Guttman et al. 2004) to confer endocytic vesicle-mediated protein trafficking events including endocytosis, transcytosis, and recycling to facilitate recycling of the "old" apical ES proteins (e.g., β1-integrin, laminins, nectin-2, nectin-3, afadin) for the assembly of "new" apical ES when step 8 spermatids arise at stage VIII of the cycle. However, cofilin-1 remains considerably expressed at the apical ES (Guttman et al. 2004) when drebrin E (Li et al. 2011) and Arp3 (Lie et al. 2010) are virtually not expressed at stage VIII. This downregulation of drebrin E is necessary since both drebrin E and cofilin-1 competes to bind onto the actin microfilaments at the same site (Zhao et al. 2006; Grintsevich and Reisler 2014). Thus, cofilin-1 further severs and depolymerizes actin microfilaments at the apical ES, causing truncation of actin microfilaments, preparing the eventual release of fully developed spermatids (i.e., spermatozoa) at spermiation, possibly in conjunction with the action of MMP-2 (Siu and Cheng 2004). It is obvious that the model depicted in Fig. 17.3 will be updated as more data are available in future years.

17.7 Drebrin E and BTB Remodeling During the Epithelial Cycle

The BTB is one of the tightest blood-tissue barriers in the mammalian body. However, unlike other tissue barriers such as the blood-brain barrier and the blood-retina barrier (Cheng 2012), the BTB undergoes extensive remodeling during the epithelial cycle to accommodate the transport of germ cells of >40-μm in diameter across the barrier. For instance, preleptotene spermatocytes derived from type B spermatogonia that first appear in stage VII tubules and are connected in clones by intercellular bridges must be transported across the BTB at stage VIII, so that spermatocytes can be prepared for meiosis that takes place in stage XIV tubules in the rat testis (Xiao et al. 2014; Parvinen 1982; Leblond and Clermont 1952). Thus, extensive junction restructuring including actin- and MT-based cytoskeleton reorganization is required to accommodate these events. Drebrin is known to be associated with cell junctions in a number of epithelial cells (e.g., MDBK cells (Madin and Darby bovine kidney epithelial cells) and A431 cells (human epithelial carcinoma cells) in which drebrin is localized to actin microfilament bundles at the cell-cell contact sites at AJ (Peitsch et al. 1999, 2001; Keon et al. 2000). Drebrin also binds to the C-terminal cytoplasmic domain of connexin 43, involving in the maintenance of gap junction function (Butkevich et al. 2004). In the rat testis, drebrin E is also found at the basal ES, a testis-specific AJ which also coexists with GJ, at the BTB based on studies by immunohistochemistry and immunofluorescence analysis (Li et al. 2011). Drebrin E is highly expressed at the BTB at stages IV–V of the epithelial cycle, but it diminishes somewhat at stage VII and considerably weakened by stage VIII (Li et al. 2011), coinciding with the restructuring of BTB. This apparent downregulation is necessary for the transport of preleptotene spermatocytes across the barrier. It is of interest to note that the expression of Arp3 at the basal ES/BTB is most intensely in stage VIII tubules (Lie et al. 2010). Since drebrin E binds to the same site as cofilin-1 in the actin microfilament and their binding is thus mutually exclusive (Zhao et al. 2006; Grintsevich and Reisler 2014), the transient downregulation of drebrin E at stage VIII may be necessary to facilitate cofilin-1 to exert its actin severing/depolymerization intrinsic activity on the actin microfilaments at the basal ES. These changes thus lead to the formation of more truncated actin microfilaments at the site, in conjunction with a robust expression of Arp3 at the BTB in stage VIII tubules (Lie et al. 2010), thus enhancing branched actin polymerization, thereby converting the F-actin network from a bundled to a branched/unbundled configuration. These changes, in turn, facilitate endocytic vesicle-mediated protein trafficking events so that BTB proteins (e.g., occludin, ZO-1, N-cadherin, β-catenin) from the "old" BTB located at the apical region of the preleptotene spermatocytes can undergo endocytosis, transcytosis, and recycling for the assembly of the "new" BTB behind these germ cells under transport at the barrier (Fig. 17.4). Other biomolecules, such as cytokines (e.g., TGF-ßs, TNF-α) and testosterone and/or its receptor which are known to modulate endocytic vesicle-mediated protein trafficking events (Yan et al. 2008; Xia et al. 2009), are also involved in BTB remodeling as depicted in Fig. 17.4.

17.8 Future Perspectives and Concluding Remarks

Based on recent advances in the field, we conclude that drebrin plays an important role in regulating both apical ES and basal ES/BTB in the testis. Drebrin is also crucial for the transport of germ cells across the BTB and the seminiferous epithelium during the epithelial cycle as noted in both Figs. 17.3 and 17.4. However, many

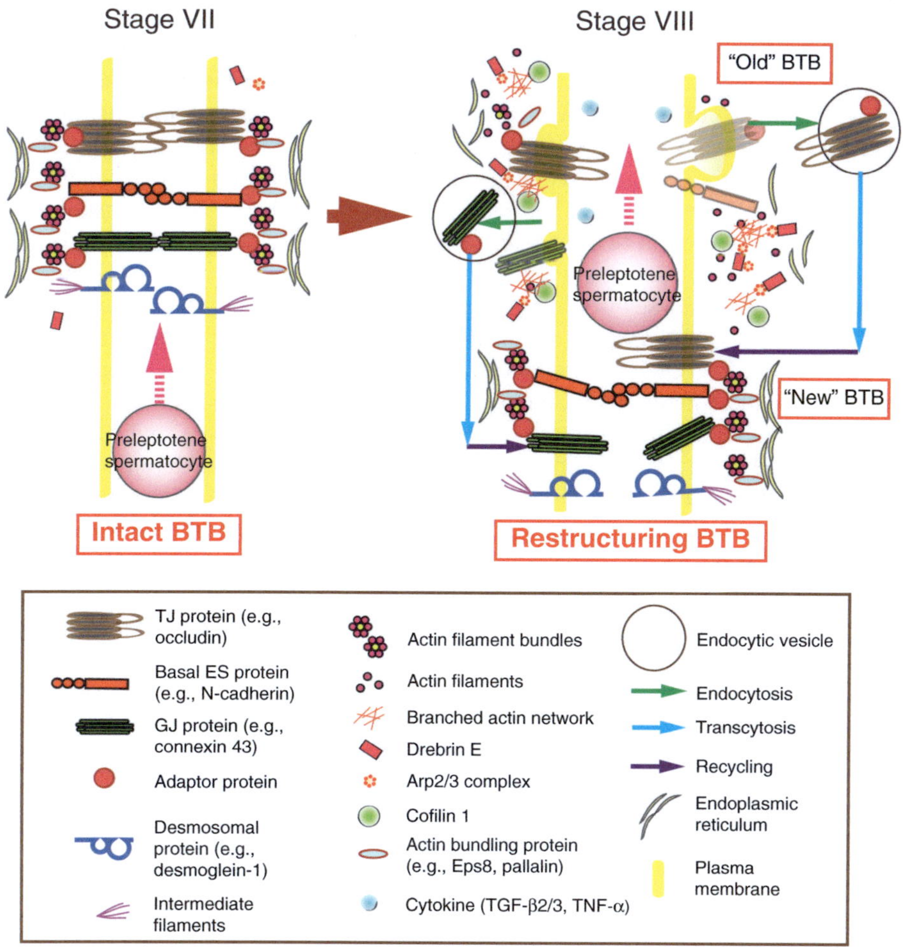

Fig. 17.4 A hypothetical model regarding the involvement of drebrin E in the remodeling of the BTB to facilitate the transport of preleptotene spermatocytes across immunological barrier. At stage VII, the BTB is intact which is supported by the actin microfilament bundles at the basal ES due to the upregulation of actin-bundling proteins such as Eps8, palladin, ezrin, plastin 3, and others (see *left panel*). However, preleptotene spermatocytes transformed from type B spermatogonia at stage VII begin to be transported across the immunological barrier at late stage VII. In late stage VII–early stage VIII, an upregulation of drebrin E recruits Arp3 to the site, so that the intrinsic branched actin polymerization activity of the Arp3 converts actin microfilaments from a bundled to an unbundled/branched configuration; this is also associated with a surge of cofilin-1 which further assists the breakdown of actin microfilaments coupled with a downregulation on the expression of actin-bundling proteins such as Eps8, palladin, ezrin, and plastin 3. These changes thus promote endocytic vesicle-mediated protein trafficking events including endocytosis, transcytosis, and recycling. Thus, "old" BTB above the preleptotene spermatocytes can be transcytosed and recycled to assemble "new" BTB below these spermatocytes to maintain the BTB integrity while preleptotene spermatocytes are being transported across the barrier

open questions remain unanswered. Since the intrinsic actin-binding capacity of drebrin is modulated by its phosphorylation and dephosphorylation, the kinases, phosphatases, and other signaling molecule(s) that are involved in drebrin activation at the ES remain to be identified. Studies in recent years have shown that cyclin-dependent kinase 5 (Cdk5)-p35, a Ser/Thr protein kinase, can phosphorylate drebrin to modulate neuronal migration (Tanabe et al. 2014). Phosphatase and tensin homolog (PTEN), a known binding partner of drebrin (Table 17.1), was also shown to dephosphorylate drebrin at Ser-647 in the brain to regulate neuronal activity (Kreis et al. 2013). It will be of interest to determine if these proteins or other testis-specific kinases and phosphatases would play a role in modulating the function of drebrin. Furthermore, fascin 1, an actin filament-bundling protein in the testis, is a component of apical ES (Gungor-Ordueri et al. 2014a). Interestingly, its stage-specific expression pattern in the seminiferous epithelium closely mimics drebrin E in which fascin 1 is predominantly expressed at the apical ES at stage VII but quickly diminishes by stage VIII (Gungor-Ordueri et al. 2014a). An earlier study has reported that the actin-bundling activity of fascin can be inhibited by drebrin in filopodia (Sasaki et al. 1996). Is it possible that drebrin E modulates or fine-tunes the actin-bundling activity of fascin 1 at the apical ES? Also, what are the phenotypes following an inactivation of drebrin E in the testis? Additionally, the physiological relationship of drebrin and cofilin-1 in the testis will need to be better defined. These questions need to be addressed in future studies.

Acknowledgments The work was supported in part by grants from the National Institutes of Health (R01 HD056034 and U54 HD029990 Project 5, to CYC).

References

Amann RP (2008) The cycle of the seminiferous epithelium in humans: a need to revisit? J Androl 29:469–487. doi:10.2164/jandrol.107.004655

Amann RP, Howards SS (1980) Daily spermatozoal production and epididymal spermatozoal reserves of the human male. J Urol 124:211–215

Aoki C, Kojima N, Sabaliauskas N, Shah L, Ahmed TH, Oakford J, Ahmed T, Yamazaki H, Hanamura K, Shirao T (2009) Drebrin a knockout eliminates the rapid form of homeostatic synaptic plasticity at excitatory synapses of intact adult cerebral cortex. J Comp Neurol 517(1):105–121. doi:10.1002/cne.22137

Bardin CW, Cheng CY, Musto NA, Gunsalus GL (1988) The Sertoli cell. In: Knobil E, Neill JD, Ewing LL, Greenwald GS, Markert CL, Pfaff DW (eds) The physiology of reproduction, vol 1. Raven Press, New York, NY, pp 933–974

Bernstein BW, Bamburg JR (2010) ADF/cofilin: a functional node in cell biology. Trends Cell Biol 20:187–195. doi:10.1016/j.tcb.2010.01.001

Billig H, Furuta I, Rivier C, Tapanainen J, Parvinen M, Hsueh AJ (1995) Apoptosis in testis germ cells: developmental changes in gonadotropin dependence and localization to selective tubule stages. Endocrinology 136:5–12

Bowman GD, Nodelman IM, Hong Y, Chua NH, Lindberg U, Schutt CE (2000) A comparative structural analysis of the ADF/cofilin family. Proteins 41:374–384

Butkevich E, Hulsmann S, Wenzel D, Shirao T, Duden R, Majoul I (2004) Drebrin is a novel connexin-43 binding partner that links gap junctions to the submembrane cytoskeleton. Curr Biol 14:650–658. doi:10.1016/j.cub.2004.03.063
Cheng CY (2012) In: Cheng CY (ed) Biology and regulation of blood-tissue barriers. Landes Bioscience/Springer Science+Business Media, LLC, Austin, TX, pp 1–361
Cheng CY, Mruk DD (2002) Cell junction dynamics in the testis: Sertoli-germ cell interactions and male contraceptive development. Physiol Rev 82:825–874
Cheng CY, Mruk DD (2010) A local autocrine axis in the testes that regulates spermatogenesis. Nat Rev Endocrinol 6:380–395
Cheng CY, Mruk DD (2011) Regulation of spermiogenesis, spermiation and blood-testis barrier dynamics: novel insights from studies on Eps8 and Arp3. Biochem J 435:553–562
Cheng CY, Mruk DD (2012) The blood-testis barrier and its implications for male contraception. Pharmacol Rev 64:16–64. doi:10.1124/pr.110.002790
Cheng CY, Mruk DD (2015) Biochemistry of Sertoli cell/germ cell junctions, germ cell transport, and spermiation in the seminiferous epithelium. In: Griswold MD (ed) Sertoli cell biology, 2nd edn. Elsevier, Amsterdam, pp 333–383. doi:10.1016/B978-0-12-417047-6.00012.0
Cheng CY, Mruk D, Silvestrini B, Bonanomi M, Wong CH, Siu MK, Lee NP, Lui WY, Mo MY (2005) AF-2364 [1-(2,4-dichlorobenzyl)-1H-indazole-3-carbohydrazide] is a potential male contraceptive: a review of recent data. Contraception 72:251–261. doi:10.1016/j.contraception.2005.03.008
Cheng CY, Lie PPY, Wong EWP, Mruk DD, Silvestrini B (2011) Adjudin disrupts spermatogenesis via the action of some unlikely partners: Eps8, Arp2/3 complex, drebrin E, PAR6 and 14-3-3. Spermatogenesis 1:291–297
Clermont Y (1972) Kinetics of spermatogenesis in mammals: seminiferous epithelium cycle and spermatogonial renewal. Physiol Rev 52:198–235
Clermont Y, Leblond CP (1955) Spermiogenesis of man, monkey, ram and other mammals as shown by the periodic acid-Schiff technique. Am J Anat 96:229–253. doi:10.1002/aja.1000960203
Condeelis J (2001) How is actin polymerization nucleated in vivo? Trends Cell Biol 11: 288–293
De La Cruz EM, Gardel ML (2015) Actin mechanics and fragmentation. J Biol Chem 290:17137–17144. doi:10.1074/jbc.R115.636472
DesMarais V, Macaluso F, Condeelis J, Bailly M (2004) Synergistic interaction between the Arp2/3 complex and cofilin drives stimulated lamellipod extension. J Cell Sci 117:3499–3510
Ehmcke J, Schlatt S (2006) A revised model for spermatogonial expansion in man: lessons from non-human primates. Reproduction 132:673–680
Ferhat L (2012) Potential role of drebrin a, an f-actin binding protein, in reactive synaptic plasticity after pilocarpine-induced seizures: functional implications in epilepsy. Int J Cell Biol 2012:474351. doi:10.1155/2012/474351
Franca LR, Auharek SA, Hess RA, Dufour JM, Hinton BT (2012) Blood-tissue barriers: morphofunctional and immunological aspects of the blood-testis and blood-epididymal barriers. Adv Exp Med Biol 763:237–259
Garg P, Verma R, Cook L, Soofi A, Venkatareddy M, George B, Mizuno K, Gurniak C, Witke W, Holzman LB (2010) Actin-depolymerizing factor cofilin-1 is necessary in maintaining mature podocyte architecture. J Biol Chem 285(29):22676–22688. doi:10.1074/jbc.M110.122929
Geraldo S, Khanzada UK, Parsons M, Chilton JK, Gordon-Weeks PR (2008) Targeting of the F-actin-binding protein drebrin by the microtubule plus-tip protein EB3 is required for neuritogenesis. Nat Cell Biol 10(10):1181–1189. doi:10.1038/ncb1778
Gordon-Alonso M, Rocha-Perugini V, Alvarez S, Ursa A, Izquierdo-Useros N, Martinez-Picado J, Munoz-Fernandez MA, Sanchez-Madrid F (2013) Actin-binding protein drebrin regulates HIV-1-triggered actin polymerization and viral infection. J Biol Chem 288(39):28382–28397. doi:10.1074/jbc.M113.494906
Grima J, Silvestrini B, Cheng CY (2001) Reversible inhibition of spermatogenesis in rats using a new male contraceptive, 1-(2,4-dichlorobenzyl)-indazole-3-carbohydrazide. Biol Reprod 64(5):1500–1508

Grintsevich EE, Reisler E (2014) Drebrin inhibits cofilin-induced severing of F-actin. Cytoskeleton 71(8):472–483. doi:10.1002/cm.21184

Gungor-Ordueri NE, Celik-Ozenci C, Cheng CY (2014a) Fascin 1 is an actin filament-bundling protein that regulates ectoplasmic specialization dynamics in the rat testis. Am J Physiol Endocrinol Metab 307:E738–E753

Gungor-Ordueri NE, Tang EI, Celik-Ozenci C, Cheng CY (2014b) Ezrin is an actin binding protein that regulates Sertoli cell and spermatid adhesion during spermatogenesis. Endocrinology 155(10):3981–3995. doi:10.1210/en.2014-1163

Gurniak CB, Perlas E, Witke W (2005) The actin depolymerizing factor n-cofilin is essential for neural tube morphogenesis and neural crest cell migration. Dev Biol 278:231–241. doi:10.1016/j.ydbio.2004.11.010

Guttman JA, Obinata T, Shima J, Griswold M, Vogl AW (2004) Non-muscle cofilin is a component of tubulobulbar complexes in the testis. Biol Reprod 70:805–812. doi:10.1095/biolreprod.103.022723

Hayashi K, Ishikawa R, Kawai-Hirai R, Takagi T, Taketomi A, Shirao T (1999) Domain analysis of the actin-binding and actin-remodeling activities of drebrin. Exp Cell Res 253(2):673–680. doi:10.1006/excr.1999.4663

Hermo L, Pelletier RM, Cyr DG, Smith CE (2010) Surfing the wave, cycle, life history, and genes/proteins expressed by testicular germ cells. Part 5: intercellular junctions and contacts between germ cells and Sertoli cells and their regulatory interactions, testicular cholesterol, and genes/proteins associated with more than one germ cell generation. Microsc Res Tech 73:409–494

Hess RA, de Franca LR (2008) Spermatogenesis and cycle of the seminiferous epithelium. Adv Exp Med Biol 636:1–15

Ichetovkin I, Grant W, Condeelis J (2002) Cofilin produces newly polymerized actin filaments that are preferred for dendritic nucleation by the Arp2/3 complex. Curr Biol 12(1):79–84. doi:10.1016/s0960-9822(01)00629-7

Ishikawa R, Hayashi K, Shirao T, Xue Y, Takagi T, Sasaki Y, Kohama K (1994) Drebrin, a development-associated brain protein from rat embryo, causes the dissociation of tropomyosin from actin filaments. J Biol Chem 269:29928–29933

Johnson L, Petty CS, Neaves WB (1980) A comparative study of daily sperm production and testicular composition in humans and rats. Biol Reprod 22:1233–1243

Jung G, Kim EJ, Cicvaric A, Sase S, Groger M, Hoger H, Sialana FJ, Berger J, Monje FJ, Lubec G (2015) Drebrin depletion alters neurotransmitter receptor levels in protein complexes, dendritic spine morphogenesis and memory-related synaptic plasticity in the mouse hippocampus. J Neurochem 134:327–339. doi:10.1111/jnc.13119

Keon BH, Jedrzejewski PT, Paul DL, Goodenough DA (2000) Isoform specific expression of the neuronal F-actin binding protein, drebrin, in specialized cells of stomach and kidney epithelia. J Cell Sci 113:325–336

Kojima N, Shirao T (2007) Synaptic dysfunction and disruption of postsynaptic drebrin-actin complex: a study of neurological disorders accompanied by cognitive deficits. Neurosci Res 58:1–5. doi:10.1016/j.neures.2007.02.003

Kojima N, Shirao T, Obata K (1993) Molecular cloning of a developmentally regulated brain protein, chicken drebrin A and its expression by alternative splicing of the drebrin gene. Mol Brain Res 19:101–114

Kojima N, Hanamura K, Yamazaki H, Ikeda T, Itohara S, Shirao T (2010) Genetic disruption of the alternative splicing of drebrin gene impairs context-dependent fear learning in adulthood. Neuroscience 165:138–150. doi:10.1016/j.neuroscience.2009.10.016

Kousaka K, Kiyonari H, Oshima N, Nagafuchi A, Shima Y, Chisaka O, Uemura T (2008) Slingshot-3 dephosphorylates ADF/cofilin but is dispensable for mouse development. Genesis 46:246–255. doi:10.1002/dvg.20389

Kreis P, Hendricusdottir R, Kay L, Papageorgiou IE, van Diepen M, Mack T, Ryves J, Harwood A, Leslie NR, Kann O, Parsons M, Eickholt BJ (2013) Phosphorylation of the actin binding protein Drebrin at S647 is regulated by neuronal activity and PTEN. PLoS One 8:e71957. doi:10.1371/journal.pone.0071957 de Kretser DM, Kerr JB (1988) The cytology of the testis. In: Knobil E, Neill JB, Ewing LL, Greenwald GS, Markert CL, Pfaff DW (eds) The Physiology of Reproduction, vol 1. Raven Press, New York, NY, pp 837–932

Lappalainen P, Kessels MM, Cope MJ, Drubin DG (1998) The ADF homology (ADF-H) domain: a highly exploited actin-binding module. Mol Biol Cell 9:1951–1959

Leblond C, Clermont Y (1952) Definition of the stages of the cycle of the seminiferous epithelium in the rat. Ann N Y Acad Sci 55:548–573

Li MWM, Mruk DD, Lee WM, Cheng CY (2009) Cytokines and junction restructuring events during spermatogenesis in the testis: An emerging concept of regulation. Cytokine Growth Factor Rev 20:329–338

Li MW, Xiao X, Mruk DD, Lam YL, Lee WM, Lui WY, Bonanomi M, Silvestrini B, Cheng CY (2011) Actin-binding protein drebrin E is involved in junction dynamics during spermatogenesis. Spermatogenesis 1:123–136. doi:10.4161/spmg.1.2.16393

Li N, Mruk DD, Cheng CY (2015a) Actin binding proteins in blood-testis barrier function. Curr Opin Endocrinol Diabetes Obes 22:238–247

Li N, Mruk DD, CKC W, Han D, Lee WM, Cheng CY (2015b) Formin 1 regulates ectoplamic specialization in the rat testis through its actin nucleation and bundling activity. Endocrinology 156:2969–2983. doi:10.1210/en2015-1161

Li N, Mruk DD, Wong CKC, Lee WM, Han D, Cheng CY (2015c) Actin bundling protein plastin 3 is a regulator of ectoplasmic specialization (ES) dynamics during spermatogenesis in the rat testis. FASEB J 29:3788–3805. doi:10.1096/fj14-267997

Lie PPY, Mruk DD, Lee WM, Cheng CY (2009) Epidermal growth factor receptor pathway substrate 8 (Eps8) is a novel regulator of cell adhesion and the blood-testis barrier integrity in the seminiferous epithelium. FASEB J 23:2555–2567

Lie PPY, Chan AYN, Mruk DD, Lee WM, Cheng CY (2010) Restricted Arp3 expression in the testis prevents blood-testis barrier disruption during junction restructuring at spermatogenesis. Proc Natl Acad Sci U S A 107:11411–11416

Lui WY, Lee WM, Cheng CY (2003) Sertoli-germ cell adherens junction dynamics in the testis are regulated by RhoB GTPase via the ROCK/LIMK signaling pathway. Biol Reprod 68:2189–2206

Majoul I, Shirao T, Sekino Y, Duden R (2007) Many faces of drebrin: from building dendritic spines and stabilizing gap junctions to shaping neurite-like cell processes. Histochem Cell Biol 127:355–361. doi:10.1007/s00418-007-0273-y

Mammoto A, Sasaki T, Asakura T, Hotta I, Imamura H, Takahashi K, Matsuura Y, Shirao T, Takai Y (1998) Interactions of drebrin and gephyrin with profilin. Biochem Biophys Res Commun 243(1):86–89. doi:10.1006/bbrc.1997.8068

Meng Y, Zhang Y, Tregoubov V, Janus C, Cruz L, Jackson M, WY L, MacDonald JF, Wang JY, Falls DL, Jia Z (2002) Abnormal spine morphology and enhanced LTP in LIMK-1 knockout mice. Neuron 35:121–133

Merriam EB, Millette M, Lumbard DC, Saengsawang W, Fothergill T, Hu X, Ferhat L, Dent EW (2013) Synaptic regulation of microtubule dynamics in dendritic spines by calcium, F-actin, and drebrin. J Neurosci 33:16471–16482. doi:10.1523/JNEUROSCI.0661-13.2013

Mikati MA, Grintsevich EE, Reisler E (2013) Drebrin-induced stabilization of actin filaments. J Biol Chem 288:19926–19938. doi:10.1074/jbc.M113.472647

Mruk DD, Cheng CY (2004a) Cell-cell interactions at the ectoplasmic specialization in the testis. Trends Endocrinol Metab 15:439–447

Mruk DD, Cheng CY (2004b) Sertoli-Sertoli and Sertoli-germ cell interactions and their significance in germ cell movement in the seminiferous epithelium during spermatogenesis. Endocr Rev 25:747–806

Mruk DD, Wong CH, Silvestrini B, Cheng CY (2006) A male contraceptive targeting germ cell adhesion. Nat Med 12:1323–1328. doi:10.1038/nm1420

Niwa R, Nagata-Ohashi K, Takeichi M, Mizuno K, Uemura T (2002) Control of actin reorganization by Slingshot, a family of phosphatases that dephosphorylate ADF/cofilin. Cell 108:233–246

O'Donnell L (2014) Mechanisms of spermatogenesis and spermiation and how they are disturbed. Spermatogenesis 4:e979623. doi:10.4161/21565562.2014.979623

O'Donnell L, Nicholls PK, O'Bryan MK, McLachlan RI, Stanton PG (2011) Spermiation: the process of sperm release. Spermatogenesis 1:14–35

Ohashi K (2015) Roles of cofilin in development and its mechanisms of regulation. Develop Growth Differ 57:275–290. doi:10.1111/dgd.12213

Ohta Y, Kousaka K, Nagata-Ohashi K, Ohashi K, Muramoto A, Shima Y, Niwa R, Uemura T, Mizuno K (2003) Differential activities, subcellular distribution and tissue expression patterns of three members of Slingshot family phosphatases that dephosphorylate cofilin. Genes Cells 8:811–824

Orth JM (1982) Proliferation of Sertoli cells in fetal and postnatal rats: A quantitative autoradiographic study. Anat Rec 203:485–492

Parvinen M (1982) Regulation of the seminiferous epithelium. Endocr Rev 3:404–417

Peitsch WK, Grund C, Kuhn C, Schnolzer M, Spring H, Schmelz M, Franke WW (1999) Drebrin is a widespread actin-associating protein enriched at junctional plaques, defining a specific microfilament anchorage system in polar epithelial cells. Eur J Cell Biol 78:767–778. doi:10.1016/s0171-9335(99)80027-2

Peitsch WK, Hofmann I, Pratzel S, Grund C, Kuhn C, Moll I, Langbein L, Franke WW (2001) Drebrin particles: components in the ensemble of proteins regulating actin dynamics of lamellipodia and filopodia. Eur J Cell Biol 80:567–579. doi:10.1078/0171-9335-00194

Pelletier RM (2011) The blood-testis barrier: the junctional permeability, the proteins and the lipids. Prog Histochem Cytochem 46:49–127. doi:10.1016/j.proghi.2011.05.001

Qian X, Mruk DD, Wong EWP, Lie PPY, Cheng CY (2013) Palladin is a regulator of actin filament bundles at the ectoplasmic specialization in the rat testis. Endocrinology 154: 1907–1920

Qian X, Mruk DD, Cheng YH, Tang EI, Han D, Lee WM, Wong EW, Cheng CY (2014) Actin binding proteins, spermatid transport and spermiation. Semin Cell Dev Biol 30:75–85. doi:10.1016/j.semcdb.2014.04.018

Rehm K, Panzer L, van Vliet V, Genot E, Linder S (2013) Drebrin preserves endothelial integrity by stabilizing nectin at adherens junctions. J Cell Sci 126:3756–3769. doi:10.1242/jcs.129437

Robaire B (2003) Advancing towards a male contraceptive: a novel approach from an unexpected direction. Trends Pharmacol Sci 24:326–328

Russell LD (1977a) Movement of spermatocytes from the basal to the adluminal compartment of the rat testis. Am J Anat 148:313–328

Russell LD (1977b) Observations on rat Sertoli ectoplasmic (junctional) specializations in their association with germ cells of the rat testis. Tissue Cell 9:475–498

Russell LD (1979) Observations on the inter-relationships of Sertoli cells at the level of the blood-testis barrier: evidence for formation and resorption of Sertoli-Sertoli tubulobulbar complexes during the spermatogenic cycle of the rat. Am J Anat 155:259–279

Russell LD, Peterson RN (1985) Sertoli cell junctions: morphological and functional correlates. Int Rev Cytol 94:177–211

Rust MB (2015) ADF/cofilin: a crucial regulator of synapse physiology and behavior. Cell Mol Life Sci 72:3521–3529. doi:10.1007/s00018-015-1941-z

Sasaki Y, Hayashi K, Shirao T, Ishikawa R, Kohama K (1996) Inhibition by drebrin of the actin-bundling activity of brain fascin, a protein localized in filopodia of growth cones. J Neurochem 66:980–988

Schlatt S, Ehmcke J (2014) Regulation of spermatogenesis: an evolutionary biologist's perspective. Semin Cell Dev Biol 29:2–16. doi:10.1016/j.semcdb.2014.03.007

Sharma S, Grintsevich EE, Phillips ML, Reisler E, Gimzewski JK (2011) Atomic force microscopy reveals drebrin induced remodeling of f-actin with subnanometer resolution. Nano Lett 11:825–827. doi:10.1021/nl104159v

Sharpe RM, McKinnell C, Kivlin C, Fisher JS (2003) Proliferation and functional maturation of Sertoli cells, and their relevance to disorders of testis function in adulthood. Reproduction 125:769–784

Shiraishi-Yamaguchi Y, Sato Y, Sakai R, Mizutani A, Knopfel T, Mori N, Mikoshiba K, Furuichi T (2009) Interaction of Cupidin/Homer2 with two actin cytoskeletal regulators, Cdc42 small GTPase and Drebrin, in dendritic spines. BMC Neurosci 10:25. doi:10.1186/1471-2202-10-25

Shirao T (1995) The roles of microfilament-associated proteins, drebrins, in brain morphogenesis: a review. J Biochem 117:231–236

Shirao T, Gonzalez-Billault C (2013) Actin filaments and microtubules in dendritic spines. J Neurochem 126:155–164. doi:10.1111/jnc.12313

Shirao T, Obata K (1985) Two acidic proteins associated with brain development in chick embryo. J Neurochem 44:1210–1216

Shirao T, Obata K (1986) Immunochemical homology of 3 developmentally regulated brain proteins and their developmental change in neuronal distribution. Brain Res 394:233–244

Shirao T, Kojima N, Kato YS, Obata K (1988) Molecular cloning of a cDNA for the developmentally regulated brain protein, drebrin. Brain Res 464:71–74

Siu MKY, Cheng CY (2004) Interactions of proteases, protease inhibitors, and the β1 integrin/laminin γ3 protein complex in the regulation of ectoplasmic specialization dynamics in the rat testis. Biol Reprod 70:945–964

Su WH, Mruk DD, Lie PPY, Lui WY, Cheng CY (2012) Filamin A is a regulator of blood-testis barrier assembly during postnatal development in the rat testis. Endocrinology 153: 5023–5035

Su WH, Mruk DD, Cheng CY (2013) Regulation of actin dynamics and protein trafficking during spermatogenesis - insights into a complex process. Crit Rev Biochem Mol Biol 48:153–172

Takahashi H, Koshimizu U, Miyazaki J, Nakamura T (2002) Impaired spermatogenic ability of testicular germ cells in mice deficient in the LIM-kinase 2 gene. Dev Biol 241:259–272. doi:10.1006/dbio.2001.0512

Takahashi H, Funakoshi H, Nakamura T (2003) LIM-kinase as a regulator of actin dynamics in spermatogenesis. Cytogenet Genome Res 103:290–298. doi:10.1159/000076815

Tanabe K, Yamazaki H, Inaguma Y, Asada A, Kimura T, Takahashi J, Taoka M, Ohshima T, Furuichi T, Isobe T, Nagata K, Shirao T, Hisanaga S (2014) Phosphorylation of drebrin by cyclin-dependent kinase 5 and its role in neuronal migration. PLoS One 9:e92291. doi:10.1371/journal.pone.0092291

Tang EI, Mok KW, Lee WM, Cheng CY (2015) EB1 Regulates tubulin and actin cytoskeletal networks at the sertoli cell blood-testis barrier in male rats: an in vitro study. Endocrinology 156:680–693. doi:10.1210/en.2014-1720

Terakawa Y, Agnihotri S, Golbourn B, Nadi M, Sabha N, Smith CA, Croul SE, Rutka JT (2013) The role of drebrin in glioma migration and invasion. Exp Cell Res 319:517–528. doi:10.1016/j.yexcr.2012.11.008

Vacca B, Bazellieres E, Nouar R, Harada A, Massey-Harroche D, Le Bivic A (2014) Drebrin E depletion in human intestinal epithelial cells mimics Rab8a loss of function. Hum Mol Genet 23(11):2834–2846. doi:10.1093/hmg/ddt670

Vogl A, Pfeiffer D, Mulholland D, Kimel G, Guttman J (2000) Unique and multifunctional adhesion junctions in the testis: ectoplasmic specializations. Arch Histol Cytol 63:1–15

Vogl AW, Vaid KS, Guttman JA (2008) The Sertoli cell cytoskeleton. Adv Exp Med Biol 636:186–211

Vogl AW, Young JS, Du M (2013) New insights into roles of tubulobulbar complexes in sperm release and turnover of blood-testis barrier. Int Rev Cell Mol Biol 303:319–355

Vogl AW, Du M, Wang XY, Young JS (2014) Novel clathrin/actin-based endocytic machinery associated with junction turnover in the seminiferous epithelium. Semin Cell Dev Biol 30:55–64

Weaver AM, Young ME, Lee WL, Cooper JA (2003) Integration of signals to the Arp2/3 complex. Curr Opin Cell Biol 15:23–30

Weber JE, Russell LD, Wong V, Peterson RN (1983) Three dimensional reconstruction of a rat stage V Sertoli cell: II. Morphometry of Sertoli-Sertoli and Sertoli-germ cell relationships. Am J Anat 167:163–179

Wong EWP, Cheng CY (2009) Polarity proteins and cell-cell interactions in the testis. Int Rev Cell Mol Biol 278:309–353

Wong EW, Mruk DD, Cheng CY (2008a) Biology and regulation of ectoplasmic specialization, an atypical adherens junction type, in the testis. Biochim Biophys Acta 1778:692–708. doi:10.1016/j.bbamem.2007.11.006

Wong EWP, Mruk DD, Cheng CY (2008b) Biology and regulation of ectoplasmic specialization, an atypical adherens junction type, in the testis. Biochem Biophys Acta 1778:692–708

Worth DC, Daly CN, Geraldo S, Oozeer F, Gordon-Weeks PR (2013) Drebrin contains a cryptic F-actin-bundling activity regulated by Cdk5 phosphorylation. J Cell Biol 202:793–806. doi:10.1083/jcb.201303005

Xia W, Wong EW, Mruk DD, Cheng CY (2009) TGF-β3 and TNFα perturb blood-testis barrier (BTB) dynamics by accelerating the clathrin-mediated endocytosis of integral membrane proteins: a new concept of BTB regulation during spermatogenesis. Dev Biol 327:48–61. doi:10.1016/j.ydbio.2008.11.028

Xiao X, Mruk DD, Wong CK, Cheng CY (2014) Germ cell transport across the seminiferous epithelium during spermatogenesis. Physiology 29:286–298. doi:10.1152/physiol.00001.2014

Xu W, Stamnes M (2006) The actin-depolymerizing factor homology and charged/helical domains of drebrin and mAbp1 direct membrane binding and localization via distinct interactions with actin. J Biol Chem 281:11826–11833. doi:10.1074/jbc.M510141200

Yamazaki H, Kojima N, Kato K, Hirose E, Iwasaki T, Mizui T, Takahashi H, Hanamura K, Roppongi RT, Koibuchi N, Sekino Y, Mori N, Shirao T (2014) Spikar, a novel drebrin-binding protein, regulates the formation and stabilization of dendritic spines. J Neurochem 128:507–522. doi:10.1111/jnc.12486

Yan HHN, Mruk DD, Lee WM, Cheng CY (2007) Ectoplasmic specialization: a friend or a foe of spermatogenesis? BioEssays 29:36–48

Yan HH, Mruk DD, Lee WM, Cheng CY (2008) Blood-testis barrier dynamics are regulated by testosterone and cytokines via their differential effects on the kinetics of protein endocytosis and recycling in Sertoli cells. FASEB J 22:1945–1959. doi:10.1096/fj.06-070342

Zhao L, Ma QL, Calon F, Harris-White ME, Yang F, Lim GP, Morihara T, Ubeda OJ, Ambegaokar S, Hansen JE, Weisbart RH, Teter B, Frautschy SA, Cole GM (2006) Role of p21-activated kinase pathway defects in the cognitive deficits of Alzheimer disease. Nat Neurosci 9:234–242. doi:10.1038/nn1630

Chapter 18
Drebrin at Junctional Plaques

Wiebke K. Ludwig-Peitsch

Abstract Adhesion, segregation, and cellular plasticity are regulated by actin filaments anchored at the plaques of adherens junctions, sites of mechanical stabilization, and interfaces of multiple signaling networks. Drebrins were originally identified in neuronal cells, but the isoform drebrin E was also detected at adherens junctions of a wide range of non-neuronal cells, including polarized epithelia, endothelia, and fibroblasts. Here the protein is enriched at actin filament bundles associated with junctional plaques. Polarized epithelial cells contain two types of actin-associated complexes, one comprising drebrin but not vinculin and the other involving vinculin, but not drebrin. At gap junctions drebrin interacts with connexin 43, stabilizes this protein at membranes, and links it to the actin cytoskeleton. In vivo drebrin is widespread in diverse non-neuronal tissues of epithelial, endothelial, and smooth muscle origin, but not ubiquitous. In intestinal cells it is involved in cell compaction, linking of actin filaments to microtubules and formation and stabilization of the terminal web. Upregulation of drebrin was noted in several types of cancers, e.g., basal cell carcinomas for which it may serve as marker, liver metastases of colon carcinomas, and bladder cancer, suggesting that it is involved in regulating actin dynamics during tumor development, progression, and metastasis.

Keywords Actin • Adherens junction • Basal cell carcinoma • Connexin 43 • Drebrin • Microtubule • Vinculin

18.1 Introduction

The ability of cells to communicate and adhere to each other is indispensable for the formation and maintenance of multicellular organisms. Adhesion, segregation, and cellular plasticity are regulated by actin filaments associated with specific domains of the plasma membrane. Among these, adherens junctions (*zonulae adhaerentes*) are the most prominent structures anchoring actin filaments. In adherens junctions

---

W.K. Ludwig-Peitsch
Department of Dermatology, Vivantes Klinikum im Friedrichshain,
Landsberger Allee 49, 10249 Berlin, Germany
e-mail: wiebke.ludwig-peitsch@vivantes.de

T. Shirao, Y. Sekino (eds.), *Drebrin*, Advances in Experimental Medicine
and Biology 1006, DOI 10.1007/978-4-431-56550-5_18

intercellular contacts are mediated by classical cadherins, calcium-dependent transmembrane proteins that may form homophilic interactions with identical cadherins on neighboring cells, or heterophilic interactions with other classical cadherins on different types of cells. At their cytoplasmic tail, cadherins are anchored in a plaque in which they bind to β-catenin, protein p120, and plakoglobin. In turn, β-catenin is linked to α-catenin, which connects adherens junctions to the actin cytoskeleton. In addition to their mechanically stabilizing function, adherens junctions have emerged as important regulators of tissue morphogenesis and plasticity and as the interface of multiple signaling networks (Takeichi 2014). These networks are required for coupling and detachment of cadherin-catenin complexes from actin filaments, allowing cell sorting, migration, and invasion. Remodeling of adherens junctions induces reshaping of cells in various ways, including, e.g., apical constriction of polarized cells driven by actomyosin contraction. The plasticity of actin filaments is regulated by Rho GTPases and their effectors as well by a large number of actin-binding proteins accumulated at adherens junctions and other sites of the actin cytoskeleton, including drebrin.

18.2 Detection of Drebrin in Diverse Non-Neuronal Cells

Drebrins were initially identified in the brain and considered to be, by and large, brain-specific. However, brain specificity has only been documented for the adult isoform drebin A. Signals of mRNA corresponding to embryonic drebrin isoforms have been reported in several non-neuronal tissues and cell cultures (Fisher et al. 1994). Furthermore, drebrin has been identified in F-actin overlay assays among the actin-binding proteins of fibroblasts and carcinoma cells (Luna et al. 1997).

The first studies investigating the occurrence and distribution of drebrin outside the brain in greater detail were performed by our group (Peitsch et al. 1999) and by Keon et al. (2000). In immunoblot experiments, drebrin was detected in a wide range of non-neuronal cell lines, including cultures from simple epithelia such as human primary liver carcinoma (PLC), human cervix adenocarcinoma (HeLa), human mammary carcinoma (MCF-7) and human colonic adenocarcinoma (CaCo-2), primary human keratinocytes, human SV80 and bovine B1 fibroblasts and endothelial cell cultures from human umbilical vein (HUVEC), human umbilical artery (HUAEC), and human dermal microvasculature (HDMEC) (Peitsch et al. 1999, 2005; Keon et al. 2000). However, according to the observations of both groups, the human squamous carcinoma line A-431 did not contain any drebrin at the level of sensitivity applied (Peitsch et al. 1999; Keon et al. 2000).

Immunoblotting of murine and bovine tissues revealed drebrin-positive reactions in bovine muzzle epithelium, lingual mucosa, kidney, and aortic endothelium (Peitsch et al. 1999) as well as in the stomach and kidney of adult mice (Keon et al. 2000). Furthermore, lower amounts of drebrin were found in murine colon and bladder and trace amounts in the heart, lung, liver, and epididymis (Keon et al. 2000). By contrast, the protein was not detectable in murine duodenum and small intestine and in bovine skeletal muscle (Peitsch et al. 1999; Keon et al. 2000). Taken together, these results indicate that drebrin is widespread but not ubiquitous outside the nervous system.

The protein detected by immunoblot was unambiguously identified as drebrin in coimmunoprecipitation experiments followed by mass spectrometry (Peitsch et al. 1999; Keon et al. 2000). The isoform expressed in non-neuronal cells was determined to be drebrin E (Keon et al. 2000).

18.3 Drebrin Enrichment at Plaques of Adherens Junctions

Immunostaining of different epithelial and endothelial monolayer cell cultures showed drebrin reactions along cell–cell junctions and, occasionally, small drebrin-positive cytoplasmic dots in the cytoplasm (Peitsch et al. 1999, 2005; Keon et al. 2000). When the cells were double-labeled with drebrin antibodies and phalloidin to visualize actin filaments, colocalization of drebrin and actin was observed at adherens junctions (*zonulae adhaerentes*, Fig. 18.1a, a′), whereas cytoplasmic actin filament bundles ("stress fibers") were free of drebrin (Fig. 18.1b, b′; Peitsch et al. 1999; Keon et al. 2000).

The localization of drebrin in comparison to other actin-binding proteins enriched at cell–cell contacts was analyzed by double-label confocal fluorescence laser scanning microscopy. Double immunostaining of epithelial cells for drebrin and the

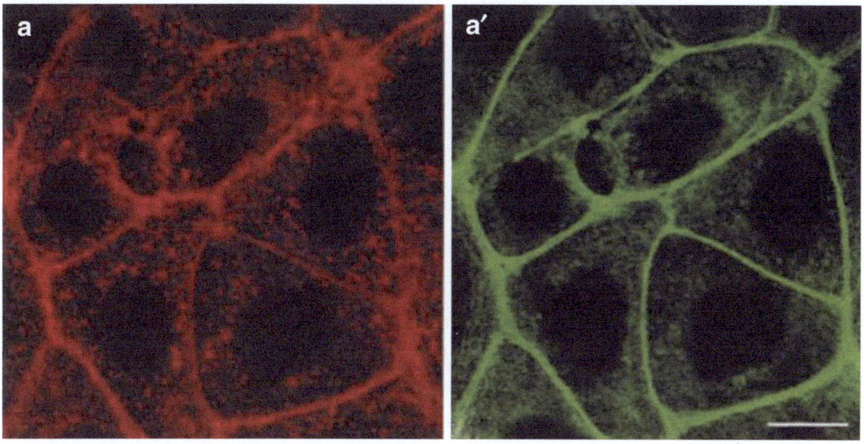

Fig. 18.1 Double-label fluorescence confocal laser microscopy of cultured MDBK cells, labeled with antibodies against drebrin (**a**, **b**, **c**, *red*) in combination with phalloidin (**a**′, **b**′, *green*) or with desmoplakin antibodies (**c**, *green*). Different focal plains of the same cell are shown in (**a**) and (**b**). In the subapical region (**a**), drebrin is enriched along the plasma membrane, colocalizing with actin, whereas the stress fibers of actin seen in a more central plain (**b**) are devoid of drebrin. Staining patterns observed with drebrin and desmoplakin antibodies are mutually exclusive (**c**). Bars: 10 μm. Source: Peitsch et al. (1999)

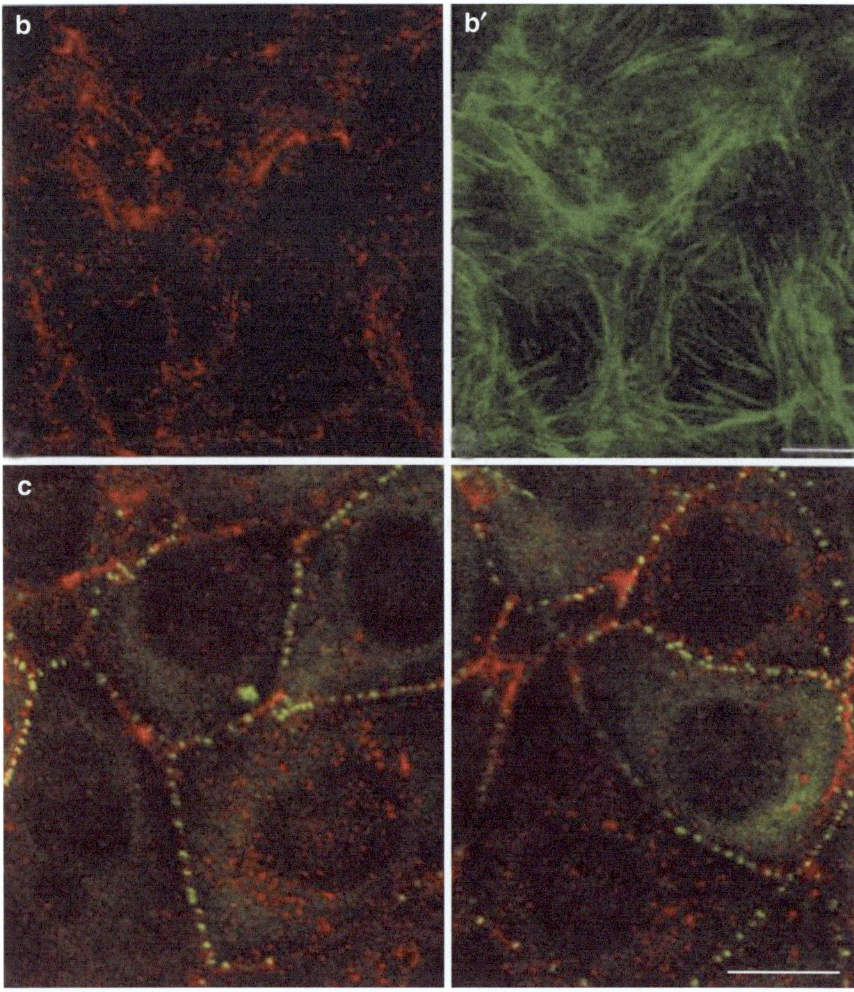

Fig. 18.2 (continued)

desmosomal marker protein desmoplakin revealed differential distribution patterns, with desmoplakin-positive punctate staining at desmosomes and drebrin reactions in distinct streaks and dot- or whisker-like structures along cell boundaries (Fig. 18.1c). Plaque proteins of adherens junctions, e.g., α- and β-catenin, protein p120 and α-actinin, colocalized with drebrin to some extent (Peitsch et al. 1999). The same observation was made when the drebrin-negative cell line A-431 was stably transfected with drebrin-EGFP constructs. In transfectants containing moderate or low amounts of drebrin-EGFP, the fusion protein was recruited to adherens junctions where it was partially codistributed with the plaque proteins (Fig. 18.2a–a″, b–b″) and showed some overlap with E-cadherin (Fig. 18.2c–c″; Peitsch et al. 2005). By contrast, transfectants overexpressing drebin-EGFP formed numerous

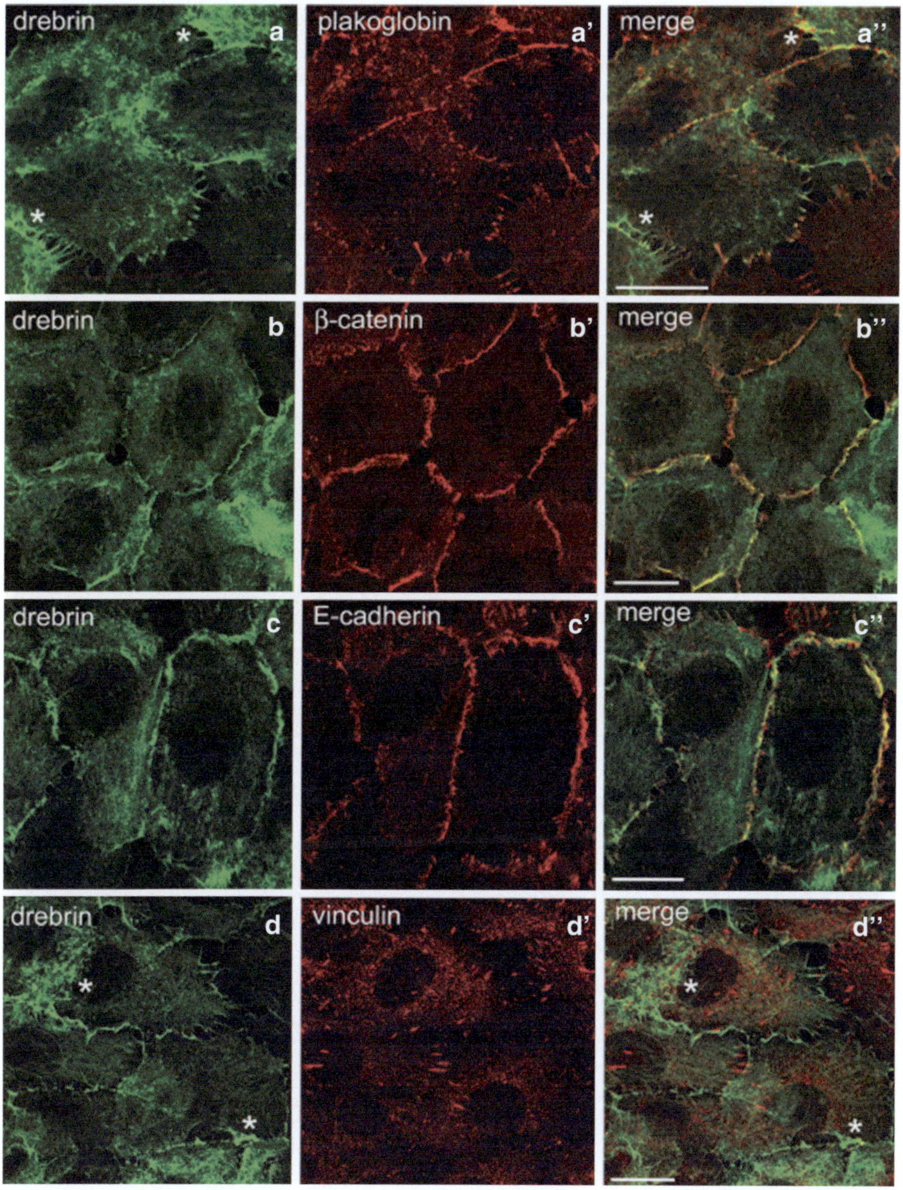

Fig. 18.2 Accumulation of drebrin-EGFP fusion proteins close to adherens junctions. A-431 cells that do not contain endogenous drebrin were stably transfected with drebin-EGFP-C1 constructs (*green* in **a**–**d**), immunostained with antibodies against components of adherens junctions (*red* in **a**′–**d**′) and analyzed by confocal laser scanning microscopy. The drebrin constructs are concentrated along cell–cell contacts, but some drebrin-transfected cells also exhibit numerous small drebrin-rich cell processes (*asterisks* in **a**, **a**″, and **d**″). Plakoglobin (**a**′, a″), β-catenin (**b**′, b″) and E-cadherin (**c**′, c″) partially colocalize with drebrin-EGFP along intercellular contacts. Vinculin antibodies label focal contacts (**d**′), which are free of drebrin (**d**″, merge). Bars: 20 μm. Source: Peitsch et al. (2005)

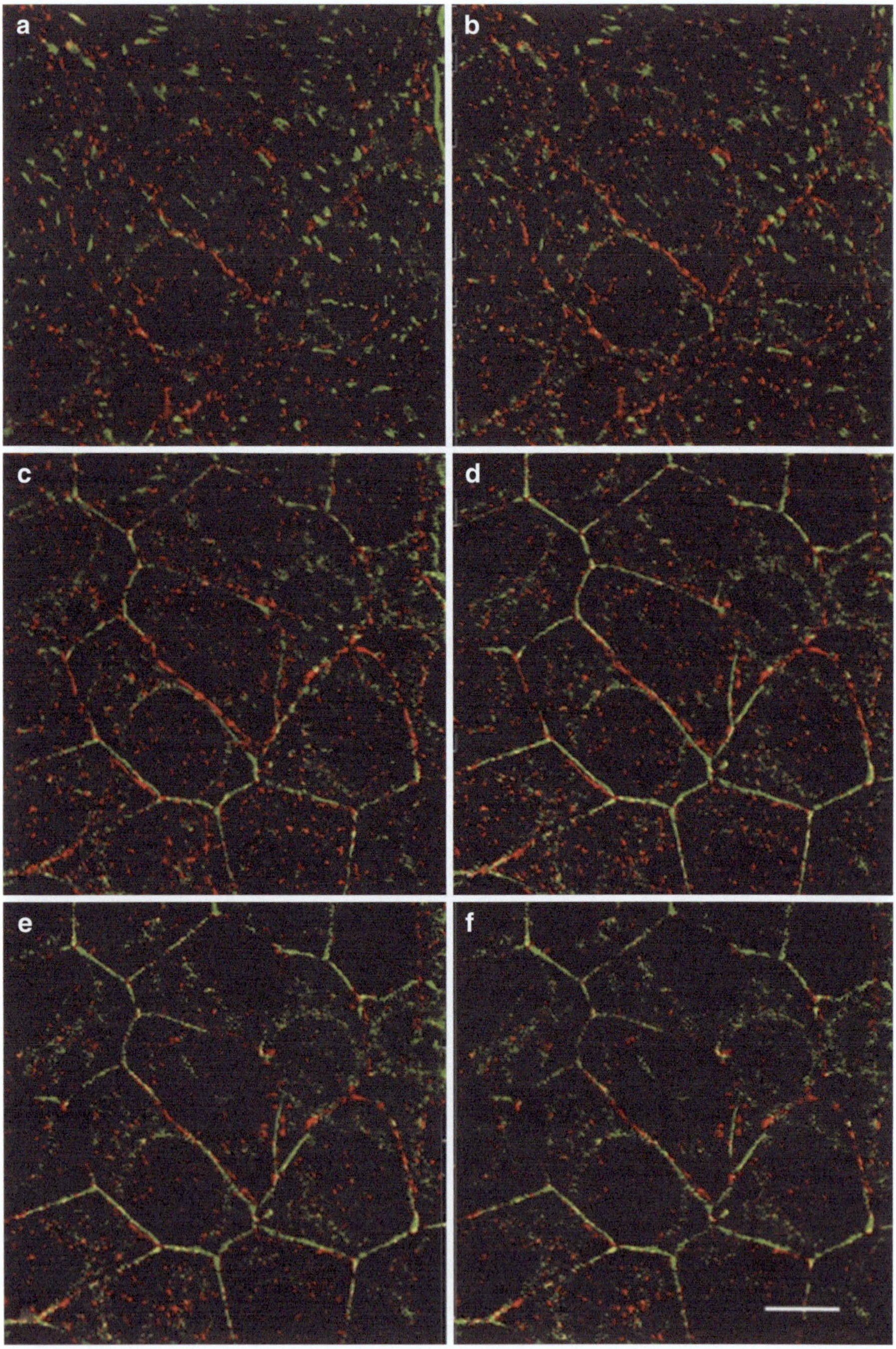
a
b
c
d
e
f

actin- and drebrin-rich cell processes (Peitsch et al. 2005, 2006). Similar observations were made upon stable transfection of Madin–Darby canine kidney (MDCK) cells (Keon et al. 2000), indicating that drebrin accumulates at junctional plaques of non-neuronal cells when expressed in moderate or low levels, whereas overexpression of the protein induces extensive morphological changes characterized by formation of cell processes, as noted in neuronal cells and fibroblasts (Shirao et al. 1992; Hayashi and Shirao 1999).

Remarkably, antibodies against drebrin and vinculin, an actin-binding protein occurring both at adherens junctions and at cell-matrix adhesions, reacted in mutually exclusive pattern (Figs. 18.2d–d″ and 18.3). When cells derived from simple epithelia, e.g., Madin–Darby bovine kidney (MDBK) or PLC cells, were double labeled with drebrin and vinculin antibodies and investigated by confocal laser scanning microscopy, vinculin was enriched at focal adhesions in the near-bottom plane (Fig. 18.3a, b). More apically, it appeared in linear cortical structures at adherens junctions (Fig. 18.3c–f). Drebrin immunoreactions were also seen along adherens junctions, but the vinculin-positive domains were consistently drebrin-negative (Fig. 18.3a–f). Focal adhesions were free of drebrin, even if it has been reported that expression of drebrin in cDNA-transformed fibroblasts results in increased assembly of vinculin in adhesion plaques (Ikeda et al. 1996). Thus, polarized epithelial cells appear to contain two actin-associated complexes, one comprising drebrin but no—or very little—vinculin and the other involving vinculin, but not drebrin. Whether this differential localization of drebrin and vinculin correlates with the specific patterns of other plaque components and is of functional relevance remains to be investigated.

As a component of endothelial adherens junctions, drebrin plays a crucial role in maintaining endothelial integrity under vascular flow conditions (Rehm et al. 2013; see Chap. 23). Drebrin knockdown was demonstrated to cause weakening of adherens junctions and loss of nectin from these sites. Stabilization of nectin at adherens junctions is achieved through a chain of interactions requiring binding of drebrin to actin, interactions between drebrin and afadin through their polyproline and PR1-2 domains, and interactions between afadin and nectin (Rehm et al. 2013). These processes are described in greater detail in Chap. 23.

Fig. 18.3 Confocal laser scanning microscopy of MDBK cells double-stained with antibodies to drebrin (*red*) and vinculin (*green*), focussing from the bottom (**a**) through central regions (**b**) to more apical plains containing the adherens junction (**c**–**f**). Vinculin is accumulated in focal adhesions at the bottom, in *puncta adhaerentia* in central regions (**b**) and at adherens junctions (**c**–**f**) in the apical region. Drebrin reactions appear in streaks or whisker-like structures along cell–cell contacts in all optical plains and sometimes additionally in small cytoplasmic dots (**a**–**f**). The distribution of drebrin and vinculin is strikingly different, suggesting existence of two actin filament systems, one associated with drebrin and the other with vinculin. A deconvolution procedure was used to increase the resolution on the Z-axis. Bars: 10 μm. Source: Peitsch et al. (1999)

18.4 Functions of Drebrin in Polarized Cells of the Gastrointestinal Tract

When immunoblot analysis was performed to examine the expression pattern of drebrin in murine tissues, particularly high amounts of the protein were detected in stomach (Keon et al. 2000). Immunofluorescence microscopy revealed specific enrichment of drebrin in acid-secreting parietal cells (Keon et al. 2000; Chew et al. 2005). Drebrin-positive reactions were particularly pronounced at the apical membrane of the intracellular canaliculi and in microvilli (Keon et al. 2000; Chew et al. 2005). The protein was differentially distributed in parietal cells along the gland axis, indicating that it may serve as a marker of parietal cell differentiation and supporting the notion that different populations of parietal cells are functionally heterogeneous (Chew et al. 2005). In parietal cells, activation of HCl secretion by cAMP-dependent agonists such as histamine involves distinct morphological changes, in particular, formation and elongation of actin-rich microvilli within intracellular canaliculi, which are the sites of acid secretion. Overexpression of epitope-tagged drebrin E induced formation of microspikes and F-actin-rich ringlike structures in cultured parietal cells and suppressed cAMP-dependent acid secretory responses (Chew et al. 2005). In MDCK cells, coexpression of epitope-tagged drebrin and the Rho family GTPase Cdc42, known to induce filopodial extension, resulted in additional elongation of the microspikes. However, coexpression of drebrin and dominant negative Cdc42 did not preclude drebrin-induced microspike formation, indicating that drebrin can induce the formation of actin-rich membrane projections both by Cdc42-dependent and by Cdc42-independent mechanisms (Chew et al. 2005).

Intestinal epithelial cells differentiate in the course of their migration along the crypt-villus axis, become elongated, and form an apical brush border of microvilli which are anchored in a dense actin-based network, the so-called terminal web. In highly polarized columnar human intestinal cells of line CaCo2/TC7 and in human enterocytes drebrin was found to be accumulated in puncta along the lateral membrane and in a dense network in the subapical region (Bazellières et al. 2012). Depletion of drebrin E affected neither cell polarity nor the composition and function of tight junctions. However, cell compaction and elongation processes were significantly impaired, i.e., drebrin-depleted cells were flatter and larger than controls. Drebrin knockdown blocked accumulation of actin in the subapical network, indicating that drebrin is crucial for formation and stabilization of the terminal web (Bazellières et al. 2012). Furthermore, myosin-IIB and βII-spectrin were downregulated, and apical accumulation of these two proteins was blocked upon drebrin depletion. The microvilli of drebrin-depleted cells were considerably less dense and disorganized, indicating that drebrin is required for brush border formation.

In polarized epithelial cells, a dense microtubular network runs parallel to the terminal web. Drebrin binds to the microtubule-plus-end-binding protein EB3 and tethers actin filaments and microtubules together by this way in growth cones (Geraldo et al. 2008). In CaCo2 cells transfected with an EB3-GFP construct drebrin, myosin-IIB and βII-spectrin coimmunoprecipitated with EB3 (Bazellières et al. 2012). However, in drebrin-depleted CaCo2, neither myosin-IIB nor βII-spectrin could be coprecipitated with this protein, proving that drebrin is indispensable for formation of the complex. In addition, accumulation of EB3 in the subapical region was disturbed upon drebrin depletion. Taken together, these observations suggest that the complex of drebrin, EB3, myosin-IIB, and βII-spectrin links F-actin and microtubule networks in the subapical region of polarized intestinal cells. Drebrin contributes to stabilizing the actin-based terminal web and promotes cell compaction dependent on myosin-IIB but not on EB3 (Bazellières et al. 2012).

Drebrin depletions also modify the distribution of several other apical marker proteins in CaCo2 cells, including the digestive enzyme dipeptidyl peptidase IV (DPPIV) which is specific for the brush border (Vacca et al. 2014). Instead of accumulating at the apical surface, DPPIV was redistributed to an enlarged subapical compartment related to lysosomes upon drebrin knockdown.

Mutations in the MYO VB gene encoding the motor protein MyoVb cause microvillar inclusion disease, a rare genetic disease characterized by intestinal malabsorption. In the apical recycling pathway, MyoVb interacts with Rab8a, which serves as a linker to transport vesicles. Conditional knockout of Rab8a leads to redistribution of apical membrane proteins into a lysosomal compartment and decrease in their protein expression levels. Interestingly, CaCo2 cells depleted of drebrin E shared several features with Rab8a-depleted cells and with the phenotype observed in microvillar inclusion disease (Vacca et al. 2014). In summary, these findings suggest that drebrin redirects the apical recycling pathway of intestinal epithelial cells to the lysosomes and that this protein is an important regulator of apical trafficking.

18.5 Drebrin as Binding Partner of Connexin 43 at Gap Junctions

Gap junctions are transmembrane channels that mediate intercellular communication by allowing the passage of ions, metabolites, and second messengers. They are formed by hexameric assemblies of six connexins, which align head to head between neighboring cells. Connexins are composed of four transmembrane helices, two extracellular loops kept together by disulfide linkages, and three variable cytoplasmic domains. Their functional importance is underscored by the fact that connexin defects can cause a wide range of human diseases. Connexin 43 is the main connexin of gap junctions in heart and in several other tissues; its null mutations in mice lead to lethal cardiac malformations.

In a screen for interaction partners of connexin 43, drebrin was identified as connexin 43 binding protein in murine brain membrane fractions (Butkevich et al. 2004). Drebrin was recovered in a pull-down assay with a GST-fusion protein encompassing the COOH-terminal domain of connexin 43. Double-label immunofluorescence microscopy revealed colocalization of drebin and connexin 43 in astrocytes and Vero cells, a kidney epithelial cell line from African green monkey. Drebrin was found to be accumulated in the plasma membrane fraction of Vero cells and to coimmunoprecipitate with connexin 43 from this fraction (Butkevich et al.

2004). Moreover, interactions between both proteins were confirmed by live-cell fluorescence resonance energy transfer (FRET) analyses of COS cells transfected with CFP-drebrin and connexin 43-YFP. Silencing of drebrin by siRNA resulted in internalization and degradation of connexin 43, and whole-cell voltage clamp recordings showed a marked decrease of cell–cell permeability upon drebrin knock-down. Conversely, overexpression of drebrin led to stabilization of gap junctions. Taken together, these findings indicate that drebrin stabilizes connexin 43 at the plasma membrane and thereby facilitates cell–cell coupling (Butkevich et al. 2004).

Based on these findings, it can be speculated that drebrin serves as a linker between connexin 43 and F-actin. In order to examine the relationship between the E-cadherin/α- and β-catenin/actin system and the connexin 43/drebrin/actin system, Vero cells were cotransfected with GFP-E-cadherin, connexin 43-CFP, and YFP-drebrin, and the fluorescent signals were linearly separated. The initial contact between two cells was found to be mediated by E-cadherin. Connexin 43 was recruited to cell–cell contacts only after a stable interface had been established. At this later stage, drebrin colocalized more with connexin 43 than with E-cadherin. This colocalization was maintained even after treatment with Latrunculin B, suggesting that actin filaments are not crucial for the interaction between drebrin and connexin 43 once stable gap junctions have been formed (Butkevich et al. 2004).

Drebrin may interact simultaneously with connexin 43 and actin via its N-terminal ADF domain and its proline-rich binding domains. Therefore, the connexin 43/drebrin complex may serve as an interface between extracellular signals and cellular responses mediated by rearrangement of the submembranous actin cytoskeleton (Majoul et al. 2007).

The composition and function of gap junctions may be modified by components of the extracellular matrix, in particular, by laminin-integrin interactions. When the impact of laminin-111 on connexin 43 was studied in mouse embryonic stem cells, laminin-111 was found to reduce intercellular communication through gap junctions and to enhance cell proliferation (Suh et al. 2012). Laminin-111 stimulated the dissociation of the connexin 43 /ZO-1 complex, followed by disruption of the connexin 43/drebrin and the connexin 43/actin complexes. This process was reversed by inhibition of RhoA. In summary, laminin-111 appears to decrease gap junction intercellular communication via increased RhoA-dependent phosphorylation of connexin 43, resulting in instability of the connexin 43/drebrin complex and increased degradation of connexin 43 (Suh et al. 2012).

18.6 Drebrin Accumulations at Junctional Plaques of Diverse Cell Types and Tissues In Vivo

To examine the distribution of drebrin in a broad range of non-neuronal cell types and organs, sections of bovine and murine tissues, including lactating udder, testis, kidney, aorta, vena cava, tongue, snout, lung, retina, heart, liver, stomach and intestine, and human tissues, including epidermis, scalp, umbilical cord, testis, and colon carcinoma, were immunostained with drebrin antibodies. In bovine udder, myoepithelial cells were markedly immunoreactive, but drebrin staining was also noted along cell–cell contacts of secretory cells. Bovine intestinal epithelium and human colon adenocarcinoma were similarly immunostained. In bovine and human testis, cell–cell contacts of Sertoli cells were particularly rich in drebrin (Peitsch et al. 1999).

In human, bovine and murine kidneys drebrin antibodies stained both glomeruli and certain tubular epithelial cells, in particular, acid-secreting type A intercalated cells and distal tubule epithelia (Peitsch et al. 1999; Keon et al. 2000, see also Chap. 20). In acid-secreting intercalated cells, drebrin was specifically accumulated at the apical plasma membrane. In distal tubule epithelia, the protein was concentrated at the luminal surface and at interdigitations of the basolateral membranes.

Stratified epithelia, e.g., squamous epithelium of bovine snout and lingual mucosa as well as the pseudostratified bronchial epithelium, showed variable drebrin reactions. The protein was sometimes enriched at cell–cell contacts, however, with variable intensity, and occasionally predominant or exclusive staining of the basal layer was noted (Peitsch et al. 1999).

Normal human epidermis was essentially drebrin-negative (Fig. 18.4a). Drebrin was, however, concentrated at the ducts of eccrine sweat glands, particularly along

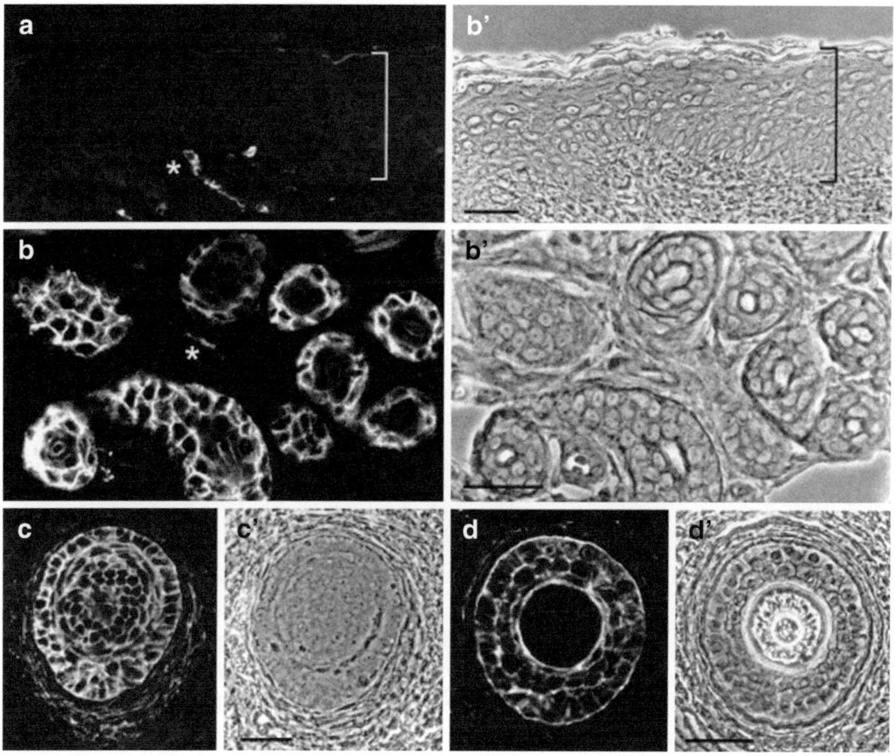

Fig. 18.4 Distribution of drebrin in normal human skin. Normal human epidermis (bracket in **a**) is virtually devoid of drebrin, but drebrin-positive reactions are noted in the endothelial cells of small blood vessels (*asterisks* in **a** and **b**). Prominent staining is observed in the ducts of eccrine sweat glands (**b**), particularly along the cell–cell contacts. Drebrin is also concentrated at the intercellular junctions of hair follicles, both in the bulb (**c**) and in the outer root sheath (**d**). (**a**′–**d**′): phase contrast micrographs. Bars: 25 μm. Source: Peitsch et al. (2005)

their cell–cell junctions (Fig. 18.4b). In addition, intercellular junctions of hair follicles were intensely drebrin-positive both in the bulbs (Fig. 18.4c) and, in upper parts, in the companion layer and in the outer root sheath (Fig. 18.4d), whereas the inner root sheath did not contain significant amounts of drebrin (Peitsch et al. 2005).

Furthermore, drebrin was sometimes detected in the endothelium of large blood vessels, including bovine aorta and vena cava and human umbilical vein and artery. Smooth muscle cells of the vascular walls as well as vasa vasorum were also drebrin-positive, but the intensity of the reaction was rather variable. In bovine retina the most intense drebrin immunoreaction was observed in the inner plexiform layer, while the zonula limitans externa, comprising densely aligned junctions between photoreceptor and Müller glia cells, was drebrin-negative (Peitsch et al. 1999).

Bovine and human skeletal and cardiac muscles were consistently negative for drebrin, in accordance with results from immunoblot analyses. Moreover, bovine and human hepatocytes did not display any drebrin immunoreactivity at the level of sensitivity applied (Peitsch et al. 1999).

Corresponding to the findings obtained in cultured cells, drebrin is a common but not ubiquitous component of cell–cell junctions in diverse human, bovine, and murine cell types and tissues in vivo. The principle underlying the distribution of drebrin in mammalian cells and tissues is still unclear, and it is so far unexplained why certain polarized cells are particularly rich in drebrin whereas others are devoid of the protein.

18.7 Role of Drebrin During Carcinogenesis

Since expression of several junction-associated proteins is altered during development and progression of tumors, the distribution of drebrin was examined in skin cancers and their precursors (Peitsch et al. 2005). With an annual incidence around 100 per 100,000 inhabitants, basal cell carcinomas are the most common skin carcinomas of the Caucasian population. They are considered as semimalignant tumors because they locally invade and destroy tissues but metastasize only extremely rarely. Basal cell carcinomas consistently showed intense drebrin-positive immunoreactions, which were sometimes enhanced in the periphery of the tumors (Fig. 18.5a). Immunoelectron microscopy revealed enrichment of drebrin in the actin filament network underneath the plaques of adherens junctions, even if β-catenin examined for control was located even closer to the membrane.

Both basal cell carcinomas and two types of benign neoplasms, trichoblastoma and trichoepithelioma, are thought to be derived from follicular germinative cells. These tumors are sometimes difficult to distinguish by routine histology. When drebrin immunostaining was compared in the three kinds of tumors, basal cell carcinomas showed intense and homogeneous immunoreactions mainly at cell–cell boundaries. By contrast, trichoblastomas and trichoepitheliomas were only weakly

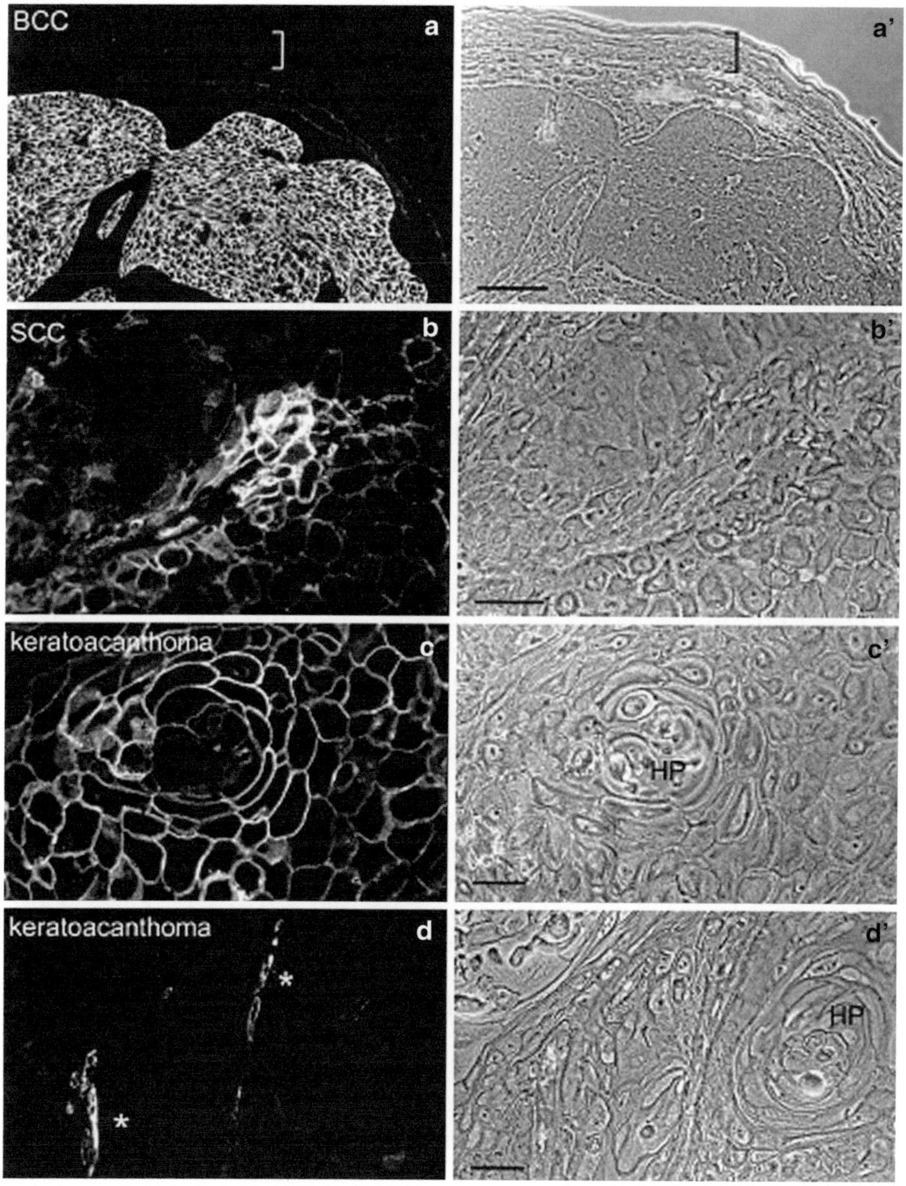

Fig. 18.5 Immunolocalization of drebrin in epithelial skin tumors. In basal cell carcinoma (BCC, **a**), intense drebrin staining is seen throughout the tumor, whereas the overlying epidermis appears drebrin-negative (*brackets*). In squamous cell carcinoma (SCC, **b**) and keratoacanthomas (**c**, **d**) drebrin antibodies react in heterogeneous patterns, showing both strongly drebrin-positive (**b**, **c**) and drebrin-negative compartments (**b**, **d**). Two different parts of the same keratoacanthoma, both containing horn pearls (HP), are displayed in (**c**) and (**d**). Asterisks: drebrin-positive blood vessels. (**a′**–**d′**): phase contrast images. Bars: 25 μm. Source: Peitsch et al. (2005)

and heterogeneously stained (Mizutani et al. 2014). Therefore, drebrin may be a useful marker for differentiating basal cell carcinomas from the two types of benign neoplasms.

In contrast to the strong and mostly homogeneous drebrin immunostaining observed in basal cell carcinomas, squamous carcinoma cells exhibited heterogeneous drebrin staining (Peitsch et al. 2005). Certain parts of the tumors showed rather intense reactions, whereas others appeared to be drebrin-negative (Fig. 18.5b). Drebrin-positive areas were often seen at the invasive tumor front or near horn pearls of differentiated squamous carcinomas, but there was no correlation between expression of drebrin and proliferation or differentiation. The same phenomenon as in squamous cell carcinomas, i.e., mosaicism of drebrin-rich and drebrin-free tumor regions, was noted in keratoacanthomas (Fig. 18.5c, d) and in precancerous lesions such as actinic keratoses. According to immunoblot analyses, normal human skin and scalp contained comparatively low amounts of drebrin, whereas the protein was markedly upregulated in basal and squamous cell carcinomas as well as in other tumors such as melanoma metastases and leiomyosarcomas (Peitsch et al. 2005).

Moreover, there is evidence that drebrin is involved in dissemination of colorectal carcinoma. In proteome profiles comparing the human colon adenocarcinoma cell line HCT-116 to its metastatic derivative E1 with iTRAQ labeling technology, drebrin was noted to be markedly overexpressed in the metastatic derivative (Lin et al. 2014). This finding was confirmed by immunoblot as well as by immunohistochemistry of colorectal cancer tissues with matched lymph node and liver metastasis tissues. Primary colorectal carcinomas were drebrin-negative in the majority of cases. By contrast, all lymph node and liver metastases examined stained positive for drebrin (Lin et al. 2014).

Drebrin was also demonstrated to promote the development and progression of bladder carcinoma by interacting with progranulin. The autocrine growth factor progranulin is known to induce proliferation, transformation, and invasion of several types of cancer, including urothelial cancer. In a pull-down assay with recombinant progranulin and extracts from bladder cancer cells, drebrin was identified as a novel progranulin-interacting protein, a finding confirmed by coimmunoprecipitation (Xu et al. 2015). Drebrin was shown to play a critical role for progranulin-dependent activation of the Akt and MAPK pathways which induce migration and invasion. Silencing of the protein resulted in impaired migration, decreased invasion, reduced colony forming in a soft agar assay, and impaired tumor formation in mouse xenograft models (Xu et al. 2015). In addition, expression of drebrin was found to be upregulated in bladder cancers compared to normal bladder tissue. In summary, these data indicate that drebrin exerts an important function in regulating progranulin activity in bladder carcinomas (Xu et al. 2015).

Development and progression of tumors is associated with rearrangements of the actin cytoskeleton during tumor cell migration, invasion, and metastasis. The finding that drebrin is upregulated in several kinds of tumors or their metastases implies that it plays an important role in regulating actin dynamics during these processes.

18.8 Conclusion

Drebrin E is a widespread component of the actin filament system in non-neuronal cells in vitro and in vivo. In polarized epithelial and endothelial cells, it has to be added to the list of actin-binding proteins regulating actin dynamics at adherens junction. At gap junctions drebrin interacts with connexin 43. Upregulation of drebrin in several kinds of carcinomas implies a role in regulating actin dynamics during tumor development, progression, and metastasis. Since it is markedly enriched and homogeneously distributed in particular tumors such as basal cell carcinomas, it may serve as marker for their detection and differential diagnosis.

References

Bazellières E, Massey-Harroche D, Barthélémy-Requin M, Richard F, Arsanto JP, Le Bivic A (2012) Apico-basal elongation requires a drebrin-E-EB3 complex in columnar human epithelial cells. J Cell Sci 125:919–931

Butkevich E, Hülsmann S, Wenzel D, Shirao T, Duden R, Majoul I (2004) Drebrin is a novel connexin-43 binding partner that links gap junctions to the submembrane cytoskeleton. Curr Biol 14:650–658

Chew CS, Okamoto CT, Chen X, Thomas R (2005) Drebrin E2 is differentially expressed and phosphorylated in parietal cells in the gastric mucosa. Am J Physiol Gastrointest Liver Physiol 289:G320–G331

Fisher LW, McBride OW, Filpula D, Ibaraki K, Young MR (1994) Human drebrin (DBNl): cDNA sequence, mRNA tissue distribution and chromosomal localization. Neurosci Res Commun 14:35–42

Geraldo S, Khanzada UK, Parsons M, Chilton JK, Gordon-Weeks PR (2008) Targeting of the F-actin-binding protein drebrin by the microtubule plus-tip protein EB3 is required for neuritogenesis. Nat Cell Biol 10:1181–1189

Hayashi K, Shirao T (1999) Change in the shape of dendritic spines caused by overexpression of drebrin in cultured cortical neurons. J Neurosci 19:3918–3925

Ikeda K, Kaub PA, Asada H, Uyemura K, Toya S, Shirao T (1996) Stabilization of adhesion plaques by the expression of drebrin A in fibroblasts. Brain Res Dev Brain Res 91:227–236

Keon BH, Jedrzejewski PT, Paul DL, Goodenough DA (2000) Isoform specific expression of the neuronal F-actin binding protein, drebrin, in specialized cells of stomach and kidney epithelia. J Cell Sci 113:325–336

Lin Q, Tan HT, Lim TK, Khoo A, Lim KH, Chung MC (2014) iTRAQ analysis of colorectal cancer cell lines suggests Drebrin (DBN1) is overexpressed during liver metastasis. Proteomics 14:1434–1443

Luna EJ, Pestonjamasp KN, Cheney RE, Strassel CP, TH L, Chia CP, Hitt AL, Fechheimer M, Furthmayr H, Mooseker MS (1997) Actin-binding membrane proteins identified by F-actin blot overlays. Soc Gen Physiol Ser 52:3–18

Majoul I, Shirao T, Sekino Y, Duden R (2007) Many faces of drebrin: from building dendritic spines and stabilizing gap junctions to shaping neurite-like cell processes. Histochem Cell Biol 127:355–361

Mizutani Y, Iwamoto I, Kanoh H, Seishima M, Nagata K (2014) Expression of drebrin, an actin binding protein, in basal cell carcinoma, trichoblastoma and trichoepithelioma. Histol Histopathol 29:757–766

Peitsch WK, Grund C, Kuhn C, Schnölzer M, Spring H, Schmelz M, Franke WW (1999) Drebrin is a widespread actin-associating protein enriched at junctional plaques, defining a specific microfilament anchorage system in polar epithelial cells. Eur J Cell Biol 78:767–778

Peitsch WK, Hofmann I, Bulkescher J, Hergt M, Spring H, Bleyl U, Goerdt S, Franke WW (2005) Drebrin, an actin-binding, cell-type characteristic protein: induction and localization in epithelial skin tumors and cultured keratinocytes. J Investig Dermatol 125:761–774

Peitsch WK, Bulkescher J, Spring H, Hofmann I, Goerdt S, Franke WW (2006) Dynamics of the actin-binding protein drebrin in motile cells and definition of a juxtanuclear drebrin-enriched zone. Exp Cell Res 312:2605–2618

Rehm K, Panzer L, van Vliet V, Genot E, Linder S (2013) Drebrin preserves endothelial integrity by stabilizing nectin at adherens junctions. J Cell Sci 126:3756–3769

Shirao T, Kojima N, Obata K (1992) Cloning of drebrin A and induction of neurite-like processes in drebrin-transfected cells. Neuroreport 3:109–112

Suh HN, Kim MO, Han HJ (2012) Laminin-111 stimulates proliferation of mouse embryonic stem cells through a reduction of gap junctional intercellular communication via RhoA-mediated Cx43 phosphorylation and dissociation of Cx43/ZO-1/drebrin complex. Stem Cells Dev 21:2058–2070

Takeichi M (2014) Dynamic contacts: rearranging adherens junctions to drive epithelial remodelling. Nat Rev Mol Cell Biol 15:397–410

Vacca B, Bazellières E, Nouar R, Harada A, Massey-Harroche D, Le Bivic A (2014) Drebrin E depletion in human intestinal epithelial cells mimics Rab8a loss of function. Hum Mol Genet 23:2834–2346

Xu SQ, Buraschi S, Morcavallo A, Genua M, Shirao T, Peiper SC, Gomella LG, Birbe R, Belfiore A, Iozzo RV, Morrione A (2015) A novel role for drebrin in regulating progranulin bioactivity in bladder cancer. Oncotarget 6:10825–10839

Chapter 19
Juxtanuclear Drebrin-Enriched Zone

Wiebke K. Ludwig-Peitsch

Abstract Drebrin E contributes to remodeling of the actin cytoskeleton and formation of cell processes. Therefore, its role in cell migration was studied in prototypes of motile cells with prominent lamellipodia such as murine B16F1 melanoma and Swiss 3T3 cells and in human SV80 fibroblasts. Confocal microscopy revealed absence of drebrin from the tips of lamellipodia but enrichment in the tail of the cells, in retraction zones and in a specific juxtanuclear actin filament compartment, named "drebrin-enriched zone." A similar subset of juxtanuclear actin filaments is characterized by the actin-binding protein SWAP-70, but drebrin and SWAP-70 localized to different compartments, suggesting the existence of novel distinct subdomains within the actin filament system. In cells overexpressing drebrin-EGFP, numerous long, branched cell processes were formed which slowly retracted and extended. However, in stable transfectants containing lower amounts of the fusion protein, drebrin-EGFP was recruited to the same sites as the endogenous protein during cell migration, i.e., to retracting membrane domains and into the juxtanuclear drebrin-enriched zone. In the leading edges of SV80 cells, characterized by pronounced actin microspikes, drebrin was concentrated along posterior portions of the microspikes, together with tropomyosin, with which it competes for actin binding. Drebrin knockdown by siRNA did not impact forward migration or ruffling. Taken together, these findings suggest that during cell migration drebrin is involved in retraction processes but not in lamellipodia formation. The novel, sizable juxtanuclear drebrin-enriched zone remains to be characterized in detail with respect to its molecular assembly and functions.

Keywords Actin-binding protein • Drebrin • Drebrin-enriched zone • Lamellipodium • Migration • SWAP-70 • Tropomyosin

W.K. Ludwig-Peitsch
Department of Dermatology, Vivantes Klinikum im Friedrichshain,
Landsberger Allee 49, 10249 Berlin, Germany
e-mail: wiebke.ludwig-peitsch@vivantes.de

T. Shirao, Y. Sekino (eds.), *Drebrin*, Advances in Experimental Medicine and Biology 1006, DOI 10.1007/978-4-431-56550-5_19

19.1 Introduction

Cell migration is a multistep process, involving polarization of the cell and formation of a forward-oriented protrusion with a leading edge; establishment of cell-substrate adhesions near the protrusions; extensions and contractions of the cell body executed by actin and myosin II; and retraction of the tail and disaggregation of cell-substrate adhesions at the rear (Small and Resch 2005). The protruding membrane of motile cells is often organized into a lamellipodium, containing branched actin filament bundles and a particular subset of actin-binding proteins that regulate the turnover of actin filaments (Rottner and Stradal 2011; Krause and Gautreau 2014). Within the lamellipodium, actin is particularly accumulated in microspikes, dense bundles often extending into filopodia. To achieve forward movement, actin filaments are elongated by polymerization at their fast growing, "barbed" ends. In parallel, depolymerization occurs at the "pointed" ends. This process, termed "treadmilling," is accurately regulated by a specific subset of actin-binding proteins. Actin polymerization is mediated by the Arp2/3 complex which also induces branching of actin filaments. The Arp2/3 complex is activated by members of the WASP family and inhibited by Arpin. Other important regulators of treadmilling are the ADF/cofilins, which possess severing, polymerization, and depolymerization activity. Actin-bundling proteins enriched in the actin network of lamellipodia include fascin, filamin, and α-actinin. Many of these proteins have proline-rich motifs via which they interact with profilin that serves as a "shuttle" for ATP-G-actin during actin polymerization. The later steps during migration, contraction of the cell body and tail retraction, are mediated by interactions of actin with myosin II.

19.2 Drebrin in Motile Cells: Recruitment to Retracting Regions and to a Specialized Juxtanuclear Zone

In polarized cells with a motile phenotype such as astrocytic glioma cells, drebrin was found to accumulate at the bases of filopodia (Peitsch et al. 2001). To examine whether drebrin was involved in the formation of lamellipodia and in cell migration, the protein was studied in murine B16F1 melanoma and murine Swiss 3T3 cells which represent typical examples of motile cells. These cells form lamellipodia and migrate spontaneously when seeded on laminin. Surprisingly, immunolocalization experiments revealed absence of drebrin from lamellipodia but enrichment of the protein in the retracting tail, in other retracting areas of the plasma membrane and in a specific zone of the juxtanuclear actin cytoskeleton, named "drebrin-enriched zone" (Fig. 19.1, Peitsch et al. 2006). A similar compartment of the juxtanuclear actin filament system is characterized by the actin-binding protein SWAP-70 (Hilpela et al. 2003). However, when cells were double-labeled with antibodies to drebrin and SWAP-70, both proteins were found to be differentially distributed,

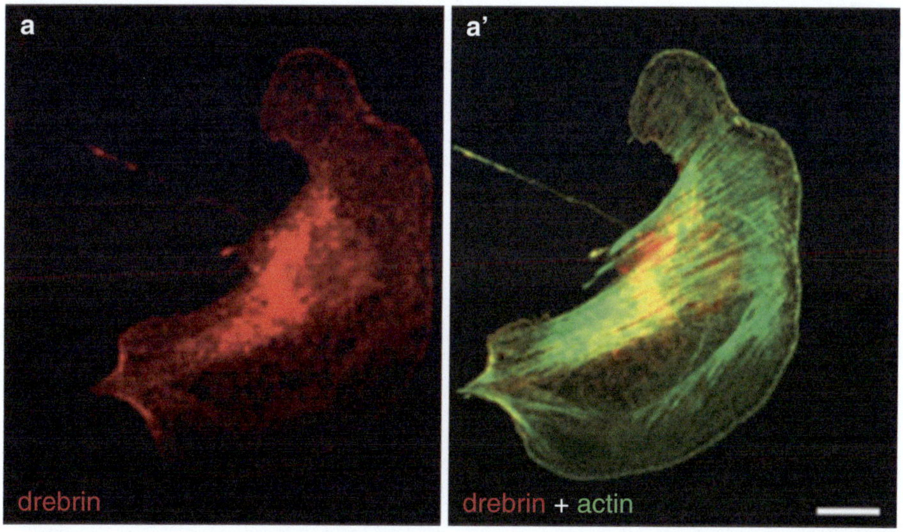

Fig. 19.1 Immunolocalization of endogenous drebrin in murine B16F1 melanoma cells seeded on laminin. Drebrin (*red* in (**a**) and (**a′**)) is absent from lamellipodia but accumulated in the posterior cell portion and in juxtanuclear aggregates of actin filaments (*green* in (**a**), phalloidin staining), the so-called drebrin-enriched zone. Bar: 10 μm. Source: Peitsch et al. (2006)

suggesting the existence of novel distinct subdomains within the actin filament system.

Transient transfection of B16F1 cells with drebrin-EGFP constructs induced the formation of multiple long cell processes and shorter, secondary processes branching therefrom when the protein was substantially overexpressed. These processes had the ability to slowly retract and elongate. However, cells expressing drebrin-EGFP in lower amounts had the motile phenotype typical of B16F1 cells. In these cells drebrin-EGFP was concentrated in the same regions as the endogenous protein, i.e., in retraction zones and in the perinuclear drebrin-enriched zone (Peitsch et al. 2006). Such cells were used to generate stable drebrin-EGFP transfectants. Compared to non-transfected B16F1 cells, they contained at least twentyfold higher amounts of drebrin, making them suitable as "knock-in" models.

Live cell confocal microscopy of stably drebrin-EGFP transfected B16F1 cells showed absence of drebrin from lamellipodia during cell migration but recruitment of the protein to the juxtanuclear drebrin-enriched zone, to retracting areas and to the tail (Fig. 19.2, Peitsch et al. 2006). Sometimes the protein was additionally accumulated in small, highly dynamic "dots" behind the lamellipodium that were reminiscent of "actin rosettes" described in motile cells.

To investigate the influence of drebrin on the speed of migration, non-transfected B16F1 cells containing low amounts of drebrin were compared to stably drebrin-EGFP expressing lines. However, the levels of drebrin had no impact on the migration distance, speed, and direction or on the velocity of wound closure in scratch wounding assays (Peitsch et al. 2006).

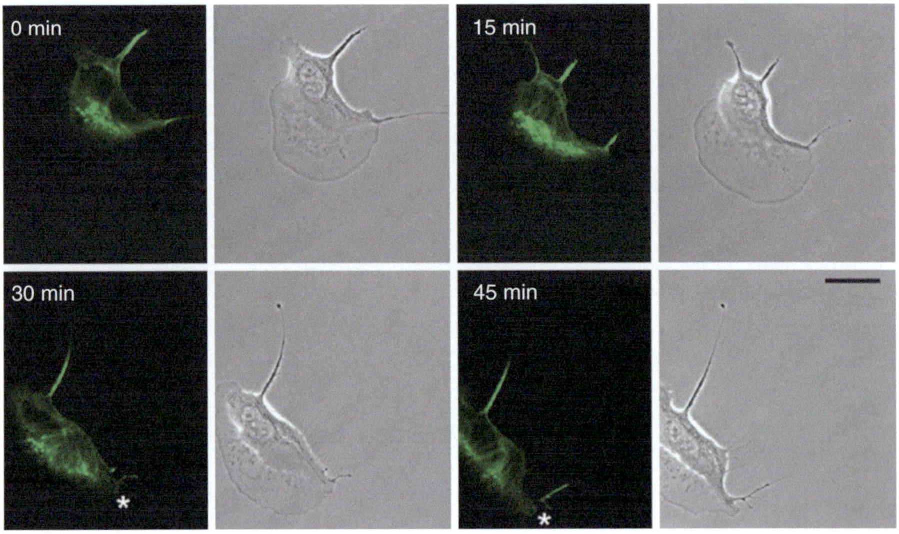

Fig. 19.2 Confocal live cell microscopy of stably drebrin-EGFP-transfected B16F1 cells, showing a time series of a cell spontaneously migrating on laminin. Drebrin-EGFP is concentrated in the tail and in the juxtanuclear drebrin-enriched zone and specifically recruited to newly retracting membrane regions (asterisks). Bar: 20 μm. Source: Peitsch et al. (2006)

To verify the observations made in murine B16F1 in another cell type from a different species, drebrin was studied in human SV80 cells which also exhibit a polarized phenotype and form lamellipodia when seeded on laminin. Double-labeling of these cells for drebrin and actin revealed prominent lamellipodia traversed by actin microspikes. Drebrin was markedly enriched along these microspikes, but only their posterior parts. In addition, intense drebrin reactions were noted in the tail and in the juxtanuclear actin-rich zone, here sometimes in granular structures (Peitsch et al. 2006). Double-label confocal microscopy with antibodies against drebrin and endosomal markers suggested that the drebrin-positive granules were not identical with endosomes but localized in close proximity to them, reminiscent of "actin tails" associated with some types of endocytic vesicles. Interestingly, a homologue of drebrin, the mammalian actin-binding protein 1 (mAbp1), was reported to be involved in endocytosis (Kessels et al. 2001). Furthermore, drebrin was isolated from a specific subset of Golgi vesicles in rat livers (Fucini et al. 2000). The small GTP-binding protein ADP-ribosylation factor (ARF) regulates interactions of actin with the Golgi apparatus. Activation of ARF was shown to induce assembly of distinct pools of actin and actin-binding proteins on Golgi membranes. The two abundant actin-binding proteins of the ARF-dependent actin complex were identified as drebrin and spectrin. Drebrin was found to be a specific component of the cytochalasin D-sensitive, ARF-dependent actin pool (Fucini et al. 2000). Although drebrin and mAbp1 share sequence homology through their N-terminal actin-binding domains, both proteins were segregated into two distinct ARF-dependent actin complexes, and mAbp1 was detected exclusively in the cytochalasin D-sensitive actin pool which was devoid of drebrin (Fucini et al. 2002). The exact nature of the drebrin-containing perinuclear granules observed in SV80 cells and their relation to the Golgi apparatus remains to be elucidated.

When the distribution of drebrin in SV80 cells was compared to that of other actin-binding proteins typical of lamellipodia and actin filament spikes, the Apr2/3 complex, α-actinin, filamin, and cofilin were enriched at the tip and in the middle of the lamellipodium, anterior to drebrin, whereas nonmuscle myosin IIA and IIB were concentrated at the transition between the lamellipodium and the lamella, mostly posterior to drebrin (Peitsch et al. 2006). Interestingly, tropomyosin colocalized with drebrin and actin along the posterior parts of microspikes (Fig. 19.3a), but both proteins did not coprecipitate with each other. Thus, the front region of SV80 cells can be subdivided in three zones, i.e., a zone at the front comprising the Arp2/3 complex, cofilin, α-actinin, and filamin; a zone containing the posterior part of the actin microspikes, drebrin, and tropomyosin; and a transition zone between the lamellipodium and the lamella harboring nonmuscle myosin II (Fig. 19.3b).

Drebrin and tropomyosin are known to compete with each other for binding of actin filaments (Ishikawa et al. 1994). Tropomyosin stabilizes actin filaments by inhibiting their depolymerization by ADF/cofilins and their nucleation by the Arp2/3 complex (Cooper 2002), whereas drebrin increases their plasticity. Absence of tropomyosin from the tips of leading edges was also reported in other motile cells, compatible with the notion that Arp2/3 and cofilins effect high turnover of actin filaments in this region, whereas tropomyosin decelerates this process in posterior parts. Further experiments will be required to fully elucidate the interplay between drebrin and tropomyosin at these sites.

When SV80 cells were transfected with drebrin-EGFP constructs, some transient transfectants overexpressing the protein formed numerous long, branched cell processes, as noted in B16F1 cells. In transfectants containing less drebrin, distribution of the EGFP fusion proteins was identical to that of endogenous drebrin. Cells with the latter phenotype were selected for the generation of stably transfected lines, which showed spontaneous migration on laminin substrates, accompanied by pronounced ruffling in the front of lamellipodia. Drebrin fusion proteins were concentrated at the posterior portions of microspikes, in the juxtanuclear drebrin-enriched zone and in retraction zones (Fig. 19.4, Peitsch et al. 2006). Silencing of drebrin with siRNA did not impact the speed of migration or the extent of ruffling. Furthermore, the distribution of actin and other actin-binding proteins, including tropomyosin, α-actinin, fascin, filamin, the Arp2/3 complex, cofilin, profilin, and nonmuscle myosin II, was not affected by drebrin knockdown. These observations may be explained by the fact that drebrin-depleted SV80 cells still contain residual amounts of drebrin that may be sufficient to maintain the normal phenotype and migration behavior. Alternatively, it is conceivable that the reduction of drebrin is compensated by other actin-binding proteins.

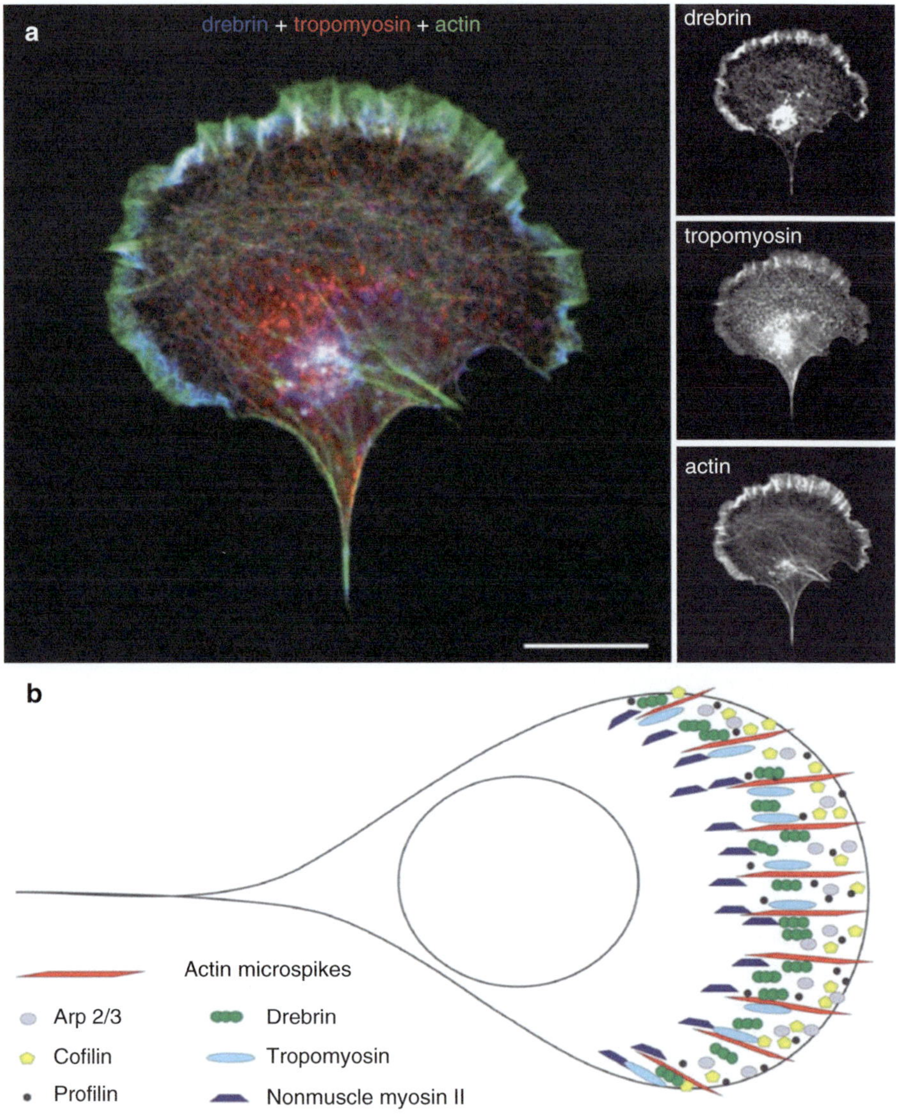

Fig. 19.3 Drebrin and tropomyosin define a specific zone of actin microspikes within posterior lamellipodia. In SV80 cells triple-labeled for drebrin (*blue*), tropomyosin (*red*) and actin (*green*), drebrin and tropomyosin colocalize along the posterior portion of actin microspikes, whereas the tips of the spikes are devoid of both proteins (**a**). Bars: 10 μm. Based on the distinctive distribution patterns of actin-binding proteins, the lamellipodium of SV80 cells can be subdivided in a front zone, containing the Arp2/3 complex, cofilin, α-actinin, and filamin (the latter two not shown), a zone comprising drebrin and tropomyosin along the posterior portion of the actin microfilaments and a zone characterized by nonmuscle myosin II at the transition to the lamella (**b**). Source: Peitsch et al. (2006)

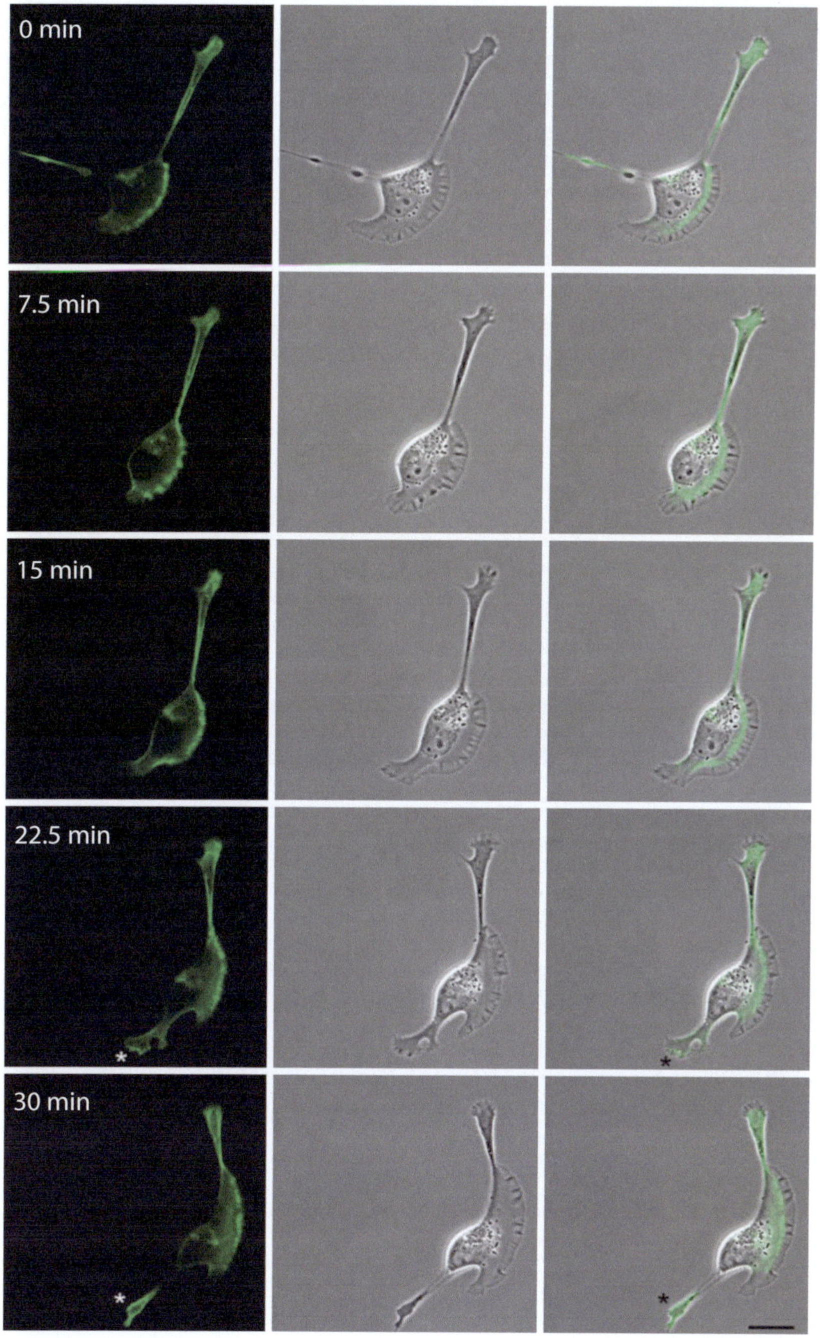

Fig. 19.4 Live cell confocal microscopy of stably drebrin-EGFP-transfected SV80 fibroblasts migrating spontaneously on laminin and showing intense ruffling at the front. Drebrin-EGFP is recruited to the rear of actin microspikes, but never to their tips, and to newly retracting membrane zones (*asterisks*). Bar: 20 μm. Source: Peitsch et al. (2006)

19.3 Conclusion

In prototypes of motile cells such as murine B16F1 melanoma and Swiss 3T3 cells, drebrin is absent from lamellipodia but concentrated in the tail of the cells, in retraction zones, and in a specific juxtanuclear actin filament compartment, the "drebrin-enriched zone." In SV80 fibroblasts it additionally accumulates in the posterior portion of actin microspikes together with tropomyosin, defining a specific zone of the posterior lamellipodium. Since silencing of drebrin does not alter forward movement, the protein appears to be involved in retraction rather than protrusion. The novel, sizable juxtanuclear drebrin-enriched zone will have to be characterized in greater detail with respect to its molecular composition and cell biological functions.

References

Cooper JA (2002) Actin dynamics: tropomyosin provides stability. Curr Biol 12:R523–R525

Fucini RV, Navarrete A, Vadakkan C, Lacomis L, Erdjument-Bromage H, Tempst P, Stamnes M (2000) Activated ADP-ribosylation factor assembles distinct pools of actin on Golgi membranes. J Biol Chem 275:18824–18829

Fucini RV, Chen JL, Sharma C, Kessels MM, Stamnes M (2002) Golgi vesicle proteins are linked to the assembly of an actin complex defined by mAbp1. Mol Biol Cell 13:621–631

Hilpela P, Oberbanscheidt P, Hahne P, Hund M, Kalhammer G, Small JV, Bähler M (2003) SWAP-70 identifies a transitional subset of actin filaments in motile cells. Mol Biol Cell 14:3242–3253

Ishikawa R, Hayashi K, Shirao T, Xue Y, Takagi T, Sasaki Y, Kohama K (1994) Drebrin, a development-associated brain protein from rat embryo, causes the dissociation of tropomyosin from actin filaments. J Biol Chem 269:29928–29933

Kessels MM, Engqvist-Goldstein AE, Drubin DG, Qualmann B (2001) Mammalian Abp1, a signal-responsive F-actin-binding protein, links the actin cytoskeleton to endocytosis via the GTPase dynamin. J Cell Biol 153:351–366

Krause M, Gautreau A (2014) Steering cell migration: lamellipodium dynamics and the regulation of directional persistence. Nat Rev Mol Cell Biol 15:577–590

Peitsch WK, Hofmann I, Prätzel S, Grund C, Kuhn C, Moll I, Langbein L, Franke WW (2001) Drebrin particles: components in the ensemble of proteins regulating actin dynamics of lamellipodia and filopodia. Eur J Cell Biol 80:567–579

Peitsch WK, Bulkescher J, Spring H, Hofmann I, Goerdt S, Franke WW (2006) Dynamics of the actin-binding protein drebrin in motile cells and definition of a juxtanuclear drebrin-enriched zone. Exp Cell Res 312:2605–2618

Rottner K, Stradal TE (2011) Actin dynamics and turnover in cell motility. Curr Opin Cell Biol 23:569–578

Small JV, Resch GP (2005) The comings and goings of actin: coupling protrusion and retraction in cell motility. Curr Opin Cell Biol 17:517–523

Chapter 20
Drebrin in Renal Glomeruli

Wiebke K. Ludwig-Peitsch

Abstract The central function of renal glomeruli is plasma ultrafiltration for primary urine production. The glomerular filtration barrier consists of a fenestrated endothelium, the glomerular basement membrane and podocytes, mesenchymal-like cells with actin filament-rich protrusions, the "foot processes." Their architecture and function are maintained and regulated by actin and several actin-binding proteins, mutations of which can be causative of glomerular diseases. Since initial immunostaining experiments had demonstrated intense drebrin reactions in renal glomeruli, the distribution of this protein was studied in detail in the kidneys of diverse mammalian species. Double-label confocal laser scanning microscopy revealed drebrin enrichment in mesangial cells of human, bovine, murine, and rat kidneys. In Thy-1.1 nephritic rat glomeruli, the protein was concentrated in mesangial cell processes and upregulated during their formation and remodeling. In adult human and bovine kidneys, drebrin was additionally accumulated in the foot processes of podocytes, a finding confirmed by immunoelectron microscopy. By contrast, podocytes of rodent glomeruli contained significant amounts of drebrin only during early developmental stages. In cultured murine podocytes induced to form cell processes, however, drebrin was concentrated in these protrusions, partly in colocalization with other actin-binding proteins. Protein extracts from human and bovine kidneys comprised 20 S-complexes of drebrin and actin, so-called drebrosomes. In summary, drebrin has to be added to the list of actin-binding proteins regulating actin dynamics of mesangial cell processes and foot processes of podocytes. It will be important to determine its role in hereditary and acquired glomerulopathies.

Keywords Actin-binding protein • Drebrin • Glomerulus • Kidney • Mesangial cell • Podocyte • Synaptopodin • Thy-1.1 nephritis

---

W.K. Ludwig-Peitsch
Department of Dermatology, Vivantes Klinikum im Friedrichshain,
Landsberger Allee 49, 10249 Berlin, Germany
e-mail: wiebke.ludwig-peitsch@vivantes.de

T. Shirao, Y. Sekino (eds.), *Drebrin*, Advances in Experimental Medicine and Biology 1006, DOI 10.1007/978-4-431-56550-5_20

20.1 Introduction

The major function of renal glomeruli is plasma ultrafiltration for production of primary urine. The structure responsible for this function is the glomerular filtration barrier, composed of a fenestrated endothelial cell layer on the inner surface, the glomerular basement membrane, and podocytes on the outer surface (Fig. 20.2a). Podocytes are mesenchymal-like cells derived from epithelial precursors with numerous actin filament-rich protrusions, the "foot processes," which are attached to the basement membrane. Their architecture and function is largely dependent on the actin cytoskeleton. At the "foot sole," the actin filament system is linked to the basement membrane via transmembrane proteins such as integrins, α-dystroglycan, syndecan-4, and type XVII collagen, which are associated with several actin-binding proteins, e.g., vinculin, paxillin, and talin (Sachs and Sonnenberg 2013). In their apical domain, the foot processes of adjacent podocytes are interconnected by the slit diaphragm, a zipper-like structure harboring specific transmembrane proteins such as nephrin, P-cadherin, FAT, Neph 1-3, and JAM-4, which are coupled to the actin cytoskeleton via a set of cytoplasmic plaque proteins including α- and β-catenin, plakoglobin, α-actinin, protein ZO-1, synpo, densin, CD2-associated protein (CD2AP), and Nck2 (Faul et al. 2007; Holthöfer 2007). At the molecular level, the slit diaphragm has considerable similarities to adherens junctions; therefore, it has also been designated as modified adherens junction (Reiser et al. 2000). In the most apical part of the foot processes, the membrane glycoprotein podocalyxin forms a complex with NHERF and ezrin, via which it is linked to actin (Nielsen and McNagny 2009).

Dynamics of the actin cytoskeleton play important roles for podocyte development, maintenance of the glomerular filtration barrier, and pathogenesis of glomerular diseases (He et al. 2013). Mutations of actin-binding proteins accumulated in podocyte foot processes, including α-actinin-4 (Kaplan et al. 2000), CD2AP (Kim et al. 2003), synaptopodin (Huber et al. 2006), myosin (Arrondel et al. 2002), and Rho GDIα (Togawa et al. 1999), lead to disturbance of the glomerular filtration barrier and renal dysfunction both in transgenic mouse models and in human genetic diseases.

The architecture of the glomerulus is supported by mesangial cells, smooth muscle-like pericytes building a framework to maintain the structural integrity of the glomerular tuft, and its capillary network (Kriz et al. 1995). The glomerular basement membrane and the mesangial cells form a biomechanical unit which is capable of generating wall tension against a higher intraglomerular pressure and of changing the geometry of glomerular capillaries by mesangial contraction or relaxation. Mesangial cells express several types of receptors for vasoactive agents such as angiotensin-II, arginine vasopressin, atrial natriuretic peptide, and endothelin on their cell surface. Therefore, they can influence glomerular hemodynamics by modifying the glomerular vascular resistance. They use different signaling pathways to cross-communicate with endothelial cells and podocytes under physiological conditions, during glomerular injury and repair (Dimke et al. 2015). Alterations in their actin cytoskeleton are characteristic for a spectrum of glomerular diseases, in particular diabetic glomerulopathy.

When the distribution of drebrin was examined in diverse non-neuronal cell types and tissues, striking accumulations of this protein were noted in cell protrusions of motile cells (Peitsch et al. 2001) as well as in renal glomeruli, the cells of which are rich in protrusions (Peitsch et al. 1999, 2003).

20.2 Localization of Drebrin in Renal Glomeruli

To characterize the drebrin-positive cells within the glomeruli, double-label immunofluorescence and confocal laser scanning microscopy were performed for drebrin in combination with marker proteins of podocytes and mesangial cells. When sections of human and bovine kidney were double-stained with antibodies to drebrin and synaptopodin, a marker protein of podocytes enriched in their foot processes, both drebrin and synaptopodin antibodies reacted intensely in podocytes and colocalized in their cell periphery (Fig. 20.2a–c) (Peitsch et al. 2003). In addition, drebrin-positive immunoreactions were noted in mesangial cells which are synaptopodin-negative (asterisk in Fig. 20.2c).

The subcellular localization of drebrin was further analyzed by immunoelectron microscopy of human renal glomeruli (Fig. 20.1). Silver-enhanced immunogold labeling demonstrated considerable accumulation of drebrin in the foot processes of podocytes, compared to the cell bodies (Fig. 20.1c, d) (Peitsch et al. 2003). Similarly, antibodies to synaptopodin strongly reacted in foot processes (Fig. 20.1b). Essentially identical observations were made with monoclonal and polyclonal drebrin antibodies as well as in immunogold-labeled bovine glomeruli.

Surprisingly, immunofluorescence microscopic examination of rodent kidneys revealed some remarkable differences compared to human and bovine kidneys. In the glomeruli of rats and mice, antibodies to drebrin and synaptopodin reacted in mutually exclusive patterns (Fig. 20.2d–f). Podocytes in the periphery of the glomeruli were positive for synaptopodin but not for drebrin, whereas cells in the center of the glomeruli reacted with drebrin but not with synaptopodin antibodies (Peitsch et al. 2003). Double-staining of rodent glomeruli with antibodies to drebrin and desmin, an intermediate filament protein of mesangial cells, confirmed that the centrally located drebrin-positive cells belong to the mesangium (Fig. 20.3) (Peitsch et al. 2003). The observation that drebrin is virtually specific for mesangial cells in rodent kidneys was made with the monoclonal antibody M2F6 against drebrin and with two different polyclonal drebrin antisera.

The finding of intense drebrin reactions in mesangial cells is in line with the observation of drebrin reactions in vascular smooth muscle cells, including pericytes

(Peitsch et al. 1999), and with the concept of mesangial cells as vascular smooth muscle derivatives (Holthöfer et al. 1995). At present the reason for the species-specific differences in the distribution of drebrin observed in rodents, humans, and cows is unclear.

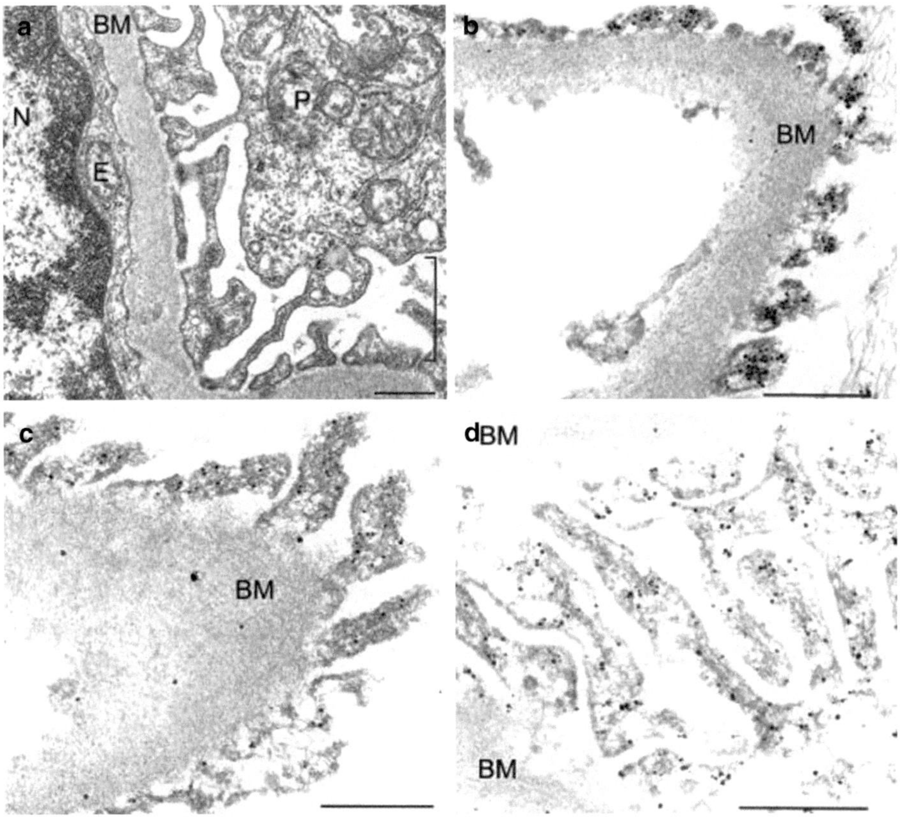

Fig. 20.1 Electron (**a**) and immunoelectron microscopy (**b–d**) of ultrathin sections through human renal glomeruli, showing the region of the glomerular filtration barrier. In the electron micrograph (**a**), the endothelial cell layer (*E* endothelial cell, *N* nucleus) can be seen on the inner surface of the glomerular basement membrane (BM) and the foot processes of podocytes (P) on the outer surface (bracket: zone of the foot processes). Immunoelectron microscopy of similar regions, using antibodies to synaptopodin (**b**) and drebrin (**c**, serum gp dreb254.2; **d**, monoclonal antibody M2F6) in combination with silver-enhanced immunogold labeling, shows marked enrichment of both proteins in the foot processes of podocytes. Bars: 500 nm. Source: Peitsch et al. (2003)

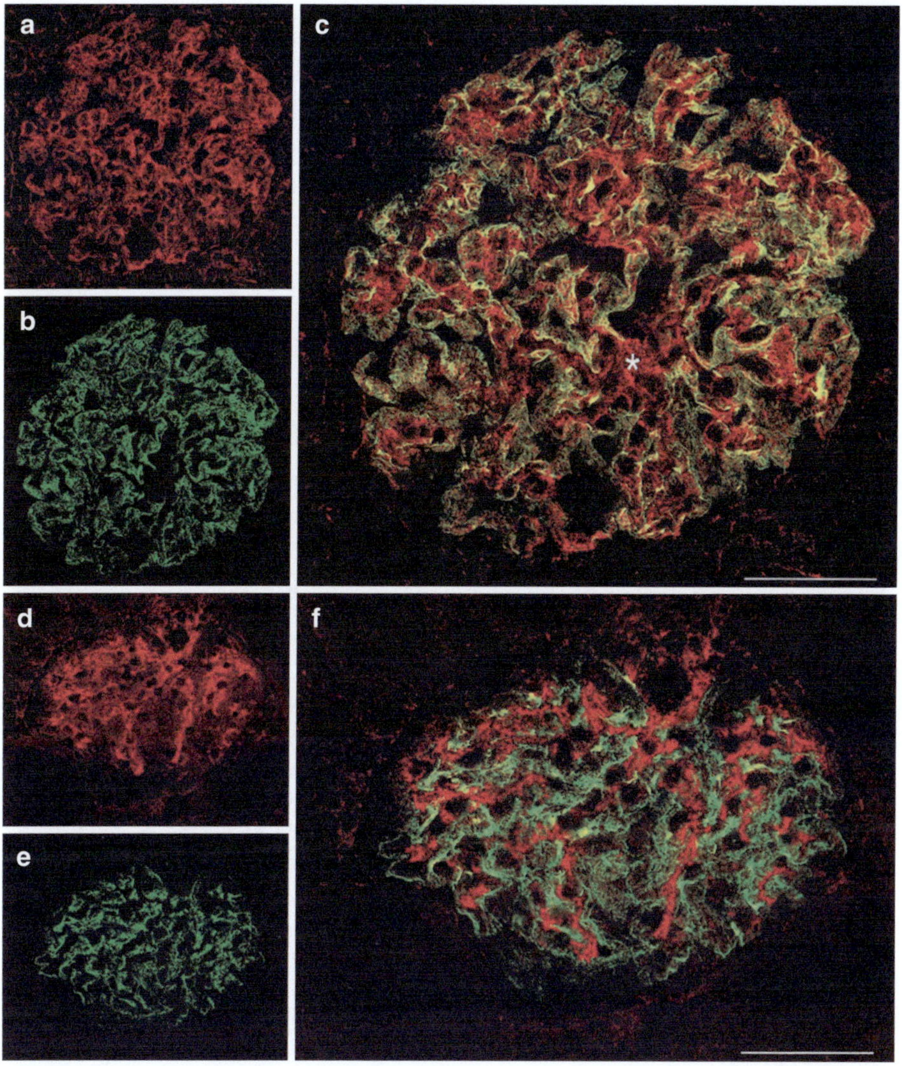

Fig. 20.2 Confocal laser scanning microscopy of sections through glomeruli of human (**a**–**c**) and rat (**d**–**f**) kidneys, stained with antibodies to drebrin (*red* in **a**, **c**, **d**, and **f**) and synaptopodin (*green* in **b**, **c**, **e**, and **f**) as marker protein for podocytes. In human glomeruli, intense drebin staining is observed in many glomerular cells (**a**, **c**), including the synaptopodin-positive podocytes (**b**, **c**). Drebrin and synaptopodin colocalize at the periphery of podocytes (**c**, *yellow*). In addition, drebrin-positive reactions are noted in synaptopodin-negative mesangial cells (**c**, asterisk). By contrast, in rat glomeruli drebrin (**d**, **f**) and synaptopodin (**e**) occur in different cell types. Drebrin-positive immunostaining can be seen in the centrally located mesangial cells, whereas synaptopodin antibodies intensely react with podocytes in the periphery of the glomerulus. Bars: 50 μm. Source: Peitsch et al. (2003)

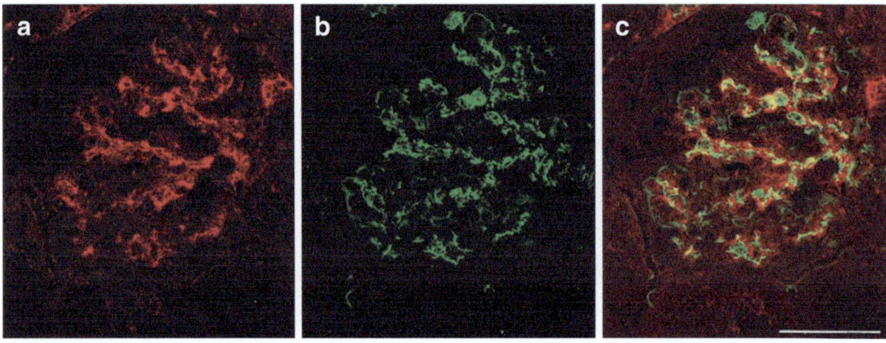

Fig. 20.3 Confocal laser scanning microscopy of a cryostat section through a rat glomerulus, double-labeled with antibodies to drebrin (*red* in **a** and **c**) and desmin (*green* in **b** and **c**). Both antibodies stain mesangial cells which are located rather in the center of the glomerulus. Bar: 50 μm. Source: Peitsch et al. (2003)

20.3 Drebrin in Developing Glomeruli

As the repertoire of cytoskeletal proteins in mesangial cells and podocytes can vary substantially during development and dependent on environmental conditions, drebrin was studied in developing rodent glomeruli. The precursor structures of glomeruli are known to pass through four stages, the renal vesicle, the S-shaped body, the capillary loop stage, and finally, the maturing glomerulus. In embryonic mice at days E16.5 and E19.5, drebrin immunoreactions were observed along intercellular contacts of renal vesicles. In S-shaped bodies and developing glomeruli in the capillary loop stage, drebrin antibodies labeled the cell borders of presumptive podocytes as well as the mesangial and endothelial cells. In glomeruli of the capillary loop stage, the protein was often codistributed with synaptopodin which is known to occur for the first time in presumptive podocytes in this stage. In maturing glomeruli, however, the differentiating synaptopodin-positive podocytes were mostly negative for drebrin, and drebrin-positive immunoreactions were noted almost exclusively in the mesangium and the capillaries. Therefore, drebrin seems to be synthesized in the precursors of murine podocytes in early and intermediate developmental stages but becomes rapidly downregulated during podocyte maturation. The mechanism underlying this downregulation observed in rodents but not in humans will have to be further elucidated.

20.4 Distribution of Drebrin in Cultured Podocytes

Cultured murine podocytes induced to differentiate and form cell processes contain drebrin E, as shown by immunoblot analysis with monoclonal and different polyclonal drebrin antibodies (Peitsch et al. 2003). Immunostaining of these cells revealed pronounced enrichment of drebrin in their cell processes and sometimes along the plasma membrane (Peitsch et al. 2003). This finding was confirmed by

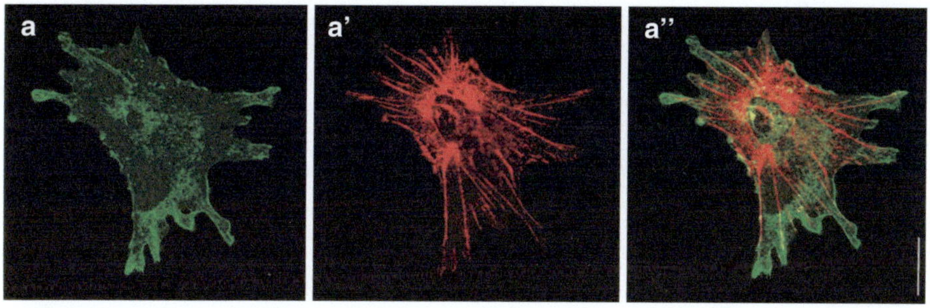

Fig. 20.4 Double-label confocal laser scanning microscopy of drebrin and synaptopodin on cultured murine podocytes. Drebrin (*green* in **a** and **a**'') is enriched in the cell processes of the podocytes, whereas antibodies to synaptopodin (*red* in **a**' and **a**'') stain cytoplasmic actin filament bundles (stress fibers). Bar: 25 μm. Source: Peitsch et al. (2003)

immunoelectron microscopy, showing accumulations of drebrin in the cortical cytoplasm, particularly along the plasma membrane and in filopodia-like cell protrusions.

When the distribution of drebrin in cultured murine podocytes was compared to that of actin and different actin-binding proteins, drebrin was accumulated in cell processes, whereas synaptopodin antibodies stained stress fibers of actin (Fig. 20.4a–a''). The actin-binding protein α-actinin was enriched along stress fibers and in cell protrusions, in the latter often colocalizing with drebrin. In addition, the actin-binding protein ezrin was codistributed with drebrin in cell processes. By contrast, drebrin and vinculin were arranged in mutually exclusive patterns, and the vinculin-positive focal adhesions appeared free of drebrin, as previously reported in other cell types (Peitsch et al. 1999).

Taken together with the observations made in embryonic and adult murine glomeruli, drebrin constitutes an example of proteins synthesized during embryonic development or regeneration, reduced or lost in adult tissues but re-expressed in cell cultures. Accumulations of drebrin in the foot processes of podocytes and in the stellate processes of mesangial cells provide renal examples of drebrin enrichment in cell processes, corresponding to observations in diverse other cell types including neurons (Peitsch et al. 2001 and references therein). Remarkably, foot processes of podocytes and cell processes of neurons are rich in drebrin but also in synaptopodin (Mundel et al. 1997), one of several features of similarity between these two cell types.

20.5 Drebrin Complexes in Kidney Extracts

To characterize the physical state of the drebrin-containing material in the kidney, sucrose gradient centrifugation was performed with kidney extracts obtained under different conditions. When bovine kidneys were extracted with rather high detergent concentrations (1% Triton X-100, 0.2% SDS, and 2 mM EDTA), drebrin-containing particles of ca. 13 S were identified (Peitsch et al. 2001). Using a milder, digitonin-containing buffer, drebrin occurred in relatively large particles sized 20 S both in bovine and in human kidneys (Peitsch et al. 2003). These 20 S-complexes, named "drebrosomes," also contained some amounts of actin.

20.6 Drebrin in Thy-1.1 Nephritic Rats

In the Thy-1.1 nephritis model, monoclonal antibodies against the membrane protein Thy-1.1 are injected into rats where they bind to the surface of glomerular mesangial cells. This leads to complement-dependent damage of the mesangium and severe disturbance of the glomerular capillary network. However, despite initial mesangiolysis and dramatic changes of the glomerular structure, most of the glomeruli can be restored. Therefore, Thy-1.1 nephritis serves as a model for investigating glomerular repair.

Mesangial cells of injured glomeruli express α-smooth muscle actin, which does not occur in healthy glomeruli. In Thy-1.1 nephritic rat, α-smooth muscle actin was found to be upregulated with a peak at day 5 and to gradually decline thereafter (Ichimura et al. 2006). At day 5, almost all mesangial cells exhibited immunoreactivity for α-smooth muscle actin along the entire cell periphery. At day 8, mesangial cells containing α-smooth muscle actin were preferentially accumulated close to the glomerular basement membrane, and α-smooth muscle actin was restricted to their cytoplasmic processes radiating toward it and touching it with their tips (Ichimura et al. 2006). Drebrin immunoreactivity was detected in the mesangial cells of both normal and Thy-1.1 nephritic glomeruli. In the latter, it was enriched in the cell processes of mesangial cells both in central and in peripheral regions of the glomeruli. At day 8, it colocalized with α-smooth muscle actin in peripheral cell processes. Subsequently, drebrin remained accumulated at these sites while α-smooth muscle actin decreased. Immunoblot analysis revealed marked upregulation of drebrin from day 3 to day 14, whereas at day 30 drebrin amounts had returned to control levels. These findings suggest that drebrin is involved in the formation and maintenance of mesangial cell processes during remodeling of Thy-1.1 nephritic glomeruli and that it is particularly upregulated in stages of intense cell process formation (Ichimura et al. 2006).

20.7 Conclusion

Actin and actin-binding proteins are essential for the formation of the cell processes of podocytes and mesangial cells, for the regulation of their shape and maintenance of their stability. For example, collapsing glomerulopathies involve disruption of the actin cytoskeleton of podocytes. Similarly, damage of the actin cytoskeleton of mesangial cells occurs in different types of glomerulopathies, including diabetic glomerulopathy. In these conditions the expression of several actin-binding proteins is altered. The role of drebrin and drebrin complexes will have to be examined for the various situations of actin filament-membrane interactions in glomeruli as well as in the different types of hereditary and acquired glomerulopathies.

References

Arrondel C, Vodovar N, Knebelmann B, Grünfeld JP, Gubler MC, Antignac C, Heidet L (2002) Expression of the nonmuscle myosin heavy chain IIA in the human kidney and screening for MYH9 mutations in Epstein and Fechtner syndromes. J Am Soc Nephrol 13:65–74

Dimke H, Maezawa Y, Quaggin SE (2015) Crosstalk in glomerular injury and repair. Curr Opin Nephrol Hypertens 24:231–238

Faul C, Asanuma K, Yanagida-Asanuma E, Kim K, Mundel P (2007) Actin up: regulation of podocyte structure and function by components of the actin cytoskeleton. Trends Cell Biol 17:428–437

He FF, Chen S, Su H, Meng XF, Zhang C (2013) Actin-associated proteins in the pathogenesis of podocyte injury. Curr Genom 14:477–484

Holthöfer H (2007) Molecular architecture of the glomerular slit diaphragm: lessons learnt for a better understanding of disease pathogenesis. Nephrol Dial Transplant 22:2124–2128

Holthöfer H, Sainio K, Miettinen A (1995) The glomerular mesangium: studies of its developmental origin and markers in vivo and in vitro. APMIS 103:354–366

Huber TB, Kwoh C, Wu H, Asanuma K, Gödel M, Hartleben B, Blumer KJ, Miner JH, Mundel P, Shaw AS (2006) Bigenic mouse models of focal segmental glomerulosclerosis involving pairwise interaction of CD2AP, Fyn, and synaptopodin. J Clin Invest 116:1337–1345

Ichimura K, Kurihara H, Sakai T (2006) Involvement of mesangial cells expressing alpha-smooth muscle actin during restorative glomerular remodeling in Thy-1.1 nephritis. J Histochem Cytochem 54:1291–1301

Kaplan JM, Kim SH, North KN, Rennke H, Correia LA, Tong HQ, Mathis BJ, Rodríguez-Pérez JC, Allen PG, Beggs AH, Pollak MR (2000) Mutations in ACTN4, encoding alpha-actinin-4, cause familial focal segmental glomerulosclerosis. Nat Genet 24:251–256

Kim JM, Wu H, Green G, Winkler CA, Kopp JB, Miner JH, Unanue ER, Shaw AS (2003) CD2-associated protein haploinsufficiency is linked to glomerular disease susceptibility. Science 300:1298–1300

Kriz W, Elger M, Mundel P, Lemley KV (1995) Structure-stabilizing forces in the glomerular tuft. J Am Soc Nephrol 5:1731–1739

Mundel P, Heid HW, Mundel TM, Krüger M, Reiser J, Kriz W (1997) Synaptopodin: an actin-associated protein in telencephalic dendrites and renal podocytes. J Cell Biol 139:193–204

Nielsen JS, McNagny KM (2009) The role of podocalyxin in health and disease. J Am Soc Nephrol 20:1669–1676

Peitsch WK, Grund C, Kuhn C, Schnölzer M, Spring H, Schmelz M, Franke WW (1999) Drebrin is a widespread actin-associating protein enriched at junctional plaques, defining a specific microfilament anchorage system in polar epithelial cells. Eur J Cell Biol 78:767–778

Peitsch WK, Hofmann I, Prätzel S, Grund C, Kuhn C, Moll I, Langbein L, Franke WW (2001) Drebrin particles: components in the ensemble of proteins regulating actin dynamics of lamellipodia and filopodia. Eur J Cell Biol 80:567–579

Peitsch WK, Hofmann I, Endlich N, Prätzel S, Kuhn C, Spring H, Gröne HJ, Kriz W, Franke WW (2003) Cell biological and biochemical characterization of drebrin complexes in mesangial cells and podocytes of renal glomeruli. J Am Soc Nephrol 14:1452–1463

Reiser J, Kriz W, Kretzler M, Mundel P (2000) The glomerular slit diaphragm is a modified adherens junction. J Am Soc Nephrol 11:1–8

Sachs N, Sonnenberg A (2013) Cell-matrix adhesion of podocytes in physiology and disease. Nat Rev Nephrol 9:200–210

Togawa A, Miyoshi J, Ishizaki H, Tanaka M, Takakura A, Nishioka H, Yoshida H, Doi T, Mizoguchi A, Matsuura N, Niho Y, Nishimune Y, Nishikawa SI, Takai Y (1999) Progressive impairment of kidneys and reproductive organs in mice lacking Rho GDIalpha. Oncogene 18:5373–5380

Chapter 21
Drebrin's Role in the Maintenance of Endothelial Integrity

Kerstin Rehm and Stefan Linder

Abstract The human endothelium forms a permeable barrier between the blood stream and surrounding tissues, strictly governing the passage of immune cells, fluids and metabolites. The regulation of cell–cell contact dynamics between endothelial cells is essential for this function and thus for the maintenance of vascular integrity. Intercellular adhesion within the endothelium is mainly dependent on adherens junctions, composed of cell–cell adhesion proteins such as VE-cadherin and nectin, and their associated proteins. Recent research points to a critical role of the actin cytoskeleton in endothelial integrity, by providing anchorage of adhesion complexes to the cell cortex. We could show that the F-actin-binding protein drebrin is a critical regulator of endothelial integrity, by linking nectin to the cortical actin cytoskeleton. In particular, the knockdown of drebrin leads to functional impairment of endothelial cells, characterized by rupturing of endothelial monolayers cultured under conditions mimicking vascular flow. This weakening of cell–cell contacts upon drebrin depletion is based on the destabilization of nectin at adherens junctions, followed by internalization and degradation in lysosomes. Conducting interaction studies, we showed that drebrin binds to nectin's interaction partner afadin, thus linking the nectin/afadin system to the cortical F-actin network. Drebrin, containing binding sites for both afadin and F-actin, is thus uniquely equipped to stabilize nectin at adherens junctions, thereby preserving endothelial integrity. Collectively, these results contribute to the current understanding of cell–cell junction regulation, introducing a new function of drebrin as a stabilizer of endothelial integrity.

Keywords Actin cytoskeleton • Adherens junctions • Afadin • Drebrin • Endothelial integrity • Nectin

---

Parts of this book chapter have been published, in modified form, in the doctoral thesis of Kerstin Rehm "Drebrin preserves endothelial integrity by stabilizing nectin at adherens junctions," 2013.

K. Rehm • S. Linder (✉)
Institute for Medical Microbiology, Virology and Hygiene, University Medical Center Eppendorf, Martinistr. 52, 20246 Hamburg, Germany
e-mail: s.linder@uke.de

T. Shirao, Y. Sekino (eds.), *Drebrin*, Advances in Experimental Medicine and Biology 1006, DOI 10.1007/978-4-431-56550-5_21

21.1 Drebrin in Monolayer-Forming Cells

Cell monolayers fulfill critical roles in multicellular organisms (Harris et al. 2012). Monolayers act by separating different tissues or compartments from each other, forming semipermeable barriers that can withstand mechanical forces. This is especially important in the case of epithelial and endothelial cells. The endothelium, in particular, serves as a semi-selective barrier between the bloodstream and the surrounding tissue, controlling the passage of leukocytes and solutes. Depending on the location of the endothelium in the vascular tree, the respective degree of permeability is quite variable. In this respect, specific and regulated anchorage between individual cells within monolayers is of major importance. Accordingly, dysregulation of cell–cell junctions compromises endothelial integrity and can thus lead to pathological scenarios such as thrombogenesis (Vestweber 2007; Muller 2003). In recent years, it has become increasingly clear that the various systems of intercellular cell–cell junctions (see Sect. 21.1.1) depend on the filamentous (F)-actin cytoskeleton for their stabilization. The regulation of F-actin dynamics is thus of major importance for the coherence and adaptability of the endothelium (Lampugnani 2010). This implies critical roles for F-actin-associated proteins in the maintenance of vascular integrity.

In this context, Peitsch et al. described drebrin E in a variety of cell types and tissues, also including epithelial and endothelial cells (Peitsch et al. 1999). Interestingly, drebrin levels are strongly increased in epidermal skin tumors, where it localizes to junctional areas. Of note, even at moderate levels of expression, such as in cultured keratinocytes, drebrin is enriched at adherens junctions (Peitsch et al. 2005). In line with these findings, confluent endothelial monolayers show a pronounced expression of drebrin E, with the majority being enriched at the cell cortex, where it associates with F-actin filaments, and preferentially near adherens junctions (Peitsch et al. 1999).

In our study, we were able to confirm the localization of drebrin to endothelial cell–cell junctions in human umbilical vein endothelial cells (HUVEC) and focused in the following on the as yet unclear function of drebrin in endothelial junction regulation (Rehm 2013). In particular, we studied the dynamic behavior of endothelial monolayers that were cultivated under constant flow conditions of 15 dyne/cm^2, thus mimicking flow conditions in medium-sized vessels (Paz et al. 2012). As expected, both control and drebrin-depleted cells aligned themselves along the direction of flow. However, in contrast to control monolayers that retained their integrity, monolayers of drebrin-depleted cells showed ruptures after 3 days, pointing to a role of drebrin in the maintenance of endothelial integrity especially under shear stress, probably through regulation of cell–cell junctions (Rehm et al. 2013) (Fig. 21.1).

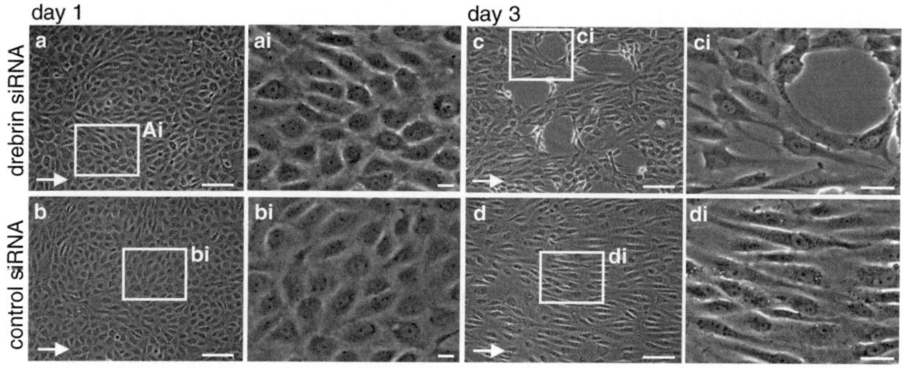

Fig. 21.1 Monolayers of HUVEC depleted for drebrin show ruptures when cultured under constant flow. Images of HUVEC monolayers, treated with drebrin-specific (**a**, **c**) or control siRNA (**b**, **d**) and seeded in Microslides. Cells were submitted to constant fluid shear stress for 1–3 days. The enlarged images (**ai**–**di**) show boxed regions of (**a**–**d**). *Arrows* indicate the direction of flow. Note the rupture of drebrin knockdown monolayers at day 3. Bars, 10 μm in (**ai**–**di**), 100 μm in (**a**–**d**) (Rehm et al. 2013)

21.1.1 Different Sets of Cell–Cell Junctions Are Involved in Maintaining Monolayer Integrity

Within monolayers, cells are connected by a number of different junctional systems, which regulate endothelial barrier function, tissue integrity, as well as cell–cell communication (Wallez and Huber 2008; Lampugnani and Dejana 1997). These junctional systems, including tight junctions, adherens junctions, gap junctions, and desmosomes, have been studied especially in epithelial cells. In these cells, the different junctions follow a well-defined distribution along the intercellular cleft, with tight junctions being the most apical component, followed by adherens junctions and desmosomes. However, in endothelial cells, the junction types are less well organized and do not exhibit such a clear progression along the contact zone (Engelhardt and Wolburg 2004). Moreover, desmosomes are missing in endothelial cells, and instead, typical endothelial proteins such as platelet endothelial cell adhesion molecule-1 (PECAM-1) and intercellular adhesion molecule-2 (ICAM-2) are found, which contribute to cell–cell adhesion and also to angiogenesis or leukocyte extravasation (Muller et al. 1993; DeLisser et al. 1994; Lyck et al. 2003). Still, junctions of the epithelium and endothelium are highly homologous, which often allows a careful transfer of observations made in epithelial cell models also to endothelial cells.

21.1.1.1 Tight Junctions

In epithelial cells, tight junctions are found at the most apical position (Dejana et al. 2009). Their main function is the formation of an impermeable barrier for soluble molecules. In endothelial cells, the necessity for permeability control is not as absolute and can vary, depending on the location within the vascular tree (Bazzoni and Dejana 2004). The most prominent components of tight junctions are occludins, claudins, and junctional adhesion molecules (JAMs), with the latter also occurring in cells that do not form junctions (Bazzoni 2003). Occludins have four membrane-spanning regions, with their N- and C-termini both being intracellular. The 15 claudins identified so far show a similar architecture, with claudin-5 being the endothelial-specific member (Morita et al. 1999). Occludin and claudin can bind to different zonula occludentes (ZO) proteins (Tsukita et al. 1999; Tsukita and Furuse 1999), which primarily serve as linkers to the F-actin cytoskeleton and to other proteins (Stevenson et al. 1986).

21.1.1.2 Gap Junctions

In addition to the junctional complexes that serve as anchoring structures, gap junctions mediate communication between neighboring cells (Simon and Goodenough 1998). They constitute clusters of few to hundreds of intercellular channels, which are permeable for ions and metabolites, excluding molecules that exceed 1 kDa in size (Alexander and Goldberg 2003). Each channel is formed by transmembrane proteins that belong to the connexin (Cx) family, which consists of 20 members in humans, with Cx43, Cx40, and Cx37 being expressed in the endothelium. Connexins assemble into hexameric clusters, forming a hemichannel (connexon) in the plasma membrane, which aligns with another connexon of an adjacent cell, thus forming a pore that connects their cytoplasms (Bazzoni and Dejana 2004). This way, adjoined cells can share second messengers or metabolites and can give coordinated responses to extracellular stimuli. Of note, Butkevich et al. found drebrin at gap junctions of green monkey kidney epithelial cells. Moreover, siRNA-mediated depletion of drebrin led to impaired cell–cell coupling and internalization of connexin-43, thus destabilizing gap junctions (Butkevich et al. 2004).

21.1.1.3 Adherens Junctions

Adherens junctions (AJ) are molecular assemblies of proteins, which provide intercellular adhesion. Among the most prominent components are cadherins, with VE (vascular endothelial)-cadherin being the only one of over 350 cadherins that is expressed exclusively in endothelial cells (Hulpiau and van Roy 2011). VE-cadherin is part of the subfamily of classical cadherins, which all share six conserved extracellular cadherin domains responsible for calcium-dependent dimerization (Boggon et al. 2002). Besides their main function of mediating adhesion, cadherins play a role in intracellular signaling, as their cytoplasmic tail region interacts with a variety of proteins. Of major importance is p120-catenin, which binds to the juxtamembrane region of cadherins, preventing their clathrin-related endocytosis through stabilization at the membrane (Davis et al. 2003; Chiasson et al. 2009). Upon the release of p120-catenin, it is able to translocate into the nucleus to regulate transcription, displaying a dual role in the cell that is common for many junction-associated proteins (Cavallaro and Dejana 2011; Yap et al. 1997). Another prominent binding partner of cadherins is β-catenin, which was thought to link cadherins to F-actin through α-catenin (Gates and Peifer 2005). However, this model has been challenged, for example, through FRAP experiments showing that F-actin has a more dynamic behavior than α-catenin, which excludes the possibility of a stable complex (Yamada et al. 2005; Drees et al. 2005). Accordingly, Drees et al. showed that α-catenin either binds to β-catenin/cadherin or to F-actin but not to both at the same time (Drees et al. 2005; Weis and Nelson 2006).

Nectins constitute a further group of AJ-localized intercellular adhesion molecules (Lopez et al. 1995). They work independently of Ca^{2+} and consist of a cytoplasmic tail region, a single transmembrane region and three extracellular immunoglobulin-like loops involved in cis- and trans-oligomerization. Comparable to cadherins, nectins initially form lateral homo (or hetero-)-cis-dimers, followed by trans-interactions with a dimer of the adjacent cell (Satoh-Horikawa et al. 2000; Reymond et al. 2001). Currently, four members of the nectin family have been identified (nectin-1, 2, 3, and 4), with several respective splice variants (Satoh-Horikawa et al. 2000; Reymond et al. 2001). The major nectin isoforms in endothelial cells, including HUVEC, are nectin-2 and nectin-3 (Lopez et al. 1998). The cytoplasmic tail of nectins contains the motif E/A-X-Y-V, which binds the Postsynaptic density protein-95/Drosophila disc large tumor suppressor/Zonula occludens-1 protein (PDZ) domain of their typical binding partner afadin (Mandai et al. 1997).

Afadin (or l-Afadin) is a multivalent adaptor protein featuring a variety of functional domains, two Ras-associated domains (RA), one dilute (DIL) domain, a fork head-associated (FHA) domain, a PDZ domain, three proline-rich regions (PR), and an F-actin-binding C-terminus, allowing interactions with a variety of other proteins (Takahashi et al. 1999; Takai et al. 2008) (Fig. 21.2). Afadin is thought to serve as a connector between the two main adhesion systems, nectins, and cadherins, by binding nectin and several of the cadherin-associated proteins. Among others, it can bind to α-catenin, which is mainly localized at cadherin-based junctions and to ponsin, which then binds vinculin (Pokutta et al. 2002; Mandai et al. 1999).

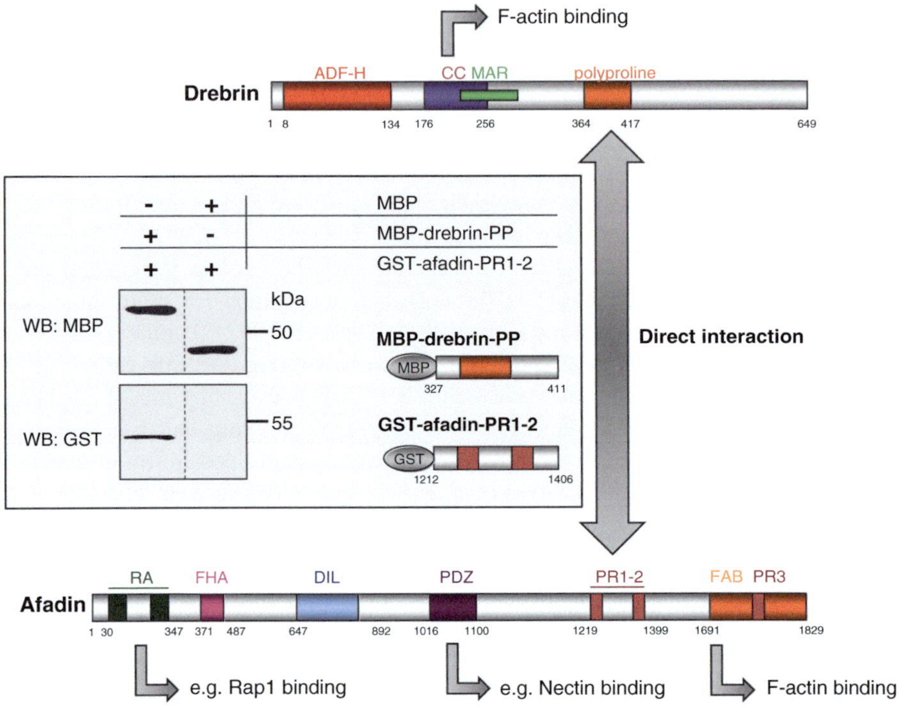

Fig. 21.2 Drebrin and afadin interact through their polyproline and PR1-2 regions. *Upper scheme*: Drebrin domain structure, including ADF-homology region (aa 8–134), coiled coil region involved in homodimerization and F-actin binding (Peitsch et al. 2001) (CC, aa 176–256), minimal actin-remodeling region (Hayashi et al. 1999) (MAR, aa 233–317), and polyproline region (PP, aa 364–417). *Lower scheme*: Afadin domain structure, including RA regions involved in Rap1 binding (aa 30–347), FHA region (aa 371–487), DIL region (aa 647–892), PDZ region involved in nectin binding (aa 1016–1100), PR1-2 region containing two polyproline stretches (aa 1219–1399), and FAB region involved in F-actin binding (aa 1691–1829), containing a third proline-rich stretch (PR3). *Middle box*: Illustration of direct interaction of drebrin's PP and afadin's PR1–2 regions. Western blots of pulldown assay using drebrin–PP fused to MBP (MBP-drebrin-PP) or MBP as control immobilized on amylose resin beads, incubated with afadin-PR1-2 fused to GST (GST-afadin-PR1-2). ± indicates presence of respective components in experiments. Western blots developed with indicated antibodies (Rehm 2013; Rehm et al. 2013)

21.2 Junctional Integrity Depends on the F-actin Cytoskeleton

The cytoskeleton of cells is essential for their morphology, migration, and cytokinesis and also for intracellular transport, endo- and exocytosis, as well as a large variety of other processes (Pollard and Borisy 2003). It is comprised of three systems, microtubules, intermediate filaments, and F-actin, each associated with numerous specific accessory and regulatory proteins (Dudek and Garcia 2001). Actin is the central cytoskeletal element of endothelial cells, which comprises up to 15% of total protein content. In confluent endothelial cells, F-actin is found primarily beneath the plasma membrane, forming the pool of "cortical F-actin" and interacting with cell–cell adhesion complexes through respective adaptor proteins. The presence of this circumferential cortical F-actin ring has been shown to be essential for adhesion in general; if disrupted by Latrunculin A or cytochalasin D, junction integrity is lost (Shen and Turner 2005; Ivanov et al. 2005; Yamazaki et al. 2007; Quinlan and Hyatt 1999).

The formation of cell–cell junctions is usually initiated between migrating cells, which form characteristic lamellipodia with unevenly distributed adhesion molecules at their leading edges (Yap et al. 1997; Hoelzle and Svitkina 2012). This initial assembly of nascent cell–cell contacts is independent of F-actin. However, once this contact is established, the further maturation of junctions relies on an intact cortical actin cytoskeleton (Chu et al. 2006; Chu et al. 2004). Furthermore, linkage to F-actin also influences the strength of the respective intercellular adhesion (Mege et al. 2006). Quickly after initial adhesion site formation at the lamellipodia, the actin-related protein (Arp) 2/3 complex is deactivated, or potentially repressed by an enrichment of α-catenin (Drees et al. 2005; Pokutta and Weis 2007). Consequently, other actin-binding proteins, among them vasodilator-stimulated phosphoprotein) and Mena, are subsequently recruited through α-catenin, leading to the formation of thicker F-actin cables needed to stabilize the developing junctions (Drees et al. 2005; Krause et al. 2003; Scott et al. 2006).

21.2.1 Adaptor Proteins Link Cortical F-actin to Junctional Proteins

The cortical actin cytoskeleton is crucial for the maintenance of cell–cell junctions. However, many proteins that form intercellular oligomers do not directly interact with F-actin. Therefore, the presence of linker proteins connecting intercellular adhesion proteins to F-actin is especially important.

Prominent examples of such adaptor proteins are α-catenin, VASP, Epithelial protein lost in neoplasm (EPLIN), or afadin, all of which are involved in anchoring adhesion systems to F-actin. Their importance is underlined by different studies, which show how their loss leads to severe defects in junction formation and maintenance (Vasioukhin et al. 2000; Kwiatkowski et al. 2010; Pappas and Rimm 2006; Abe and Takeichi 2008). The underlying reasons for these defects in proper junction formation are varied, depending on the function of the respective protein. Colon carcinoma cell lines deficient in α-catenin were not able to maintain their adhesive properties, due to an inability of forming cadherin-mediated contacts (Pappas and Rimm 2006). VASP activity was shown to be necessary both for F-actin accumulation and its assembly at cell–cell contacts, and sequestering VASP through relocating it to mitochondria led to defects in F-actin organization at contact sites (Scott et al. 2006), while VASP knockdown led to enhanced endothelial permeability (Schlegel et al. 2008; Reinhard et al. 2001). In afadin-depleted epithelial cells, adherens junctions do not form due to defective recruitment of cadherins to initial adherens junction sites, underlining the importance of afadin during initial junction formation (Tachibana et al. 2000). EPLIN is another linker of the cadherin–catenin complex to F-actin and is also needed for the stabilization of cortical actin fibers. Accordingly, siRNA-mediated knockdown of EPLIN in colon adenocarcinoma cell lines led to misorganization of the cortical F-actin, with E-cadherin showing only a punctate accumulation at cell–cell contact zones (Abe and Takeichi 2008; Chervin-Petinot et al. 2012).

21.2.2 Drebrin Acts as a Stabilizer for Nectins, by Linking Afadin to Cortical F-actin

It is well described that cadherins are linked to the F-actin cytoskeleton. Also, the roles of the different adaptor proteins through which this linkage is established have been explored in depth (Pappas and Rimm 2006; Abe and Takeichi 2008; Tachibana et al. 2000). In contrast, knowledge on how other adherens junction components, in particular nectins, are stabilized in the junctional area is scarce. This is in part based on the fact that nectins are not well characterized in the endothelium in general and are usually only mentioned in the context of initial junction formation (Takahashi et al. 1999; Mizoguchi et al. 2002; Rikitake et al. 2012; Reymond et al. 2000; Ikeda et al. 1999; Mueller et al. 2003). In our work, we could show that drebrin acts by stabilizing the nectin/afadin system through linking it to the cortical F-actin network. These data revealed a novel function of drebrin in the endothelium, as an adaptor protein for the nectin/afadin system (Rehm et al. 2013). In addition, our work showed that nectins are not only necessary for the formation of nascent cell–cell contacts but are also of major importance for the maintenance of established adherens junctions, especially under conditions of enhanced shear stress such as vascular flow (Fig. 21.1).

In our study, we could show that siRNA-mediated depletion of drebrin leads to functional impairments of endothelial monolayers, as demonstrated by a decrease of transendothelial electrical resistance (TER) and also by rupturing of HUVEC monolayers cultured under constant unidirectional flow conditions (Fig. 21.1). The observed weakening of cell–cell contacts upon drebrin depletion is characterized by a specific loss of nectin-2 from adherens junctions, whereas other junctional proteins such as connexin-43, VE-cadherin, occludin, or PECAM-1 remain apparently unaffected (Rehm et al. 2013). Further experiments showed that, in the absence of drebrin, nectin is endocytosed and subsequently degraded in lysosomes, pointing to an important role of drebrin in the stabilization of nectin at the junctional area. The importance of drebrin for nectin's presence at junctions was further underlined by rescue experiments, where reexpression of siRNA-insensitive drebrin led to recovery of nectin at junctions.

In co-immunoprecipitation experiments, we could show that drebrin does not interact with nectin directly but with its most prominent intracellular binding partner, afadin. Direct binding of drebrin and afadin is mediated through their polyproline and PR1–2 regions, as shown by glutathione S-transferase (GST)-pulldown experiments using bacterially expressed domain constructs of both proteins fused to GST or maltose-binding protein (MBP) (Fig. 21.2). Moreover, confocal microscopy studies revealed strong binding between drebrin and afadin also in a cellular context, as drebrin's polyproline region fused to a mitochondrial targeting signal is sufficient to relocalize endogenous afadin to the outer membrane of mitochondria. The close association of afadin and drebrin was further demonstrated by FRAP experiments, which showed that afadin's mobility at the junctional area is enhanced under drebrin depletion. The fact that afadin remains at junctions, in contrast to nectin, is most probably due to accessory binding by other proteins, such as ZO-1 or α-catenin (Tachibana et al. 2000; Takai and Nakanishi 2003).

Furthermore, we could demonstrate that drebrin maintains junctional integrity through its ability to link the nectin/afadin system to the cortical F-actin network. Being equipped with an F-actin-binding module (CC-region) and the afadin-binding polyproline region, drebrin is able to anchor afadin to the cortical actin cytoskeleton. As afadin can bind to nectin through its PDZ region, this results in a chain of protein interactions, with the sequence F-actin-drebrin-afadin-nectin. Nectin is thus indirectly stabilized by drebrin at junctions through drebrin's dual ability to bind afadin and F-actin (Fig. 21.3).

In order to verify that the linkage of nectin to F-actin is indeed essential for monolayer integrity, we generated a variety of minimal rescue constructs and expressed them in cells under knockdown of both drebrin and afadin. Remarkably, minimal constructs containing only afadin's PDZ region coupled to drebrin´s F-actin-binding region, or to lifeact, were able to fully rescue nectin at adherens junctions. These results showed that drebrin functions indeed by providing indirect linkage for nectin to F-actin. At the same time, the construct containing lifeact instead of drebrin's F-actin-binding region demonstrated that only binding to F-actin per se is necessary and that the drebrin-provided anchorage is not based on a specific property of the drebrin–actin interface. Still, drebrin is uniquely equipped to stabilize nectin at endothelial junctions, as it contains binding sites for both afadin and F-actin. This enables drebrin to function as a linker for the nectin/afadin system to the cortical actin cytoskeleton and thus preserve endothelial integrity. Collectively, these results contribute to the current understanding of cell–cell junction regulation in the endothelium, and especially under vascular flow. In particular, the newly identified interaction between drebrin and afadin is shown to be crucial for junctional integrity, not only for junction formation but also during the maintenance of junctions under steady-state conditions (Rehm et al. 2013).

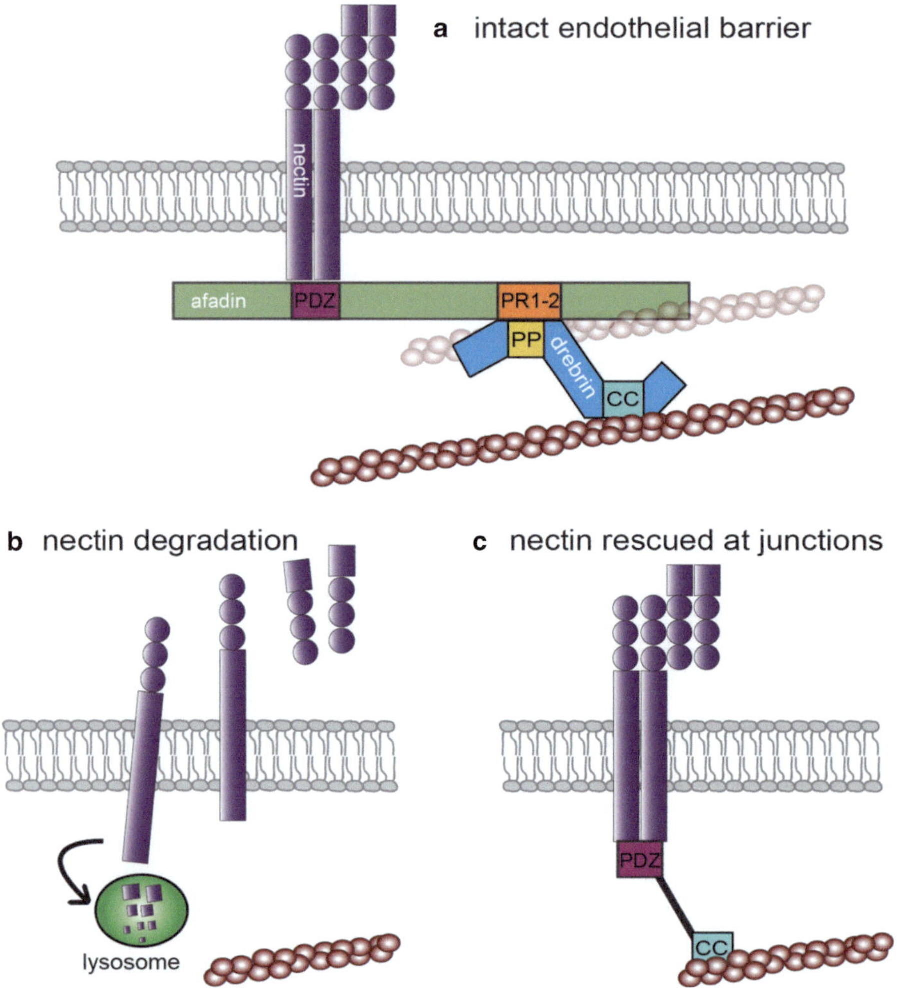

Fig. 21.3 Drebrin binds afadin and anchors nectin to the F-actin cytoskeleton in endothelial cells. (**a**) Drebrin binds to F-actin with high affinity through its coiled coil (CC) region and binds via its polyproline (PP) region to afadin's PR1–2 regions, while afadin's PDZ domain binds nectin. Nectin is thus stabilized at the junctional region, preserving endothelial integrity. (**b**) The absence of drebrin and/or afadin leads to the loss of nectin's indirect anchorage to the actin cytoskeleton. Nectins are subsequently internalized and degraded in lysosomes, resulting in impaired endothelial integrity. (**c**) Nectin can be artificially stabilized at junctions even in the absence of both afadin and drebrin upon the overexpression of constructs containing the afadin PDZ region and drebrin's CC region, reestablishing proper anchorage to the actin cytoskeleton

References

Abe K, Takeichi M (2008) EPLIN mediates linkage of the cadherin catenin complex to F-actin and stabilizes the circumferential actin belt. Proc Natl Acad Sci U S A 105(1):13–19. doi:10.1073/pnas.0710504105

Alexander DB, Goldberg GS (2003) Transfer of biologically important molecules between cells through gap junction channels. Curr Med Chem 10(19):2045–2058

Bazzoni G (2003) The JAM family of junctional adhesion molecules. Curr Opin Cell Biol 15(5):525–530

Bazzoni G, Dejana E (2004) Endothelial cell-to-cell junctions: molecular organization and role in vascular homeostasis. Physiol Rev 84(3):869–901. doi:10.1152/physrev.00035.2003

Boggon TJ, Murray J, Chappuis-Flament S, Wong E, Gumbiner BM, Shapiro L (2002) C-cadherin ectodomain structure and implications for cell adhesion mechanisms. Science 296(5571):1308–1313. doi:10.1126/science.1071559

Butkevich E, Hulsmann S, Wenzel D, Shirao T, Duden R, Majoul I (2004) Drebrin is a novel connexin-43 binding partner that links gap junctions to the submembrane cytoskeleton. Curr Biol 14(8):650–658. doi:10.1016/j.cub.2004.03.063

Cavallaro U, Dejana E (2011) Adhesion molecule signalling: not always a sticky business. Nat Rev Mol Cell Biol 12(3):189–197. doi:10.1038/nrm3068

Chervin-Petinot A, Courcon M, Almagro S, Nicolas A, Grichine A, Grunwald D, Prandini MH, Huber P, Gulino-Debrac D (2012) Epithelial protein lost in neoplasm (EPLIN) interacts with alpha-catenin and actin filaments in endothelial cells and stabilizes vascular capillary network in vitro. J Biol Chem 287(10):7556–7572. doi:10.1074/jbc.M111.328682

Chiasson CM, Wittich KB, Vincent PA, Faundez V, Kowalczyk AP (2009) p120-catenin inhibits VE-cadherin internalization through a Rho-independent mechanism. Mol Biol Cell 20(7):1970–1980. doi:10.1091/mbc.E08-07-0735

Chu YS, Thomas WA, Eder O, Pincet F, Perez E, Thiery JP, Dufour S (2004) Force measurements in E-cadherin-mediated cell doublets reveal rapid adhesion strengthened by actin cytoskeleton remodeling through Rac and Cdc42. J Cell Biol 167(6):1183–1194. doi:10.1083/jcb.200403043

Chu YS, Eder O, Thomas WA, Simcha I, Pincet F, Ben-Zéev A, Perez E, Thiery JP, Dufour S (2006) Prototypical type I E-cadherin and type II cadherin-7 mediate very distinct adhesiveness through their extracellular domains. J Biol Chem 281(5):2901–2910. doi:10.1074/jbc.M506185200

Davis MA, Ireton RC, Reynolds AB (2003) A core function for p120-catenin in cadherin turnover. J Cell Biol 163(3):525–534. doi:10.1083/jcb.200307111

Dejana E, Tournier-Lasserve E, Weinstein BM (2009) The control of vascular integrity by endothelial cell junctions: molecular basis and pathological implications. Dev Cell 16(2):209–221. doi:10.1016/j.devcel.2009.01.004

DeLisser HM, Newman PJ, Albelda SM (1994) Molecular and functional aspects of PECAM-1/CD31. Immunol Today 15(10):490–495. doi:10.1016/0167-5699(94)90195-3

Drees F, Pokutta S, Yamada S, Nelson WJ, Weis WI (2005) Alpha-catenin is a molecular switch that binds E-cadherin-beta-catenin and regulates actin-filament assembly. Cell 123(5):903–915. doi:10.1016/j.cell.2005.09.021

Dudek SM, Garcia JG (2001) Cytoskeletal regulation of pulmonary vascular permeability. J Appl Physiol 91(4):1487–1500

Engelhardt B, Wolburg H (2004) Mini-review: Transendothelial migration of leukocytes: through the front door or around the side of the house? Eur J Immunol 34(11):2955–2963. doi:10.1002/eji.200425327

Gates J, Peifer M (2005) Can 1000 reviews be wrong? Actin, alpha-Catenin, and adherens junctions. Cell 123(5):769–772. doi:10.1016/j.cell.2005.11.009

Harris AR, Peter L, Bellis J, Baum B, Kabla AJ, Charras GT (2012) Characterizing the mechanics of cultured cell monolayers. Proc Natl Acad Sci U S A 109(41):16449–16454. doi:10.1073/pnas.1213301109

Hayashi K, Ishikawa R, Kawai-Hirai R, Takagi T, Taketomi A, Shirao T (1999) Domain analysis of the actin-binding and actin-remodeling activities of drebrin. Exp Cell Res 253(2):673–680. doi:10.1006/excr.1999.4663

Hoelzle MK, Svitkina T (2012) The cytoskeletal mechanisms of cell–cell junction formation in endothelial cells. Mol Biol Cell 23(2):310–323. doi:10.1091/mbc.E11-08-0719

Hulpiau P, van Roy F (2011) New insights into the evolution of metazoan cadherins. Mol Biol Evol 28(1):647–657. doi:10.1093/molbev/msq233

Ikeda W, Nakanishi H, Miyoshi J, Mandai K, Ishizaki H, Tanaka M, Togawa A, Takahashi K, Nishioka H, Yoshida H, Mizoguchi A, Nishikawa S, Takai Y (1999) Afadin: a key molecule essential for structural organization of cell–cell junctions of polarized epithelia during embryogenesis. J Cell Biol 146(5):1117–1132

Ivanov AI, Hunt D, Utech M, Nusrat A, Parkos CA (2005) Differential roles for actin polymerization and a myosin II motor in assembly of the epithelial apical junctional complex. Mol Biol Cell 16(6):2636–2650. doi:10.1091/mbc.E05-01-0043

Krause M, Dent EW, Bear JE, Loureiro JJ, Gertler FB (2003) Ena/VASP proteins: regulators of the actin cytoskeleton and cell migration. Annu Rev Cell Dev Biol 19:541–564. doi:10.1146/annurev.cellbio.19.050103.103356

Kwiatkowski AV, Maiden SL, Pokutta S, Choi HJ, Benjamin JM, Lynch AM, Nelson WJ, Weis WI, Hardin J (2010) In vitro and in vivo reconstitution of the cadherin-catenin-actin complex from Caenorhabditis elegans. Proc Natl Acad Sci U S A 107(33):14591–14596. doi:10.1073/pnas.1007349107

Lampugnani MG (2010) Endothelial adherens junctions and the actin cytoskeleton: an 'infinity net'? J Biol 9(3):16. doi:10.1186/jbiol232

Lampugnani MG, Dejana E (1997) Interendothelial junctions: structure, signalling and functional roles. Curr Opin Cell Biol 9(5):674–682

Lopez M, Eberle F, Mattei MG, Gabert J, Birg F, Bardin F, Maroc C, Dubreuil P (1995) Complementary DNA characterization and chromosomal localization of a human gene related to the poliovirus receptor-encoding gene. Gene 155(2):261–265

Lopez M, Aoubala M, Jordier F, Isnardon D, Gomez S, Dubreuil P (1998) The human poliovirus receptor related 2 protein is a new hematopoietic/endothelial homophilic adhesion molecule. Blood 92(12):4602–4611

Lyck R, Reiss Y, Gerwin N, Greenwood J, Adamson P, Engelhardt B (2003) T-cell interaction with ICAM-1/ICAM-2 double-deficient brain endothelium in vitro: the cytoplasmic tail of endothelial ICAM-1 is necessary for transendothelial migration of T cells. Blood 102(10):3675–3683. doi:10.1182/blood-2003-02-0358

Mandai K, Nakanishi H, Satoh A, Obaishi H, Wada M, Nishioka H, Itoh M, Mizoguchi A, Aoki T, Fujimoto T, Matsuda Y, Tsukita S, Takai Y (1997) Afadin: A novel actin filament-binding protein with one PDZ domain localized at cadherin-based cell-to-cell adherens junction. J Cell Biol 139(2):517–528

Mandai K, Nakanishi H, Satoh A, Takahashi K, Satoh K, Nishioka H, Mizoguchi A, Takai Y (1999) Ponsin/SH3P12: an l-afadin- and vinculin-binding protein localized at cell–cell and cell-matrix adherens junctions. J Cell Biol 144(5):1001–1017

Mege RM, Gavard J, Lambert M (2006) Regulation of cell–cell junctions by the cytoskeleton. Curr Opin Cell Biol 18(5):541–548. doi:10.1016/j.ceb.2006.08.004

Mizoguchi A, Nakanishi H, Kimura K, Matsubara K, Ozaki-Kuroda K, Katata T, Honda T, Kiyohara Y, Heo K, Higashi M, Tsutsumi T, Sonoda S, Ide C, Takai Y (2002) Nectin: an adhesion molecule involved in formation of synapses. J Cell Biol 156(3):555–565. doi:10.1083/jcb.200103113

Morita K, Sasaki H, Furuse M, Tsukita S (1999) Endothelial claudin: claudin-5/TMVCF constitutes tight junction strands in endothelial cells. J Cell Biol 147(1):185–194

Mueller S, Rosenquist TA, Takai Y, Bronson RA, Wimmer E (2003) Loss of nectin-2 at Sertoli-spermatid junctions leads to male infertility and correlates with severe spermatozoan head and midpiece malformation, impaired binding to the zona pellucida, and oocyte penetration. Biol Reprod 69(4):1330–1340. doi:10.1095/biolreprod.102.014670

Muller WA (2003) Leukocyte-endothelial-cell interactions in leukocyte transmigration and the inflammatory response. Trends Immunol 24(6):327–334
Muller WA, Weigl SA, Deng X, Phillips DM (1993) PECAM-1 is required for transendothelial migration of leukocytes. J Exp Med 178(2):449–460
Pappas DJ, Rimm DL (2006) Direct interaction of the C-terminal domain of alpha-catenin and F-actin is necessary for stabilized cell–cell adhesion. Cell Commun Adhes 13(3):151–170. doi:10.1080/15419060600726142
Paz NG, Walshe TE, Leach LL, Saint-Geniez M, D'Amore PA (2012) Role of shear-stress-induced VEGF expression in endothelial cell survival. J Cell Sci 125:831–843. doi:10.1242/jcs.084301
Peitsch WK, Grund C, Kuhn C, Schnolzer M, Spring H, Schmelz M, Franke WW (1999) Drebrin is a widespread actin-associating protein enriched at junctional plaques, defining a specific microfilament anchorage system in polar epithelial cells. Eur J Cell Biol 78(11):767–778
Peitsch WK, Hofmann I, Pratzel S, Grund C, Kuhn C, Moll I, Langbein L, Franke WW (2001) Drebrin particles: components in the ensemble of proteins regulating actin dynamics of lamellipodia and filopodia. Eur J Cell Biol 80(9):567–579
Peitsch WK, Hofmann I, Bulkescher J, Hergt M, Spring H, Bleyl U, Goerdt S, Franke WW (2005) Drebrin, an actin-binding, cell-type characteristic protein: induction and localization in epithelial skin tumors and cultured keratinocytes. J Invest Dermatol 125(4):761–774. doi:10.1111/j.0022-202X.2005.23793.x
Pokutta S, Weis WI (2007) Structure and mechanism of cadherins and catenins in cell–cell contacts. Annu Rev Cell Dev Biol 23:237–261. doi:10.1146/annurev.cellbio.22.010305.104241
Pokutta S, Drees F, Takai Y, Nelson WJ, Weis WI (2002) Biochemical and structural definition of the l-afadin- and actin-binding sites of alpha-catenin. J Biol Chem 277(21):18868–18874. doi:10.1074/jbc.M201463200
Pollard TD, Borisy GG (2003) Cellular motility driven by assembly and disassembly of actin filaments. Cell 112(4):453–465
Quinlan MP, Hyatt JL (1999) Establishment of the circumferential actin filament network is a prerequisite for localization of the cadherin-catenin complex in epithelial cells. Cell Growth Differ 10(12):839–854
Rehm K (2013) Drebrin preserves endothelial integrity by stabilizing nectin at adherens junctions. Dissertation/ Doctoral thesis at the Universität Hamburg, Institut für Medizinische Mikrobiologie, Virologie und Hygiene at the Universitätsklinikum Hamburg-Eppendorf
Rehm K, Panzer L, van Vliet V, Genot E, Linder S (2013) Drebrin preserves endothelial integrity by stabilizing nectin at adherens junctions. J Cell Sci 126(Pt 16):3756–3769. doi:10.1242/jcs.129437
Reinhard M, Jarchau T, Walter U (2001) Actin-based motility: stop and go with Ena/VASP proteins. Trends Biochem Sci 26(4):243–249
Reymond N, Borg JP, Lecocq E, Adelaide J, Campadelli-Fiume G, Dubreuil P, Lopez M (2000) Human nectin3/PRR3: a novel member of the PVR/PRR/nectin family that interacts with afadin. Gene 255(2):347–355
Reymond N, Fabre S, Lecocq E, Adelaide J, Dubreuil P, Lopez M (2001) Nectin4/PRR4, a new afadin-associated member of the nectin family that trans-interacts with nectin1/PRR1 through V domain interaction. J Biol Chem 276(46):43205–43215. doi:10.1074/jbc.M103810200
Rikitake Y, Mandai K, Takai Y (2012) The role of nectins in different types of cell–cell adhesion. J Cell Sci 125(Pt 16):3713–3722. doi:10.1242/jcs.099572
Satoh-Horikawa K, Nakanishi H, Takahashi K, Miyahara M, Nishimura M, Tachibana K, Mizoguchi A, Takai Y (2000) Nectin-3, a new member of immunoglobulin-like cell adhesion molecules that shows homophilic and heterophilic cell–cell adhesion activities. J Biol Chem 275(14):10291–10299
Schlegel N, Burger S, Golenhofen N, Walter U, Drenckhahn D, Waschke J (2008) The role of VASP in regulation of cAMP- and Rac 1-mediated endothelial barrier stabilization. Am J Physiol Cell Physiol 294(1):C178–C188. doi:10.1152/ajpcell.00273.2007
Scott JA, Shewan AM, den Elzen NR, Loureiro JJ, Gertler FB, Yap AS (2006) Ena/VASP proteins can regulate distinct modes of actin organization at cadherin-adhesive contacts. Mol Biol Cell 17(3):1085–1095. doi:10.1091/mbc.E05-07-0644

Shen L, Turner JR (2005) Actin depolymerization disrupts tight junctions via caveolae-mediated endocytosis. Mol Biol Cell 16(9):3919–3936. doi:10.1091/mbc.E04-12-1089

Simon AM, Goodenough DA (1998) Diverse functions of vertebrate gap junctions. Trends Cell Biol 8(12):477–483

Stevenson BR, Siliciano JD, Mooseker MS, Goodenough DA (1986) Identification of ZO-1: a high molecular weight polypeptide associated with the tight junction (zonula occludens) in a variety of epithelia. J Cell Biol 103(3):755–766

Tachibana K, Nakanishi H, Mandai K, Ozaki K, Ikeda W, Yamamoto Y, Nagafuchi A, Tsukita S, Takai Y (2000) Two cell adhesion molecules, nectin and cadherin, interact through their cytoplasmic domain-associated proteins. J Cell Biol 150(5):1161–1175. doi:10.1083/jcb.150.5.1161

Takahashi K, Nakanishi H, Miyahara M, Mandai K, Satoh K, Satoh A, Nishioka H, Aoki J, Nomoto A, Mizoguchi A, Takai Y (1999) Nectin/PRR: an immunoglobulin-like cell adhesion molecule recruited to cadherin-based adherens junctions through interaction with Afadin, a PDZ domain-containing protein. J Cell Biol 145(3):539–549

Takai Y, Nakanishi H (2003) Nectin and afadin: novel organizers of intercellular junctions. J Cell Sci 116(Pt 1):17–27

Takai Y, Miyoshi J, Ikeda W, Ogita H (2008) Nectins and nectin-like molecules: roles in contact inhibition of cell movement and proliferation. Nat Rev Mol Cell Biol 9(8):603–615. doi:10.1038/nrm2457

Tsukita S, Furuse M (1999) Occludin and claudins in tight-junction strands: leading or supporting players? Trends Cell Biol 9(7):268–273

Tsukita S, Furuse M, Itoh M (1999) Structural and signalling molecules come together at tight junctions. Curr Opin Cell Biol 11(5):628–633

Vasioukhin V, Bauer C, Yin M, Fuchs E (2000) Directed actin polymerization is the driving force for epithelial cell–cell adhesion. Cell 100(2):209–219

Vestweber D (2007) Adhesion and signaling molecules controlling the transmigration of leukocytes through endothelium. Immunol Rev 218:178–196. doi:10.1111/j.1600-065X.2007.00533.x

Wallez Y, Huber P (2008) Endothelial adherens and tight junctions in vascular homeostasis, inflammation and angiogenesis. Biochim Biophys Acta 1778(3):794–809. doi:10.1016/j.bbamem.2007.09.003

Weis WI, Nelson WJ (2006) Re-solving the cadherin-catenin-actin conundrum. J Biol Chem 281(47):35593–35597. doi:10.1074/jbc.R600027200

Yamada S, Pokutta S, Drees F, Weis WI, Nelson WJ (2005) Deconstructing the cadherin-catenin-actin complex. Cell 123(5):889–901. doi:10.1016/j.cell.2005.09.020

Yamazaki D, Oikawa T, Takenawa T (2007) Rac-WAVE-mediated actin reorganization is required for organization and maintenance of cell–cell adhesion. J Cell Sci 120(Pt 1):86–100. doi:10.1242/jcs.03311

Yap AS, Brieher WM, Gumbiner BM (1997) Molecular and functional analysis of cadherin-based adherens junctions. Annu Rev Cell Dev Biol 13:119–146. doi:10.1146/annurev.cellbio.13.1.119

Chapter 22
Regulation of Skeletal Myoblast Differentiation by Drebrin

Robert S. Krauss

Abstract Myoblast differentiation is a complex process. As myoblasts differentiate into myofibers, they acquire a cell type-specific transcriptional program, irreversibly exit the cell cycle, and dramatically change their morphology. The morphological changes include cell elongation, alignment, and fusion into syncytial myofibers. Several lines of evidence suggest that these events may be co-regulated. However, the mechanisms that coordinate major alterations in a cell's transcriptome and its shape are not well understood. Muscle-specific transcription is controlled by proteins of the MyoD family, transcription factors whose activity is regulated by specific signal transduction pathways, including the p38 MAP kinase pathway. In a search for genes that might play a role in linking myogenic signal transduction, cytoskeletal regulation, and myoblast differentiation, *Dbn1* (encoding the actin regulator drebrin) was identified. *Dbn1* expression is induced during myoblast differentiation, in a p38 MAP kinase- and MyoD- dependent manner. RNAi-mediated depletion of drebrin, or treatment with a chemical drebrin inhibitor, resulted in a similar phenotype in myoblasts: defective differentiation, with low levels of early and late differentiation markers and inefficient production of myofibers. Drebrin localizes at sites of cell-cell contact and cell extensions, locations that are also enriched for F-actin. Drebrin may be important in linking transcriptional and morphological aspects of myoblast differentiation.

Keywords Drebrin • Myogenesis • Skeletal muscle • MyoD • Myogenin • Myoblast • Cell-cell fusion • p38 MAP kinase • Cell differentiation

R.S. Krauss
Icahn School of Medicine at Mount Sinai, New York, NY 10029, USA
e-mail: Robert.Krauss@mssm.edu

T. Shirao, Y. Sekino (eds.), *Drebrin*, Advances in Experimental Medicine and Biology 1006, DOI 10.1007/978-4-431-56550-5_22

22.1 Skeletal Muscle Development and Myoblast Differentiation

In vertebrates, skeletal muscles of the trunk and limbs arise from somites, blocks of columnar epithelial cells that form in an anterior-to-posterior manner from paraxial mesoderm. Muscle progenitor cells are derived from the dorsal region of the maturing somite, the dermomyotome (Tajbakhsh and Buckingham 2000; Pownall et al. 2002; Charge and Rudnicki 2004; Biressi et al. 2007). In response to signals from adjacent axial tissues and surface ectoderm, some dermomyotomal progenitors become committed to the muscle lineage and form the myotome, a set of differentiated muscle cells that underlies the dermomyotome. Subsequent embryonic, fetal, and postnatal stages of myogenesis involve additional muscle progenitors that migrate from the dermomyotome and ultimately establish the trunk and limb musculature (Gros et al. 2005; Kassar-Duchossoy et al. 2005; Relaix et al. 2005). These progenitor cells express the transcription factors Pax3 and Pax7 (Kassar-Duchossoy et al. 2005; Relaix et al. 2005).

At the appropriate developmental times and locations, somite-derived muscle progenitor cells are specified to become lineage-committed (i.e., "determined") myoblasts. Myoblasts are proliferative muscle precursor cells that, again at the appropriate developmental times and locations, differentiate into multinucleated myofibers (Tapscott 2005; Rudnicki et al. 2008) (Fig. 22.1). The processes of determination and differentiation in the skeletal muscle lineage are orchestrated by the family of myogenic basic helix-loop-helix (bHLH) transcription factors, which includes Myf5, MyoD, myogenin, and MRF4 (Tapscott 2005; Rudnicki et al. 2008). *Myf5* and *MyoD* are directly regulated by Pax3 and Pax7, and they serve redundant roles in determination of cells to the muscle lineage. Upon generation of appropriate numbers of myoblasts, Myf5 and MyoD induce expression of *Myog* (encoding myogenin), which is essential for myoblast differentiation. MRF4 and MyoD itself also participate in myoblast differentiation.

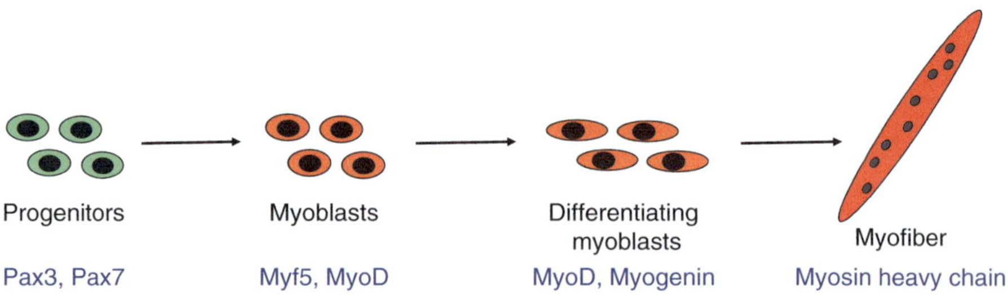

Fig. 22.1 Developmental progression in the skeletal muscle lineage. Pax3/Pax7-expressing progenitor cells become committed to the muscle lineage by expression of Myf5 and/or MyoD. Such lineage-committed cells are myoblasts. Induction of myogenin by Myf5 and/or MyoD is associated with the onset of myoblast differentiation. The differentiation process includes expression of proteins of the contractile apparatus, such as myosin heavy chain, and fusion into highly elongated, multinucleated myofibers

Expression of the myogenic bHLH factors is almost completely skeletal muscle-specific, and MyoD, in particular, may be viewed as a master regulator of the lineage. Expression of MyoD in many non-muscle cell types converts such cells to the skeletal muscle program, revealing its ability to drive tissue-specific transcription and cell differentiation (Tapscott 2005). MyoD's ability to function in this manner occurs in conjunction with non-muscle-specific factors, such as its heterodimeric bHLH partners, the E proteins, Mef2 family members, transcriptional co-activators, and chromatin remodeling factors (Tapscott 2005; Rudnicki et al. 2008). These and other transcription factors orchestrate the process of myoblast differentiation.

Myoblast differentiation is a complex and multistage process. As myoblasts differentiate into myofibers, they acquire a cell type-specific transcriptional program, irreversibly exit the cell cycle, and change their morphology. The morphological changes include cell elongation, alignment, cell-cell fusion into syncytia that can include hundreds of nuclei, and positioning and organization of sarcomeres (Tajbakhsh and Buckingham 2000; Pownall et al. 2002; Charge and Rudnicki 2004; Biressi et al. 2007) (Fig. 22.1). It is extremely likely that the changes that occur in myoblasts during differentiation are co-regulated. However, the mechanisms that coordinate major alterations in a cell's transcriptome and shape are not well understood. Myoblast differentiation can be studied in vitro with both primary myoblasts isolated from mice and myoblast cell lines, such as C2C12 cells. The availability of such culture systems allows for the types of detailed molecular and cell biological analyses that are required to address this issue.

22.2 Regulation of Myoblast Differentiation by Signal Transduction Pathways

Transcriptional regulation directed by the myogenic bHLH factors lies at the core of skeletal myogenesis. However, the differentiation process is also controlled by extracellular signaling factors. These include insulin-like growth factor 1 (IGF-1), fibroblast growth factors (FGFs), Wnts, transforming growth factor β (TGFβ) superfamily members, and additional factors (Tajbakhsh and Buckingham 2000; Pownall et al. 2002; Charge and Rudnicki 2004; Biressi et al. 2007). Several intracellular signal transduction pathways promote myoblast differentiation, including the PI-3 kinase/Akt, FAK, calcium/calcineurin, and p38α/β mitogen-activated protein (MAP) kinase pathways (Jiang et al. 1998, 1999; Tamir and Bengal 2000; Wu et al. 2000; Horsley and Pavlath 2002; Schulz and Yutzey 2004; Quach and Rando 2006; Quach et al. 2009).

The p38α/β MAP kinase (hereafter called simply "p38") pathway is of particular interest. Although total levels of p38 protein remain constant during myoblast differentiation in vitro, p38 activity rises progressively during this process. Inhibition of p38 activity with small molecule inhibitors such as SB203580 prevents induction of select muscle-specific genes and blocks differentiation of myoblasts in vitro (Cuenda and Cohen 1999; Zetser et al. 1999; Wu et al. 2000). Mice with a mutation in *p38a* develop small, immature myofibers, and myoblasts isolated from such animals have defects in cell cycle exit and differentiate inefficiently (Perdiguero et al. 2007).

The activity of MyoD and other promyogenic transcription factors is tightly controlled at the posttranslational level by signal transduction pathways (Tapscott 2005; Guasconi and Puri 2009). p38 directly phosphorylates several substrates that promote MyoD-dependent expression of specific genes. Among these substrates are E proteins, which efficiently heterodimerize with MyoD when phosphorylated; Mef2, leading to enhanced transcriptional activity; BAF60c, which remodels chromatin at muscle-specific genes; and KSRP, which promotes stability of myogenic mRNAs (Wu et al. 2000; Simone et al. 2004; Briata et al. 2005; de Angelis et al. 2005; Lluis et al. 2005; Rampalli et al. 2007). Interestingly, although treatment of cultured myoblasts with the p38 inhibitor SB230580 strongly blocks formation of myotubes, p38 activity is only necessary for induction of a subset of MyoD target genes (Bergstrom et al. 2002). This subset is enriched for structural genes, including those that encode various troponin isoforms and myosin heavy chain (*Myh3*). Few of the p38-dependent MyoD target genes that promote myoblast differentiation have been identified.

MAP kinases are usually activated in response to extracellular signaling cues, and many such stimuli are known to activate p38 (Cuenda and Rousseau 2007). Several distinct pathways appear to participate in p38 activation during myoblast differentiation (Sorci et al. 2004; Takaesu et al. 2006; Chen et al. 2007b; Kang et al. 2008). One pathway that accounts for a major proportion of differentiation-associated p38 activation is stimulated in response to cell-cell contact and adhesion between myoblasts. Upon ligation of the cell-cell adhesion molecule N- cadherin, the intracellular region of the cadherin-associated coreceptor Cdo nucleates formation of a signaling complex that includes the small GTPase, Cdc42, and components of the p38 MAP kinase cascade (Takaesu et al. 2006; Kang et al. 2008; Lu and Krauss 2010). This leads to Cdc42-dependent activation of p38 and promotion of myoblast differentiation.

22.3 Potential Roles of Cytoskeleton Regulation in Myoblast Differentiation

It is obvious that myoblasts and mature myofibers have very different cytoskeletal architectures. Several lines of evidence argue that actin- and tubulin-based cytoskeletal structures do not simply change to meet the requirements of precursor and differentiated cell types; rather, active changes to these structures may play a regulatory role in the differentiation process. For example, lamellipodia and filopodia, functional cellular structures that require actin remodeling, form dynamically during myoblast differentiation in vitro (Abramovici and Gee 2007; Kim et al. 2008; Nowak et al. 2009). Additionally, disruption of the F-actin or microtubule cytoskeletons by several different means blocks various aspects of myoblast differentiation (Chang et al. 2002; O'Connor et al. 2008; Nowak et al. 2009).

One critical event during muscle cell differentiation that obviously involves a change in morphology is myoblast fusion. Recent work has begun to illuminate how signal transduction pathways intersect with cytoskeletal regulation to promote myoblast fusion, resulting in multinucleated myofibers (Jansen and Pavlath 2008; Richardson et al. 2008). Myoblast fusion proceeds through discernable stages of cell-cell recognition, adhesion, alignment, and membrane fusion. Genetic analyses in *Drosophila*, zebrafish, and mice, plus in vitro studies with cultured myoblasts, have revealed that a largely evolutionarily conserved mechanism underlies myoblast fusion. Adhesive cell surface receptors signal via the guanine nucleotide exchange factor (GEF) Mbc/Dock1 and its target the small GTPase Rac (Chen et al. 2007a; Richardson et al. 2008; Onel and Renkawitz-Pohl 2009). Rac regulates actin cytoskeletal dynamics, and fusion is dependent on the WAVE, WASP, and Arp2/3 actin-remodeling complexes (Chen et al. 2007a; Richardson et al. 2008; Onel and Renkawitz-Pohl 2009). Brag2/Loner, a GEF for the GTPase Arf6, is also required for fusion (Chen et al. 2007a; Richardson et al. 2008; Onel and Renkawitz-Pohl 2009). While most of this work has been performed in the fruit fly, zebrafish morphants and/or mouse mutants in *Dock1* and *Rac1* display myoblast fusion defects (Chen et al. 2007a; Moore et al. 2007; Srinivas et al. 2007; Laurin et al. 2008; Pajcini et al. 2008; Sohn et al. 2009; Vasyutina et al. 2009). Additional signaling factors not implicated in *Drosophila* myoblast fusion are important for this process to occur efficiently in mice or cultured mammalian myoblasts, and many of these are known cytoskeletal regulators, including integrin β1 and its associated proteins FAK and talin, M-cadherin, Cdc42, and the microtubule-binding protein EB3 (Straube and Merdes 2007; Richardson et al. 2008; Conti et al. 2009; Vasyutina et al. 2009).

Cross regulation of the transcriptional and morphological aspects of myoblast differentiation is very likely to occur. Depletion in cultured mammalian myoblasts of factors required for fusion in *Drosophila* (e.g., Dock1, Brag2) not only blocks formation of multinucleated cells but also decreases the percentage of cells that express muscle-specific markers (Pajcini et al. 2008). A similar phenomenon is seen when myoblasts are treated with the F-actin-remodeling inhibitor latrunculin B or depleted of the microtubule-binding protein EB1 (Nowak et al. 2009; Zhang et al. 2009). Furthermore, at least some aspects of the cytoskeletal rearrangement and morphological changes that occur during differentiation are likely to be mediated by transcriptional induction of regulators of these processes, as expression of MyoD in fibroblasts induces not only expression of muscle-specific genes but also elongation and fusion into multinucleate myotubes (Tapscott 2005). Taken together, acquisition of the muscle-specific transcriptional program and alterations in cytoskeletal architecture both play key roles during differentiation of myoblasts into myofibers, and it is likely that these processes are coordinately regulated.

22.4 Drebrin Promotes Myoblast Differentiation

We sought to identify genes that might play a role in linking myogenic signal transduction, cytoskeletal regulation, and myoblast differentiation. To do so, we performed microarray analyses on C2C12 myoblasts cultured in growth medium (GM) or in differentiation medium (DM) for various time points, with or without inclusion of the p38 MAP kinase inhibitor, SB203580 (Mancini et al. 2011). Among the genes we identified was *Dbn1*, encoding drebrin.

22.4.1 Regulation of Dbn1 *Expression During Myogenic Differentiation*

Dbn1 mRNA was expressed at very low levels in proliferating myoblasts (in GM), but it was induced at the time the cultures were shifted to DM (when cells are nearly confluent and there is substantial cell-cell contact). *Dbn1* mRNA levels continued to increase in abundance over 48 h in DM (Mancini et al. 2011). Expression was biphasic, as *Dbn1* mRNA levels decreased again at 72 h in DM. Drebrin protein levels largely paralleled the mRNA levels, although downregulation was not seen at 72 h in DM, suggesting the protein is relatively stable in these differentiated cultures. Importantly, *Dbn1* mRNA and protein levels were strongly dampened in parallel cultures treated with SB203580, indicating that *Dbn1* induction during differentiation was dependent on p38 activity (Mancini et al. 2011). Similar results were seen with both C2C12 cells and with primary mouse myoblasts. The isoform of drebrin expressed by these cells was drebrin E. A portion of the drebrin protein was phosphorylated, although this did not seem to be regulated in a differentiation-dependent manner.

Even though *Dbn1* is expressed in most cell types, its p38-dependent induction during myoblast differentiation suggests it may be a direct target of the myogenic transcriptional network. Consistent with this notion, chromatin immunoprecipitation studies by Forcales et al. showed that MyoD binds to the *Dbn1* locus during myoblast differentiation (Forcales et al. 2012). Furthermore, RNAi-mediated depletion of BAF60c, a direct p38 substrate that is required for chromatin remodeling at MyoD-dependent genes, led to a decrease in *Dbn1* expression during differentiation.

In addition to signaling by p38, the activity of the non-receptor tyrosine kinase, FAK, is also important for *Dbn1* expression during myoblast differentiation (Quach et al. 2009). Primary mouse myoblasts expressing a dominant negative form of FAK failed to induce *Dbn1* efficiently and also displayed defective maturation and fusion. FAK is activated by integrin signaling and probably has many roles in myogenesis (Quach and Rando 2006; Quach et al. 2009). It is worth noting that FAK is important for expression of an upstream regulator of p38, Cdo, and a functional downstream target of p38, MyoD (Han et al. 2011). Therefore, FAK and p38 signaling may be interlinked.

Gene profiling on cells directly isolated from mice has provided additional evidence for *Dbn1* induction during myogenesis. Satellite cells are skeletal muscle stem cells and in adult mice are characterized by a long-term quiescent state (Chang and Rudnicki 2014). Satellite cells are activated during injury or muscle disease and are a source of myoblasts during regeneration. Satellite cells are also the source of myoblasts for the enormous muscle growth that occurs during the first three weeks of mouse life. Pallafacchina et al. performed transcriptome analyses on such cells isolated from 1-week-old mice, normal adult mice, and adult *mdx* mice, an animal model of muscular dystrophy in which muscles undergo cycles of degeneration and regeneration. Consistent with in vitro observations of *Dbn1* induction during myogenic differentiation, *Dbn1* mRNA levels were significantly higher in cells isolated from both 1-week-old mice and adult *mdx* mice than from normal adult mice (Pallafacchina et al. 2010).

22.4.2 Regulation of Myoblast Differentiation by Drebrin

To investigate a role for drebrin in myoblast differentiation, multiple approaches were pursued. First, we used RNAi-mediated knockdown of drebrin in C2C12 cells and primary mouse myoblasts (Mancini et al. 2011). In both cell types, differentiation-dependent induction of drebrin was substantially dampened. When these cells were placed into DM, drebrin-depleted myoblasts formed myotubes that were much smaller and thinner than those formed by control cells (Fig. 22.2). Additionally, the percentage of nuclei present in multinucleated cells (i.e., the fusion index) formed

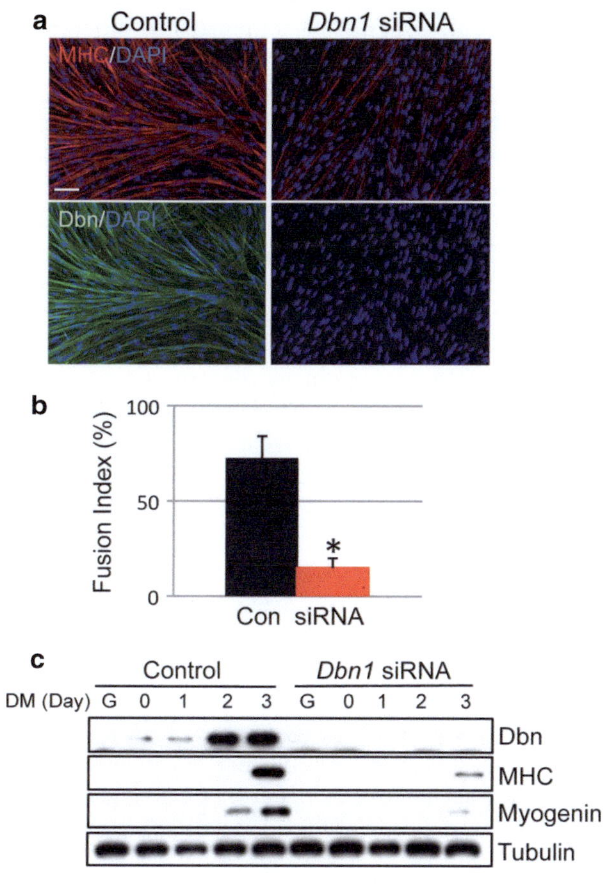

Fig. 22.2 RNAi-mediated depletion of drebrin inhibits primary myoblast fusion. (**a**) C2C12 cells were transfected with non-silencing (control) or *Dbn1*-specific small interfering RNAs (siRNAs). After 3 days in DM, cultures were fixed and triple-stained for myosin heavy chain (MHC; *red*), drebrin (Dbn; *green*), and DAPI (*blue*). Bar: 20 μM. (**b**) Quantification of fusion index in cultures from (**a**). Con, control siRNA. Values represent means of triplicate determinations ±SD. *$p < 0.01$. (**c**) Parallel cultures to those in (**a**) were analyzed by Western blotting for drebrin, myogenin, and MHC production. Tubulin was used as a loading control

by drebrin-depleted myoblasts was much lower than in control cells (Mancini et al. 2011). Strikingly, expression of the early differentiation marker, myogenin, and the late differentiation marker, myosin heavy chain, were both strongly reduced by knockdown of drebrin (Fig. 22.2). Cell proliferation and apoptosis were also measured in primary myoblasts with RNAi-mediated reduction of drebrin levels. Such cultures showed a modest increase in cell proliferation and in apoptosis (Mancini et al. 2011). Although the increases in these processes may contribute to the defective differentiation in drebrin-depleted cells, they were insufficient to explain the much greater effect on myotube formation. These results indicate that (1) drebrin is important for myotube formation in vitro and (2) drebrin's role in the differentiation process appears to occur relatively early, as myogenin induction occurs prior to elongation or fusion of myoblasts.

We also assessed the effect of a small molecule inhibitor of drebrin on myoblast differentiation. *N*-[4-[3,5-*Bis*(trifluoromethyl)-1*H*-pyrazol-1-yl]phenyl]-4-methyl-1,2,3-thiadiazole-5-carboxamide (BTP2; also called YM-58483) binds directly to drebrin, and it inhibits actin cytoskeletal alterations that occur upon drebrin overexpression in cultured Chinese hamster ovary cells (Trevillyan et al. 2001; Mercer et al. 2010). Treatment of C2C12 cells and primary mouse myoblasts with BTP2 had very similar effects as RNAi-mediated depletion of drebrin from these cell types (Mancini et al. 2011). Cells stimulated to differentiate in the presence of BTP2 formed fewer and smaller myotubes than vehicle-treated cells. They also displayed lower levels of both myogenin and myosin heavy chain.

To establish that drebrin was the critical target of BTP2, we sought to rescue differentiation of BTP2-treated myoblasts through exogenous expression of a BTP2-resistent form of drebrin. C2C12 cells were engineered to stably express either wild-type drebrin E or a mutant form of drebrin E that is unable to bind to BTP2 (the K270M/K271M mutant). Both constructs were tagged with GFP, and cells that expressed GFP alone were used as controls; the levels of exogenous drebrin were similar to those achieved endogenously during differentiation. While wild-type drebrin was unable to rescue myotube formation in BTP2-treated cultures, the drebrin K270M/K271M mutant did so to nearly the level seen in GFP-expressing control cells (Mancini et al. 2011). These results argue persuasively that drebrin is indeed the critical target of BTP2 in its ability to inhibit myogenic differentiation.

The effectiveness and specificity of BTP2 as a drebrin inhibitor in myoblast differentiation permitted analysis when in the differentiation time course drebrin is required (i.e., defined as the window of BTP2 sensitivity). A C2C12 cell differentiation assay was used, in which myotube formation and myosin heavy chain expression increased quantitatively over a 5-day time course subsequent to addition of DM to the cultures. It was found that BTP2 ceased to have any effect if added 60 h after the cultures were transferred to DM (Mancini et al. 2011). Therefore, drebrin functions to promote myogenic differentiation relatively early in this process, and it is largely dispensable for the later phase characterized by increased levels of myosin heavy chain and formation of large myotubes. This result is consistent with the results described above, wherein BTP2 treatment and drebrin depletion resulted in diminished expression of the critical early regulator of differentiation, myogenin.

22.4.3 Subcellular Localization of Drebrin in Myoblasts

To determine where drebrin is localized in myoblasts, both indirect immunofluorescence and GFP-tagged drebrin were studied; similar results were obtained. In C2C12 myoblasts proliferating at low density, overall the level of drebrin was low, but it was enriched in cell projections and at cell cortices, locations that also accumulated F-actin (Mancini et al. 2011). After 24 h in DM, when the cultures were at higher density and preparing for initial fusion events, drebrin was concentrated at sites of cell-cell contact and also at the tips of nascent myotubes, sites where fusion often occurs.

22.4.4 How Might Drebrin Promote Myoblast Differentiation?

It is of interest to speculate on the mechanisms whereby drebrin promotes myoblast differentiation. Drebrin interacts with several proteins also shown to regulate various aspects of myogenesis. Among these proteins are the microtubule plus-tip-binding protein, EB3; the scaffold protein, Homer; and the chemokine receptor CXCR4. The drebrin-EB3 interaction links F-actin and dynamic microtubules during formation of neuronal growth cones and growth of axons (Geraldo et al. 2008; Mizui et al. 2009). Interestingly, EB3 expression is induced during myoblast differentiation with similar kinetics to *Dbn1*, and it plays a role in myoblast fusion (Straube and Merdes 2007). Homer proteins promote calcineurin/NFAT pathway activity, which is known to enhance myoblast differentiation and myotube formation (Horsley and Pavlath 2002; Stiber et al. 2005). Intriguingly, BTP2 was identified in a screen for antagonists of NFAT activity, only subsequently being discovered to bind drebrin (Trevillyan et al. 2001). Homer proteins also interact with Cdc42, a signaling hub that is a positive regulator of differentiation and fusion (Kang et al. 2008; Shiraishi-Yamaguchi et al. 2009; Vasyutina et al. 2009). Finally, drebrin binds the intracellular region of CXCR4 (Pérez-Martínez et al. 2010). This chemokine receptor promotes myoblast and myocyte migration, activities thought to be important for fusion (Bae et al. 2008; Griffin et al. 2010).

It is sensible to speculate that one or more of these interactions are important for drebrin's role in myoblast differentiation. However, it is extremely likely that drebrin's ability to remodel actin is central to its function in this context. The colocalization of drebrin with F-actin structures in early differentiating myoblasts is consistent with this notion, and some of these interactions may be a way for drebrin to achieve the appropriate subcellular localization for its actions in actin regulation. As described above, F-actin regulation is central to myoblast fusion, but also important for overall differentiation in ways that remain mysterious. How drebrin's actin-regulating function contributes to its ability to regulate calcium signaling is still unclear, but calcium signaling has many roles in myogenesis, and this aspect of drebrin function may be critical in myoblasts. Finally, a *C. elegans* drebrin-related protein (DBN-1) is expressed in body wall muscle cells, where it is a regulatory component of the contractile apparatus (Butkevich et al. 2015). Drebrin and drebrin-related proteins may therefore have functions subsequent to the terminal differentiation of skeletal muscle cells.

22.5 Perspectives

Numerous observations about drebrin indicate it is a positive regulator of myoblast differentiation. First, *Dbn1* expression is induced upon stimulation of myoblasts to differentiate. *Dbn1* is a direct target gene of the myogenic master regulator, MyoD, and *Dbn1* induction requires p38 activity, a known promyogenic enzyme that promotes posttranslational activation of MyoD. Second, RNAi-mediated depletion of drebrin and drebrin inhibition with BTP2 both result in a similar phenotype in myoblasts: defective differentiation, with low levels of early and late differentiation markers and inefficient production of myotubes. Future mechanistic studies should focus on how drebrin is required in this process. Cytoskeletal regulation is important in several aspects of myoblast differentiation, but the underlying mechanisms are not well understood. Furthermore, how cytoskeletal alterations are co-regulated with changes in transcription is unclear. Given that drebrin's best studied function is as an actin regulator, yet its inhibition in myoblasts effects even early transcriptional changes associated with differentiation, it is possible that it plays a role in linking these processes. Finally, drebrin's role in developmental and regenerative myogenesis in vivo is unexplored. It will be important to study mice with a targeted mutation of *Dbn1* to address this issue.

References

Abramovici H, Gee SH (2007) Morphological changes and spatial regulation of diacylglycerol kinase-zeta, syntrophins, and Rac1 during myoblast fusion. Cell Motil Cytoskeleton 64:549–567

de Angelis L, Zhao J, Andreucci JJ, Olson EN, Cossu G, McDermott JC (2005) Regulation of vertebrate myotome development by the p38 MAP kinase-MEF2 signaling pathway. Dev Biol 283:171–179

Bae GU, Gaio U, Yang YJ, Lee HJ, Kang JS, Krauss RS (2008) Regulation of myoblast motility and fusion by the CXCR4-associated sialomucin, CD164. J Biol Chem 283:8301–8309

Bergstrom DA, Penn BH, Strand A, Perry RL, Rudnicki MA, Tapscott SJ (2002) Promoter-specific regulation of MyoD binding and signal transduction cooperate to pattern gene expression. Mol Cell 9:587–600

Biressi S, Molinaro M, Cossu G (2007) Cellular heterogeneity during vertebrate skeletal muscle development. Dev Biol 308:281–293

Briata P, Forcales SV, Ponassi M, Corte G, Chen CY, Karin M, Puri PL, Gherzi R (2005) p38-dependent phosphorylation of the mRNA decay-promoting factor KSRP controls the stability of select myogenic transcripts. Mol Cell 20:891–903

Butkevich E, Bodensiek K, Fakhri N, von Roden K, Schaap IAT, Majoul I, Schmidt CF, Klopfenstein DR (2015) Drebrin-like protein DBN-1 is a sarcomere component that stabilizes actin filaments during muscle contraction. Nat Commun 6:7523. doi:10.1038/ncomms8523

Chang NC, Rudnicki MA (2014) Satellite cells: the architects of skeletal muscle. Curr Top Dev Biol 107:161–181

Chang W, Webster DR, Salam AA, Gruber D, Prasad A, Eiserich JP, Bulinski JC (2002) Alteration of the C-terminal amino acid of tubulin specifically inhibits myogenic differentiation. J Biol Chem 277:30690–30698

Charge SB, Rudnicki MA (2004) Cellular and molecular regulation of muscle regeneration. Physiol Rev 84:209–238

Chen EH, Grote E, Mohler W, Vignery A (2007a) Cell-cell fusion. FEBS Lett 581:2181–2193

Chen SE, Jin B, Li YP (2007b) TNF-α regulates myogenesis and muscle regeneration by activating p38 MAPK. Am J Phys Cell Physiol 292:C1660–C1671

Conti FJ, Monkley SJ, Wood MR, Critchley DR, Müller U (2009) Talin 1 and 2 are required for myoblast fusion, sarcomere assembly and the maintenance of myotendinous junctions. Development 136:3597–3606

Cuenda A, Cohen P (1999) Stress-activated protein kinase-2/p38 and a rapamycin- sensitive pathway are required for C2C12 myogenesis. J Biol Chem 274:4341–4346

Cuenda A, Rousseau S (2007) p38 MAP-kinases pathway regulation, function and role in human diseases. Biochim Biophys Acta 1773:1358–1375

Forcales SV, Albini S, Giordani L, Malecova B, Cignolo L, Chernov A, Coutinho P, Saccone V, Consalvi S, Williams R et al (2012) Signal-dependent incorporation of MyoD–BAF60c into Brg1-based SWI/SNF chromatin-remodelling complex. EMBO J 31:301–316

Geraldo S, Khanzada UK, Parsons M, Chilton JK, Gordon-Weeks PR (2008) Targeting of the F-actin-binding protein drebrin by the microtubule plus-tip protein EB3 is required for neuritogenesis. Nat Cell Biol 10:1181–1189

Griffin CA, Apponi LH, Long KK, Pavlath GK (2010) Chemokine expression and control of muscle cell migration during myogenesis. J Cell Sci 123:3052–3060

Gros J, Manceau M, Thomé V, Marcelle C (2005) A common somitic origin for embryonic muscle progenitors and satellite cells. Nature 435:954–958

Guasconi V, Puri PL (2009) Chromatin: the interface between extrinsic cues and the epigenetic regulation of muscle regeneration. Trends Cell Biol 19:286–294

Han J-W, Lee H-J, Bae G-U, Kang J-S (2011) Promyogenic function of Integrin/FAK signaling is mediated by Cdo, Cdc42 and MyoD. Cell Signal 23(7):1162–1169

Horsley V, Pavlath GK (2002) NFAT: ubiquitous regulator of cell differentiation and adaptation. J Cell Biol 156:771–774

Jansen KM, Pavlath GK (2008) Molecular control of mammalian myoblast fusion. Methods Mol Biol 475:115–133

Jiang BH, Zheng JZ, Vogt PK (1998) An essential role of phosphatidylinositol 3- kinase in myogenic differentiation. Proc Natl Acad Sci U S A 95:14179–14183

Jiang BH, Aoki M, Zheng JZ, Li J, Vogt PK (1999) Myogenic signaling of phosphatidylinositol 3-kinase requires the serine-threonine kinase Akt/protein kinase B. Proc Natl Acad Sci U S A 96:2077–2081

Kang J-S, Bae G-U, Yi M-J, Yang Y-J, Oh J-E, Takaesu G, Zhou YT, Low BC, Krauss RS (2008) A Cdo/Bnip-2/Cdc42 signaling pathway regulates p38α/β MAPK activity and myogenic differentiation. J Cell Biol 182:497–507

Kassar-Duchossoy L, Giacone E, Gayraud-Morel B, Jory A, Gomès D, Tajbakhsh S (2005) Pax3/Pax7 mark a novel population of primitive myogenic cells during development. Genes Dev 19:1426–1431

Kim SJ, Kim S, Shin H, Uhm CS (2008) Intercellular interaction observed by atomic force microscopy. Ultramicroscopy 108:1148–1151

Laurin M, Fradet N, Blangy A, Hall A, Vuori K, Côté JF (2008) The atypical Rac activator Dock180 (Dock1) regulates myoblast fusion in vivo. Proc Natl Acad Sci U S A 105:15446–15451

Lluis F, Ballestar E, Suelves M, Esteller M, Munoz-Canoves P (2005) E47 phosphorylation by p38 MAPK promotes MyoD/E47 association and muscle-specific gene transcription. EMBO J 24:974–984

Lu M, Krauss RS (2010) N-cadherin ligation, but not Sonic hedgehog binding, initiates Cdo-dependent p38α/β MAPK signaling in skeletal myoblasts. Proc Natl Acad Sci U S A 107:4212–4217

Mancini A, Sirabella D, Zhang W, Yamazaki H, Shirao T, Krauss RS (2011) Regulation of myotube formation by the actin-binding factor drebrin. Skelet Muscle 1:36

Mercer JC, Qi Q, Mottram LF, Law M, Bruce D, Iyer A, Morales JL, Yamazaki H, Shirao T, Peterson BR et al (2010) Chemico-genetic identification of drebrin as a regulator of calcium responses. Int J Biochem Cell Biol 42:337–345

Mizui T, Kojima N, Yamazaki H, Katayama M, Hanamura K, Shirao T (2009) Drebrin E is involved in the regulation of axonal growth through actin-myosin interactions. J Neurochem 109:611–622

Moore CA, Parkin CA, Bidet Y, Ingham PW (2007) A role for the Myoblast city homologues Dock1 and Dock5 and the adaptor proteins Crk and Crk-like in zebrafish myoblast fusion. Development 134:3145–3153

Nowak SJ, Nahirney PC, Hadjantonakis AK, Baylies MK (2009) Nap1-mediated actin remodeling is essential for mammalian myoblast fusion. J Cell Sci 122:3282–3293

O'Connor RS, Steeds CM, Wiseman RW, Pavlath GK (2008) Phosphocreatine as an energy source for actin cytoskeletal rearrangements during myoblast fusion. J Physiol 586:2841–2853

Onel SF, Renkawitz-Pohl R (2009) FuRMAS: triggering myoblast fusion in Drosophila. Dev Dyn 238:1513–1525

Pajcini KV, Pomerantz JH, Alkan O, Doyonnas R, Blau HM (2008) Myoblasts and macrophages share molecular components that contribute to cell-cell fusion. J Cell Biol 180:1005–1019

Pallafacchina G, François S, Regnault B, Czarny B, Dive V, Cumano A, Montarras D, Buckingham M (2010) An adult tissue-specific stem cell in its niche: a gene profiling analysis of in vivo quiescent and activated muscle satellite cells. Stem Cell Res 4(2):77–91

Perdiguero E, Ruiz-Bonilla V, Gresh L, Hui L, Ballestar E, Sousa-Victor P, Baeza-Raja B, Jardí M, Bosch-Comas A, Esteller M et al (2007) Genetic analysis of p38 MAP kinases in myogenesis: fundamental role of p38α in abrogating myoblast proliferation. EMBO J 26:1245–1256

Pérez-Martínez M, Gordón-Alonso M, Cabrero JR, Barrero-Villar M, Rey M, Mittelbrunn M, Lamana A, Morlino G, Calabia C, Yamazaki H et al (2010) F-actin- binding protein drebrin regulates CXCR4 recruitment to the immune synapse. J Cell Sci 123:1160–1170

Pownall ME, Gustafsson MK, Emerson CP Jr (2002) Myogenic regulatory factors and the specification of muscle progenitors in vertebrate embryos. Annu Rev Cell Dev Biol 18:747–783

Quach NL, Rando TA (2006) Focal adhesion kinase is essential for costamerogenesis in cultured skeletal muscle cells. Dev Biol 293:38–52

Quach NL, Biressi S, Reichardt LF, Keller C, Rando TA (2009) Focal Adhesion Kinase Signaling Regulates the Expression of Caveolin 3 and β1 Integrin, Genes Essential for Normal Myoblast Fusion. Mol Biol Cell 20:3422–3435

Rampalli S, Li L, Mak E, Ge K, Brand M, Tapscott SJ, Dilworth FJ (2007) p38 MAPK signaling regulates recruitment of Ash2L-containing methyltransferase complexes to specific genes during differentiation. Nat Struct Mol Biol 14:1150–1156

Relaix F, Rocancourt D, Mansouri A, Buckingham M (2005) A Pax3/Pax7-dependent population of skeletal muscle progenitor cells. Nature 435:948–953

Richardson BE, Nowak SJ, Baylies MK (2008) Myoblast fusion in fly and vertebrates: new genes, new processes and new perspectives. Traffic 9:1050–1059

Rudnicki MA, Le Grand F, McKinnell I, Kuang S (2008) The Molecular Regulation of Muscle Stem Cell Function. Cold Spring Harb Symp Quant Biol 73:323–331

Schulz RA, Yutzey KE (2004) Calcineurin signaling and NFAT activation in cardiovascular and skeletal muscle development. Dev Biol 266:1–16

Shiraishi-Yamaguchi Y, Sato Y, Sakai R, Mizutani A, Knöpfel T, Mori N, Mikoshiba K, Furuichi T (2009) Interaction of Cupidin/Homer2 with two actin cytoskeletal regulators, Cdc42 small GTPase and Drebrin, in dendritic spines. BMC Neurosci 10:25

Simone C, Forcales SV, Hill DA, Imbalzano AN, Latella L, Puri PL (2004) p38 pathway targets SWI-SNF chromatin-remodeling complex to muscle-specific loci. Nat Genet 36:738–743

Sohn RL, Huang P, Kawahara G, Mitchell M, Guyon J, Kalluri R, Kunkel LM, Gussoni E (2009) A role for nephrin, a renal protein, in vertebrate skeletal muscle cell fusion. Proc Natl Acad Sci U S A 106:9274–9279

Sorci G, Riuzzi F, Arcuri C, Giambanco I, Donato R (2004) Amphoterin stimulates myogenesis and counteracts the antimyogenic factors basic fibroblast growth factor and S100B via RAGE binding. Mol Cell Biol 24:4880–4894

Srinivas BP, Woo J, Leong WY, Roy S (2007) A conserved molecular pathway mediates myoblast fusion in insects and vertebrates. Nat Genet 39:781–786

Stiber JA, Tabatabaei N, Hawkins AF, Hawke T, Worley PF, Williams RS, Rosenberg P (2005) Homer modulates NFAT-dependent signaling during muscle differentiation. Dev Biol 287:213–224

Straube A, Merdes A (2007) EB3 regulates microtubule dynamics at the cell cortex and is required for myoblast elongation and fusion. Curr Biol 17:1318–1325

Tajbakhsh S, Buckingham M (2000) The birth of muscle progenitor cells in the mouse: spatiotemporal considerations. Curr Top Dev Biol 48:225–268

Takaesu G, Kang JS, Bae GU, Yi MJ, Lee CM, Reddy EP, Krauss RS (2006) Activation of p38α/β MAPK in myogenesis via binding of the scaffold protein JLP to the cell surface protein Cdo. J Cell Biol 175:383–388

Tamir Y, Bengal E (2000) Phosphoinositide 3-kinase induces the transcriptional activity of MEF2 proteins during muscle differentiation. J Biol Chem 275:34424–34432

Tapscott SJ (2005) The circuitry of a master switch: myod and the regulation of skeletal muscle gene transcription. Development 132:2685–2695

Trevillyan J, Chiou XG, Chen YW, Ballaron SJ, Sheets MP, Smith ML, Wiedeman PE, Warrior U, Wilkins J, Gubbins EJ et al (2001) Potent inhibition of NFAT activation and T cell cytokine production by novel low molecular weight pyrazole compounds. J Biol Chem 276:48118–48126

Vasyutina E, Martarelli B, Brakebusch C, Wende H, Birchmeier C (2009) The small G-proteins Rac1 and Cdc42 are essential for myoblast fusion in the mouse. Proc Natl Acad Sci U S A 106:8935–8940

Wu Z, Woodring PJ, Bhakta KS, Tamura K, Wen F, Feramisco JR, Karin M, Wang JY, Puri PL (2000) p38 and extracellular signal-regulated kinases regulate the myogenic program at multiple steps. Mol Cell Biol 20:3951–3964

Zetser A, Gredinger E, Bengal E (1999) p38 mitogen-activated protein kinase pathway promotes skeletal muscle differentiation. Participation of the Mef2c transcription factor. J Biol Chem 274:5193–5200

Zhang T, Zaal KJ, Sheridan J, Mehta A, Gundersen GG, Ralston E (2009) Microtubule plus-end binding protein EB1 is necessary for muscle cell differentiation, elongation and fusion. J Cell Sci 122:1401–1409

Chapter 23
The Role of Drebrin in Cancer Cell Invasion

Anna E. Dart and Phillip R. Gordon-Weeks

Abstract Cancer progression is characterized by the capacity of malignant cells to exploit an innate migratory ability in order to invade adjacent tissues, enter the vasculature and eventually metastasize to secondary organs. It is this spread of cancer cells that is the major cause of death in cancer patients. Understanding the basic biology of how cancer cells generate an invasive phenotype will be crucial to the identification of drug targets with the aim of impeding tumour dissemination. Ten years on from its initial description in neuronal cells, drebrin expression was found in a wide variety of non-neuronal cells that importantly included cancer cell lines. Since then mounting evidence suggests that drebrin may be a key player in the advancement of several diverse cancer types where its expression is frequently upregulated. Cancer cell motility and invasion are crucial elements in the metastatic cascade and involve dramatic changes in cellular morphology that are associated with dynamic remodelling of the cytoskeleton. Interestingly, it now appears that drebrin could deliver this role during cancer development.

Keywords Drebrin • Cancer • EMT • Migration • Invasion • Metastasis

23.1 Introduction

Cancer metastasis is a disease arising from deregulated cellular behaviour whereby malignant cells are able to use their intrinsic motility machinery to escape their tissue of origin, invade other tissues and vasculature and ultimately disseminate to distant sites (Fig. 23.1). Minimally, for malignant cells to metastasize, these cells must first detach from the primary tumour, gain a migratory phenotype and breach the extracellular matrix (ECM) of basement membranes and stroma, which physically divide the tumour from the blood and lymph circulation. From here, the cancer cells must transmigrate across the endothelium to enter the bloodstream or lymph in

---

A.E. Dart (✉) • P.R. Gordon-Weeks
MRC Centre for Developmental Neurobiology, New Hunt's House, King's College London, Guy's Campus, London SE1 1UL, UK
e-mail: anna.dart@kcl.ac.uk

T. Shirao, Y. Sekino (eds.), *Drebrin*, Advances in Experimental Medicine and Biology 1006, DOI 10.1007/978-4-431-56550-5_23

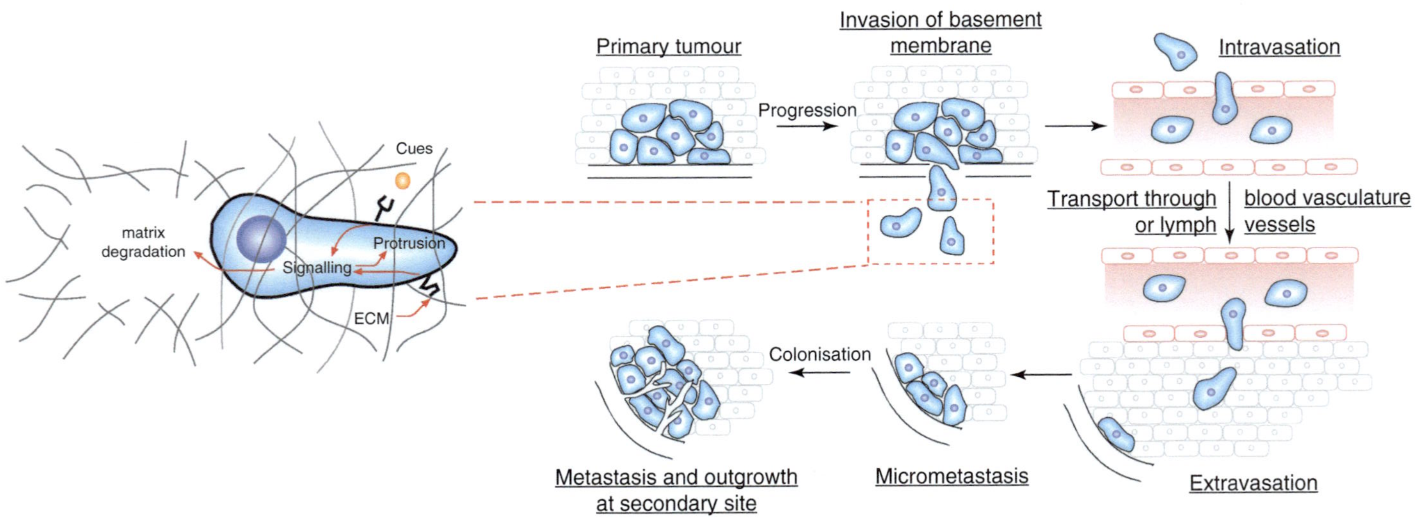

Fig. 23.1 The metastatic cascade. Schematic representation of the multistep process of cancer metastasis, which is characterized by the physical translocation of malignant cells from the primary tumour to a distal organ. Cancer cells within the initial tumour must progressively acquire enhanced motile and invasive traits allowing them to degrade underlying basement membranes, infiltrate the extracellular matrix (ECM) of the surrounding tissue and migrate towards blood and lymphatic vessels. From here, tumour cells intravasate to enter the circulation and be disseminated to secondary sites. Upon reaching microvessels of distant tissues, cancer cells exit the circulation via extravasation and begin to colonize these foreign niches by adaptation to the microenvironment. Active proliferation through recruitment of a blood supply ensures a small cluster of malignant cells can develop from a micrometastasis to a macrometastasis. In order to seed metastases, primary tumour cells must have the capacity to invade ECM, and this crucially requires a change in cell morphology driven by cytoskeletal reorganization. Cues emanating extracellularly trigger a diverse array of intracellular signalling pathways that generate protrusive structures, dynamic cycling of focal adhesions and proteolysis of the matrix for efficient, directed invasion

a process known as intravasation. Only those cells that survive transport through the circulation can undertake transmigration or extravasation through the endothelial cell lining into the stroma of the targeted tissue. Once in this microenvironment, malignant cells can switch to a proliferative phenotype and colonize the secondary niche (Chaffer and Weinberg 2011). Inherent in the progression of cancer from tumour initiation to invasion are dramatic transformations of cell shape primarily resultant from dynamic changes in the cell cytoskeleton (Fig. 23.1).

To facilitate invasion, the cancer cell undergoes a gradual acquisition of traits through an epithelial to mesenchymal transition (EMT). Its diverse cellular responses are tightly controlled by intracellular signalling cascades evoked by extracellular stimuli, such as growth factors, chemokines and cytokines, in the ECM. Indeed it is becoming clear that cancer cells frequently divert extrinsic signalling pathways associated with physiological processes to support tumour cell dissemination (Roussos et al. 2011). In turn, the signalling pathways triggered by receptor activation can generate an array of invasive properties, namely, (1) cytoskeletal remodelling and actomyosin contractility to promote the formation of protrusive structures for polarized migration, (2) initiation of proteolysis by matrix-degrading enzymes to degrade ECM components for efficient invasion, (3) matrix remodelling by secretion of ECM proteins and (4) dynamic turnover of focal adhesions between the integrins of the cell and the ECM components (Fig. 23.1; enlarged schematic of a cancer cell).

As such, cancer pathogenesis can be seen as a multistage process culminating in metastasis, the foremost cause of cancer-related mortality (Chaffer and Weinberg 2011). Identifying the broad spectrum of fundamental motility mechanisms and their associated molecular triggers which ultimately underlie the critical steps of invasion and metastasis will allow us to gain an understanding of both the basic biology and the pathology of cancer. Moreover, this knowledge will be crucial in the discovery of new diagnostic methods and targets for cancer therapy.

For just over 10 years following its original identification, drebrin was generally believed to be a neuronal-specific protein. However by 1999, drebrin E was described as being expressed in a range of non-neuronal cell types and tissues of diverse origin that importantly included several cancer-derived cell lines (Peitsch et al. 1999; Shirao et al. 1994). Since then drebrin has been characterized in non-neuronal cell types from multiple organs and recognized as being integral to the developmental stages of these organs. It is perhaps then not surprising that drebrin is emerging as an important player in cancer cell biology.

23.2 Drebrin Expression in Cancer Types

Drebrin E was initially observed, by immunoblotting, to be present in the human cell lines: metastatic-derived breast carcinoma MCF-7, primary liver carcinoma PLC and primary cervix adenocarcinoma HeLa (Peitsch et al. 1999). Subsequently, drebrin expression has been identified in a range of cancer cell lines encompassing cancer types derived from both simple epithelia and astrocytes. These include human colon adenocarcinoma Caco-2 and glioma U333 CG/343 (Keon et al. 2000; Peitsch et al. 2001). Early profiling of cancer cell lines revealed that drebrin was undetectable in a cell line derived from an epidermoid carcinoma A-431. This raised the intriguing possibility that drebrin may not be universally expressed in all cancer types with an absence in tumours derived from stratified epithelia (Keon et al. 2000). However, this could have been an oversimplification of the expression profile, as drebrin was later detected in many cancers originating in the skin, a stratified epithelium. These include basal cell carcinoma (BCC), squamous cell carcinoma (SCC) and a subcutaneous metastasis of a melanoma (Mizutani et al. 2014; Peitsch et al. 2005). In normal human skin, drebrin is albeit absent from the epidermis and only heavily enriched in eccrine sweat glands and hair follicles, regions where BCC is thought to derive from. Moreover, it has been noted that the drebrin distribution in SCC is often variable with a mosaic patterning (Peitsch et al. 2005). Taken altogether, these observations suggest that the discrepancy in drebrin expression in cancers of stratified epithelia may be dependent on the specific cancer stem cell of origin. It would be of great interest for future studies to systematically investigate the drebrin expression pattern in an expansive panel of simple versus stratified epithelial cancer cell types.

The number of human cancer types now known to express drebrin is extensive and wide-ranging. To date, drebrin has been detected in gliomas (Terakawa et al. 2013), skin cancers (Mizutani et al. 2014; Peitsch et al. 2005), breast cancer (Yagoub et al. 2014), bladder cancer (Xu et al. 2015), leiomyosarcoma smooth muscle tumours (Peitsch et al. 2005), recurrent non-small cell lung cancer (Mitra et al. 2011), lymphoblastic leukaemia (Vaskova et al. 2011), mantle cell lymphoma (Wang et al. 2010), prostate adenocarcinoma (Dart et al. 2017), colorectal cancer lymph node-positive (Kwon et al. 2004) and liver-positive (Lin et al. 2014) metastasis and pancreatic cancer (Iacobuzio-Donahue et al. 2002). Nonetheless, drebrin expression is unlikely to be limited to these cancer types as it is yet to be fully characterized, and thus, the full extent of drebrin expression in human cancers is undoubtedly underestimated.

Regardless of the cancer type, a unifying theme is the upregulation of drebrin, conceivably indicating that this event may be a common factor in malignancy. Illustrating this is the greater than fivefold change in drebrin expression in pancreatic cancer cell lines and tissues compared to normal pancreas tissue (Iacobuzio-Donahue et al. 2002), and the marked increase in different skin cancers contrasted with the virtual absence in non-tumourigenic skin disorders and normal skin (Peitsch et al. 2005). As reports of drebrin in cancer have accrued, insights into potential mechanisms of aberrant upregulation are also emerging. Recently, drebrin has been identified as a direct transcriptional target of the protein SOX11 (Wang et al. 2010). This is particularly pertinent to tumourigenesis since the transcription factor SOX11 is itself upregulated in several malignancies including ovarian cancer (Brennan et al. 2009), gliomas (Weigle et al. 2005) and mantle cell lymphoma (Wang et al. 2010). Importantly in validation of drebrin being modulated by SOX11 in cancer is the finding that SOX11 expression is highly correlated with that of drebrin in both mantle cell lymphoma primary cells and cell lines (Wang et al. 2010). Post-transcriptional modifications can also control gene expression, and one such modulation is through the activity of microRNAs. There is growing evidence that microRNAs are abnormally expressed or mutated in various types of human cancer, and in turn these changes can impact on the targets they regulate (Jansson and Lund 2012). Specifically, some microRNAs that act as tumour suppressors undergo epigenetic silencing during carcinogenesis, an example of which is the downregulation of miR-517a in bladder cancer and hepatocellular carcinoma (Liu et al. 2013; Yoshitomi et al. 2011). Significantly, ectopic restoration of miR-517a in bladder cancer cell lines results in a downregulation of drebrin (Yoshitomi et al. 2011) suggesting that for some cancers at least, the dysregulation of miR-517a could be instrumental in inducing drebrin gene expression.

Upregulation of a protein in cancer can often be viewed as a predictor of malignancy. This may also be the case for drebrin as its expression has now been correlated with malignancy in recurrent non-small cell lung cancer and colorectal and bladder cancer tissues and prostate adenocarcinoma tissues (Dart et al. 2017; Lin et al. 2014; Mitra et al. 2011; Terakawa et al. 2013; Xu et al. 2015). A tissue microarray (TMA) of human bladder cancer has shown that drebrin expression is prominently upregulated in a large number of malignant tissues ranging from highly to poorly differentiated bladder cancer in comparison to normal bladder tissue (Xu et al. 2015). Of particular note in this study was the finding that the drebrin staining intensity was substantially elevated in higher-grade (T2, T3 and T4) bladder carcinoma tissues over low-grade (Ta, T1) and normal bladder tissues (Xu et al. 2015). Colorectal cancer exhibits a strong propensity to metastasize to the liver, and drebrin has been identified as a differentially expressed protein between the proteomic profiles of the poorly metastatic human colon adenocarcinoma cell line HCT-116 and its liver metastatic derivative, E1 (Lin et al. 2014). Significantly, these two cell lines are derived from the same genetic background suggesting that drebrin upregulation could be both causal and indicative of a metastatic phenotype. Furthermore, drebrin overexpression has additionally been validated in a clinical context for colorectal cancer as immunohistochemical analysis of patient sections has revealed drebrin is absent from primary adenocarcinoma tumours whilst present in matched lymph node and liver metastases (Lin et al. 2014).

However, a negative relationship between drebrin expression and disease is not always absolute for all cancers. In clinical samples of human glioblastoma, the link between drebrin levels and malignancy is less well defined with only 40.3% of examined specimens displaying drebrin expression and, moreover, no association with patient outcome (Terakawa et al. 2013). As alluded to by the authors of this work, analyses which consider overall protein levels in a randomly selected tissue sample may not accurately reflect the impact of a protein on tumour progression and/or invasiveness, and hence, more detailed examinations of the heterogeneity of protein expression within a tumour may be more informative. Reinforcing this idea is the observation that in drebrin-positive glioblastoma samples, drebrin expression was substantially higher in the periphery of the tumours compared to central regions, perhaps denoting the more invasive nature of these outermost cells (Terakawa et al. 2013). Collectively, these reports advocate the potential of drebrin as a novel biomarker and highlight how its expression could carry prognostic information (Iacobuzio-Donahue et al. 2002; Mitra et al. 2011; Mizutani et al. 2014; Vaskova et al. 2011; Xu et al. 2015). In a similar vein, there is accumulating evidence to suggest that the presence of autoantibodies is symptomatic of cancer progression, and hence, their identification in sera from cancer patients could be used as a diagnostic tool. In light of this, it is interesting to note that drebrin has been recognized as an immunogenic tumour-associated autoantigen in prostate cancer patients (Fossa et al. 2000).

Mutations in proteins are often causal in the development of many types of cancer, and therefore perhaps not unexpectedly, such mutations have been identified in drebrin. Specifically, two point mutations have been found in cases of breast cancer: glutamate to lysine at position 278 (E278K) and glutamate to glutamine at position 640 (E640Q) (Sjoblom et al. 2006). These correspond to single nucleotide changes that fall in the helical (Hel) and blue box (BB) domains of drebrin, respectively (Fig. 23.2) (Worth et al. 2013). However, the implications for the functional and structural properties of the protein and how these would ultimately impact on breast cancer progression are presently unknown. If one were to speculate on the nature of the effect of these mutations, it would be important to highlight that the E278K

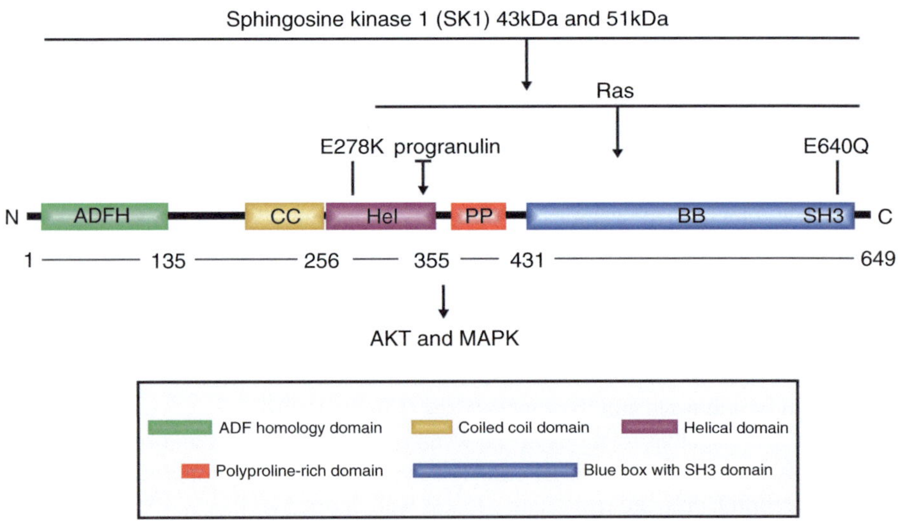

Fig. 23.2 Drebrin signalling in cancer. Schematic representation of the domain structure of drebrin highlighting key drebrin-mediated signalling elements associated with cancer. Two point mutations in drebrin have been identified in breast cancer patients at positions E278K and E640Q. The growth factor progranulin interacts with drebrin in a region encompassing the end of the Hel domain and the start of the PP domain. This interaction triggers downstream activation, by phosphorylation, of Akt and MAPK in bladder cancer cells (Xu et al. 2015). Both isoforms of sphingosine kinase 1 (SK1) have been reported to interact with drebrin in breast cancer cells, although the binding site is yet unknown (Yagoub et al. 2014). Lastly, Ras, the most common oncogene in human cancer, has been shown to drive drebrin-directed signalling in neuronal cells, but an interaction has so far not been demonstrated in cancer cells (Biou et al. 2008). Amino acid numbering is for human drebrin E

mutation lies within a region of drebrin sufficient for cell surface localization and filopodia formation (Xu and Stamnes 2006), whilst the E640Q is positioned in an assumed SH3 domain (Hayashi and Shirao 1999; Lappalainen et al. 1998). At present, it seems likely that such mutations could influence the downstream signalling cascades of drebrin, resulting in changes to cytoskeletal dynamics that culminate in a pro-migratory phenotype. Future studies will need to address if similar mutations can be identified in other cancer types and whether they can potentially confer a gain of function upon drebrin.

23.3 Drebrin Localization in Cancer Cells

Aberrant cell migration is essential for cancer cells to invade and metastasize, and importantly this involves the precise control of changes in cell shape. Typically, the malignant cell will become polarized in the direction of migration by generating protrusive structures in the leading edge cell membrane. Current thinking is that these structures can detect extracellular cue gradients in the surrounding matrix to orchestrate a cellular response whereby the structures in a favourable position can then guide cell movement (Lidke et al. 2005). These protrusive structures can take the form of filopodia, lamellipodia and invadopodia which are all morphologically and structurally different. Nonetheless, these structures have in common a requirement for spatially and temporally regulated dynamic cytoskeletal remodelling that is indicative of an EMT transition. Consistent with a role for drebrin in promoting a motile phenotype, it has been observed accumulating discretely in cellular protrusions or membrane ruffles at the leading edge of lamellipodia and also at the base of filopodia in the human glioma cell line U333 (Peitsch et al. 2001). Notably, drebrin was absent from stress fibres (Peitsch et al. 2001), the contractile actomyosin bundles that are often a prominent cytoskeletal feature of cancer cells on two-dimensional (2D) substrates (Tojkander et al. 2012). The disparity between drebrin localization in filopodia and not in stress fibres is likely explained by the difference in F-actin organization within these structures. In filopodia, F-actin is arranged in parallel bundles, whereas in stress fibres, the F-actin is antiparallel proposing that drebrin can exclusively recognize parallel F-actin bundles (Worth et al. 2013).

More recently, drebrin expression levels have been shown to induce morphological changes in a range of human glioma cell lines (Terakawa et al. 2013). Overexpression of drebrin in the Glioma cell line U87 triggered a spindle/stellate cell shape accompanied by a larger number of cellular projections. In contrast, depletion of drebrin from another glioma cell line A172 (which has significantly high drebrin expression in comparison to other glioma cell lines) produced a more rounded, spread phenotype that had substantially less cellular protrusions concomitant with an increase in stress fibre formation (Terakawa et al. 2013). This result is consistent with current thinking that stress fibres are more conspicuous in stationary cells on 2D surfaces suggesting that they may be inhibitory to cell migration in three-dimensional (3D) environments (Tojkander et al. 2012). Furthermore, in line with earlier studies, drebrin was also focally localized at the projecting tips and membrane ruffles of lamellipodia where it co-localized with F-actin in these glioma cell lines (Terakawa et al. 2013). Similarly, drebrin is detected in F-actin-enriched spikes at the membrane edge of growth factor-stimulated 5637 bladder cancer cells (Xu et al. 2015).

Besides the leading edge protrusions of motile cells, drebrin has often been observed in a juxta-nuclear cytoplasmic position within non-neuronal cells, including cancer cells, such as glioma U333 cells (Peitsch et al. 2001). Whilst the exact nature of this cytoplasmic pool of drebrin has not been fully characterized, some reports indicate drebrin can be found in complex with actin on the cisternae of Golgi and associated budding vesicles in rat liver fractions (Fucini et al. 2000). Although there is no direct data to point towards this potential drebrin accumulation at Golgi impacting on the malignancy of cells, it is significant to note that a proteomics approach looking at the protein content of breast tumour cell membranes revealed that drebrin was heavily enriched in integral membranes of four breast cancer cell lines MDA-MB-468, T-47D, BT-474 and MCF-7, three of which derive from metastatic sites (Adam et al. 2003).

Another unique distribution pattern of drebrin in non-neuronal cells can be seen in polarized epithelial cancers. In these cancer types, which include human primary liver carcinoma and epithelial skin tumours, drebrin is strongly enriched at cell-cell boundaries and, more specifically, at intercellular adherens junctions (Keon et al. 2000; Peitsch et al. 1999, 2005). In both basal and squamous cell carcinomas, drebrin decorates the submembranous actin cortex of cell-cell borders where it co-localizes with F-actin. Similarly, in the PLC cell line, drebrin intensely labels apical plasma membranes between neighbouring cells but is notably absent from cell free borders (Keon et al. 2000). Furthermore, overexpression of drebrin in a cell line derived from a human epidermoid carcinoma A-431 that ordinarily does not express drebrin produces the corresponding localization of drebrin at intercellular borders, where it partially co-localizes with adherens junction constituents, such as β-catenin and E-cadherin. Drebrin localization did not however overlap with vinculin, a component of focal adhesions. Additionally, these drebrin-GFP-A-431 cells frequently formed small cellular processes that became more pronounced and filopodial-like upon EGF stimulation, features indicative of an EMT transition (Peitsch et al. 2005). In normal epithelium, drebrin may be required to control actin filament attachment to adherens junctions, but in terms of tumourigenesis, the relevance of junctional drebrin is unclear. Yet, it is tempting to speculate that this localization could become misregulated to provide a protrusive force at cell-free borders and perhaps a contractile/retractive force at cell-cell boundaries enabling extrusion of a single cell from its surrounding neighbours within a tumour mass (Slattum and Rosenblatt 2014).

All studies so far of drebrin involvement in cancer have exclusively looked at malignant cells on 2D substrates, and as such, it remains to be seen whether these specific localizations of drebrin translate to invading cancer cells in 3D microenvironments. One such protrusive structure widely observed in in vitro 3D ECM models are F-actin-rich projections known as invadopodia, which contain a high density of proteases capable of degrading the matrix. Although drebrin has not been definitively localized within this structure, it has been shown by mass spectrometry to bind Tks5, a component of invadopodia important in linking actin assembly to matrix proteolysis (Stylli et al. 2009). Furthermore, TIRF microscopy has highlighted that in a metastatic breast cancer cell line, the formation of the initial primary invadopodial process is accompanied by transient filament-like extensions reminiscent of filopodia (Artym et al. 2011). This finding intimates it is plausible drebrin could be functioning in the actin dynamics associated with invadopodia. At present however, this remains to be investigated.

23.4 Drebrin Signalling in Cancer Cells

Delineating the signalling cascades that are subverted to afford cancer cells an invasive advantage is an active field of cancer research that will ultimately provide drug targets. This is also true of drebrin-mediated signalling pathways during cancer progression as many large proteomic and mass spectrometry-based screens are frequently revealing drebrin as an interaction partner. One such example is the novel identification of drebrin as a sphingosine kinase 1-binding protein in MCF-7 epithelial breast cancer cells (Yagoub et al. 2014). Sphingosine kinase 1 (SK1) is a signalling enzyme that catalyzes the formation of sphingosine-1-phosphate (S1P) through the phosphorylation of the lipid sphingosine. The dysregulation of the intricate balance between these two lipid forms is expected to influence the tumourigenic potential of SK1. Furthermore, the tumourigenic effects of SK1 are well recognized in breast cancer, specifically stimulating the growth and oestrogen receptor regulation and responsiveness of breast cancer cells (Sukocheva et al. 2003). Importantly, drebrin has been shown to interact with both major transcriptional isoforms of SK1, SK1 43kDa and SK1 51 kDa, with the 51 kDa variant being more abundant in breast cancer cells. Although the function of drebrin in SK1 signalling is yet to be defined, it seems likely it could be central in the acquisition of cancer cell migratory phenotypes by providing a bridge between the membrane of the cell and the F-actin network. Strengthening this concept is the finding that SK1 has been implicated in EGF-mediated MCF-7 cell motility (Sarkar et al. 2005).

In another signalling pathway that impacts on cancer cell motility and invasiveness of bladder cancer, a secreted growth factor glycoprotein known as progranulin associates with drebrin to regulate F-actin reorganization (Xu et al. 2015). Mass spectrometry mapped the progranulin-interactive drebrin fragment to a sequence that ranges from the end of the Hel region to the beginning of the polyproline-rich region (PP) (Fig. 23.2) (Xu et al. 2015). Under steady-state conditions, drebrin was demonstrated to interact weakly with progranulin in two malignant bladder cell lines, 5637 and T24. The kinetics following progranulin stimulation showed that the drebrin complex with progranulin was lost after 5 min but then markedly increased after 30 min, implying that the interaction between drebrin and progranulin occurs subsequent to progranulin internalization from the cell membrane (Xu et al. 2015). Progranulin-dependent effects on bladder and prostate cancer cell migration and invasion involve the downstream activation of Akt and MAPK pathways. It now appears that progranulin induces the phosphorylation of targets Akt and ERK1/2 via drebrin since depletion of drebrin from bladder cancer cell lines effectively prevents the activation of these two signalling cascades (Xu et al. 2015).

Although very little evidence has directly placed drebrin into signalling networks that give rise to oncogenic attributes, we may be able to infer drebrin-mediated signalling patterns from studies of drebrin function in neuronal systems. An example of this is the regulation of the drebrin-driven effects on dendritic spine actin cytoskeleton and morphology by Ras (Biou et al. 2008). This modulation requires the C-terminal domain of drebrin, specifically a region encompassing amino acids 233–317 (personal communication RC Malenka), and implies that a direct interaction between drebrin and Ras may occur. Importantly, Ras is a common oncogene in a wide variety of cancer subtypes (Pylayeva-Gupta et al. 2011). Thus, it is entirely possible that oncogenic and hence constitutively activated Ras could be causal in the acquisition of an invasive phenotype by promoting the F-actin-binding activity of drebrin. Further studies will need to determine the relationship between drebrin and Ras, initially determining if there is a direct interaction, but it presently stands as an exciting prospect.

23.5 Drebrin Involvement in Tumourigenesis

For a tumour to develop, it requires a transformation in cellular adhesivity that disrupts the normal tissue architecture and creates a microenvironment conducive to proliferation and angiogenesis to deliver a blood supply for continued tumour growth. Therefore, the ability of cancer cells to proliferate without adhesion to ECM proteins is considered a hallmark of tumourigenicity and can be modelled in vitro using anchorage-independent cell growth assays. Interestingly, drebrin depletion in a malignant bladder carcinoma cell line UMUC-3 has been shown to impair the colony-forming capacity of these cells in soft agar (Xu et al. 2015) and is suggestive of drebrin expression being an important determinant of tumourigenic and metastatic potential, at least for bladder cancer.

So far only one study has provided evidence that drebrin could underlie tumour development and aggressiveness in vivo. In a mouse xenograft model of bladder cancer, malignant UMUC-3 cells with shRNA-driven reduction of drebrin expression substantially decreased both the occurrence and size of tumours (Xu et al. 2015). However, no definitive relationship between drebrin expression and a proliferative or survival advantage for the cancer cell has been demonstrated in vitro, and by itself the mechanistic insight as to how drebrin could drive tumour formation and maintenance in vivo is unclear.

23.6 Drebrin Involvement in Cancer Cell Migration and Invasion

The capacity of cancer cells to colonize a metastatic tumour site is predicated on their ability to escape the confines of the primary tumour through acquisition of migratory and invasive programmes. It is now emerging that drebrin could be crucial to both these elements in several cancer types (Terakawa et al. 2013; Xu et al. 2015). Progranulin-mediated migration of malignant bladder cells appears to be regulated by drebrin expression, as suppression of drebrin protein levels in two bladder cancer cell lines, 5637 and T24, produces an inhibition of motility upon progranulin stimulation (Xu et al. 2015). A role for drebrin in glioma cell migration has also been demonstrated using an in vitro radial cell migration assay whereby depletion of drebrin attenuated migration rates in several glioma cell lines. Interestingly, these experiments were performed on a range of different ECM proteins (laminin, collagen type IV, vitronectin and fibronectin), all of which displayed a dependency on drebrin expression (Terakawa et al. 2013). This is particularly striking given that diverse modes of cell migration are often determined by the ECM composition, ostensibly as a result of the varying nature of the cell: substratum adhesions formed. Hence, this finding perhaps proposes a unifying role for drebrin in cancer cell motility.

Cell migration is intrinsically linked to invasion, and as such it would be expected that any effect of protein misregulation on the migratory capacity of a cell would be directly correlative with that of its invasive capacity. This appears to be true of drebrin in cancer. With the application of 3D in vitro Transwell invasion assays utilizing Matrigel as the matrix constituent, it has been shown that drebrin depletion can decrease invasive rates of U87 and A172 glioma cell lines (Terakawa et al. 2013) and prostate adenocarcinoma cell lines (Dart et al. 2017) as well as progranulin-induced invasion of 5637 bladder cancer cells (Xu et al. 2015). Further evidence demonstrating that the balance of drebrin expression is crucial to the generation of a malignant phenotype comes from overexpression of drebrin in the glioma cell line U87 which results in a greater than ninefold increase in the invasiveness through a Matrigel matrix (Terakawa et al. 2013).

Whilst these studies have provided strong support for a role for drebrin in migration and invasion of several cancer cell types, they do not directly elucidate the drebrin-driven signalling pathway(s) exploited to provide the cancer cell its invasive attributes. One possible hypothesis proposed by Terakawa et al. (2013) is that in some way drebrin is able to modulate the interaction of the cancer cell with the ECM substratum, which principally derives from experiments showing that the glioma cell line U87 exhibits increased adhesion to ECM macromolecules upon drebrin depletion (Terakawa et al. 2013). However, this theory is inconsistent with reports that drebrin displays a strikingly dissimilar localization pattern to that of vinculin, an archetypal protein of focal adhesions. Nonetheless, despite a lack of current evidence, it may be that drebrin could regulate the focal adhesion turnover of migrating cancer cells indirectly through an effect on cytoskeletal dynamics. This should be fully explored by seeking functional interactions of drebrin with fundamental focal adhesion proteins, such as FAK and paxillin (Xu et al. 2015) (Ikeda et al. 1995, 1996).

23.7 Future Perspectives

Given the presence of drebrin in human cells with a distinctive and specialized need for actin plasticity, such as neurons, parietal cells of the stomach and podocytes of the kidneys (Aoki et al. 2005; Keon et al. 2000; Peitsch et al. 2003), it was perhaps not unexpected to find cancer cells might too be modulating the expression of this actin-binding protein as part of an adaptive strategy to facilitate increased motility and invasion. Evidence is now rapidly accruing to suggest this is the case. Taken together recent data would propose that drebrin becomes upregulated in cancerous cells undergoing EMT and that this misregulation of drebrin expression is associated with a greater invasive potential. However, reports pertaining to a role for drebrin overexpression in cancer progression are still limited, leaving many unanswered questions. Whilst dysfunctional levels of drebrin may be a common consequence of tumourigenesis, we cannot yet say whether it is the primary driver of tumour formation or simply a secondary effect of a sequence of other genetic or epigenetic alterations. An attractive area of research for the future would be to investigate the concept that drebrin expression may be fundamental to the cellular origins of at least some cancer types.

Another path of potential exploration will be to find and characterize binding partners of drebrin as well as signalling cascades activated downstream of drebrin that could help elucidate its potential functionality in cancer. It will also be necessary to determine how drebrin precisely coordinates the cytoskeleton in the varied modes of directed migration and invasion observed during cancer metastasis. Much research of late has focused on the involvement of the actin cytoskeleton and its regulation in cancer cell invasion and metastasis, but it is important to emphasize that the actin cytoskeleton does not work in isolation and its interplay with the microtubule cytoskeleton will ultimately influence the migratory outcome of a cancer cell. This perhaps explains the recent surge in literature highlighting the microtubule system as a crucial factor in cancer cell invasion and metastasis (reviewed in Fife et al. (2014)). In addition, recent reports have underlined the necessity of the mutual regulation of actin and microtubule cytoskeletons indicating that targeting this crosstalk could be an attractive approach for developing novel combinatorial therapeutics (reviewed in Fife et al. (2014)). Therefore, is the ability of drebrin to bind actin and the microtubule +TIP-binding protein EB3, thereby coupling actin filaments to microtubules (Geraldo et al. 2008), necessary for cancer cell invasion and metastasis? Furthermore, the key to metastasis is directional, polarized motility which is acquired in response to extracellular cues. Drebrin has already been shown to be involved in regulating progranulin-dependent signalling in bladder cancer, but importantly, we need to identify if other extracellular chemical gradients mobilize drebrin to influence cytoskeletal dynamics.

The capacity of cancer cells to disseminate is intimately linked to the microenvironment in which they find themselves, and moreover, the interaction of cancer cells with stromal cells and other constituents surrounding the tumour is important for disease development. Taking this into consideration, drebrin expression has also been detected in fibroblasts and endothelial cells (Keon et al. 2000; Peitsch et al. 1999; Xu et al. 2015) indicating it could also facilitate invasion indirectly via other components of the tumour microenvironment. This aspect of drebrin functionality influencing cancer progression certainly warrants further study.

Collectively the current research presents drebrin as not only a necessary but more importantly a feasible drug target. It will be crucial in the future to determine the exact mechanism of action of drebrin in order to generate appropriate target selectivity. As it stands, efforts directed towards the actin-binding capacity of drebrin could prove to be an effective therapeutic.

References

Adam PJ, Boyd R, Tyson KL, Fletcher GC, Stamps A, Hudson L, Poyser HR, Redpath N, Griffiths M, Steers G, Harris AL, Patel S, Berry J, Loader JA, Townsend RR, Daviet L, Legrain P, Parekh R, Terrett JA (2003) Comprehensive proteomic analysis of breast cancer cell membranes reveals unique proteins with potential roles in clinical cancer. J Biol Chem 278:6482–6489

Aoki C, Sekino Y, Hanamura K, Fujisawa S, Mahadomrongkul V, Ren Y, Shirao T (2005) Drebrin A is a postsynaptic protein that localizes in vivo to the submembranous surface of dendritic sites forming excitatory synapses. J Comp Neurol 483:383–402

Artym VV, Matsumoto K, Mueller SC, Yamada KM (2011) Dynamic membrane remodeling at invadopodia differentiates invadopodia from podosomes. Eur J Cell Biol 90:172–180

Biou V, Brinkhaus H, Malenka RC, Matus A (2008) Interactions between drebrin and Ras regulate dendritic spine plasticity. Eur J Neurosci 27:2847–2859

Brennan DJ, Ek S, Doyle E, Drew T, Foley M, Flannelly G, O'Connor DP, Gallagher WM, Kilpinen S, Kallioniemi OP, Jirstrom K, O'Herlihy C, Borrebaeck CA (2009) The transcription factor Sox11 is a prognostic factor for improved recurrence-free survival in epithelial ovarian cancer. Eur J Cancer 45:1510–1517

Chaffer CL, Weinberg RA (2011) A perspective on cancer cell metastasis. Science 331:1559–1564

Dart AE, Worth DC, Muir G, Chandra A, Morris JD, McKee C, Verrill C, Bryant RJ, Gordon-Weeks PR (2017) The drebrin/EB3 pathway drives invasive activity in prostate cancer. Oncogene. doi:10.1038/onc.2017.45

Fife CM, McCarroll JA, Kavallaris M (2014) Movers and shakers: cell cytoskeleton in cancer metastasis. Br J Pharmacol 171:5507–5523

Fossa A, Siebert R, Aasheim HC, Maelandsmo GM, Berner A, Fossa SD, Paus E, Smeland EB, Gaudernack G (2000) Identification of nucleolar protein No55 as a tumour-associated autoantigen in patients with prostate cancer. Br J Cancer 83:743–749

Fucini RV, Navarrete A, Vadakkan C, Lacomis L, Erdjument-Bromage H, Tempst P, Stamnes M (2000) Activated ADP-ribosylation factor assembles distinct pools of actin on golgi membranes. J Biol Chem 275:18824–18829

Geraldo S, Khanzada UK, Parsons M, Chilton JK, Gordon-Weeks PR (2008) Targeting of the F-actin-binding protein drebrin by the microtubule plus-tip protein EB3 is required for neuritogenesis. Nat Cell Biol 10:1181–1189

Hayashi K, Shirao T (1999) Change in the shape of dendritic spines caused by overexpression of drebrin in cultured cortical neurons. J Neurosci 19:3918–3925

Iacobuzio-Donahue CA, Maitra A, Shen-Ong GL, van Heek T, Ashfaq R, Meyer R, Walter K, Berg K, Hollingsworth MA, Cameron JL, Yeo CJ, Kern SE, Goggins M, Hruban RH (2002) Discovery of novel tumor markers of pancreatic cancer using global gene expression technology. Am J Pathol 160:1239–1249

Ikeda K, Shirao T, Toda M, Asada H, Toya S, Uyemura K (1995) Effect of a neuron-specific actin-binding protein, drebrin A, on cell-substratum adhesion. Neurosci Lett 194:197–200

Ikeda K, Kaub PA, Asada H, Uyemura K, Toya S, Shirao T (1996) Stabilization of adhesion plaques by the expression of drebrin A in fibroblasts. Brain Res Dev Brain Res 91:227–236

Jansson MD, Lund AH (2012) MicroRNA and cancer. Mol Oncol 6:590–610

Keon BH, Jedrzejewski PT, Paul DL, Goodenough DA (2000) Isoform specific expression of the neuronal F-actin binding protein, drebrin, in specialized cells of stomach and kidney epithelia. J Cell Sci 113(Pt 2):325–336

Kwon HC, Kim SH, Roh MS, Kim JS, Lee HS, Choi HJ, Jeong JS, Kim HJ, Hwang TH (2004) Gene expression profiling in lymph node-positive and lymph node-negative colorectal cancer. Dis Colon Rectum 47:141–152

Lappalainen P, Kessels MM, Cope MJ, Drubin DG (1998) The ADF homology (ADF-H) domain: a highly exploited actin-binding module. Mol Biol Cell 9:1951–1959

Lidke DS, Lidke KA, Rieger B, Jovin TM, Arndt-Jovin DJ (2005) Reaching out for signals: filopodia sense EGF and respond by directed retrograde transport of activated receptors. J Cell Biol 170:619–626

Lin Q, Tan HT, Lim TK, Khoo A, Lim KH, Chung MC (2014) iTRAQ analysis of colorectal cancer cell lines suggests Drebrin (DBN1) is overexpressed during liver metastasis. Proteomics 14:1434–1443

Liu RF, Xu X, Huang J, Fei QL, Chen F, Li YD, Han ZG (2013) Down-regulation of miR-517a and miR-517c promotes proliferation of hepatocellular carcinoma cells via targeting Pyk2. Cancer Lett 329:164–173

Mitra R, Lee J, Jo J, Milani M, McClintick JN, Edenberg HJ, Kesler KA, Rieger KM, Badve S, Cummings OW, Mohiuddin A, Thomas DG, Luo X, Juliar BE, Li L, Mesaros C, Blair IA, Srirangam A, Kratzke RA, McDonald CJ, Kim J, Potter DA (2011) Prediction of postoperative recurrence-free survival in non-small cell lung cancer by using an internationally validated gene expression model. Clin Cancer Res 17:2934–2946

Mizutani Y, Iwamoto I, Kanoh H, Seishima M, Nagata K (2014) Expression of drebrin, an actin binding protein, in basal cell carcinoma, trichoblastoma and trichoepithelioma. Histol Histopathol 29:757–766

Peitsch WK, Grund C, Kuhn C, Schnolzer M, Spring H, Schmelz M, Franke WW (1999) Drebrin is a widespread actin-associating protein enriched at junctional plaques, defining a specific microfilament anchorage system in polar epithelial cells. Eur J Cell Biol 78:767–778

Peitsch WK, Hofmann I, Pratzel S, Grund C, Kuhn C, Moll I, Langbein L, Franke WW (2001) Drebrin particles: components in the ensemble of proteins regulating actin dynamics of lamellipodia and filopodia. Eur J Cell Biol 80:567–579

Peitsch WK, Hofmann I, Endlich N, Pratzel S, Kuhn C, Spring H, Grone HJ, Kriz W, Franke WW (2003) Cell biological and biochemical characterization of drebrin complexes in mesangial cells and podocytes of renal glomeruli. J Am Soc Nephrol 14:1452–1463

Peitsch WK, Hofmann I, Bulkescher J, Hergt M, Spring H, Bleyl U, Goerdt S, Franke WW (2005) Drebrin, an actin-binding, cell-type characteristic protein: induction and localization in epithelial skin tumors and cultured keratinocytes. J Invest Dermatol 125:761–774

Pylayeva-Gupta Y, Grabocka E, Bar-Sagi D (2011) RAS oncogenes: weaving a tumorigenic web. Nat Rev Cancer 11:761–774

Roussos ET, Condeelis JS, Patsialou A (2011) Chemotaxis in cancer. Nat Rev Cancer 11:573–587

Sarkar S, Maceyka M, Hait NC, Paugh SW, Sankala H, Milstien S, Spiegel S (2005) Sphingosine kinase 1 is required for migration, proliferation and survival of MCF-7 human breast cancer cells. FEBS Lett 579:5313–5317

Shirao T, Hayashi K, Ishikawa R, Isa K, Asada H, Ikeda K, Uyemura K (1994) Formation of thick, curving bundles of actin by drebrin A expressed in fibroblasts. Exp Cell Res 215:145–153
Sjoblom T, Jones S, Wood LD, Parsons DW, Lin J, Barber TD, Mandelker D, Leary RJ, Ptak J, Silliman N, Szabo S, Buckhaults P, Farrell C, Meeh P, Markowitz SD, Willis J, Dawson D, Willson JK, Gazdar AF, Hartigan J, Wu L, Liu C, Parmigiani G, Park BH, Bachman KE, Papadopoulos N, Vogelstein B, Kinzler KW, Velculescu VE (2006) The consensus coding sequences of human breast and colorectal cancers. Science 314:268–274
Slattum GM, Rosenblatt J (2014) Tumour cell invasion: an emerging role for basal epithelial cell extrusion. Nat Rev Cancer 14:495–501
Stylli SS, Stacey TT, Verhagen AM, Xu SS, Pass I, Courtneidge SA, Lock P (2009) Nck adaptor proteins link Tks5 to invadopodia actin regulation and ECM degradation. J Cell Sci 122:2727–2740
Sukocheva OA, Wang L, Albanese N, Pitson SM, Vadas MA, Xia P (2003) Sphingosine kinase transmits estrogen signaling in human breast cancer cells. Mol Endocrinol 17:2002–2012
Terakawa Y, Agnihotri S, Golbourn B, Nadi M, Sabha N, Smith CA, Croul SE, Rutka JT (2013) The role of drebrin in glioma migration and invasion. Exp Cell Res 319:517–528
Tojkander S, Gateva G, Lappalainen P (2012) Actin stress fibers--assembly, dynamics and biological roles. J Cell Sci 125:1855–1864
Vaskova M, Kovac M, Volna P, Angelisova P, Mejstrikova E, Zuna J, Brdicka T, Hrusak O (2011) High expression of cytoskeletal protein drebrin in TEL/AML1pos B-cell precursor acute lymphoblastic leukemia identified by a novel monoclonal antibody. Leuk Res 35:1111–1113
Wang X, Bjorklund S, Wasik AM, Grandien A, Andersson P, Kimby E, Dahlman-Wright K, Zhao C, Christensson B, Sander B (2010) Gene expression profiling and chromatin immunoprecipitation identify DBN1, SETMAR and HIG2 as direct targets of SOX11 in mantle cell lymphoma. PLoS One 5:e14085
Weigle B, Ebner R, Temme A, Schwind S, Schmitz M, Kiessling A, Rieger MA, Schackert G, Schackert HK, Rieber EP (2005) Highly specific overexpression of the transcription factor SOX11 in human malignant gliomas. Oncol Rep 13:139–144
Worth DC, Daly CN, Geraldo S, Oozeer F, Gordon-Weeks PR (2013) Drebrin contains a cryptic F-actin-bundling activity regulated by Cdk5 phosphorylation. J Cell Biol 202:793–806
Xu W, Stamnes M (2006) The actin-depolymerizing factor homology and charged/helical domains of drebrin and mAbp1 direct membrane binding and localization via distinct interactions with actin. J Biol Chem 281:11826–11833
Xu SQ, Buraschi S, Morcavallo A, Genua M, Shirao T, Peiper SC, Gomella LG, Birbe R, Belfiore A, Iozzo RV, Morrione A (2015) A novel role for drebrin in regulating progranulin bioactivity in bladder cancer. Oncotarget 6(13):10825–10839
Yagoub D, Wilkins MR, Lay AJ, Kaczorowski DC, Hatoum D, Bajan S, Hutvagner G, Lai JH, Wu W, Martiniello-Wilks R, Xia P, McGowan EM (2014) Sphingosine kinase 1 isoform-specific interactions in breast cancer. Mol Endocrinol 28:1899–1915
Yoshitomi T, Kawakami K, Enokida H, Chiyomaru T, Kagara I, Tatarano S, Yoshino H, Arimura H, Nishiyama K, Seki N, Nakagawa M (2011) Restoration of miR-517a expression induces cell apoptosis in bladder cancer cell lines. Oncol Rep 25:1661–1668

Erratum to: Drebrin - From Structure and Function to Physiological and Pathological Roles

Tomoaki Shirao and Yuko Sekino

Erratum to:
T. Shirao, Y. Sekino (eds.), *Drebrin,* Advances in Experimental Medicine and Biology 1006, DOI 10.1007/978-4-431-56550-5

This book was initially published with error in the title of Part IV, which has now been corrected.

The correct title has been updated in the part page and in the table of contents.

The updated original online version of book can be found at https://doi.org/10.1007/978-4-431-56550-5

T. Shirao, Y. Sekino (eds.), *Drebrin*, Advances in Experimental Medicine and Biology 1006, DOI 10.1007/978-4-431-56550-5_24

Index

T. Shirao, Y. Sekino (eds.), *Drebrin*, Advances in Experimental Medicine and Biology 1006, DOI 10.1007/978-4-431-56550-5

MIX
Papier aus verantwortungsvollen Quellen
Paper from responsible sources
FSC® C105338

If you have any concerns about our products,
you can contact us on
ProductSafety@springernature.com

In case Publisher is established outside the EU,
the EU authorized representative is:
Springer Nature Customer Service Center GmbH
Europaplatz 3, 69115 Heidelberg, Germany

Printed by Libri Plureos GmbH
in Hamburg, Germany